Lecture Notes in

T0238137

Springer
Berlin
Heidelberg
New York
Hong Kong
London
Milan
Paris
Tokyo

The Editorial Policy for Edited Volumes

The series *Lecture Notes in Physics* (LNP), founded in 1969, reports new developments in physics research and teaching - quickly, informally but with a high degree of quality. Manuscripts to be considered for publication are topical volumes consisting of a limited number of contributions, carefully edited and closely related to each other. Each contribution should contain at least partly original and previously unpublished material, be written in a clear, pedagogical style and aimed at a broader readership, especially graduate students and nonspecialist researchers wishing to familiarize themselves with the topic concerned. For this reason, traditional proceedings cannot be considered for this series though volumes to appear in this series are often based on material presented at conferences, workshops and schools.

Acceptance

A project can only be accepted tentatively for publication, by both the editorial board and the publisher, following thorough examination of the material submitted. The book proposal sent to the publisher should consist at least of a preliminary table of contents outlining the structure of the book together with abstracts of all contributions to be included. Final acceptance is issued by the series editor in charge, in consultation with the publisher, only after receiving the complete manuscript. Final acceptance, possibly requiring minor corrections, usually follows the tentative acceptance unless the final manuscript differs significantly from expectations (project outline). In particular, the series editors are entitled to reject individual contributions if they do not meet the high quality standards of this series. The final manuscript must be ready to print, and should include both an informative introduction and a sufficiently detailed subject index.

Contractual Aspects

Publication in LNP is free of charge. There is no formal contract, no royalties are paid, and no bulk orders are required, although special discounts are offered in this case. The volume editors receive jointly 30 free copies for their personal use and are entitled, as are the contributing authors, to purchase Springer books at a reduced rate. The publisher secures the copyright for each volume. As a rule, no reprints of individual contributions can be supplied.

Manuscript Submission

The manuscript in its final and approved version must be submitted in ready to print form. The corresponding electronic source files are also required for the production process, in particular the online version. Technical assistance in compiling the final manuscript can be provided by the publisher's production editor(s), especially with regard to the publisher's own LaTeX macro package which has been specially designed for this series.

LNP Homepage (springerlink.com)

On the LNP homepage you will find:
−The LNP online archive. It contains the full texts (PDF) of all volumes published since 2000. Abstracts, table of contents and prefaces are accessible free of charge to everyone. Information about the availability of printed volumes can be obtained.
−The subscription information. The online archive is free of charge to all subscribers of the printed volumes.
−The editorial contacts, with respect to both scientific and technical matters.
−The author's / editor's instructions.

B. Grammaticos Y. Kosmann-Schwarzbach
T. Tamizhmani (Eds.)

Discrete Integrable Systems

Springer

Editors

Basil Grammaticos
GMPIB, Université Paris VII
Tour 24-14, 5e étage, case 7021
2 place Jussieu
75251 Paris Cedex 05, France

Yvette Kosmann-Schwarzbach
Centre de Mathématiques
École Polytechnique
91128 Palaiseau, France

Thamizharasi Tamizhmani
Department of Mathematics
Kanchi Mamunivar Centre
for Postgraduate Studies
Pondicherry, India

B. Grammaticos, Y. Kosmann-Schwarzbach, T. Tamizhmani (Eds.), *Discrete Integrable Systems*, Lect. Notes Phys. **644** (Springer, Berlin Heidelberg 2004), DOI 10.1007/b94662

Bibliographic information published by Die Deutsche Bibliothek Die Deutsche Bibliothek lists this publication in the Deutsche Nationalbibliografie; detailed bibliographic data is available in the Internet at <http://dnb.ddb.de>

ISSN 0075-8450
ISBN 978-3-642-05978-0   e-ISBN 978-3-540-40357-9

Springer-Verlag is a part of Springer Science+Business Media

springeronline.com

© Springer-Verlag Berlin Heidelberg 2010
Printed in Germany

Data conversion: PTP-Berlin Protago-TeX-Production GmbH
Cover design: *design & production*, Heidelberg

Printed on acid-free paper
54/3141/ts - 5 4 3 2 1 0

Preface

This volume contains ten courses based on the lectures delivered at the International School on Discrete Integrable Systems, organized by CIMPA, UNESCO and the Government of Pondicherry, which was held in Pondicherry, India, from February 2 to February 14, 2003. These expository articles constitute both an introduction to and a survey of recent results on discrete integrable systems. This volume constitutes a companion volume to Lecture Notes in Physics 638, *Integrability of Nonlinear Systems*, edited by Y. Kosmann-Schwarzbach, B. Grammaticos and K. M. Tamizhmani, Springer 2004. Both volumes will be of great use to students and researchers interested in a rapidly expanding field.

CIMPA, Centre International de Mathématiques Pures et Appliquées (International Center for Pure and Applied Mathematics) is a non-profit international organization based in Nice, France, and is sponsored by the French Government and by UNESCO, in partnership with several scholarly societies. Its aim is to promote international cooperation in higher education and research in mathematics and related subjects, for the benefit of developing countries.

Forty-four young scientists in mathematics and physics, including six from outside India, participated in this school. Fourteen of them were women.

This school was made possible by the financial and material support of both CIMPA and the government of Pondicherry, which was directly involved in the organization of the school at every stage. Its scientific directors and organizers were Basil Grammaticos (Université Paris VII, France), Yvette Kosmann-Schwarzbach (École Polytechnique, France) and Thamizharasi Tamizhmani (Kanchimamunivar Centre for Postgraduate Studies, KMCPGS, Pondicherry, India).

We offer sincere thanks to His Excellency the Lt. Governor of Pondicherry, K. R. Malkani, for inaugurating the school. We thank the Honourable Chief Minister, N. Rangasamy, the Honourable Speaker, D. Ramachandran, the Honourable Education Minister, K. Lakshminarayanan, the Chief Secretary, Dr. R. Padmanabhan, IAS, the Secretary to Education, Mrs. M. Sathiyavathy, IAS, and the Director of Education, Mr. Theva Neethi Dhas, for their kind support and encouragement throughout the school. We also thank M. Michel Séguy, Consul Général de France à Pondichéry, Prof. Michel Waldschmidt (Université Paris VI and President of the Société

Mathématique de France), CIMPA representative, Prof. V.T. Patil, Vice-Chancellor, Pondicherry University, Prof. Martin Kruskal and Prof. M. Lakshmanan, who delivered addresses at the opening session.

We sincerely acknowledge the financial support received from ICTP, IMU, NBHM, DST (Delhi), DST (Pondicherry), and the Departments of Tourism and of Education of the Government of Pondicherry.

We also thank the Principal of the Avvaiyar Government College for Women, Karaikal, the Director of the KMCPGS, as well as the staff and students of the Department of Mathematics of KMCPGS for their valuable assistance. We are grateful to Mr. M. Sivaprakasam and Dr. K. M. Tamizhmani for their support in the administration of the school and to Dr. B. E. Schwarzbach for his precious editorial help.

The editors acknowledge with thanks the encouragement of Dr. Christian Caron, physics editor at Springer-Verlag, and the assistance of his staff, in particular the technical expertise of Miss Sandra Thoms.

December 2003   *The Editors*

Introduction

Discrete integrable systems burst upon the physics and mathematics communities with hurricane force less than fifteen years ago. When integrable systems, a lost heritage of the mathematicians of the early twentieth century, were rediscovered, the revival was at first limited to the case of continuous dynamical systems, described by either differential or partial differential equations. A variety of techniques were devised for the study of both finite- and infinite-dimensional integrable systems. Already in the late 70s there were deviations from the strict differential framework. R. Hirota, in a series of articles that went essentially unnoticed, because he was so far in advance of his time, produced discrete analogues of many of the evolution partial differential equations known at that time. Another direction was explored by M. Ablowitz and his collaborators. Trying to find solutions for some of the known integrable partial differential equations by numerical methods, they were led to propose integrators in the form of difference schemes which were integrable in their own right. Thus, both Hirota and Ablowitz dealt with discrete integrable systems. Curiously, the memory of these innovations was lost, except among a handful of researchers.

Why did the development of the domain of discrete integrability have to wait almost a quarter of a century after the pioneering work of M. D. Kruskal and his collaborators on the Korteweg-de Vries equation and the subsequent developments in the theory of solitons for continuous integrable systems? Two facts could explain this delay. The first is that there did not exist, till the early

90s, any method for detecting the integrable character of a discrete system. The only procedure available was to search for numerical solutions with a chaotic behavior, and to declare the system integrable if no such solution could be found. But this method was far from being a valid predictor of integrability. The second reason is that not enough discrete integrable systems were yet known for the domain to be self-sustaining. Although Hirota had identified a large number of discrete systems, few people were aware of them at that time and almost nobody outside Japan.

Then the situation gradually changed. Semi-discrete and discrete systems started appearing in field-theoretical models in the work of M. Jimbo and T. Miwa. Then H. Capel and his school tackled the problem of constructing integrable lattice equations directly, thus providing some of the first systematic tools for the study of discrete equations. All of a sudden the number of discrete integrable systems approached a critical mass and the time was thus ripe for the introduction of the first discrete integrability criterion. This test of integrability was the property of singularity confinement which was proposed by B. Grammaticos, A. Ramani and V. Papageorgiou in 1991. Singularity confinement is a property characteristic of discrete equations integrable by spectral methods: in this case, every spontaneously appearing singularity disappears after a few iteration steps. The same property was also observed, under the name "orbits of pole-like behaviour," by N. Joshi who surmised that it should play a role analogous to that of the Painlevé property for continuous integrability. This volume is introduced by a course by Kruskal on the Painlevé property and the Painlevé equations.

The singularity confinement criterion was further refined in collaboration with K. M. Tamizhmani, but it became clear that it was not sufficient to serve as a predictor of integrability. In fact it furnishes no information about the rate of growth of the solutions of the discrete system, and such information is essential because a fast rate is incompatible with integrability. Another boost given to the study of discrete intregrable systems was the discovery of discrete Painlevé equations. The full story is told in the course contributed by Grammaticos and Ramani to this book. Decisive progress in this domain was obtained with the derivation by F. Nijhoff and Papageorgiou of the discrete P_{II}, a difference equation obtained as the similarity reduction of the discrete modified Korteweg-de Vries equation, thus establishing a perfect parallel to the continuous case. Soon afterwards, Ramani, Grammaticos and J. Hietarinta discovered the multiplicative discrete Painlevé equations, known as q-Painlevé equations. A third variety of discrete Painlevé equation, the elliptic-discrete type, was only found several years later, when H. Sakai provided the definitive classification of discrete Painlevé equations. The properties of these systems have been the object of numerous studies. Foremost among those is the construction of special solutions, the work of the Paris-Pondicherry collaboration of Grammaticos, Ramani, K. M. Tamizhmani and Thamizharasi Tamizhmani, which is described in their course. We must also mention the most interesting approach of K. Kajiwara and Y. Ohta who, together with their collaborators, addressed the difficult question of the representation of the solution of discrete Painlevé equations in terms of Casorati determinants.

In the domain of integrable lattices, the resurrection of the Hirota-Miwa equation (discrete Kadomtsev-Petviashvilii), coupled with the extension of the Sato theory to discrete systems, has led to a host of results of capital importance. They are reviewed in the courses of Ohta and of R. Willox and J. Satsuma. In Ohta's article, the bilinear formulation of hierarchies of discrete soliton equations is derived from algebraic identities satisfied by determinants and Pfaffians, while Willox and Satsuma present the algebro-geometric properties of the hierarchies of integrable partial differential equations, their discretizations and their reductions. The course of Y. Suris extends the notions of Lagrangian and Hamiltonian mechanics to discrete time, and describes a discrete analogue of Liouville integrability. He examines several models which are well-known examples of integrable systems in classical mechanics, and produces their integrable discretization. P. Winternitz and his team have studied the symmetries of discrete equations. For differential equations, symmetries play a major role, in particular in the search for solutions. In the case of discrete systems, the problem is considerably more difficult and Winternitz's course contains a state-of-the-art account of the question. The course of A. Bobenko is devoted to another aspect of discrete integrable systems, their link to geometry. In this setting, the introduction of discrete surfaces makes the determination of integrable lattices possible by requiring nothing more than a condition of consistency on the labeling of their vertices. Finally, a very special aspect or rather a special limit of discrete systems is the object of the course of T. Tokihiro. This limit allows an algorithmic construction, starting from an integrable discrete system, of a generalised cellular automaton analogue, while preserving integrability. The systems thus obtained are called ultra-discrete, since, with appropriate initial choices, the dependent variable can be constrained to a discrete set of values. These lectures are complemented by a course by Y. Pomeau on time in physics and the question of reversibility.

The most important fact that emerges from a decade of intense activity exploring discrete integrable systems is that they are fundamental objects because they contain all other integrable systems as their limits. Moreover, the analogy between their properties and those of continuous systems is nearly perfect, a fact that has greatly facilitated their study. When, ten years ago, the study of discrete integrable systems had just begun, Satsuma claimed that "only crazy people are interested in discrete systems". By now a solid arsenal of methods for the study of such systems exists. This book testifies to the amazing progress accomplished in this domain.

The Editors

Table of Contents

Discrete Differential Geometry. Integrability as Consistency

Discrete Lagrangian Models

List of Contributors

A. I. Bobenko
Institut für Mathematik, Fakultät 2,
Technische Universität Berlin,
Str. des 17. Juni 136,
10623 Berlin, Germany
bobenko@math.tu-berlin.de

B. Grammaticos
GMPIB,
Université Paris VII,
Tour 24-14, 5ᵉ étage, case 7021,
75251 Paris, France
grammati@paris7.jussieu.fr

Y. Kosmann-Schwarzbach
Centre de Mathématiques,
École Polytechnique,
91128 Palaiseau, France
yks@math.polytechnique.fr

M. D. Kruskal
Department of Mathematics,
Rutgers University,
New Brunswick, NJ 08903, USA
kruskal@math.rutgers.edu

Y. Ohta
Department of Mathematics,
Kobe University,
Rokko, Kobe 657-8501, Japan
ohta@math.kobe-u.ac.jp

Y. Pomeau
Laboratoire de Physique Statistique
de l'École Normale Supérieure,
24 rue Lhomond,
75231 Paris Cedex 05, France
pomeau@physique.ens.fr

A. Ramani
CNRS UMR 7644,
Centre de Physique Théorique,
École Polytechnique,
91128 Palaiseau, France
ramani@cpht.polytechnique.fr

J. Satsuma
Graduate School of
Mathematical Sciences,
University of Tokyo,
3-8-1 Komaba,
Meguro-ku, Tokyo 153-8914, Japan
satsuma
@poisson.ms.u-tokyo.ac.jp

Yu. B. Suris
Institut für Mathematik,
Technische Universität Berlin,
Str. des 17. Juni 136,
10623 Berlin, Germany
suris@sfb288.math.tu-berlin.de

K. M. Tamizhmani
Departement of Mathematics,
Pondicherry University,
Kalapet,
Pondicherry 605014, India
tamizh@yahoo.com

T. Tamizhmani
Department of Mathematics,
Kanchi Mamunivar Centre for
Postgraduate Studies,
Pondicherry 605008, India
arasi55@yahoo.com

T. Tokihiro
Graduate School of
Mathematical Sciences,
University of Tokyo,
3-8-1 Komaba,
Meguro-ku, Tokyo 153-8914, Japan
toki@poisson.ms.u-tokyo.ac.jp

R. Willox
Graduate School of
Mathematical Sciences,
University of Tokyo,
3-8-1 Komaba,
Meguro-ku, Tokyo 153-8914, Japan
and
Theoretical Physics,
Free University of Brussels (VUB),
Pleinlaan 2,
1050 Brussels, Belgium
willox
@poisson.ms.u-tokyo.ac.jp

P. Winternitz
Centre de Recherches
Mathématiques,
Université de Montréal,
C.P. 6128-A, Montréal,
Québec, H3C 3J7, Canada
wintern@crm.umontreal.ca

Three Lessons on the Painlevé Property and the Painlevé Equations

M. D. Kruskal[1], B. Grammaticos[2], and T. Tamizhmani[3]

[1] Department of Mathematics, Rutgers University, New Brunswick, NJ 08903, USA, kruskal@math.rutgers.edu
[2] GMPIB, Université Paris VII, Tour 24-14, 5^{e}étage, case 7021, 75251 Paris, France, grammati@paris7.jussieu.fr
[3] Department of Mathematics, Kanchi Mamunivar Centre for Postgraduate Studies, Pondicherry 605008, India, arasi55@yahoo.com

Abstract. While this school focuses on discrete integrable systems we feel it necessary, if only for reasons of comparison, to go back to fundamentals and introduce the basic notion of the Painlevé property for continuous systems together with a critical analysis of what is called the Painlevé test. The extension of the latter to what is called the poly-Painlevé test is also introduced. Finally we devote a lesson to the proof that the Painlevé equations do have the Painlevé property.

1 Introduction

A course on integrability often starts with introducing the notion of soliton and how the latter emerges in integrable partial differential equations. Here we will focus on simpler systems and consider only ordinary differential equations. Six such equations play a fundamental role in integrability theory, the six Painlevé equations [1]:

$$x'' = 6x^2 + t \qquad\qquad \mathrm{P_I}$$

$$x'' = 2x^3 + tx + a \qquad\qquad \mathrm{P_{II}}$$

$$x'' = \frac{x'^2}{x} - \frac{x'}{t} + \frac{1}{t}(ax^2 + b) + cx^3 + \frac{d}{x} \qquad \mathrm{P_{III}}$$

$$x'' = \frac{x'^2}{2x} + \frac{3x^3}{2} + 4tx^2 + 2(t^2 - a)x - \frac{b^2}{2x} \qquad \mathrm{P_{IV}}$$

$$x'' = x'^2\left(\frac{1}{2x} + \frac{1}{x-1}\right) - \frac{x'}{t} + \frac{(x-1)^2}{t^2}\left(ax + \frac{b}{x}\right) + c\frac{x}{t} + \frac{dx(x+1)}{x-1} \quad \mathrm{P_V}$$

$$x'' = \frac{x'^2}{2}\left(\frac{1}{x} + \frac{1}{x-1} + \frac{1}{x-t}\right) - x'\left(\frac{1}{t} + \frac{1}{t-1} + \frac{1}{x-t}\right)$$
$$+ \frac{x(x-1)(x-t)}{2t^2(t-1)^2}\left(a - \frac{bt}{x^2} + c\frac{t-1}{(x-1)^2} + \frac{(d-1)t(t-1)}{(x-t)^2}\right) \quad \mathrm{P_{VI}}$$

Here the dependent variable x is a function of the independent variable t, while a, b, c, and d are parameters (constants). These are second order equations in normal form (solved for x''), rational in x' and x.

M.D. Kruskal, B. Grammaticos, and T. Tamizhmani, Three Lessons on the Painlevé Property and the Painlevé Equations, Lect. Notes Phys. **644**, 1–15 (2004)
http://www.springerlink.com/

These may look like more or less random equations, but that is not the case. Apart from some simple transformations they cannot have a form other than shown above. They are very special.

The equations form a hierarchy. Starting from the highest we can, through appropriate limiting processes, obtain the lower ones (after some rescalings and changes of variables):

$$\begin{array}{ccc} P_{VI} \longrightarrow & P_V \longrightarrow & P_{IV} \\ & \downarrow & \downarrow \\ & P_{III} \longrightarrow P_{II} & \longrightarrow P_I \end{array}$$

Note that P_{IV} and P_{III} are at the same level since they can both be obtained from P_V. What makes these equations really special is the fact that they possess the Painlevé property [2].

2 The Painlevé Property and the Naive Painlevé Test

The Painlevé property can be loosely defined as the absence of movable branch points. A glance at the Painlevé equations above reveals the fact that some of them possess *fixed* branch points. Equation P_{III} for instance has $t = 0$ as (fixed) singular point. At such points one can expect bad behaviour, branching, of the solutions. In order to study this one has to go to the complex plane of the independent variable. This is a most interesting feature. Typically when the six Painlevé and similar equations arise from physical applications, the variables are real and t represents physical time, which is quintessentially real. The prototypical example that springs to mind is the "Kowalevski top" [3]. It is surprising that the behaviour of the solution for complex values of t should be relevant.

Kovalevskaya set out to study the integrability of a physical problem, namely the motion of an ideal frictionless top in a uniform gravitational field, spinning around a fixed point in three dimensions, using what today we call singularity-analysis techniques. The equations of motion of a moving Cartesian coordinate system based on the principal axes of inertia with the origin at its fixed point, known as Euler's equations, are:

$$A\frac{dp}{dt} = (B - C)qr + Mg(\gamma y_0 - \beta z_0)$$

$$B\frac{dq}{dt} = (C - A)pr + Mg(\alpha z_0 - \gamma x_0)$$

$$C\frac{dr}{dt} = (A - B)pq + Mg(\beta x_0 - \alpha y_0) \qquad (2.1)$$

$$\frac{d\alpha}{dt} = \beta r - \gamma q$$

$$\frac{d\beta}{dt} = \gamma p - \alpha r$$

$$\frac{d\gamma}{dt} = \alpha q - \beta p$$

where (p, q, r) are the components of angular velocity, (α, β, γ) the direction cosines of the force of gravity, (A, B, C) the moments of inertia, (x_0, y_0, z_0) the centre of mass of the system, M the mass of the top, and g the acceleration of gravity. Complete integrability of the system requires four integrals of motion. Three such integrals are straightforward: the geometric constraint

$$\alpha^2 + \beta^2 + \gamma^2 = 1 \tag{2.2}$$

the total energy

$$Ap^2 + Bq^2 + Cr^2 - 2Mg(\alpha x_0 + \beta y_0 + \gamma z_0) = K_1 \tag{2.3}$$

and the projection of the angular momentum on the direction of gravity

$$A\alpha p + B\beta q + C\gamma r = K_2 \tag{2.4}$$

A fourth integral was known only in three special cases:

i) Spherical: $A = B = C$ with integral $px_0 + qy_0 + rz_0 = K$,

ii) Euler: $x_0 = y_0 = z_0 = 0$ with integral $A^2p^2 + B^2q^2 + C^2r^2 = K$, and

iii) Lagrange: $A = B$ and $x_0 = y_0 = 0$ with integral $Cr = K$.

In each of these cases the solutions of the equations of motion were given in terms of elliptic functions and were thus meromorphic in time t. Kovalevskaya set out to investigate the existence of other cases with solutions meromorphic in t, and found the previously unknown case

$$A = B = 2C \quad \text{and} \quad z_0 = 0 \tag{2.5}$$

with integral

$$[C(p+iq)^2 + Mg(x_0+iy_0)(\alpha+i\beta)][C(p-iq)^2 + Mg(x_0-iy_0)(\alpha-i\beta)] = K \tag{2.6}$$

This case has been dubbed the *Kowalevski top* in her honour.

Using (2.6) Kovalevskaya was able to show that the solution can be expressed as the inverse of a combination of hyperelliptic integrals. Such inverses are not meromorphic in general, but it turns out that the symmetric combinations of hyperelliptic integrals involved in the solution of the Kowalevski top do have meromorphic inverses, called hyperelliptic functions.

Going back to the question of singularities and the Painlevé property, we require that the solutions be free of movable singularities other than poles. (Poles can be viewed as nonsingular values of ∞ on the "complex sphere," the compact closure of the complex plane obtained by adjoining the point at infinity.)

Fixed singularities do not pose a major problem. Linear equations can have only the singularities of their coefficients and thus these singularities are fixed. The case of fixed singularities of nonlinear equations can also be dealt with. Consider for example the $t = 0$ branch point of P_{III}. The change of variable $t = e^z$ removes the fixed singularity by moving it to ∞ (without creating any new singularity in the finite plane). The same or something similar can be done for all the Painlevé equations. Thus we can rationalise ignoring fixed singularities.

The simplest singularities are poles. Consider the equation $x' = x^2$, an extremely simple nonlinear equation. Its solution is $x = -1/(t - t_0)$, with a pole of residue -1 at the point t_0. So no problem arises in this case. (But what about essential singularities? Consider the function $x = ae^{1/(t-t_0)}$, which satisfies the equation $(x''/x - x'^2/x^2)^2 + 4x'^3/x^3 = 0$. This function has no branching but its movable singularity is an essential one, not a pole.) Painlevé himself decreed that any movable singularities should be no worse than poles, i.e. no movable branch points or essential singularities should be present.

Next we can ask for a method to investigate whether there are movable singularities other than poles, the "Painlevé test". There exists a standard practice for the investigation of the Painlevé property which we call the naive Painlevé test [4]. It is not really satisfactory but we can consider it as a useful working procedure. We present an example like P_I but generalised somewhat to

$$x'' = 6x^2 + f(t) \qquad (2.7)$$

where $f(t)$ is an analytic function of its argument in some region. If the solutions are not singlevalued then the equation does not possess the Painlevé property. We use the test to find a condition (on f) for the equation to have all its solutions singlevalued. We look for branched solutions in a straightforward way. Assume $x \sim a(t - t_0)^p$ which is branched unless p is an integer. We look for something like a Laurent series with a leading term (or even Taylor series, depending on the exponents) and write

$$x \sim a_0(t - t_0)^{p_0} + a_1(t - t_0)^{p_1} + \cdots \quad \text{with} \quad \Re p_0 < \Re p_1 < \cdots$$

Looking for branching in such an expansion can be done algorithmically. This is an asymptotic series; we do not care (in the present context) whether it converges. We do not say that this is a solution, only that it is asymptotic to a solution. Since $\Re p_0 < \Re p_1 < \cdots$, the first term is dominant as $t \to t_0$. If there are two codominant leading terms (two terms with the same $\Re p_0$ at dominant order), then even the leading behaviour is bad; however this situation does not arise in practice. If two terms have complex conjugate p_i's at orders other than the dominant one, this violates the condition for asymptoticity, but still the formalism goes through.

We substitute the series into the differential equation and differentiate term by term (though this is generally not allowed for asymptotic series, it is all right here because we are operating at a formal level) and find

$$a_0 p_0 (p_0 - 1)(t - t_0)^{p_0 - 2} + \cdots = 6(a_0^2 (t - t_0)^{2p_0} + \cdots) + f(t_0) + (t - t_0) f'(t_0) + \cdots$$

Using the principle of dominant balance [11] we try to balance as many dominant terms as possible. Looking at the possibly dominant exponents we have

$$p_0 - 2 : 2p_0 : 0$$

and two must be equal and dominate the third for a balance (or all three may be equal). There are three ways to equate a pair of these exponents:

First way. $2p_0 = 0$. Then $p_0 = 0$ and the ignored exponent $p_0 - 2$ is -2, which dominates (has real part less than) the two assumed dominant exponents. So this is not a possible case.

Second way. $p_0 - 2 = 0$. Then $p_0 = 2$ and the ignored exponent $2p_0$ is 4, which is, satisfactorily, dominated by the two assumed dominant exponents. However, p_0 is a integer so no branched behaviour has appeared. A proper treatment would develop the series with this leading behaviour to see whether branching occurs at higher order, but we do not pursue that issue here.

Third way. $p_0 - 2 = 2p_0$. Then $p_0 = -2$ so the two balanced exponents are -4, which is, satisfactorily, less than the other exponent 0. There is no branching to dominant order, but now we will test higher order terms. We assume an expansion in integer powers and determine the coefficients one by one. (A more general procedure is to generate the successive terms recursively and see whether branching such as fractional powers or logarithms arise, as we will demonstrate almost immediately.) Assume $x = \sum_{n=-2}^{\infty} a_n (t - t_0)^n$, substitute into the equation, and obtain a recursion relation for the coefficients. If at some stage the coefficient of a_n vanishes, this is called "resonance". In the case of (2.7) we find a recurrence relation of the form

$$(n + 3)(n - 4)a_n = F_n(a_{n-1}, a_{n-2}, \cdots, a_0) \tag{2.8}$$

where F_n is a definite polynomial function of its arguments. The resonances are at $n = -3$ and 4. The one at -3 is outside the range of meaningful values of n but was to be expected: a formal resonance at $p_0 - 1$ is always present (unless $p_0 = 0$), because infinitesimal perturbation of the free constant t_0 in the leading term gives the derivative with respect to t_0 and thereby the formally dominant power $p_0 - 1$; this is called the "universal resonance". From (2.8) we see that a_4 drops out and so is not determined. Since the only free parameter in the solution we had till now was t_0, it is natural for the second order equation to have another, here a_4. (Of course there is no guarantee that what we find within the assumptions we made, in particular on the dominant balance, will be a general solution.) Since the left side of (2.8) vanishes, the right side must also vanish if a power series is to work. There is no guarantee for this. If the right side is not zero the test fails: the equation does not have the Painlevé property. (More properly, the test succeeds: it succeeds in showing that the equation fails to have the property.) However it turns out that for P_I this condition is indeed satisfied. But what about the generalised equation (2.7)?

We will now present the more general way to set up the recursion to generate a series that is not prejudiced against the actual appearance of terms exhibiting branching when it occurs. What we do, analogous to what Picard did to solve an ordinary differential equation near an ordinary point, is to integrate the equation formally, obtain an integral equation to view as a recursion relation, and iterate it. If we look at the dominant terms of the equation near the singularity (for "the third way" above) we have $x'' = 6x^2 + \cdots$, and these terms we can integrate explicitly after multiplying by $2x'$:

$$x'^2 = 4x^3 + 2xf(t) - 2\int_{t_1}^{t} xf'(t)\, dt$$

No confusion should result from the convenient impropriety of using t for both the variable of integration and the upper limit of integration. The lower limit of choice would have been t_0 but since x behaves dominantly like a double pole there the integrand would not be integrable, so we choose some arbitrary other point t_1 instead.

A second integration of the dominant terms is now possible. For this we take the square root and multiply by the integrating factor $x^{-3/2}$:

$$\frac{1}{2}x^{-3/2}x' = \left(1 + \frac{1}{2x^2}f(t) - \frac{1}{2}x^{-3}\int_{t_1}^{t} xf'(t)\, dt\right)^{1/2} \tag{2.9}$$

One sees immediately that near a singularity the expression in parentheses behaves like "1+ small terms" and can thus be expanded formally. There exists a whole theory of manipulation of formal series but it is not widely identified and taught as such; it is used naively most of the time but, fortunately, in a correct way. Integrating (2.9) and expanding we find

$$-x^{-1/2} = \int_{t_0}^{t}\left(1 + \frac{1}{2x^2}f(t) - \frac{1}{2}x^{-3}\int_{t_1}^{t} xf'(t)dt\right)^{1/2} dt = (t-t_0) + O((t-t_0)^5) \tag{2.10}$$

This time we can integrate from t_0 because the integrand is finite there. From (2.10) we find immediately that the dominant behaviour of x is $(t - t_0)^{-2}$, the double pole as expected. Starting from this we can iterate (2.10) (raised to the -2 power) and obtain an expansion for x with leading term. The only term that might create a problem is $\int xf'(t)\, dt$ which, because of the double pole leading term in the expansion of x, might have a residue and contribute a logarithm. In order to investigate this we expand in the neighbourhood of t_0: $f'(t) = f'(t_0) + f''(t_0)(t - t_0) + \cdots$. The term $f''(t_0)(t - t_0)$ times the double pole, when integrated, gives rise to a logarithm. This multivaluedness is incompatible with the Painlevé property. Thus $f''(t_0)$ must vanish if we are to have the Painlevé property. Since t_0 is an arbitrary point this means that $f''(t) = 0$ and f must be linear. (We can take $f(t) = at + b$ but it is then straightforward to transform it to just $f(t) = t$.) So the only equation of the form (2.7) that has the Painlevé property is P$_\text{I}$.

The technique of integrating dominant terms and generating expansions can be used to analyse the remaining Painlevé equations. P_{II} and P_{III} have simple poles ($x = \infty$), but for the latter $x = 0$ is also singular. Thus here we must consider not only poles but also zeros and ensure that these are pure zeros without logarithms appearing. The Painlevé equations are very special in the sense that they do indeed satisfy the Painlevé property. What is less clear is why they appear so often in applications.

What we presented above is the essence of the naive Painlevé test. Without assuming anything we can seek dominant balances and for each one generate a series for the solution, finding the possible logarithms (and fractional or complex powers) naturally. The main difficulty is in finding all possible dominant behaviours. Some equations have a dominant behaviour that is not power-like. We have seen in the example above an equation with an essential singularity for which the naive Painlevé test would not find anything troublesome. While the solution to that equation was singlevalued, it is straightforward to generate similar examples with branching. Thus, starting from the branched function $x = ae^{(t-t_0)^{-1/2}}$ we obtain the differential equation $2(x''/x - x'^2/x^2)^3 + 27x'^5/x^5 = 0$ for which, again, the naive Painlevé test can say nothing.

These arguments show that one must be very cautious when using the naive Painlevé test [5]. Nevertheless, people have been using it and obtaining results with it. If the naive Painlevé test is satisfied this means that the equation probably has the Painlevé property.

A lot of mysteries remain. While many problems (like the one of Kovalevskaya) are set in real time, one still has to look for branching in the complex plane. It is not clear why one has to look outside the real line. If one thinks of a simple one-dimensional system in Newtonian mechanics, $x'' = F(x)$ with smooth F, it is always possible to integrate it over real time but the equation is, in general, not integrable in the complex plane, nor even analytically extendable there. Another question is, "Why does an equation that passes the Painlevé test behave nicely numerically?" Still, the numerical study of an equation and the detection of chaotic behaviour is an indication that the Painlevé property is probably absent. One should think deeply about these mysteries and try to explain them.

3 From the Naive to the Poly-Painlevé Test

As we have seen the application of the naive Painlevé test makes possible the detection of multivaluedness related to logarithms. But what about fractional powers? We illustrate such an analysis with the differential equation

$$x'' = -\frac{x'^2}{x} + x^5 + \frac{1}{2}tx + \frac{\alpha}{2x} \tag{3.1}$$

We apply the naive test by assuming that the dominant behaviour is $x \sim a\tau^p$ where $\tau = t - t_0$ and $\tau \ll 1$ in the vicinity of the singularity. Furthermore we assume that $a \neq 0$. (The case $a = 0$ seems nonsensical but can be an indication of the existence of logarithms at dominant order.) Substituting into the equation we obtain the possible dominant order terms

$$ap(p-1)\tau^{p-2} \sim -ap^2\tau^{p-2} + a^5\tau^{5p} + \frac{1}{2}t_0 a\tau^p - \frac{\alpha}{2a}\tau^{-p}$$

leading to the comparison of powers $p-2 : p-2 : 5p : p : -p$. The principle of maximum balance [11] requires that two (at least) terms be equal. Balancing $p-2$ with $-p$ gives $p = 1$, which on the face of it gives a simple zero and so no singularity (though one should pursue its analysis to higher order in case a singularity arises later). However here we concentrate on the balance $p-2 = 5p$ which gives $p = -1/2$: a fractional power appears already in the leading order! In view of this result we can conclude, correctly, that the equation does not have the Painlevé property. However, computing the series we find that only half-integer powers appear to all orders. Thus if we square the solution we may find poles as the only singularities. So, while the initial equation does not have the Painlevé property, there exists a simple change of variable which transforms it to an equation that does. Indeed, multiplying (3.1) by x we find

$$xx'' + x'^2 = x^6 + \frac{1}{2}tx^2 + \frac{\alpha}{2}$$

and putting $y = x^2$ we recover the Painlevé II equation

$$y'' = 2y^3 + ty + \alpha$$

the solution of which which has simple poles with leading terms $\pm 1/(t - t_0)$ as its only singularities.

Since at each singular point we have a square root, with a branching into two branches, we have potentially an infinite number of branches. However, as we saw, this is not the case for (3.1). How can we determine, given some equation with many (even infinitely many) branch points, that something like what happened here is possible? The answer to this is the poly-Painlevé test [2,6]. While the naive Painlevé test studies the solution around just one singularity, the poly-Painlevé test considers more than one singularity at a time (hence the name). The idea is that if we start from a first singularity and make a loop around a second one and come back to the first, we may end up on a different branch of it. Thus branching may be detected through the "interaction" of singularities.

To show how this works in a first-order equation we consider equations which are mostly analytic, i.e. equations involving functions which are analytic except for some special singular points. We try to find the simplest nontrivial example. Clearly, linear and quadratic (Riccati) equations are too simple. Thus we choose the cubic equation

$$x' = x^3 + t \qquad (3.2)$$

(Abel's equation) which we have taken to be nonautonomous lest it be integrable through quadratures. Other more or less similar forms could have been considered, for instance $x' = x^3 + tx$. However it turns out that this last equation is a Riccati equation in disguise (for the variable $y = x^2$) and passes the poly-Painlevé test in a trivial way.

Now, the Painlevé test looks for any multivaluedness of a solution in the neighbourhood of a (movable) singularity; if any is found the test "fails" (actually the test succeeds, it's the equation that fails — to be integrable!), and one can go on to the poly-Painlevé test which looks for "bad" (dense) multivaluedness, generally not in the neighbourhood of a single singular point but by following a path winding around several (movable) singular points. Like the Painlevé test it relies on asymptotic expansions of the solution. This means that one must have a small parameter in which to expand.

But equation (3.2) does not contain a small parameter, and if it did, such an "external" parameter wouldn't suit our purpose. We introduce an appropriate "internal" parameter by transforming variables. One way is to look in an asymptotic region with t large (but not approaching infinity), a region where t is approximately constant. We effect this formally by introducing the change of variable $t = N + az$ where N is a large (complex) number ($N \gg 1$), a is a parameter, and z is a new variable (to be thought of as taking "finite" values). We must have az much smaller than N (which means that $a \ll N$) and we expand in the small quantity a/N. We also rescale x through $x = by$ where b is a parameter (which can be of any size, small or large or even finite). The equation now becomes

$$\frac{b}{a}\frac{dy}{dz} = b^3 y^3 + N + az$$

We try to balance the terms as much as possible: $b/a = b^3 = N$ (assuming that we are not at a pole and thus y is finite). We find $b = N^{1/3}$ and $a = N^{-2/3}$ (so a/N is indeed small). The equation now becomes

$$\frac{dy}{dz} = y^3 + 1 + \epsilon z \qquad (3.3)$$

where $\epsilon \sim N^{-5/3}$. Equation (3.3) is autonomous at leading order with a small nonautonomous perturbation. (Here we see an application of another asymptotological [11] principle: transform the problem so that you can treat it by perturbation theory.) The parameter ϵ is an internal one, just like the parameter α in the eponymous α-method of Painlevé.

In order to investigate whether (3.3) has the poly-Painlevé property we start by inverting the roles of the variables, taking z as independent and y as dependent. Introducing $q = 1 + y^3$ we have

$$\frac{dz}{dy} = \frac{1}{q + \epsilon z} = \frac{1}{q} - \frac{\epsilon z}{q^2} + \frac{\epsilon^2 z^2}{q^3} + \cdots$$

Next we expand z in powers of ϵ, $z = z_0 + \epsilon z_1 + \epsilon z_2 + \cdots$, and set up the equations for the z_i recursively. At lowest order we have

$$z_0 = c + \int_{y_0}^{y} \frac{dy}{y^3 + 1}$$

which is not quite as trivial as it appears. Decomposing the integrand into partial fractions we have

$$\frac{1}{y^3 + 1} = \frac{1}{3}\left(\frac{1}{y+1} + \frac{j}{y+j} + \frac{j^2}{y+j^2}\right)$$

where $j = e^{2\pi i/3}$. Thus the integration for z_0 leads to a sum of three logarithms. A single logarithm in the complex plane of the independent variable is defined up to a quantity $2\pi i n$, which would introduce a one-dimensional lattice of values of the integration constant. Two logarithms would lead to a two-dimensional lattice, a multivaluedness still acceptable in the poly-Painlevé spirit. In the present case of three logarithms, the integration constant c *in the complex plane* is defined up to a quantity $2\pi i(k + mj + nj^2)/3$, where k, m, n are arbitrary integers. In general such a multivaluedness involving three integers and arbitrary residues would be dense and thus unacceptable. However, since the three cube roots of unity are related through $1 + j + j^2 = 0$, the multivaluedness of c is not dense. Thus at leading order the equation (3.3) is integrable in the poly-Painlevé sense.

However, to decide the integrability of the full equation (3.2) we must continue with the poly-Painlevé test to higher orders of (3.3). We are not going to give these details here. They can be found in the course of two of the authors (MDK, BG) together with A. Ramani in the 1989 Les Houches winter school [2]. It turns out that while no bad multivaluedness is introduced at the next (first) order, the second-order contribution gives an uncertainty (in the value of the integration constant) that accumulates densely as we go around the singularities. Thus no constant of integration can be defined and the equation is not integrable according to the poly-Painlevé test.

Of course in this problem we studied the behaviour of the solutions only near infinity. So the question is whether we can apply the results obtained near infinity in all regions of the complex plane. The simple answer to this question is that if an equation violates the poly-Painlevé criterion in any region, then this means that the equation is not integrable. However if we find that the poly-Painlevé criterion is satisfied in the region we studied then we cannot conclude that it is so everywhere.

Having dealt with Abel's equation, we return to the case of the second-order equation (2.7), $x'' = 6x^2 + f(t)$, for which we have found that the Painlevé property requires $f(t) = t$. We ask whether some "mild" branching, compatible with the poly-Painlevé property, is possible for this equation. Here we shall work around some finite point and introduce the change of variables $t = t_0 + \delta z$ where $\delta \ll 1$. We scale x through $x = \alpha y$ and rewrite the equation as

$$\frac{\alpha}{\delta^2}\frac{d^2y}{dz^2} = 6\alpha^2 y^2 + f(t_0) + f'(t_0)\delta z + \cdots$$

We balance the terms by taking $\alpha/\delta^2 = \alpha^2$ or $\alpha = \delta^{-2} \gg 1$, since δ is assumed to be small. The equation can now be written as

$$\frac{d^2y}{dz^2} = 6y^2 + \delta^4[f(t_0) + f'(t_0)\delta z + \frac{1}{2}f''(t_0)\delta^2 z^2 + \cdots] \tag{3.4}$$

which is trivially integrable to leading order ($\delta = 0$).

We now treat (3.4) by perturbation analysis. We start by formally integrating it from the dominant terms as before, first multiplying it by $2\,dy/dz$:

$$\left(\frac{dy}{dz}\right)^2 = 4y^3 + 2\int_{z_1}^z y'[\delta^4 f(t_0) + \delta^5 f'(t_0)z + \delta^6 f''(t_0)z^2/2 + \cdots]\,dz$$

or equivalently

$$\frac{1}{2}y^{-3/2}\frac{dy}{dz} = \left(1 + \frac{1}{2y^3}\int_{z_1}^z y'[\delta^4 f(t_0) + \delta^5 f'(t_0)z + \delta^6 f''(t_0)z^2/2 + \cdots]\,dz\right)^{1/2}$$

The integral term is small because of the powers of δ, so the square root can be expanded as 1 plus powers of that term. Integrating the whole equation leads to

$$-y^{-1/2} = (z - z_0) + \frac{1}{4}\int_{z_0}^z \frac{1}{y^3}\left[\int_{z_1}^z y'(\cdots)\,dz\right]dz$$

So $y \sim (z - z_0)^{-2} + \cdots$ and the leading singularity is a double pole as expected. Next we iteratively construct the solution. The problems arise when we integrate y' multiplied by $f''(t_0)z^2$, resulting in a logarithm. The only way to avoid having this logarithm is to have $f''(t_0) = 0$, which as before, since t_0 is arbitrary, means that f must be linear. In this case the poly-Painlevé test has uncovered no equations that don't already satisfy the more stringent naive Painlevé test, that is, no instances of (2.7) whose solutions are free of dense branching other than P_I itself, with no branching at all.

4 The Painlevé Property for the Painlevé Equations

The Painlevé equations possess the Painlevé property, one would say, almost by definition. They were discovered by asking for necessary conditions for this property to be present. But do they really have it? Painlevé himself realized that this had to be shown. He did, in fact, produce a proof which is rather complicated (although it looks essentially correct) [7]. Moreover Painlevé treated only the P_I case, assuming that the remaining equations can be treated in a similar way (something which is not entirely clear). A simple proof thus appeared highly desirable.

In a series of papers [8,9] one of the authors (MDK) together with various collaborators has proposed a straightforward proof of the Painlevé property that can be applied to all six equations. The latest version of this proof is that obtained in collaboration with K.M. Tamizhmani [10]. In what follows we shall outline this proof in the case of P_{III} (for x as a function of z), which is a bit complicated but still tractable:

$$x'' = \frac{x'^2}{x} - \frac{x'}{z} + \frac{1}{z}(ax^2 + b) + cx^3 + \frac{d}{x}$$

To prove that P_{III} has the Painlevé property we must show that in the neighbourhood of any arbitrary movable singular point of the equation (which is not necessarily a singular point of the *solution*) the solution can be expressed as a convergent Laurent expansion with leading term. We shall examine the series up to the highest power where an arbitrary constant may enter ("the last resonance"). We shall not be concerned with the fixed singularity at $z = 0$: it suffices to put $z = e^t$ to send the fixed singular point to infinity without significantly affecting other singularities. Infinity is a bad singularity for the independent variable in all the Painlevé equations, being a limit point of poles. We are only interested here in singularities in the finite plane.

We note that the equation is singular where the dependent variable $x = 0$ (but not the solution, which has a simple Taylor series around this point). The other value of the dependent variable where the equation is singular is $x = \infty$. Any other initial value for x leads, given x', to a solution by the standard theory of ordinary differential equations. Moreover the points 0 and ∞ are reciprocal through the transformation $x \to 1/x$, which leaves the equation invariant up to some parameter changes. Thus the Laurent expansion at a pole is essentially like the Taylor expansion at a zero. This allows us to confine our study to just one of the two kinds of singularity. In order to simplify the calculations we put $a = b = 0$ (which turns out not to change anything significant) and rescale the remaining ones to $c = 1, d = 1$. We have finally the equation

$$uu'' - u'^2 + \frac{uu'}{z} = u^4 - 1 \qquad (4.1)$$

where the possible values of the dependent variable at movable singularities are $u = 0$ and $u = \infty$.

A crucial ingredient of the proof not previously sufficiently exploited is the *localness* of the Painlevé property: if in any given arbitrarily small region (of the finite plane with the origin removed) an *arbitrary* solution has no movable "bad" singularities, then it can have no bad singularities anywhere (in the similarly punctuated finite plane). Use of this localness does away with the difficulties encountered in previous proofs where one had to bound integrals over (finitely) long paths in the complex plane.

Consider some region which is a little disk around z_1 (which we assume to be neither a pole nor a zero) with radius ϵ. (As we have shown in [10]

$\epsilon = |z_1|/96$ suffices for our estimates.) As in the previous lessons we start by formally integrating our equation so as to be able to iterate; the path of integration is to be entirely contained in the little disk. We solve it recursively to obtain an asymptotic series for the solution. Near the singular points the important term on the right side (containing the terms not involving derivatives of u, namely $u^4 - 1$), is u^4 when u is large, and -1 when u is small. In order to integrate (4.1) we need an integrating factor. We start by noting that the left side has the obvious integrating factor $1/(uu')$, after multiplying by which we can write the left side as $[\ln(u'z/u)]'$ or $[(u'z/u)]'/(u'z/u)$, while the right side becomes $(u^4 - 1)/(uu')$. To render the right side integrable we would like to multiply by u'^2 and any function of u alone, while to maintain the integrability of the left side we can multiply by any function of $u'z/u$. If we could do both of these at the same time we would succeed in integrating the equation exactly, which is more than we can hope for. However, here localness enters effectively: in our little disk z is nearly constant, and we can treat it as constant up to small corrections.

Accordingly, we multiply the latest version of the equation by $(u'z/u)^2$ and integrate to

$$\left(\frac{zu'}{u}\right)^2 = z^2\left(u^2 + \frac{1}{u^2}\right) + k - 2\int_{z_1}^z z\left(u^2 + \frac{1}{u^2}\right) dz \qquad (4.2)$$

where the right side has resulted from integration by parts with k as the constant of integration.

Our aim is to show that the solution is regular everywhere in the little disk with center z_1 and radius ϵ. If u and $1/u$ are finite along the path of integration then the integral is small (because length of the integration path is of order ϵ). But what happens when u passes close to 0 or ∞? If z were constant then the solution of (4.1) would be given in terms of elliptic functions. The latter have two zeros and two poles in each elementary parallelogram. When the parameter (here the integration constant) becomes large, the poles (and the zeros) of the elliptic functions get packed closely together. Thus when we integrate we may easily pass close to an ∞ (or a zero). It is important in this case to have a more precise estimate of the integral. To this end we put a little disk around the pole z_0 and assume that on its circumference the value of $|u|$ becomes large, say A. Similarly we can treat the case where $|u|$ is small, say $1/A$, with A large as before. The integration path is now a straight line starting at z_1, till $|u|$ hits the value A (or $1/A$). Then we make a detour around the circumference of the small disk where $|u| > A$ (or $u < 1/A$) and we proceed along the straight line extrapolation of the previous path till we encounter the next singularity. In general the integration path will be a straight line from z_1 to z interspersed with several small detours.

We now make more precise estimates. For definiteness we choose to work with the case of u small, but u large is entirely similar, *mut. mut.* To solve the equation by iteration, we note that the contribution of $1/u^2$ is more important than that of u^2. The integral of z/u^2 may be an important contribution but since it is taken over a short path it is much smaller than the z^2/u^2 term outside the integral in (4.2). The precise bounds can be worked out and the choice of a small enough ϵ guarantees that the integral is indeed subdominant.

Thus (4.2) becomes $(zu'/u)^2 = z^2/u^2$ plus smaller terms or equivalently $u' = \pm(1 + \cdots)$. More precisely we have

$$u' = \pm \left(1 + u^4 + \frac{u^2}{z^2}\left(k - 2\int_{z_1}^z z(u^2 + 1/u^2)\,dz\right)\right)^{1/2} \tag{4.3}$$

Integrating we find $u = \pm(z - z_0) + \cdots$ where z_0 is the point where $u = 0$. (This makes the constant of this last integration exactly zero.) We find thus that u has a simple zero, *if* we can show that no logarithmic term appears in the recursively generated expansion. (We would have found a pole had we worked with a u which became large instead of small).

The dangerous term is the integral

$$I := \int_{z_1}^z z(u^2 + 1/u^2)\,dz$$

because near z_0 u starts with a series like a simple zero and it looks as if z/u^2 may have a nonzero residue, which would produce a logarithmic term. To see that this doesn't happen we can write

$$I' = z/u^2 + \cdots = \frac{z}{u^2}[u' - (u' - 1)] + \cdots$$

$$= \frac{z}{u^2}u' - \frac{z}{u^2}[-\frac{u^2}{z^2}I + \cdots] + \cdots$$

Moving the last explicit term to the left side, multiplying by the integrating factor $1/z$, and integrating gives for I the formula

$$I = -\frac{z}{u} + \cdots$$

We simultaneously iterate for I, u', and u from this, (4.3), and the obvious $u = \int_{z_0}^z u'\,dz$ treated as three coupled equations, and in this form it is clear that no logarithm can be generated. (This is true only for the precise z dependence of (4.1): any other dependence would have introduced a logarithm.)

This completes the proof that the special form of P_{III} (4.1) has the Painlevé property.

Open problems remain. First one has to repeat the proof for the full P_{III} without any special choice of the parameters. Then the proof should be extended to all the other Painlevé equations, including their special cases (where one or more parameters vanish). Still we expect the approach presented above to be directly applicable without any fundamental difficulty.

References

1. E.L. Ince, *Ordinary Differential Equations*, Dover, London, 1956.
2. M.D. Kruskal, A. Ramani and B. Grammaticos, NATO ASI Series C 310, Kluwer 1989, p. 321.
3. S. Kovalevskaya, Acta Math. 12 (1889) 177.
4. M.J. Ablowitz, A. Ramani and H. Segur, Lett. Nuov. Cim. 23 (1978) 333.
5. M.D. Kruskal, NATO ASI B278, Plenum 1992, p. 187.
6. M.D. Kruskal and P.A. Clarkson, Stud. Appl. Math. 86 (1992) 87.
7. P. Painlevé, Acta Math. 25 (1902) 1.
8. N. Joshi and M.D. Kruskal, in "Nonlinear evolution equations and dynamical systems" (Baia Verde, 1991), World Sci. Publishing 1992, p. 310.
9. N. Joshi and M.D. Kruskal, Stud. Appl. Math. 93 (1994), no. 3, 187.
10. M.D. Kruskal, K.M. Tamizhmani, N. Joshi and O. Costin, "The Painlevé property: a simple proof for Painlevé equation III", preprint (2004).
11. M.D. Kruskal, Asymptotology, in Mathematical Models in Physical Sciences (University of Notre Dame, 1962), S. Drobot and P.A. Viebrock, eds., Prentice-Hall 1963, pp. 17-48.

Sato Theory and Transformation Groups. A Unified Approach to Integrable Systems

Ralph Willox[1,2] and Junkichi Satsuma[1]

[1] Graduate School of Mathematical Sciences, University of Tokyo, 3-8-1 Komaba, Meguro-ku, 153-8914 Tokyo, Japan,
{willox, satsuma}@poisson.ms.u-tokyo.ac.jp
[2] Theoretical Physics, Free University of Brussels (VUB), Pleinlaan 2, 1050 Brussels, Belgium

Abstract. More than 20 years ago, it was discovered that the solutions of the Kadomtsev-Petviashvili (KP) hierarchy constitute an infinite-dimensional Grassmann manifold and that the Plücker relations for this Grassmannian take the form of Hirota bilinear identities. As is explained in this contribution, the resulting unified approach to integrability, commonly known as Sato theory, offers a deep algebraic and geometric understanding of integrable systems with infinitely many degrees of freedom. Starting with an elementary introduction to Sato theory, followed by an exposé of its interpretation in terms of infinite-dimensional Clifford algebras and their representations, the scope of the theory is gradually extended to include multi-component systems, integrable lattice equations and fully discrete systems. Special emphasis is placed on the symmetries of the integrable equations described by the theory and especially on the Darboux transformations and elementary Bäcklund transformations for these equations. Finally, reductions to lower dimensional systems and eventually to integrable ordinary differential equations are discussed. As an example, the origins of the fourth Painlevé equation and of its Bäcklund transformations in the KP hierarchy are explained in detail.

1 The Universal Grassmann Manifold

More than 20 years ago, it was discovered by Sato that the solutions of the Kadomtsev-Petviashvili (KP) hierarchy constitute an infinite-dimensional Grassmann manifold (which he called the Universal Grassmann manifold) and that the Plücker relations for this Grassmannian take the form of Hirota bilinear identities [38, 39]. The resulting "unified approach" to integrability, commonly known as *Sato theory* [36], offers a deep algebraic and geometric understanding of integrable systems with infinitely many degrees of freedom and their solutions. At the heart of the theory lies the idea that integrable systems are not isolated but should be thought of as belonging to infinite families, so-called *hierarchies* of mutually compatible systems, i.e., systems governed by an infinite set of evolution parameters in terms of which their (common) solutions can be expressed.

R. Willox and J. Satsuma, Sato Theory and Transformation Groups. A Unified Approach to Integrable Systems, Lect. Notes Phys. **644**, 17–55 (2004)
http://www.springerlink.com/ © Springer-Verlag Berlin Heidelberg 2004

1.1 The KP Equation

One could summarize the original idea of Sato as follows:
Start from an ordinary differential equation and suppose that its solutions satisfy certain dispersion relations, for a set of supplementary parameters. Then, as conditions on the coefficients of this ordinary differential equation, we obtain a set of integrable nonlinear partial differential equations.

Let us show how this recipe allows one to derive the famous KP equation from a particularly simple set of linear dispersion relations. Consider the following second-order linear differential equation, denoting derivatives $\dfrac{df}{dx}$ by f', ... ,

$$f''(x) + a(x)f'(x) + b(x)f(x) = 0 , \tag{1}$$

for which we choose two linearly independent solutions, $f_1(x)$ and $f_2(x)$. The coefficients $a(x)$ and $b(x)$ in (1) can be expressed in terms of these solutions as

$$a(x) = - \begin{vmatrix} f_1 & f_2 \\ f_1'' & f_2'' \end{vmatrix} / \tau(x) , \qquad b(x) = \begin{vmatrix} f_1' & f_2' \\ f_1'' & f_2'' \end{vmatrix} / \tau(x) . \tag{2}$$

The function $\tau(x)$ denotes the Wronski determinant

$$\tau(x) := \begin{vmatrix} f_1 & f_2 \\ f_1' & f_2' \end{vmatrix} . \tag{3}$$

If, as mentioned at the ouset, we now suppose that besides this x-dependence the solutions $f_1(x)$ and $f_2(x)$ also depend on two new parameters, y and t, $(f_i(x; y = 0, t = 0) = f_i(x)$ for $i = 1, 2)$ such that they satisfy the dispersion relations

$$\frac{\partial f_i}{\partial y} = \frac{\partial^2 f_i}{\partial x^2} , \qquad \frac{\partial f_i}{\partial t} = \frac{\partial^3 f_i}{\partial x^3} , \tag{4}$$

then the coefficients a and b will obviously also depend on these new parameters. However, this dependence will be of a much more complicated form than (4) ; $a(x; y, t)$ and $b(x; y, t)$ must satisfy certain *nonlinear* partial differential equations which will turn out to be solvable in terms of the KP equation.

Let us see what kind of partial differential equations we obtain. Denote by $\widehat{W}$ the second-order differential operator acting on $f(x)$ in (1) $(\partial_x := \frac{\partial}{\partial x})$

$$\widehat{W} := \partial_x^2 + a(x; y, t)\partial_x + b(x; y, t) , \tag{5}$$

for which, by definition,

$$\widehat{W} f_i(x; y, t) = 0 , \quad (i = 1, 2) . \tag{6}$$

By differentiating with respect to y we see that the functions f_i also solve the fourth-order differential equation,

$$\widehat{W}_y f_i + \widehat{W} \partial_y f_i \equiv \left[a_y \partial_x + b_y + \widehat{W} \partial_x^2 \right] f_i = 0 , \tag{7}$$

where a_y denotes $\frac{\partial a}{\partial y}$, etc..., which indicates that the fourth-order differential operator $\left[\widehat{W}_y + \widehat{W} \partial_x^2 \right]$ must factorize in terms of $\widehat{W}$ and a suitable second-order differential operator $\widehat{B}_2$,

$$\widehat{W}_y + \widehat{W} \partial_x^2 = \widehat{B}_2 \widehat{W} . \tag{8}$$

If we parametrize $\widehat{B}_2$ as $\widehat{B}_2 := \partial_x^2 + \alpha_2(x; y, t)\partial_x + \beta_2(x; y, t)$, its coefficients are obtained from (8),

$$\alpha_2 = 0 , \qquad \beta_2 = -2a_x , \tag{9}$$

under the following nonlinear conditions on a and b,

$$a_y = a_{2x} - 2aa_x + 2b_x$$
$$b_y = b_{2x} - 2a_x b . \tag{10}$$

Similarly, from the t derivative of (6), one obtains the following factorization of a sixth-order operator,

$$\widehat{W}_t + \widehat{W} \partial_x^3 = \widehat{B}_3 \widehat{W} , \quad \widehat{B}_3 = \partial_x^3 + \alpha_3 \partial_x^2 + \beta_3 \partial_x + \gamma_3 , \tag{11}$$
$$\alpha_3 = 0 , \quad \beta_3 = -3a_x , \quad \gamma_3 = -3a_{2x} + 3a_x a - 3b_x ,$$

under the conditions :

$$a_t = a_{3x} + 3b_{2x} - 3a(b_x + a_{2x}) - 3a_x^2 - 3ba_x + 3a^2 a_x$$
$$b_t = b_{3x} - 3ba_{2x} - 3a_x b_x - 3bb_x + 3aba_x . \tag{12}$$

Hence, by assuming a simple linear parameter-dependence for the solutions of the linear ordinary differential equation (1) – as in (4) – we obtain a far more interesting system of nonlinear partial differential equations (10) and (12) for the parameter-dependence of the coefficients of that ordinary differential equation. The system (10,12) can be seen to be equivalent to the KP equation. If we use the factorizations (8) and (11) to express the equality of the cross-derivatives $(\widehat{W}_y)_t$ and $(\widehat{W}_t)_y$, we obtain a compatibility condition for the operators $\widehat{B}_2$ and $\widehat{B}_3$,

$$(\widehat{B}_2)_t - (\widehat{B}_3)_y = \left[\widehat{B}_3, \widehat{B}_2 \right]_- , \tag{13}$$

where $[A, B]_- := AB - BA$. It is straightforward to show that (13) amounts to the *KP equation*,

$$(4u_t - 12uu_x - u_{3x})_x - 3u_{2y} = 0 , \tag{14}$$

in the field $u(x, y, t) := (-a(x; y, t))_x$.

Furthermore, since solutions a and b to (10) and (12) can be expressed in terms of the function $\tau(x; y, t)$ (2), subject to (4),

$$a(x; y, t) = \frac{-\tau_x(x; y, t)}{\tau(x; y, t)} , \quad b(x; y, t) = \frac{\tau_{2x}(x; y, t) - \tau_y(x; y, t)}{2\, \tau(x; y, t)} , \tag{15}$$

it follows that the solution $u(x, y, t)$ to the KP equation derived from $a(x, y, t)$ is also completely determined by this function,

$$u(x, y, t) = \partial_x^2 \log \tau(x; y, t) . \tag{16}$$

Exercise 1.1. Show that (13) really yields the KP equation (14).

1.2 Plücker Relations

It is the function $\tau(x; y, t)$ that will turn out to be the single most important object in Sato theory. In fact, it is directly connected to the notion of an infinite-dimensional Grassmann manifold. To demonstrate this, we start by expanding the solutions $f_1(x)$ and $f_2(x)$ of the original differential equation (1) around a common point of analyticity, say, $x = 0$ for simplicity, $(i = 1, 2)$

$$f_i(x) = \sum_{j=0}^{\infty} \zeta_j^i \frac{x^j}{j!} , \quad \zeta_j^i = \frac{d^j f_i}{dx^j}\bigg|_{x=0} , \tag{17}$$

from the coefficients of which we construct a (rank 2) $\infty \times 2$ matrix,

$$\zeta_0 := \begin{pmatrix} \zeta_0^1 & \zeta_0^2 \\ \zeta_1^1 & \zeta_1^2 \\ \zeta_2^1 & \zeta_2^2 \\ \vdots & \vdots \end{pmatrix} . \tag{18}$$

Observe that, due to (3), $\tau(x; 0, 0)$ is completely determined by the entries in this matrix. However, due to the linearity of the differential equation (1), the matrix ζ_0 itself is only defined up to right-multiplication with an element of $GL(2, \mathbb{C})$, i.e., a non-singular 2×2 matrix. The resulting change in $\tau(x)$ being but a mere multiplication with the determinant of that transformation matrix, these transformations obviously leave a, b and thus also u invariant.

Definition 1.1. *Given an n-dimensional vector space V, then the Grassmann manifold GM(m; n) (or Grassmannian for short) is defined as the set of all m-dimensional linear subspaces of V.*

Alternatively, one may also think of $GM(m; n)$ as the quotient space obtained by the right-action of the Lie group $GL(m, \mathbb{C})$ on the manifold $M(m, n)$ of all $n \times m$ matrices of rank m, $M(m, n)/GL(m, \mathbb{C})$. In particular, $GM(m; n)$ is $m(n - m)$-dimensional (see, e.g., [30] or [41] for further details on Grassmann manifolds).

By extension of these ideas, it can be shown that the set of all $\infty \times 2$ matrices ζ_0, defined up to the right-action of $GL(2, \mathbb{C})$, constitutes an infinite-dimensional Grassmann manifold, in this case denoted by $GM(2; \infty)$.

It is instructive however to dwell a little longer on the case of a finite-dimensional Grassmannian, the simplest (non-trivial) example of which is $GM(2; 4)$, i.e., the set W of all 2-dimensional planes passing through the origin of a 4-dimensional vector space V. In practice, one needs to introduce a coordinate system on this Grassmannian. If we take v_i $(i = 1, \ldots, 4)$ to be basis vectors for the vector space V $(dim(V) = 4)$, we can express a basis $\{w_1, w_2\}$ for a 2-dimensional plane by means of the coordinates ζ_{ij} of the w_j in the $\{v_i\}$ basis,

$$w_i = \sum_{j=1}^{4} \zeta_{ji} v_j . \tag{19}$$

The 4×2 matrix $(\zeta_{ij})_{4\times2} \in M(4, 2)$ is called a *frame* of W and we can think of the Grassmannian $GM(2; 4)$ as the quotient space $M(4, 2)/GL(2, \mathbb{C})$.

Now, using the minor determinants of the frame $(\zeta_{ij})_{4\times2}$, we can define the following homogeneous coordinates in 5-dimensional projective space ($\mathbb{P}^5$),

$$\begin{aligned}
\xi &= (\xi_{12} : \xi_{13} : \xi_{14} : \xi_{23} : \xi_{24} : \xi_{34}) \\
&:= \left(\begin{vmatrix} \zeta_{11} & \zeta_{12} \\ \zeta_{21} & \zeta_{22} \end{vmatrix} : \begin{vmatrix} \zeta_{11} & \zeta_{12} \\ \zeta_{31} & \zeta_{32} \end{vmatrix} : \begin{vmatrix} \zeta_{11} & \zeta_{12} \\ \zeta_{41} & \zeta_{42} \end{vmatrix} : \begin{vmatrix} \zeta_{21} & \zeta_{22} \\ \zeta_{31} & \zeta_{32} \end{vmatrix} : \begin{vmatrix} \zeta_{21} & \zeta_{22} \\ \zeta_{41} & \zeta_{42} \end{vmatrix} : \begin{vmatrix} \zeta_{31} & \zeta_{32} \\ \zeta_{41} & \zeta_{42} \end{vmatrix} \right) .
\end{aligned} \tag{20}$$

Since the action of $GL(2, \mathbb{C})$ on w_1, w_2 only results in the multiplication of each minor by the determinant of the transformation matrix, it is clear that ξ is invariant under such transformations. In this way we see that $GM(2; 4)$ can also be regarded as a 4-dimensional subvariety of $\mathbb{P}^5$. The homogeneous coordinates ξ are called the *Plücker coordinates* of this Grassmannian. It is worth observing that a Grassmann manifold $GM(m; n)$ can always be embedded in the projective space $\mathbb{P}^{\binom{n}{m}-1}$ [30, 41].

It is important to realize however that the Plücker coordinates are not independent ; they satisfy (and are in fact fully characterized by) a set of nonlinear algebraic relations which are called the *Plücker relations*. In the case of $GM(2; 4)$ there is only one such relation, which can be obtained from the Laplace expansion of the (0) determinant,

$$\begin{vmatrix} \zeta_{11} & \zeta_{12} & 0 & 0 \\ \zeta_{21} & \zeta_{22} & \zeta_{21} & \zeta_{22} \\ \zeta_{31} & \zeta_{32} & \zeta_{31} & \zeta_{32} \\ \zeta_{41} & \zeta_{42} & \zeta_{41} & \zeta_{42} \end{vmatrix} = \xi_{12}\xi_{34} - \xi_{13}\xi_{24} + \xi_{14}\xi_{23} = 0 . \tag{21}$$

Since $GM(2;4)$ is the first non trivial subvariety of the infinite-dimensional Grassmannian $GM(2;\infty)$, (21) is of course also the simplest Plücker relation for the Grassmannian $GM(2;\infty)$. We shall now see that this Plücker relation actually encodes the KP equation.

1.3 The KP Equation as a Dynamical System on a Grassmannian

Let us introduce an evolution with respect to the x-coordinate in $GM(2;\infty)$. This can be done by means of the shift matrix (see, e.g., [36] for a detailed account)

$$\Lambda := \begin{pmatrix} 0 & 1 & 0 & \cdots \\ 0 & 0 & 1 & 0 & \cdots \\ & & & \ddots & \ddots & \ddots \end{pmatrix} . \tag{22}$$

It is easily seen that

$$\exp(x\Lambda) = \begin{pmatrix} 1 & x & \frac{x^2}{2!} & \frac{x^3}{3!} & \cdots \\ 0 & 1 & x & \frac{x^2}{2!} & \cdots \\ 0 & 0 & 1 & x & \cdots \\ & & & \ddots & \ddots \end{pmatrix} , \tag{23}$$

which allows us to define the x evolution $\zeta(x)$ of the matrix ζ_0 (18) as

$$\zeta(x) := \exp(x\Lambda)\,\zeta_0 \equiv \begin{pmatrix} f_1 & f_2 \\ f_1' & f_2' \\ f_1'' & f_2'' \\ \vdots & \vdots \end{pmatrix} . \tag{24}$$

The y and t dependencies can be introduced in a similar way,

$$\zeta(x;y,t) := \exp(x\Lambda + y\Lambda^2 + t\Lambda^3)\,\zeta_0 = \begin{pmatrix} h_1 & h_2 \\ h_1' & h_2' \\ h_1'' & h_2'' \\ \vdots & \vdots \end{pmatrix} , \tag{25}$$

for functions $h_i(x,y,t)$ $(i = 1,2)$ that satisfy the conditions :

$$h_i(x,0,0) = f_i(x), \qquad \frac{\partial h_i}{\partial y} = \frac{\partial^2 h_i}{\partial x^2}, \qquad \frac{\partial h_i}{\partial t} = \frac{\partial^3 h_i}{\partial x^3}. \tag{26}$$

These conditions are however identical to (4), identifying the functions $h_i(x, y, t)$ with the $f_i(x; y, t)$, $i = 1, 2$, and thus we have succeeded in introducing the parameter dependence (4) into the Grassmannian $GM(2; \infty)$. The evolution of the function $\tau(x; y, t)$ then corresponds to the motion $\zeta(x; y, t)$ of an initial point ζ_0 on $GM(2; \infty)$, under the action of the 3-parameter transformation group $\exp(x\Lambda + y\Lambda^2 + t\Lambda^3)$.

We can now translate Plücker relation (21) for $GM(2; \infty)$ into an equation for $\tau(x; y, t)$. If we introduce the frame (25)

$$(\zeta_{ij})_{\infty \times 2} \equiv (\partial_x^{i-1} f_j)_{\infty \times 2} , \tag{27}$$

we immediately obtain the Plücker coordinate $\xi_{12} \equiv \tau(x; y, t)$. Differentiation with respect to x, y, t yields the remaining coordinates (20): $\xi_{13} = \partial_x \tau$, $\xi_{14} = \frac{1}{2}(\partial_x^2 + \partial_y)\tau$, $\xi_{23} = \frac{1}{2}(\partial_x^2 - \partial_y)\tau$, $\xi_{24} = \frac{1}{3}(\partial_x^3 - \partial_t)\tau$ and $\xi_{34} = \frac{1}{12}(\partial_x^4 + 3\partial_y^2 - 4\partial_x \partial_t)\tau$. With the help of these expressions, the Plücker relation (21) can be transformed into a quadratic relation for $\tau(x; y, t)$, which can be rewritten as

$$\left(D_x^4 - 4D_x D_t + 3D_y^2\right) \tau \cdot \tau = 0 \tag{28}$$

in terms of the Hirota operators [16],

$$D_{x_1}^{m_1} \cdots D_{x_n}^{m_n} F \cdot G = \left(\partial_{\epsilon_1}^{m_1} \cdots \partial_{\epsilon_n}^{m_n}\right) F(x_1 + \epsilon_1, \dots, x_n + \epsilon_n)$$
$$G(x_1 - \epsilon_1, \dots, x_n - \epsilon_n)\big|_{\epsilon_i = 0} \forall i. \tag{29}$$

This is nothing but the Hirota bilinear form of the KP equation (14).

Exercise 1.2. Show that the KP equation can be obtained from (28) by means of the "bilinearizing transformation" (16).

1.4 Generalization to the KP Hierarchy

If one introduces infinitely many evolution parameters, $t = (t_1, t_2, t_3, \dots)$, the functions $\tau(t_1, t_2, t_3, \dots)$ will correspond to the orbits

$$\zeta(t) := \exp(\sum_{n=1}^{\infty} t_n \Lambda^n) \, \zeta_0 \equiv \sum_{n=0}^{\infty} p_n(t) \Lambda^n \, \zeta_0 , \tag{30}$$

where, as compared to the above, the variable t_1 plays the rôle of x and t_2 and t_3 those of y and t respectively : we shall adhere to this convention from here on.

Definition 1.2. *The Schur polynomials, $p_n(t)$ ($n \in \mathbb{N}$), are defined by the following generating formula,*

$$\exp(\sum_{n=1}^{\infty} t_n z^n) \equiv \sum_{m=0}^{\infty} p_m(t) z^m. \tag{31}$$

Exercise 1.3. Calculate Schur polynomials $p_0, p_1 \cdots$ up to p_5 explicitly.

This generalization to infinitely many evolution parameters, accompanied by a careful limit $GM(m \to \infty; \infty)$, is necessary if one wants to capture all the evolutions contained in the KP hierarchy. For it can be shown that [38]

Theorem 1.1 (Sato 1981). *The solution-space of the KP hierarchy is isomorphic to the infinite-dimensional Grassmannian $GM(\infty/2; \infty)$ whose Plücker relations take the form of Hirota bilinear identities for the equations in the KP hierarchy. The evolution of a KP τ-function is defined by the motion of a point on that Grassmannian, under the action of the Abelian infinite-parameter group* (30).

In particular, the τ-function can be expressed in terms of Plücker coordinates ξ_Y for $GM(\infty/2; \infty)$ and vice versa,

$$\tau(t) = \sum_Y \xi_Y \, \chi_Y(t) \quad \text{or} \quad \xi_Y = \chi_Y(\tilde{\partial}_t)\tau(t)|_{t=0} , \tag{32}$$

where $\tilde{\partial}_t := (t_1, \frac{1}{2}t_2, \frac{1}{3}t_3, \dots)$. The symbol $\chi_Y(t)$ denotes the character-polynomials associated with the irreducible tensor representations of $GL(n)$, classified in terms of Young diagrams, Y. The interested reader is referred to [36, 40] for a more detailed explanation, proofs and for some explicit examples.

2 Wave Functions, τ-Functions and the Bilinear Identity

It is intuitively clear that if one wishes to construct the Universal Grassmann Manifold $GM(\infty/2; \infty)$ by mimicking the construction of $GM(2; 4)$, performed in the previous sections, one would have to start from a linear differential operator of infinite-order and study its deformations in terms of infinitely many auxiliary parameters. This is exactly the point where so-called *pseudo-differential operators* [36, 12] come into play.

2.1 Pseudo-differential Operators

We start by pointing out that the second-order operator $\widehat{W}$ (5) can be written as

$$\widehat{W} = W^{(2)}\partial^2 , \quad W^{(2)} := 1 + \alpha_1(x; y, t)\partial^{-1} + \alpha_2(x; y, t)\partial^{-2} , \tag{33}$$

where $\partial^m := \partial_x^m$, when $m \geq 0$, and ∂^{-1} is defined such that $\partial\partial^{-1} = \partial^{-1}\partial = 1$.

Definition 2.1. *A pseudo-differential operator $A(\partial)$ is a linear operator,*

$$A(\partial) = \sum_{j \ll +\infty} a_j(x)\partial^j , \tag{34}$$

where the symbol ∂, *as an operator, is defined by its action on a function* $f(x)$: $\partial(f) := f_x$, *and* ∂^{-1} *is defined by* $\partial\partial^{-1} = \partial^{-1}\partial \equiv 1$.

A sum of pseudo-differential operators is defined in the usual way by collecting terms, and their product is defined by the following extension of Leibniz' s rule,

$$A(\partial)B(\partial) = \sum_{i,j \ll +\infty} a_i\partial^i b_j\partial^j = \sum_{i,j \ll +\infty} \sum_{k=0}^{\infty} \binom{i}{k} a_i(b_j)_{kx}\, \partial^{i+j-k} , \tag{35}$$

where ($\forall i \in \mathbb{Z}$)

$$\binom{i}{k} = \begin{cases} \frac{i(i+1)\cdots(i+k-1)}{k!} & \text{for } k \geq 1 \\ 1 & \text{for } k = 0 \end{cases} ; \tag{36}$$

$(A(\partial))_+ := \sum_{j \geq 0} a_j\partial^i$ denotes the so-called "differential" part of a pseudo-differential operator and its complement, $A(\partial) - (A(\partial))_+$, is denoted by $(A(\partial))_-$. A pseudo-differential operator possesses a unique inverse, denoted simply by $A(\partial)^{-1}$, and its formal adjoint can be calculated from (35) :

$$A(\partial)^* := \sum_i (-1)^i \partial^i a_i = \sum_i (-1)^i \sum_{k=0}^{+\infty} \binom{i}{k} a_{i,kx} \partial^{i-k} . \tag{37}$$

2.2 The Sato Equation and the Bilinear Identity

Extending (33), we define a pseudo-differential operator,

$$W := 1 + \sum_{j=1}^{+\infty} w_j(t)\partial^{-j} , \tag{38}$$

called the *gauge* operator, whose coefficients w_j depend on infinitely many parameters, $t = (t_1 \equiv x, t_2, t_3, \cdots)$, as introduced in Sect. 1.4.

In order to generalize the line of thought running through Sect. 1.1, we need to define differential operators, B_n ($\forall n \geq 1$),

$$B_n := (W\partial^n W^{-1})_+ , \tag{39}$$

which will provide us with an extension of the factorizations (8) and (11) to arbitrary orders ($n = 2, 3, \ldots$),

$$W_{t_n} = B_n W - W\partial^n = -(W\partial^n W^{-1})_- W . \tag{40}$$

This equation is known as the *Sato equation* [36] and it provides a collective description of the t_n evolutions of the coefficient functions $w_j(t)$, i.e., the Sato equation actually encodes a "doubly infinite" sequence of partial differential equations for the $w_j(t)$, generalizing (10) and (12).

Furthermore, the gauge operator W can be used to define the operator

$$L := W \partial W^{-1} , \tag{41}$$

which will turn out to be of crucial importance, not least because it underlies the differential operators B_n,

$$B_n \equiv (L^n)_+ . \tag{42}$$

Let us now define so-called *wave functions* and *adjoint wave functions*.

Definition 2.2. *A wave function (adjoint wave function) $\Psi(t, \lambda)$ $(\Psi^*(t, \lambda))$ is defined by the expression*

$$\Psi(t, \lambda) := W(t, \lambda) \exp \xi(t, \lambda) \qquad \left(\Psi^*(t, \lambda) := W^{*-1}(t, \lambda) \exp -\xi(t, \lambda) \right) , \tag{43}$$

for

$$\xi(t, \lambda) := \sum_{n=1}^{\infty} t_n \lambda^n \tag{44}$$

and the formal Laurent series in λ,

$$W(t, \lambda) := 1 + \sum_{j=1}^{\infty} w_j(t) \lambda^{-j} , \tag{45}$$

is obtained from the gauge operator W by setting $\partial \to \lambda^{-1}$. $W^{*-1}(t, \lambda)$ can similarly be obtained from (37).

This leads to the following proposition [38, 11, 12],

Proposition 2.1 (Sato). *If a pseudo-differential operator W satisfies the Sato equation (40), then the operators L and B_n, obtained from W by means of (41) and (39), satisfy $\forall n, m$,*

$$L_{t_n} = [B_n, L]_- \tag{46}$$

$$(B_n)_{t_m} - (B_m)_{t_n} = [B_m, B_n]_- . \tag{47}$$

Furthermore, the wave function $\Psi(t, \lambda)$ and adjoint wave function $\Psi^*(t, \lambda)$ which can be derived from W satisfy the linear systems,

$$\begin{cases} L\Psi(t, \lambda) = \lambda \Psi(t, \lambda) \\ \partial_{t_n} \Psi(t, \lambda) = B_n \Psi(t, \lambda) \end{cases} \qquad \begin{cases} L^*\Psi^*(t, \lambda) = \lambda \Psi^*(t, \lambda) \\ \partial_{t_n} \Psi^*(t, \lambda) = -B_n^* \Psi^*(t, \lambda) \end{cases} , \tag{48}$$

as well as the bilinear identity,

$$Res_{\lambda=\infty} [\Psi(t, \lambda) \Psi^*(t', \lambda)] = 0 \qquad \forall t, t' . \tag{49}$$

In (49), *Res* denotes the operation $A(\lambda) := \sum_i a_i \lambda^i$, $Res_{\lambda=\infty} [A(\lambda)] := a_{-1}$.

Several remarks are in order. First of all it should be clear that (47) generalizes (13) from Sect. 1.1. Secondly, forgetting for a moment the W-origins of the operators L and B_n and simply parametrizing L as

$$L := \partial + u_1(t)\partial^{-1} + u_2(t)\partial^{-2} + \cdots , \qquad (50)$$

equations (46) and (42) will yield an infinite system of partial differential equations in the coefficients $u_j(t)$.

Exercise 2.1. Show that at $n = 2, 3$, the so-called *Lax equation* (46) yields a system of equations from which, by elimination of $u_2, u_3, \cdots$, the KP equation expressed in the field $u_1(t)$ is obtained.

Observe that, due to (41), the coefficients $u_j(t)$ can easily be connected to those of the gauge operator W, e.g., $u_1(t) = -(w_1(t))_x$, etc... More improtantly however, observe also that due to Prop. 2.1, all the equations obtained for the $w_j(t)$ or $u_j(t)$ are mutually compatible.

Corollary 2.1 ([38, 12]).

$$\partial_{t_m}\partial_{t_n}W = \partial_{t_n}\partial_{t_m}W \quad \forall m, n. \qquad (51)$$

In fact, the *KP hierarchy* is the set of nonlinear $(2+1)$-dimensional evolution equations expressed in $u_1(t)$ that can be obtained from (46) and (47). In turn, these equations can be thought of as the compatibility conditions of the linear systems (48) which underlie the KP hierarchy. The *bilinear* identity (49) however, as we shall see in the next section, encodes both the equations of the KP hierarchy in their Hirota bilinear forms as well as their associated linear formulations (48). A remark regarding the nature of the equations in the system (48) is in order here. As the "action" (as a differential operator) of ∂^{-1} is not defined on a function, the first equation in each system in (48) should be thought of as a *formal relation* linking the coefficients in the (formal) Laurent expansions on both sides of the equality, defining "$\partial^{-1}\exp\xi(t, \lambda)$" to be $\lambda^{-1}\exp\xi(t, \lambda)$. The second set of equations in each system however only consists of differential equations, whose compatibility conditions are given by (47). The compatibility of these differential equations with the formal relations in (48) is guaranteed by (46).

The converse of Prop.2.1 can be formulated as follows [39, 11, 12].

Proposition 2.2 (Sato). *If $\Psi(t, \lambda)$ and $\Psi^*(t, \lambda)$ of the form*

$$\Psi(t, \lambda) = W(\lambda)e^{\xi(t,\lambda)} , \quad W(\lambda) = 1 + w_1(t)\lambda^{-1} + \cdots \qquad (52)$$
$$\Psi^*(t, \lambda) = V(-\lambda)e^{-\xi(t,\lambda)} , \quad V(-\lambda) = 1 + v_1(t)(-\lambda)^{-1} + \cdots , \qquad (53)$$

are solutions to the bilinear equation (49), then the pseudo-differential operator $W(\partial)$ solves the Sato equation (40) and $V(-\lambda) \equiv W^{*-1}(-\lambda)$, i.e., $\Psi(t, \lambda)$ and $\Psi^*(t, \lambda)$ are, respectively, a wave function and an adjoint wave function.

2.3 τ-Functions and the Bilinear Identity

Just as the coefficients of the second-order operator $\widehat{W}$ in Sect. 1.1 could be expressed in terms of a particular (Wronski) determinant $\tau(x)$, so can the gauge operator W, and with it the entire KP hierarchy and its associated linear formulations, be expressed in terms of just a single function $\tau(t)$, the KP τ-function.

The reader is referred to [36] for a more detailed discussion of this construction, based on an mth-order extension of the approach that was adapted in Sect. 1.1, or to [39] for a Grassmannian based explanation. In Sect. 3 a third approach will be introduced, based on the representation theory of a fermionic algebra. For now we simply state the main result [39, 11, 12].

Proposition 2.3 (Sato). *There exists a function $\tau(t)$, in terms of which the coefficients of W and V in (52) and (53) can be expressed as*

$$w_j = \frac{p_j(-\tilde{\partial})\tau}{\tau} \quad \text{and} \quad v_j = \frac{p_j(\tilde{\partial})\tau}{\tau} \qquad \forall j \geq 1 \ . \tag{54}$$

The p_j are the Schur polynomials (31), and the "weighted" differential operators $\tilde{\partial}_t := (t_1, \frac{1}{2}t_2, \frac{1}{3}t_3, \dots)$ were already introduced in (32) (Sect. 1.4).

Observing that, due to the definition of the Schur polynomials (at least formally [12]),

$$\sum_{n=0}^{\infty} p_n(-\tilde{\partial})\tau(t)\lambda^{-n} = \tau(t - \varepsilon[\lambda]) \ , \tag{55}$$

one finds that $\Psi(t, \lambda)$ and $\Psi^*(t, \lambda)$ can be expressed as

$$\Psi(t, \lambda) = \frac{\tau(t - \varepsilon[\lambda])}{\tau(t)} \, e^{\xi(t,\lambda)}, \quad \Psi^*(t, \lambda) = \frac{\tau(t + \varepsilon[\lambda])}{\tau(t)} \, e^{-\xi(t,\lambda)} \ , \tag{56}$$

where $\xi(t, \lambda)$ is as in (44) and the *shift* $\varepsilon[\lambda]$ on the coordinates t stands for the infinite sequence

$$\varepsilon[\lambda] = (\frac{1}{\lambda}, \frac{1}{2\lambda^2}, \frac{1}{3\lambda^3}, \cdots) \ . \tag{57}$$

Hence, one can reformulate identity (49) in terms of $\tau(t)$ only

$$Res_{\lambda=\infty} \left[\tau(t - \varepsilon[\lambda]) \, \tau(t' + \varepsilon[\lambda]) \, e^{\xi(t-t',\lambda)} \right] = 0 \qquad \forall t, t' \ . \tag{58}$$

Actually, this identity encodes all the evolution equations which make up the KP hierarchy when written in Hirota bilinear form. This can be easily seen by changing to new variables, x and y, as in $t = x - y$ and $t' = x + y$. Direct calculation of Res in (58) then yields

$$\exp(\sum_{i=1}^{\infty} y_i D_{x_i}) \sum_{j=0}^{\infty} p_j(-2y) p_{j+1}(\tilde{D}) \tau \cdot \tau = 0 \qquad \forall y , \qquad (59)$$

where the symbol $\tilde{D}$ stands for the sequence of "weighted" Hirota D-operators, $\tilde{D} = (D_{x_1}, D_{x_2}/2, D_{x_3}/3, \dots)$. The above expression is nothing but a generating formula for all the Hirota bilinear equations in the KP hierarchy, e.g., as the coefficients of y_3 and y_4 in (59) one finds the following equations,

$$\begin{aligned} (4D_{x_1} D_{x_3} - 3D_{x_2}^2 - D_{x_1}^4) \tau \cdot \tau = 0 \\ (3D_{x_1} D_{x_4} - D_{x_2} D_{x_1}^3 - 2D_{x_2} D_{x_3}) \tau \cdot \tau = 0 , \end{aligned} \qquad (60)$$

the first of which is the KP equation in bilinear form (28).

Exercise 2.2. Derive (59) from (58).

Exercise 2.3. Compute the coefficient of y_n in formula (59) for general n, in order to find bilinear expressions for all flows x_n in the KP hierarchy. Demonstrate (60).

Observe that because of the use of the Hirota bilinear operators, equations (60) are not explicitly in $(2 + 1)$-dimensional form since the "higher weight" bilinear equation involves a Hirota operator corresponding to the "lower weight" time variable x_3. Elimination of this "lower" time variable is needed to obtain a genuine $(2 + 1)$-dimensional equation governing the x_4 flow.

To show that the bilinear identity also encompasses the linear formulation of the KP hierarchy, it is convenient to rewrite (58) as an integral identity [11],

$$\oint_{C_\lambda} \frac{d\lambda}{2\pi i} \tau(t - \varepsilon[\lambda]) \tau(t' + \varepsilon[\lambda]) e^{\xi(t-t',\lambda)} = 0 \qquad \forall t, t' , \qquad (61)$$

for a narrow loop C_λ in the complex plane around $\lambda \approx \infty$. Then, introducing 3 points ν_i $(i = 1, 2, 3)$ inside C_λ, i.e., $\forall i \, |\nu_i| > |\lambda|$, and identifying t' as $t' \equiv t - \sum_{i=1}^{3} \varepsilon[\nu_i]$, we can easily compute the integral in (61),

$$\begin{aligned} (\nu_2 - \nu_3) \, \tau(t - \varepsilon[\nu_1]) \tau(t - \varepsilon[\nu_2] - \varepsilon[\nu_3]) \\ + (\nu_3 - \nu_1) \, \tau(t - \varepsilon[\nu_2]) \tau(t - \varepsilon[\nu_3] - \varepsilon[\nu_1]) \\ + (\nu_1 - \nu_2) \, \tau(t - \varepsilon[\nu_3]) \tau(t - \varepsilon[\nu_1] - \varepsilon[\nu_2]) = 0 . \quad (62) \end{aligned}$$

Exercise 2.4. Show that $\exp \sum_{n=1}^{\infty} \frac{1}{n} (\frac{\lambda}{\nu})^n = \frac{\nu}{\nu - \lambda}$ for $|\nu| > |\lambda|$, i.e., the particular choice of t' made above turns the essential singularity at $\lambda = \infty$ in the bilinear identity into simple poles at ν_i. Use this result to calculate (62) from (61).

The quadratic expression in the τ-functions (62) is often called the *Fay identity* for the KP hierarchy, since it is connected to Fay's tri-secant formula [14] for theta functions. It first appeared in the KP context in [39] where it was pointed out that this identity actually includes all useful information about the KP hierarchy, such as the Hirota forms of the evolution equations or the underlying linear formulations. As we shall see later on, it is also closely related to the discretization procedure for the KP equation.

To show that (62) contains all information about the linear system underlying the KP hierarchy, it suffices, e.g., to take the limit $\nu_3 \to \infty$ which, at $o(\nu_3^0)$ yields,

$$(\nu_2 - \nu_1)\,[\tau(t - \varepsilon[\nu_1])\tau(t - \varepsilon[\nu_2]) - \tau(t)\tau(t - \varepsilon[\nu_1] - \varepsilon[\nu_2])]$$
$$+ \tau(t - \varepsilon[\nu_1])\,(\tau(t - \varepsilon[\nu_2]))_x - (\tau(t - \varepsilon[\nu_1]))_x\,\tau(t - \varepsilon[\nu_2]) = 0 \ . \quad (63)$$

Then, introducing the wave function, $\Psi(t) := \tau(t - \varepsilon[\nu_2])/\tau(t)\,\exp\xi(t,\nu_2)$, this equation can be reformulated as

$$\nu_1[\Psi(t) - \Psi(t - \varepsilon[\nu_1])] = \Psi(t)_x + \Psi(t)[\log\tau(t) - \log\tau(t - \varepsilon[\nu_1])]_x \ , \quad (64)$$

which, upon expansion in powers of ν_1^{-1}, which was after all the real meaning of the shifts introduced in (55), yields the infinite set of linear equations [39],

$$p_n(-\tilde\partial)\Psi = \Psi\,p_{n-1}(-\tilde\partial)(\log\tau)_x \qquad \forall n \geq 2 \ . \quad (65)$$

Analogously, one can also obtain

$$p_n(\tilde\partial)\Psi^* = -\Psi^*\,p_{n-1}(\tilde\partial)(\log\tau)_x \qquad \forall n \geq 2 \ . \quad (66)$$

These equations are nothing but a recursive formulation of the *Zakharov-Shabat* (ZS) linear system, or its adjoint form, for the KP hierarchy, i.e., of the set of (λ-independent) differential equations in (48). For a combined approach to the KP hierarchy, its linear system and its symmetries, in the same vain as the above, see [4].

Exercise 2.5. Calculate the ZS equations at $n = 2, 3$ from (65), and show that they are equivalent to those obtained from (48) at the same order, if one sets $u_1 = (\log\tau)_{2x}$ and $u_2 = \frac{1}{2}[(\log\tau)_{xt_2} - (\log\tau)_{3x}]$.

Observe that, since the ZS equations (65) do not depend explicitly on the spectral parameter associated to the wave function Ψ, here ν_2, any linear combination of wave functions will solve the same set of equations, and similarly for the adjoint case. Conversely, it can be shown that [45,3],

Proposition 2.4. *Any solution Φ of the ZS system (65), generally called a KP eigenfunction, can be expressed as a superposition of wave functions,*

$$\Phi(t) = \oint_{C_\lambda} \frac{d\lambda}{2\pi i}\,h(\lambda)\Psi(t,\lambda) \ , \quad (67)$$

$$\text{where} \qquad h(\lambda) = \frac{1}{\lambda}\,\Phi(t + \varepsilon[\lambda])\Psi^*(t,\lambda) \ . \quad (68)$$

The relevant formulae for (KP) *adjoint eigenfunctions* are [45]

$$\Phi^*(t) = \oint_{\mathcal{C}_\lambda} \frac{d\lambda}{2\pi i} \, h^*(\lambda) \Psi^*(t, \lambda) \,, \tag{69}$$

$$\text{where} \qquad h^*(\lambda) = \frac{1}{\lambda} \, \Phi^*(t - \varepsilon[\lambda]) \Psi(t, \lambda) \,. \tag{70}$$

In the above formulas, the loop $\mathcal{C}_\lambda$ is taken as in (61).

That (62) contains information on the KP evolution equations themselves can be seen from the following exercise.

Exercise 2.6. Show that subsequent limits, $\nu_2 \to \infty$, $\nu_1 \to \infty$, of (63) yield the KP equation in bilinear form (28) at $o(\nu_2^{-1}, \nu_1^{-2})$.

3 Transformation Groups

In this section we shall present a description of Sato theory which makes use of the representation theory for an infinite-dimensional Clifford algebra, or *free Fermion algebra*. This description is originally due to Date, Jimbo, Kashiwara and Miwa (see [11, 19] for a review of results or [21, 22] for a slightly different point of view). We shall see that this description not only offers an interesting perspective on the theory we presented so far, but that it also serves as an extremely convenient starting point for further extensions or generalizations of the theory. For details of proofs or derivations, the reader is referred to [30] where an elementary treatment of the case of a finite (free) Fermion algebra can be found.

3.1 The Boson-Fermion Correspondence

In terms of the usual *anti-commutator*, $[X, Y]_+ := XY + YX$, we define the Clifford algebra or free Fermion algebra.

Definition 3.1. *The algebra over $\mathbb{C}$ with generators ψ_j and ψ_j^* that satisfy the anti-commutation relations*

$$[\psi_i, \psi_j]_+ = [\psi_i^*, \psi_j^*]_+ = 0 \,, \quad [\psi_i, \psi_j^*]_+ = \delta_{i+j,0} \,, \tag{71}$$

where the indices run over the set of half integers $\mathbb{Z} + 1/2$, and $\delta_{i,j}$ is the Kronecker delta, is called the free Fermion algebra. *It will be denoted by $\mathcal{A}$.*

This algebra possesses a standard representation on a so-called fermionic *Fock space* $\mathcal{F}$ (see, e.g., [30, 21]) which can be decomposed as $\mathcal{F} = \bigoplus_{\ell \in \mathbb{Z}} \mathcal{F}_\ell$, where $\mathcal{F}_\ell$ is referred to as the ℓth charge-section of the fermionic Fock space $\mathcal{F}$. We shall not go into full detail as to how general elements $|u\rangle \in \mathcal{F}$ can be constructed. Instead, let us define the heighest-weight vectors in this representation as ($\forall \ell \in \mathbb{N} \setminus \{0\}$),

$$|\ell\rangle := \psi_{1/2-\ell} \cdots \psi_{-1/2} |0\rangle \tag{72}$$

$$|-\ell\rangle := \psi^*_{1/2-\ell} \cdots \psi^*_{-1/2} |0\rangle , \tag{73}$$

for a "cyclic vector" $|0\rangle$, sometimes called the vacuum state. The cyclic vector $|0\rangle$ is defined in terms of the genuine vacuum state $|\Omega\rangle$,

$$|0\rangle := \psi^*_{1/2} \psi^*_{3/2} \psi^*_{5/2} \cdots |\Omega\rangle , \tag{74}$$

which has the characteristic of being annihilated by the ψ_j : $\psi_j|\Omega\rangle = 0$ ($\forall j$). The *dual* $\mathcal{F}^*$ of this Fock space, is defined by means of the duality relation $\psi_j \leftrightarrow \psi^*_{-j}$. From the above, one immediately obtains that the highest-weightvectors $|\ell\rangle \in \mathcal{F}$ and $\langle \ell | \in \mathcal{F}^*$ ($\ell \in \mathbb{Z}$) are such that ($j \in \mathbb{Z} + 1/2$),

$$\psi_j|\ell\rangle = 0 \quad \text{if } j > -\ell , \quad \psi^*_j|\ell\rangle = 0 \quad \text{if } j > \ell , \tag{75}$$

$$\langle \ell|\psi_j = 0 \quad \text{if } j < -\ell , \quad \langle \ell|\psi^*_j = 0 \quad \text{if } j < \ell . \tag{76}$$

There exists a pairing $\mathcal{F}^* \times \mathcal{F} \to \mathbb{C}$, the "expectation value", such that $\langle 0|1|0\rangle = 1$. Since operators ψ_j and ψ^*_j carry "charge" $+1$ and -1 respectively, and since the charge-sectors $\mathcal{F}_\ell$ of the Fock space and its dual are othogonal for this pairing ($\langle \ell|1|\ell'\rangle = \delta_{\ell,\ell'}$), it is easily seen that only charge-0 combinations of Fermion operators, i.e., combinations with equal amounts of ψ_j's and ψ^*_j's, can yield non-zero *vacuum-expectation values*,

$$\langle 0|\psi_i\psi^*_j|0\rangle = \delta_{i+j,0}\, \theta(j < 0) , \quad \theta(j < 0) := \begin{cases} 1 \text{ if } j < 0 \\ 0 \text{ if } j > 0 \end{cases} . \tag{77}$$

General vacuum-expectation values are calculated with the help of the well known *Wick theorem*, where u_i denotes either ψ_{j_i} or $\psi^*_{j_i}$,

$$\langle 0|u_1 \cdots u_r|0\rangle = \begin{cases} 0 \\ \sum_\sigma \text{sgn}(\sigma)\langle 0|u_{\sigma(1)}u_{\sigma(2)}|0\rangle \cdots \langle 0|u_{\sigma(r-1)}u_{\sigma(r)}|0\rangle , \end{cases} \tag{78}$$

depending on whether $r \in \mathbb{N}$ is odd or even, and where the sum $\sum_\sigma$ runs over all possible permutations of the indices such that $\sigma(1) < \sigma(2), \ldots, \sigma(r-1) < \sigma(r)$ and $\sigma(1) < \sigma(3) < \cdots < \sigma(r-1)$.

Most importantly however, from the above Fermion operators it is possible to construct *bosonic* operators,

$$H_n := \sum_{j\in\mathbb{Z}+1/2} \psi_{-j}\psi^*_{j+n} , \quad n \in \mathbb{Z} \setminus \{0\} , \tag{79}$$

which satisfy the usual commutation relations, $[H_n, H_m]_- = n\delta_{n+m,0}$, and which can easily be seen to annihilate highest-weight vectors, $H_n|\ell\rangle = \langle \ell|H_{-n} = 0$, $\forall \ell \in \mathbb{Z}, n \geq 1$.

It was discovered by Date, Jimbo, Kashiwara and Miwa that, conversely, it is also possible to express fermionic operators in terms of bosonic ones [11,30].

Theorem 3.1 (Boson-Fermion correspondence). *Map* $\Phi : \mathcal{F} \to \mathbb{C}[z, z^{-1}; t]$, *from* $\mathcal{F}$ *to the space of formal power series in* $t = (t_1, t_2, \ldots)$, z *and* z^{-1} *which are polynomial in* z *and* z^{-1}, *is an isomorphism of vector spaces. In particular,* $\forall |u\rangle \in \mathcal{F}$,

$$\Phi(H_n|u\rangle) = \begin{cases} \partial_{t_n} \Phi(|u\rangle) & \text{if } n \geq 1 \\ -nt_{-n} \Phi(|u\rangle) & \text{if } n \leq -1 \end{cases}, \tag{80}$$

whereas the fermionic operators are realised in the bosonic Fock space, $\mathbb{C}[z, z^{-1}; t]$,

$$\Phi(\psi(\lambda)|u\rangle) = \Gamma(t, \lambda)\Phi(|u\rangle) , \qquad \Phi(\psi^*(\lambda)|u\rangle) = \Gamma^*(t, \lambda)\Phi(|u\rangle) , \tag{81}$$

in terms of vertex operators $\Gamma(t, \lambda)$ *and* $\Gamma^*(t, \lambda)$,

$$\Gamma(t, \lambda) := z e^{\xi(t, \lambda)} e^{-\xi(\tilde{\partial}, 1/\lambda)} \lambda^{H_0} , \tag{82}$$

$$\Gamma^*(t, \lambda) := z^{-1} e^{-\xi(t, \lambda)} e^{\xi(\tilde{\partial}, 1/\lambda)} \left(\lambda^{-1}\right)^{H_0} . \tag{83}$$

Operator λ^{H_0} acts on elements of $\mathbb{C}[z, z^{-1}; t]$ as $\lambda^{H_0} f(z; t) := f(\lambda z; t)$, and the *field operators* $\psi(k)$ and $\psi^*(k)$ that appear in (81)

$$\psi(k) := \sum_{j \in \mathbb{Z}+1/2} \psi_j k^{-j-1/2} , \quad \psi^*(k) := \sum_{j \in \mathbb{Z}+1/2} \psi_j^* k^{-j-1/2} , \tag{84}$$

should really be thought of as generating functions for the Fermion operators ψ_j, ψ_j^*. Below we list some important properties of these field operators. First of all, concerning their right-action on $\mathcal{F}^*$,

$$\langle \ell | \psi(\lambda) = \lambda^{\ell-1} \langle \ell - 1 | e^{-H(\varepsilon[\lambda])} , \tag{85}$$

$$\langle \ell | \psi^*(\lambda) = \lambda^{-\ell-1} \langle \ell + 1 | e^{H(\varepsilon[\lambda])} , \tag{86}$$

where the operator $H(t)$ is defined as

$$H(t) := \sum_{n=1}^{\infty} t_n H_n , \tag{87}$$

and secondly concerning their "evolution" with respect to such $H(t)$'s :

$$e^{H(t)} \psi(\lambda) e^{-H(t)} = e^{\xi(t, \lambda)} \psi(\lambda) , \quad e^{H(t)} \psi^*(\lambda) e^{-H(t)} = e^{-\xi(t, \lambda)} \psi^*(\lambda) . \tag{88}$$

Thirdly, the vacuum-expectation value of the product, $\psi(\lambda)\psi^*(\mu)$, of field operators can be calculated as

$$\langle 0|\psi(\lambda)\psi^*(\mu)|0\rangle = \frac{1}{\lambda - \mu} , \quad \text{if } \lambda > \mu , \tag{89}$$

$$\text{or} \qquad \langle 0|\psi^*(\mu)\psi(\lambda)|0\rangle = \frac{1}{\mu - \lambda} , \quad \text{if } \mu > \lambda , \tag{90}$$

which allows us to say that, at least in expectation-value, $\psi(\lambda)$ and $\psi^*(\mu)$ anti-commute when $\mu \neq \lambda$, $\psi(\lambda)\psi^*(\mu) = -\psi^*(\mu)\psi(\lambda)$.

3.2 Transformation Groups and τ-Functions

It was shown in Sect. 1.4 that the evolution of a τ-function is defined by the action of the infinite-parameter group (30) on a point of the Universal Grassmann manifold (UGM). This parameter group is a subgroup of the group $GL(\infty)$ of automorphisms of the UGM. In fact (citing [38]), "the automorphism group $GL(\infty)$ of the Grassmannian plays the rôle of the group of transformations [i.e., symmetry group] of the KP equation".

The group $GL(\infty)$ can be defined by means of its associated Lie algebra, $\mathfrak{gl}(\infty)$, which, it turns out, possesses a representation on the fermionic Fock space $\mathcal{F}$. Denote by $\mathbb{A}$ the set of $(\infty \times \infty)$ matrices :

$$\mathbb{A} := \{(a_{ij}) \ i,j \in \mathbb{Z} + 1/2 \mid \exists R \in \mathbb{N} : a_{ij} = 0 \ \forall |i - j| > R\} \ . \tag{91}$$

We then associate with each infinite matrix $A \in \mathbb{A}$ a quadratic expression $X_A = \sum_{i,j} a_{ij} : \psi_{-i}\psi_j^* :$ where $: \psi_k\psi_\ell^* :$ denotes the *normal ordered* product $: \psi_k\psi_\ell^* := \psi_k\psi_\ell^* - \langle 0|\psi_k\psi_\ell^*|0\rangle$. The commutator of two such X's then has all the attributes of a Lie product. Most importantly,

$$[X_A, X_B]_- = X_{[A,B]_-} + \omega(A, B) \ , \tag{92}$$

where $[A, B]_- = AB - BA$ and $\omega(A, B) := \sum_{i,j} a_{ij}b_{ji} (\theta(i < 0) - \theta(j < 0))$, with the θ's defined as in (77). This leads to the following.

Definition 3.2. *The Lie algebra* $\mathfrak{gl}(\infty)$ *and its associated group are defined as*

$$\mathfrak{gl}(\infty) := \left\{ \sum_{i,j \in \mathbb{Z}+1/2} a_{ij} : \psi_{-i}\psi_j^* : \Big| \ \exists R \in \mathbb{N} : a_{ij} = 0 \ \forall |i - j| > R \right\} \oplus \mathbb{C} \ , \tag{93}$$

$$and \qquad GL(\infty) := \{g \mid g = e^{X_1} \cdots e^{X_k}, \ X_i \in \mathfrak{gl}(\infty)\} \ . \tag{94}$$

It can be shown that elements of $GL(\infty)$ possess the fundamental property [30]

Proposition 3.1. $\forall g \in GL(\infty)$,

$$\sum_{j \in \mathbb{Z}+1/2} \psi_{-j}g \otimes \psi_j^*g = \sum_{j \in \mathbb{Z}+1/2} g\psi_{-j} \otimes g\psi_j^* \ . \tag{95}$$

This is a property which directly underlies the existence of the bilinear identity (58) (and several others) for the KP hierarchy, as we shall see shortly.

In fact, and most importantly, it can be shown that the $GL(\infty)$-orbit of the cyclic vector $|0\rangle$ naturally gives rise to the UGM. For a proof in the case of finite-dimensional Grassmannians, see [30] ; for the general construction see, e.g., [11]. More precisely,

Proposition 3.2. *The KP τ-functions are described by the $GL(\infty)$-orbit of the cyclic vector $|0\rangle$ in the Fock space $\mathcal{F}$,*

$$\tau(\boldsymbol{x}) := \langle 0|e^{H(\boldsymbol{x})}g|0\rangle , \quad g \in GL(\infty) , \tag{96}$$

with Hamiltonians $H(\boldsymbol{x})$ as defined in (87).

In particular [30],

Theorem 3.2. *A state $|u\rangle \in \mathcal{F}_0$ belongs to the $GL(\infty)$-orbit of the cyclic vector $|0\rangle$, i.e., $\exists\, g \in GL(\infty) : |u\rangle = g|0\rangle$, if and only if :*

$$\sum_{j\in\mathbb{Z}+1/2} \psi_{-j}|u\rangle \otimes \psi_j^*|u\rangle = 0 . \tag{97}$$

This means that a function $\tau(\boldsymbol{x}) \in \mathbb{C}[z, z^{-1}; \boldsymbol{x}]$ is a KP τ-function if and only if it solves the KP bilinear identity (58).

To clarify this last point let us point out that due to (85,86) and the definition of the τ-function (96), the action of $\langle 1|e^{H(\boldsymbol{x})} \otimes \langle -1|e^{H(\boldsymbol{x})}$ on (97) exactly yields the KP bilinear identity in its τ-function form (58).

A natural question to ask in the light of Prop.3.2 is, given a KP τ-function, what is the element in $GL(\infty)$ that will generate this function through the definition (96) ? This problem can be settled by first constructing an appropriate element of the Lie algebra, $\mathfrak{gl}(\infty)$, from which it is then possible to generate such a $g \in GL(\infty)$. For technical reasons we first need to introduce a shifted version of the coordinates, $\boldsymbol{x} : \bar{\boldsymbol{x}} := (x_1+\delta, x_2, x_3, \dots)$, $(\delta \in \mathbb{C})$, where δ is any number such that $\tau(\bar{\boldsymbol{x}})|_{\boldsymbol{x}=0} \neq 0$. We can then define the following expression in $\lambda, \mu \in \mathbb{C}$, close to ∞,

$$h^0(\lambda, \mu) := \left[\frac{\tau(\bar{\boldsymbol{x}} - \varepsilon[\mu] + \varepsilon[\lambda])}{(\mu - \lambda)\tau(\bar{\boldsymbol{x}})}\right]\Bigg|_{\boldsymbol{x}=0} - \frac{1}{\mu - \lambda} . \tag{98}$$

The following can be proven [48].

Proposition 3.3. *For any spectral density $h^0(\lambda, \mu)$ associated with KP τ-function by (98),*

$$X = c + e^{-\delta H_1}\, \boldsymbol{v}\, e^{\delta H_1} , \tag{99}$$

with $c \in \mathbb{C}$, $H_1 = \sum_{j\in\mathbb{Z}+1/2} \psi_{-j}\psi_{j+1}^*$ and

$$\boldsymbol{v} = \oint_{\mathcal{C}_\lambda} \frac{d\lambda}{2\pi i} \oint_{\mathcal{C}_\mu} \frac{d\mu}{2\pi i}\, h^0(\lambda, \mu)\psi(\lambda)\psi^*(\mu) , \tag{100}$$

for contours $\mathcal{C}_\lambda, \mathcal{C}_\mu$ in the complex plane around $\lambda, \mu \approx \infty$ but excluding all singularities of $h^0(\lambda, \mu)$ itself, belongs to a completion $\overline{\mathfrak{gl}}(\infty)$ of $\mathfrak{gl}(\infty)$ such that

$$\tau(\boldsymbol{x}) \equiv \langle 0|e^{H(\boldsymbol{x})}e^X|0\rangle , \tag{101}$$

for a suitable normalization constant $c \in \mathbb{C}$.

Furthermore, one can show that if the spectral density $h^0(\lambda, \mu)$ separates as

$$h^0(\lambda, \mu) = \sum_{\ell=1}^{\infty} h_\ell(\lambda) \, h_\ell^*(\mu) \, , \tag{102}$$

the τ-function can be represented by the possibly infinite determinant,

$$\tau(\boldsymbol{x}) = \det[\mathbb{I} - A] \, , \tag{103}$$

$$A_{nm} = \oint_{C_\lambda} \frac{d\lambda}{2\pi i} \, h_n(\lambda) \, e^{\xi_\lambda(\boldsymbol{x})} \oint_{C_\mu} \frac{d\mu}{2\pi i} \, \frac{h_m^*(\mu) \, e^{-\xi_\mu(\boldsymbol{x})}}{\mu - \lambda} \, . \tag{104}$$

Observe that this implies that such τ-functions can, generally, be written in the form

$$\tau(\boldsymbol{x}) = \det[(\omega_{i,j})] \, , \tag{105}$$

where (ω_{ij}) stands for a possibly infinite matrix with entries ω_{ij},

$$(\omega_{i,j})_x = \varphi_i \varphi_j^* \, , \quad (\varphi_i)_{x_n} = (\varphi_i)_{nx} \, , \quad (\varphi_j^*)_{x_n} = (-)^{n+1} (\varphi_j^*)_{nx} \, , \tag{106}$$

for some otherwise arbitrary functions $\varphi_i(\boldsymbol{x})$ and $\varphi_j^*(\boldsymbol{x})$.

Readers familiar with the $\bar{\partial}$-method [26] will notice that (103),(104) are quite similar to determinant formulas for the solutions of the KP hierarchy that one encounters there. This is by no means a coincidence. In fact, it can be shown [48] that KP τ-functions are required to satisfy linear integral equations similar to those appearing in the non-local $\bar{\partial}$-problem for the KP hierarchy [26] and hence that, at least on the operational level, both methods are indeed very similar, with each method having its particular advantages [49].

Finally, observe that the above reconstruction of an appropriate element of $\overline{\mathfrak{gl}}(\infty)$ from a given τ-function can actually be extended to the case in which only partial information on the τ-function is available : It is possible to carry out a similar reconstruction starting from an initial x_1, x_2-profile $\tau(x_1, x_2, 0, \dots)$, thus effectively solving the initial-value problem for the KP τ-functions. We shall not go into this question here, since it would lead us astray, but the interested reader is referred to [48] for further details and explanations.

Exercise 3.1. Calculate the spectral densities for the KP τ-functions $\varepsilon + x_1$ ($\varepsilon \neq 0$) and $1 + 1/(p - q) \exp(\xi(\boldsymbol{x}, p) - \xi(\boldsymbol{x}, q))$, and show that these indeed give rise to those τ-functions by Prop.3.3.

3.3 Bäcklund Transformations for the KP Hierarchy

In Sect. 3.2 it was mentioned that $GL(\infty)$ acts like a symmetry group for the equations in the KP hierarchy. Here we shall try to make that statement a bit more tangible.

Although it was not explicitly mentioned at the time, it can be seen from the discussion in Sect. 3.1 that the bosonic operators H_n (79) span a Heisenberg sub-algebra of $\mathfrak{gl}(\infty)$. By the Boson-Fermion correspondence, these H_n then correspond to symmetry generators ∂_{x_n} and x_n, responsible for the translational invariance ($x_n \to x_n + \varepsilon$) and gauge invariance ($\tau \to \tau \exp \xi(\boldsymbol{x}, k)$) of the KP bilinear equations. As these particular symmetries are rather trivial, let us take a look at some more involved symmetries that can be deduced from $GL(\infty)$.

First, let us define a τ-function for any highest weight vector $|\ell\rangle$ in $\mathcal{F}$,

$$\tau_\ell(\boldsymbol{x}) := \langle \ell | e^{H(\boldsymbol{x})} g | \ell \rangle , \quad g \in GL(\infty) . \tag{107}$$

The τ-function $\tau_0(\boldsymbol{x})$ is, of course, the one defined in (96). In fact, the choice of a particular highest-weight vector is irrelevant in the general definition of a τ-function since it can be seen that any given τ can actually be realized in any of the charge sectors by appropriately altering g.

However, from the present definition (107) and the interwining relation (95), one can obtain a quite general bilinear identity relating τ-functions defined in different charge sectors of the Fock space, but generated by the same element g of $GL(\infty) : (\forall \boldsymbol{x}, \boldsymbol{x}'; \ell, \ell' : \ell \geq \ell')$

$$Res_{\lambda=\infty} \left[\lambda^{\ell-\ell'} \tau_\ell(\boldsymbol{x} - \boldsymbol{\varepsilon}[\lambda]) \, \tau_{\ell'}(\boldsymbol{x}' + \boldsymbol{\varepsilon}[\lambda]) \, e^{\xi(\boldsymbol{x}-\boldsymbol{x}',\lambda)} \right] = 0 . \tag{108}$$

Exercise 3.2. Derive the above identity from (95). Hint : calculate the action of $\langle \ell + 1 | e^{H(\boldsymbol{x})} \otimes \langle \ell' - 1 | e^{H(\boldsymbol{x}')}$ and $|\ell\rangle \otimes |\ell'\rangle$ on that identity.

The above bilinear identity is often referred to as that for the nth-modified KP hierarchy, where $n = \ell - \ell'$, because it can be shown to correspond to the following generator for the bilinear equations in the nth-modified KP hierarchy,

$$\exp(\sum_{i=1}^{\infty} y_i D_{x_i}) \sum_{j=0}^{\infty} p_j(2\boldsymbol{y}) p_{n+j+1}(-\tilde{D}) \tau_\ell \cdot \tau_{\ell'} = 0 , \quad (n = \ell - \ell' \geq 0) . \tag{109}$$

The equations contained in this bilinear identity are in fact integrable evolution equations in their own right, i.e., irrespective of the precise origins of the τ-functions that appear in them, as will be seen for instance in Prop. 9.

As an example one can take $n = 1$ (set $\ell = m + 1$) and obtain at lowest orders,

$$(D_{x_2} - D_{x_1}^2) \tau_{m+1} \cdot \tau_m = 0 , \tag{110}$$

$$(4D_{x_3} - 3D_{x_1} D_{x_2} - D_{x_1}^3) \tau_{m+1} \cdot \tau_m = 0 , \tag{111}$$

which is the Hirota form of the 1st-modified KP equation. In fact, it can be shown [27] that the bilinear equations in the 1st-modified KP hierarchy can, equivalently, be interpreted as a bilinear representation of the ZS system (65) for the KP hierarchy if one introduces $\Psi = \tau_{m+1}/\tau_m$.

Exercise 3.3. Derive the generator for the bilinear representation of the adjoint ZS system (66), and calculate the two lowest-order bilinear equations contained in it.

The bilinear equations encoded in (109) are often referred to as the equations describing *Bäcklund transformations* for the KP hierarchy and (110,111) in particular, as *the* Bäcklund transformation for the KP equation. As we shall see this is a slight misnomer since these equations do not describe the most general Bäcklund transformation for the KP equations. System (109) actually describes what are called *Darboux* transformations for the KP τ-functions.

Definition 3.3. *Let Φ be a KP eigenfunction for a given τ-function $\tau(\boldsymbol{x})$, i.e., a solution to the Zakharov-Shabat system (65) with respect to $\tau(\boldsymbol{x})$, then the map*

$$\tau(\boldsymbol{x}) \to \tau(\boldsymbol{x}) \times \Phi(\boldsymbol{x}) \tag{112}$$

is called a Darboux transformation for $\tau(\boldsymbol{x})$.
Similarly, an adjoint Darboux transformation for $\tau(\boldsymbol{x})$ is defined by the map, $\tau(\boldsymbol{x}) \to \tau(\boldsymbol{x}) \times \Phi^*(\boldsymbol{x})$, where $\Phi^*(\boldsymbol{x})$ is an adjoint eigenfunction with respect to $\tau(\boldsymbol{x})$.

It can be shown that the Darboux transformation of a τ-function yields again a τ-function. In fact [45],

Proposition 3.4. *The following action of $GL(\infty)$ or a completion thereof,*

$$GL(\infty) \ni \quad g \longrightarrow S^{-1}\phi g \qquad \in GL(\infty)$$
$$\updownarrow \qquad \qquad \updownarrow$$
$$\tau \longrightarrow \tau \times \Phi \equiv \tilde{\tau} \ ,$$

where ϕ stands for the linear superposition of Fermion operators,

$$\phi := \oint_{\mathcal{C}_\lambda(\infty)} \frac{d\lambda}{2\pi i} \ h(\lambda) \ \psi(\lambda) \ , \tag{113}$$

is the Darboux transformation for $\tau(\boldsymbol{x})$, in terms of the eigenfunction Φ obtained from (67) by means of the density $h(\lambda)$.

The operator S^{-1} can be defined on the Fock space by

$$S^{-1}\psi_j = \psi_{j+1}S^{-1} \ , \quad S^{-1}\psi_j^* = \psi_{j-1}^*S^{-1} \tag{114}$$
$$\langle \ell|S^{-1} = \langle \ell+1| \ , \quad S^{-1}|\ell\rangle = |\ell-1\rangle \ . \tag{115}$$

In fact, it can be shown [45] that the above transformation entails the classical Darboux transformation for KP eigenfunctions [28]. More precisely, one can show that for any eigenfunction Ψ associated to τ but different from Φ, the transformation,

$$\Psi \longrightarrow \frac{\Psi_x \Phi - \Psi \Phi_x}{\Phi} \tag{116}$$

yields a non-trivial KP eigenfunction associated with the new τ-function $\tilde{\tau}$.

As we already mentioned, (109) describes the bilinear form of the ZS system for the KP hierarchy and hence a τ-function and its Darboux transform, i.e., τ and $\tilde{\tau} \equiv \tau \Phi$, should by definition also satisfy such bilinear equations. In fact, denoting iterated Darboux transformations of a τ-function by $\tau_{[k]}$ $(k = 1, 2, \cdots)$,

Proposition 3.5. *A τ-function, $\tau(\boldsymbol{x})$, and its kth Darboux iterate, $\tau_{[k]}(\boldsymbol{x})$, satisfy the kth modified KP hierarchy,*

$$Res_{\lambda=\infty} \left[\lambda^k \ \tau_{[k]}(\boldsymbol{x} - \varepsilon[\lambda]) \ \tau(\boldsymbol{x'} + \varepsilon[\lambda]) \ e^{\xi(\boldsymbol{x}-\boldsymbol{x'},\lambda)} \right] = 0 \ . \tag{117}$$

This is a consequence of the fact that the action of a Darboux transformation (or an adjoint Darboux transformation) on a τ-function is a mere generalization, in the sense of (67) or (69), of the action of the vertex operators $\Gamma(\boldsymbol{x}, \lambda)$, or $\Gamma^*(\boldsymbol{x}, \lambda)$, which arose in the context of the Boson-Fermion correspondence.

As can be seen from Prop. 3.4 however, a Darboux transformation is not the most natural Bäcklund transformation if one thinks in terms of the general action of $GL(\infty)$, the natural transformation being

Definition 3.4. *A binary Darboux transformation is a map,*

$$\tau(\boldsymbol{x}) \rightarrow \tau(\boldsymbol{x}) \times \Omega(\Phi, \Phi^*) \ , \tag{118}$$

defined in terms of a so-called eigenfunction potential, $\Omega(\Phi, \Phi^*)$, which in turn is defined by means of a total differential involving KP eigenfunctions Φ and adjoint eigenfunctions Φ^* [35],

$$d\Omega(\Phi, \Phi^*) := \sum_{n=1}^{\infty} A_n dx_n \ , \quad (A_n)_{x_m} = (A_m)_{x_n} \ , \tag{119}$$

$$A_n := n \Phi^* p_{n-1}(\tilde{\partial})\Phi - \sum_{k=1}^{n-1} \left(\Phi^* p_{n-k-1}(\tilde{\partial})\Phi \right)_{x_k} \ . \tag{120}$$

The first few A's take the form : $A_1 = \Phi \Phi^*, A_2 = \Phi_x \Phi^* - \Phi \Phi_x^*, \ldots$.

Exercise 3.4. Show that $(A_1)_{x_2} = (A_2)_x$.

A binary Darboux transformation maps τ-functions into τ-functions, as can be seen [45] from

Proposition 3.6. *A binary Darboux transformation corresponds to the following action of $GL(\infty)$ or a completion thereof,*

$$GL(\infty) \ni \quad g \longrightarrow (1 + \phi\phi^*)g \quad \in GL(\infty)$$
$$\updownarrow \qquad\qquad \updownarrow$$
$$\tau \longrightarrow \quad \tau \times \Omega \equiv \hat{\tau} \quad ; \qquad \Omega := \partial^{-1}\phi\phi^* \,,$$

where ϕ is as in Prop. 3.4 and ϕ^* denotes

$$\phi^* = \oint_{\mathcal{C}_\mu} \frac{\mathrm{d}\mu}{2\pi i} \, h^*(\mu) \, \psi^*(\mu) \,, \tag{121}$$

with a density $h^*(\mu)$ used in the decomposition (69) of Φ^* in terms of adjoint wave functions.

The notation $\Omega = \partial^{-1}\Phi\Phi^*$ can be justified by the fact that $\Omega_x \equiv \Phi\Phi^*$.

Just as was the case for Darboux transformations, it can be shown that the above binary transformation induces a transformation for KP eigenfunctions,

$$\Psi \longrightarrow \Psi - \Phi \, \frac{\Omega(\Psi, \Phi^*)}{\Omega(\Phi, \Phi^*)} \,, \tag{122}$$

mapping an eigenfunction Ψ ($\neq \Phi$) associated to τ into a non-trivial eigenfunction associated to $\hat{\tau}$.

Binary Darboux transformations possess an important property which can be viewed as a generalization of the well-known Bianchi permutation theorem for Bäcklund transformations for (1+1)-dimensional integrable systems [37],

Proposition 3.7. *Every binary Darboux transformation $\tau \to \hat{\tau} \equiv \tau \times \Omega(\Phi, \Phi^*)$ is associated to a Bianchi diagram for Darboux transformations,*

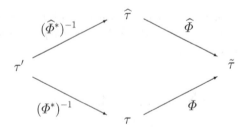

and vice versa. In this diagram a particular arrow indicates the direction of a Darboux transformation involving the eigenfunction mentioned next to it. This diagram is unique for given Φ and Φ^*, up to a constant mutiple of the τ-functions,

$$\hat{\Phi}^* \equiv \frac{\Phi^*}{\Omega(\Phi, \Phi^*)} \,, \qquad \hat{\Phi} \equiv \frac{\Phi}{\Omega(\Phi, \Phi^*)} \,. \tag{123}$$

An important property used in the proof of the above is

Lemma 3.1. *If Φ^* is an adjoint eigenfunction associated to a τ-function τ, then its inverse $(\Phi^*)^{-1}$ will be an eigenfunction for the τ-function $\tau' \equiv \tau \times \Phi^*$.*

These properties of binary Darboux transformations will be used later on, when the reductions of the KP hierarchy to ordinary differential equations are discussed.

Observe that the action of a binary Darboux transformation on a τ-function generalizes the action of the so-called *solitonic* vertex operator,

$$\Gamma(p,q) := \exp\left(\xi(\boldsymbol{x},p) - \xi(\boldsymbol{x},q)\right) \times \exp\left(-\xi(\tilde{\partial},p^{-1}) + \xi(\tilde{\partial},q^{-1})\right), \quad (124)$$

which is known to generate the KP N-soliton solutions [11,30],

$$\tau_{N-\text{sol}} = e^{c_1 \Gamma(p_1,q_1)} \ldots e^{c_N \Gamma(p_N,q_N)} \cdot 1, \quad (125)$$

whence its name. The N-soliton solutions for the KP hierarchy are obtained from an interation of binary Darboux transformations, $(1 + c_i\psi(p_i)\psi^*(q_i))$ $(c_i \in \mathbb{C})$, starting from a vacuum element $g_{[0]} \equiv 1$,

$$g_{[0]} \longrightarrow g_{[N]} = \prod_{i=1}^{N} (1 + c_i\psi(p_i)\psi^*(q_i)) \in \overline{GL}(\infty), \quad (126)$$

with the operator product ordered in descending order for the indices ; $\overline{GL}(\infty)$ denotes a completion of $GL(\infty)$ since the product $\psi(p)\psi^*(q)$ is, strictly speaking, not contained in $\mathfrak{gl}(\infty)$ but rather in a completion thereof. The corresponding N-soliton τ-function can be expressed as

$$\tau_{N-\text{sol}} = \det\left[\delta_{i,j} + \frac{c_i}{p_i - q_j} \exp\left(\xi(\boldsymbol{x},p_i) - \xi(\boldsymbol{x},q_j)\right)\right]_{1 \leq i,j \leq N}, \quad (127)$$

in accordance with the general form (105,106). Observe that the $\omega_{i,j}$ in (106) are really nothing but eigenfunction potentials, $\Omega(\varphi_i, \varphi_j^*)$.

4 Extensions and Reductions

One can think of several extensions of the above construction. On the one hand, there are possible extensions of the theory to the case of semi-discrete systems, also known as *differential-difference* systems, or to fully discrete equations. On the other hand, there are also extensions involving not so much the ranges (of types) of systems that can be described, but rather the toolbox of methods available for the study of the properties of these systems. This is especially the case when one wishes to study, for example, integrable sub-cases of the KP hierarchy, i.e., systems which possess properties similar to the full hierarchy, but whose solutions only form a subset of the solution space of the KP hierarchy. Such a selection process is what is commonly referred to as a *reduction* of the KP hierarchy.

4.1 Extensions of the KP Hierarchy

The first possible extension which springs to mind is one to so-called *multi-component* systems [38]. The theory that describes such systems is very close to standard Sato theory, since it only requires a refinement of the representation space $\mathcal{F}$.

The n-Component KP Hierarchy. The main idea underlying this extension is to sub-divide the set of Fermion operators $\{\psi_j, \psi_j^*\}$ that formed the free Fermion algebra for the KP theory. Define [19, 23], for ($\alpha = 1, 2, \ldots, n$),

$$\psi_j^{(\alpha)} := \psi_{(j-1/2)n+\alpha-1/2} \,, \quad \psi^{*(\alpha)}_j := \psi^*_{(j+1/2)n-\alpha+1/2} \,, \tag{128}$$

in terms of the operators ψ_j and ψ_j^*. This guarantees that these newly defined operators satisfy the anti-commutation relations,

$$[\psi_i^{(\alpha)}, \psi^{*(\beta)}_j]_+ = \delta_{i+j,0}\,\delta_{\alpha,\beta}\,, \quad [\psi_i^{(\alpha)}, \psi_j^{(\beta)}]_+ = [\psi^{*(\alpha)}_i, \psi^{*(\beta)}_j]_+ = 0\,, \tag{129}$$

which shows that they actually form n separate sets of Fermion operators. The fermionic Fock space $\mathcal{F}$ can therefore be sub-divided accordingly. If we define shift operators S_α (cf. (114) and (115)),

$$S_\alpha \psi_j^{(\alpha)} = \psi_{j-1}^{(\alpha)} S_\alpha\,, \quad S_\alpha \psi^{*(\alpha)}_j = \psi^{*(\alpha)}_{j+1} S_\alpha\,, \quad S_\alpha |0\rangle := \psi_{-1/2}^{(\alpha)}|0\rangle\,, \tag{130}$$

which leave all operators $\psi_j^{(\beta)}, \psi^{*(\beta)}_j$ ($\beta \neq \alpha$) invariant, we can then define so-called coloured highest-weight vectors,

$$\boldsymbol{m} := (m_1, m_2, \ldots, m_n)\,: \quad |\boldsymbol{m}\rangle := S_1^{m_1} S_2^{m_2} \cdots S_n^{m_n}|0\rangle\,, \tag{131}$$

and their appropriate duals. The Lie algebra $\mathfrak{gl}(\infty)$ and its Lie group $GL(\infty)$ can then be expresssed in terms of the new operators $\psi_j^{(\alpha)}$ and $\psi^{*(\alpha)}_j$, from which point onwards the n-component theory runs parallel to the usual 1-component theory. Introducing a Hamiltonian $H(\boldsymbol{t}) := \sum_{\alpha=1}^n \sum_{\ell=1}^\infty t_\ell^{(\alpha)} H_\ell^{(\alpha)}$, $H_\ell^{(\alpha)} := \sum_{j\in\mathbb{Z}+1/2} :\psi_{-j}^{(\alpha)}\psi^{*(\alpha)}_{j+\ell}:$, which now depends on n sets of infinitely many time variables, $\boldsymbol{t}^{(\alpha)} := (t_1^{(\alpha)}, t_2^{(\alpha)}, \ldots)$, one can define an n-component τ-function as ($g \in \overline{GL}(\infty)$),

$$\tau^{\boldsymbol{m}}(\boldsymbol{t}) := \langle \boldsymbol{m}|e^{H(\boldsymbol{t})}g|0\rangle\,, \quad \sum_{\alpha=1}^n m_\alpha = 0\,. \tag{132}$$

Observe that, due to the constraint on $\boldsymbol{m}$, these τ's are naturally defined on an $(n-1)$-dimensional lattice. They satisfy a bilinear identity of the form ($\forall \boldsymbol{t}, \boldsymbol{t}'; \boldsymbol{k}$),

$$\sum_{\alpha=1}^n Res_\lambda \Big[(-)^{\sum_{i=1}^{\alpha-1} k_i + k_i'} \lambda^{k_\alpha + k_\alpha'} \tau^{\boldsymbol{k}-\delta_\alpha}(\boldsymbol{t} - \varepsilon_\alpha[\lambda])$$

$$\times \tau^{\boldsymbol{k}'+\delta_\alpha}(\boldsymbol{t}' + \varepsilon_\alpha[\lambda]) e^{\xi_\alpha(\boldsymbol{t}-\boldsymbol{t}',\lambda)} \Big] = 0\,, \tag{133}$$

where $k \pm \delta_\alpha := (k_1, \ldots, k_\alpha \pm 1, \ldots, k_n)$, $\xi_\alpha(t, \lambda) := \sum_{\ell=1}^\infty t_\ell^{(\alpha)} \lambda^\ell$, and with shifts $\varepsilon_\alpha[\lambda^{-1}]$ that, as in (57) only affect the $t^{(\alpha)}$ component. The bilinear identity yields a generator for the so-called n-component KP hierarchy,

$$\prod_{\beta=1}^n e^{\sum_{n=1}^\infty y_n^{(\beta)} D^{(\beta)}} \sum_{\alpha=1}^n (-)^{\sum_{i=1}^{\alpha-1} k_i + k_i'}$$

$$\times \sum_{m=0}^\infty p_m(-2y^{(\alpha)}) \, p_{m-1+k_\alpha - k'_\alpha}(\tilde{D}^{(\alpha)}) \, \tau^{k'+\delta_\alpha} \cdot \tau^{k-\delta_\alpha} = 0 \; . \quad (134)$$

The simplest equation contained in this expression is

$$\forall \beta \neq \alpha \; : \qquad D_{x_1^{(\beta)}} D_{x_1^{(\alpha)}} \, \tau^k \cdot \tau^k = 2 \, \tau^{k+\delta_\beta - \delta_\alpha} \, \tau^{k - \delta_\beta + \delta_\alpha} \; , \qquad (135)$$

which, at $n = 2$, yields the (2D)-*molecule Toda* equation, i.e., the 2D, finite, non-periodic or semi-infinite Toda latttice (see, e.g., [46] for a survey of solutions for this system). Observe that the usual KP evolutions are also contained in (134). For a further discussion of the systems contained in the n-component KP hierarchy, the reader is referred to [19] or [23]. Observe also that a further possible extension of this hierarchy would be one in which the τ-functions are defined for general charge sectors (not only zero-charge), thus including all modified n-component equations as well.

The 2D-Toda Lattice Hierarchy. Taking a closer look at the derivations in the preceding sections, it becomes clear that the actual form of the KP bilinear identity is largely due to the presence of an essential singularity at $\lambda = \infty$ in the wave functions (cf. (61)). The introduction, next to this singularity, of an extra essential singular point in the complex λ-plane, say at $\lambda = 0$, leads to another interesting extension of the KP hierarchy.

This extension can be achieved by taking into account bosons $H_{n<0}$ (79) in the time evolution of the τ-functions. Set

$$H^+(x) := \sum_{n=1}^\infty x_n H_n \; , \quad H^-(y) := \sum_{n=1}^\infty y_n H_{-n} \; , \qquad (136)$$

in terms of two sets of infinitely many time variables x and y. The time-evolution of the field operators $\psi(\lambda)$ and $\psi^*(\lambda)$ is then given by :

$$e^{H^+(x)} e^{H^-(y)} \, \psi(\lambda) \, e^{-H^-(y)} e^{-H^+(x)} = \psi(\lambda) \, e^{\xi(x,\lambda) + \xi(y,\lambda^{-1})} \; ,$$
$$e^{H^+(x)} e^{H^-(y)} \, \psi^*(\lambda) \, e^{-H^-(y)} e^{-H^+(x)} = \psi^*(\lambda) \, e^{-\xi(x,\lambda) - \xi(y,\lambda^{-1})} \; , \qquad (137)$$

exhibiting the desired, essentially singular behaviour at $\lambda = \infty$ and $\lambda = 0$. The rest of the theory can then be developed in close analogy with the KP case, e.g., the left-action of the field operators on highest-weight vectors in $\mathcal{F}$ can be shown to be (cf. (85) and (86)),

$$\psi(\lambda)|\ell\rangle = \lambda^\ell e^{H^-(\varepsilon[\frac{1}{\lambda}])}|\ell+1\rangle \,, \quad \psi^*(\lambda)|\ell\rangle = \lambda^{-\ell}e^{-H^-(\varepsilon[\frac{1}{\lambda}])}|\ell-1\rangle \,. \quad (138)$$

τ-functions are then defined as

$$\tau_n(\boldsymbol{x},\boldsymbol{y}) := \langle n|e^{H^+(\boldsymbol{x})}e^{H^-(\boldsymbol{y})}\, g\, e^{-H^-(\boldsymbol{y})}|n\rangle \,, \quad (139)$$

and the corresponding bilinear identity takes the form

$$Res_{\lambda=\infty}\left[\lambda^{n-n'}\,\tau_n(\boldsymbol{x}-\boldsymbol{\varepsilon}[\lambda],\boldsymbol{y})\,\tau_{n'}(\boldsymbol{x}'+\boldsymbol{\varepsilon}[\lambda],\boldsymbol{y}')\,e^{\xi(\boldsymbol{x}-\boldsymbol{x}',\lambda)+\xi(\boldsymbol{y}-\boldsymbol{y}',\frac{1}{\lambda})}\right]=$$
$$Res_{\lambda=0}\left[\lambda^{n-n'}\,\tau_{n+1}(\boldsymbol{x},\boldsymbol{y}-\boldsymbol{\varepsilon}[\frac{1}{\lambda}])\,\tau_{n'-1}(\boldsymbol{x}',\boldsymbol{y}'+\boldsymbol{\varepsilon}[\frac{1}{\lambda}])\,e^{\xi(\boldsymbol{x}-\boldsymbol{x}',\lambda)+\xi(\boldsymbol{y}-\boldsymbol{y}',\frac{1}{\lambda})}\right].$$
$$(140)$$

Since the expression for the generator of the bilinear equations in this hierarchy is quite complicated, let us rather present an identity for the τ-functions, calculated along the lines of (62),

$$(\mu-\nu)\,\tau_{n-1}(\boldsymbol{x}-\boldsymbol{\varepsilon}[\lambda],\boldsymbol{y})\,\tau_{n+1}(\boldsymbol{x},\boldsymbol{y}-\boldsymbol{\varepsilon}[\frac{1}{\mu}]-\boldsymbol{\varepsilon}[\frac{1}{\nu}])$$
$$+(\lambda-\mu)\,\tau_n(\boldsymbol{x},\boldsymbol{y}-\boldsymbol{\varepsilon}[\frac{1}{\nu}])\,\tau_n(\boldsymbol{x}-\boldsymbol{\varepsilon}[\lambda],\boldsymbol{y}-\boldsymbol{\varepsilon}[\frac{1}{\mu}])$$
$$+(\nu-\lambda)\,\tau_n(\boldsymbol{x},\boldsymbol{y}-\boldsymbol{\varepsilon}[\frac{1}{\mu}])\,\tau_n(\boldsymbol{x}-\boldsymbol{\varepsilon}[\lambda],\boldsymbol{y}-\boldsymbol{\varepsilon}[\frac{1}{\nu}])=0 \,. \quad (141)$$

Successive limits, $(\nu\to 0, \mu\to 0, \lambda\to\infty)$, yield the 2D-Toda lattice equation in its bilinear form,

$$\frac{1}{2}D_{x_1}D_{y_1}\tau_n\cdot\tau_n = \tau_n^2 - \tau_{n+1}\tau_{n-1} \,. \quad (142)$$

Observe that, in contrast to the molecule Toda type lattices (135), this lattice is infinite in both directions of n. For a full treatment of the 2D-Toda lattice hierarchy in the spirit of Sect. 2, we refer to [43] or, for a a description of these lattices and their symmetries in terms of τ-function identities like the one just above, to [5].

The Discrete KP Equation. The idea behind the construction of the discrete KP equation [15] is the same as that used in case of the 2D-Toda lattice. In order to obtain new identities for the KP τ-functions it suffices to alter the singular behaviour of the wave functions.

Here we shall, so as to speak, resolve the essential singularity at ∞ in the KP wave function by introducing 3 new singular points, in effect simple poles near infinity, in exactly the same way as was done in (61) when we constructed the τ-function identity (62) (cf. Ex. 2.4). At the same time, we shall make a change of variables known as the *Miwa transformation* [29],

$$x \to \ell := (\ell_1, \ell_2, \ldots, \ell_L) : \qquad x = \sum_{i=1}^{L} \ell_i \, \varepsilon[a_i] \, , \qquad (\ell_i \in \mathbb{Z}) \qquad (143)$$

which relates the continuous variables x to possibly infinitely many discrete variables ℓ by means of a set of parameters $\{a_i \in \mathbb{C}\}$.

Since we are only interested in obtaining a discrete equivalent of the KP equation, introducing only 3 of such variables, i.e., $L = 3$ in (143) will prove to be sufficient. Let us, for notational simplicity, denote these variables by $\ell_1 \equiv \ell, \ell_2 \equiv m, \ell_3 \equiv n$, and the 3 lattice parameters a_i by $a_1 \equiv a, a_2 \equiv b, a_3 \equiv c$. Next we choose the 3 points ν_i close to ∞, so as to resolve the essential singularity at ∞ in (61) in exactly the same way as was done for (62). Furthermore, we choose the parameters a, b and c such that $\nu_1 = a, \nu_2 = b, \nu_3 = c$.

The Miwa transformation then turns (62) into

$$(b - c) \, \tau(\ell - 1, m, n) \, \tau(\ell, m - 1, n - 1)$$
$$+ (c - a) \, \tau(\ell, m - 1, n) \, \tau(\ell - 1, m, n - 1)$$
$$+ (a - b) \, \tau(\ell, m, n - 1) \, \tau(\ell - 1, m - 1, n) = 0 \, , \quad (144)$$

where the τ-functions now live on a 3-dimensional lattice whose vertices are described by the coordinates ℓ, m and n. This lattice equation is known as the discrete KP (dKP) equation or, sometimes, as the *Hirota-Miwa* equation [15, 29].

Just like its continuous counterpart, the dKP equation can be described as the compatibility condition of a system of linear equations,

$$\psi(\ell - 1, m - 1, n) = \frac{1}{b - a} \frac{\tau(\ell - 1, m, n) \, \tau(\ell, m - 1, n)}{\tau(\ell, m, n) \, \tau(\ell - 1, m - 1, n)}$$
$$\times \, [\, b \, \psi(\ell, m - 1, n) - a \, \psi(\ell - 1, m, n)] \, , \quad (145)$$

and two more equations obtained from (145) by cyclic permutation of the variables (ℓ, m, n) and the lattice parameters (a, b, c).

For a discussion of the dKP equation, its linear system and its Darboux transformations or binary Darboux transformations, the reader is referred to [44], where the general case of the non-autonomous dKP equation, i.e., defined on a lattice with varying lattice parameters, is discussed.

Discretizations of the n-component KP hierarchy can be found in [13, 47] where it is shown that these bear geometrical meaning as so-called quadrilateral lattices.

Finally, the procedure of adding particular singular behaviour to the KP wave functions, mixing essential singularities with poles or continuous variables with discrete ones, can be carried out in general [8, 19] and provides access to a wide variety of differential-difference and lattice systems. It also provides a powerful method by which the existence of τ-functions expressible in terms of special functions can be proved for a whole range of systems. These ideas are explored in great detail in [42].

4.2 Reductions of the KP Hierarchy

As was pointed out in the beginning of this section, one of the most interesting problems in the Sato description of the KP hierarchy or its extensions, is that of describing lower-dimensional integrable systems whose solutions form a subset of the solution space for the KP hierarchy. The process by means of which such systems are singled out is known as a *reduction* of the KP hierarchy, or of one of its extensions.

The spectrum of such reductions ranges from (2+1)-dimensional systems, through (1+1)-dimensional ones, all the way to integrable ordinary differential equations. Interesting integrable systems arise at many different stages throughout the reduction process. Let us start by describing a hierarchy of (2+1)-dimensional systems whose solutions only make up a subset of the KP solution space, but which still appear as an integrable hierarchy in its own right.

The BKP Hierarchy. A well-known example of such a sub-hierarchy is the so-called *BKP hierarchy* [10]. Its name derives from the fact that, whereas $\mathfrak{gl}(\infty)$ can be identified with the infinite-rank Kac-Moody algebra A_∞, the Lie algebra that underlies the BKP hierarchy is of B-type, (B_∞) [22].

In the Sato picture, the BKP hierarchy can be obtained from the KP hierarchy (see, e.g., [11]) by requiring that the B_n operators as defined in (39) or (42) vanish when acting on a constant,

$$B_n \, 1 = 0 \,, \qquad \forall n = 1, 3, 5, \dots \,, \tag{146}$$

(since this only yields non-trivial operators for the odd flows $x, t_1, t_3, \dots$, the even time-flows $t_2, t_4, \dots$ have to be discarded). Equivalently, one could require the L-operator (50) to satisfy the symmetry requirement,

$$\partial L + L^* \partial = 0 \,. \tag{147}$$

The BKP τ-functions are related to KP τ-functions by

$$\tau_{\mathrm{KP}}(x_1, 0, x_3, 0, x_5, \dots) \equiv \tau_{\mathrm{BKP}}^2(x_1, x_3, x_5, \dots) \,, \tag{148}$$

and they satisfy, e.g., the BKP equation which arises as the lowest member of the hierarchy of integrable evolution equations that carries the same name,

$$\left(9 D_{x_1} D_{x_5} - 5 D_{x_3}^2 - 5 D_{x_1}^3 D_{x_3} + D_{x_1}^6\right) \tau_{\mathrm{BKP}} \cdot \tau_{\mathrm{BKP}} = 0 \,. \tag{149}$$

For the derivation of the bilinear identity for BKP τ-functions and for further examples of the resulting bilinear equations, we refer the reader to [19] or [10,11] where it is also explained how to describe the BKP hierarchy using the representation theory for so-called *neutral* fermion operators.

Alternatively (see, e.g., [19]), the reduction from KP to BKP τ-functions can also be described by the following automorphism of the free fermion algebra $\mathcal{A}$,

$$\sigma_0(\psi_i) = (-)^{i-1/2}\psi_{i+1}^* , \quad \sigma_0(\psi_i^*) = (-)^{i-1/2}\psi_{i+1} , \qquad (150)$$

since B_∞ can be defined as a sub-algebra of A_∞ $(\mathfrak{gl}(\infty))$ by

$$B_\infty := \left\{ X \in A_\infty \middle| \sigma_0(X) = X \right\} . \qquad (151)$$

Similarly, the CKP hierarchy generated from the C_∞-Kac-Moody algebra is obtained from Sato theory by requiring that $L^* = -L$ for (50). D_∞ type systems are associated with B-type reductions of the 2-component KP hierarchy [19]. Recently [24,2], there has been much interest in the so-called coupled KP system [17] which is obtained from a different representation of the D_∞ algebra (see [19] or [20] for a detailed explanation of this reduction of the 2-component KP hierarchy). Of particular interest are the remarkable soliton solutions that exist for the coupled KP equation [18].

A general discussion of the above reductions for the 2D-Toda lattice can be found in [31].

Reduction from A_∞ to Affine Algebras. Instead of reductions to sub-hierarchies of the KP hierarchy that consist of (2+1)-dimensional partial differential equations, one can also describe reductions to hierarchies of (1+1)-dimensional equations. Such a dimensional reduction amounts to stating that the τ-functions of the reduced system no longer depend on a particular variable, say $x_\ell : \tau_{x_\ell} = 0$. From definition (56) and the ZS equations in (48), it is then easily seen that wave functions $\Psi(x, \lambda)$ associated with such a τ-function will satisfy the eigenvalue equation,

$$\Psi_{x_\ell} = B_\ell \Psi = \lambda^\ell \Psi , \qquad (152)$$

the right-hand-side of which is nothing other than $L^\ell \psi$. Hence one comes to realize that imposing the constraint

$$L^\ell = \left(L^\ell\right)_+ , \qquad (153)$$

i.e., requiring the ℓth-power of the pseudo-differential operator L to be a differential operator , amounts to making the τ-functions associated with this L operator x_ℓ-independent. Observe that all $L^{j\ell}$ $(j = 1, 2, \dots)$ will then be differential operators as well, and that accordingly, the τ-functions will no longer depend on the $x_{j\ell}$ variables $(j = 1, 2, \dots)$: $\tau_{x_{j\ell}} = 0$. The process of imposing the constraint (153) and the subsequent elimination of the $x_{j\ell}$-dependencies in the τ-functions is called an ℓ-reduction.

For example, as is well-known, the 2-reduction of the KP equation (28) is nothing but the Korteweg-de Vries (KdV) equation written here in Hirota bilinear form,

$$\left(4D_{x_1}D_{x_3} - D_{x_1}^4\right)\tau \cdot \tau = 0 . \qquad (154)$$

Proposition 4.1. *Besides the KP bilinear identity* (58), *ℓ-reduced KP τ-functions must also satisfy the identity,*

$$Res_{\lambda=\infty} \left[\lambda^\ell \, \tau(t - \varepsilon[\lambda]) \, \tau(t' + \varepsilon[\lambda]) \, e^{\xi(t-t',\lambda)} \right] = 0 \qquad \forall t, t' . \tag{155}$$

Combined, these two bilinear identities describe the so-called ℓ-reduced KP hierarchy.

As explained above, all equations in the ℓ-reduced KP hierarchy admit a Lax representation,

$$\begin{cases} B_\ell \Psi(t, \lambda) = \lambda^\ell \Psi(t, \lambda) \\ \partial_{t_n} \Psi(t, \lambda) = B_n \Psi(t, \lambda) \qquad (n \bmod \ell \neq 0) . \end{cases} \tag{156}$$

As we saw before, studying the symmetry group for the KP hierarchy provides much information about its τ-functions and Bäcklund transformations. It is therefore natural to ask what symmetries underly the ℓ-reduced KP hierarchies ? It can be shown [9,19] that the bilinear identity (155) is actually valid for all τ-functions generated from elements of the form $X = \sum_{i,j} a_{ij} : \psi_{-i} \psi_j^* :$ in $\mathfrak{gl}(\infty)$, that are invariant under the automorphism of the fermion algebra $\mathcal{A}$,

$$\iota_\ell : \qquad \iota_\ell(\psi_j) = \psi_{j-\ell} , \quad \iota_\ell(\psi_j^*) = \psi_{j-\ell}^* . \tag{157}$$

Condition $\iota_\ell(X) = X$ can be easily seen to be equivalent to the following constraint on the matrix elements a_{ij},

$$a_{i+\ell, j+\ell} = a_{i,j} . \tag{158}$$

Exercise 4.1. Show that the above constraint also implies that such X's commute with all bosons, $H_{j\ell}$ ($j = 1, 2, \dots$), i.e., $[H_{j\ell}, X]_- = 0$. Show that this implies that all τ-functions generated from such X's are $x_{j\ell}$-independent.

Furthermore, it can be shown [9] (or see [30] for an explicit proof in case of the KdV hierarchy, including an explicit construction of the relevant algebras) that the elements of $\mathfrak{gl}(\infty)$ (A_∞) that satisfy (158), as well as a second constraint,

$$\sum_{m=0}^{\ell-1} a_{m-1/2, m+k\ell-1/2} = 0 \qquad (\forall k \in \mathbb{Z}) , \tag{159}$$

form the affine Lie algebra $A_{\ell-1}^{(1)}$,

$$A_{\ell-1}^{(1)} := \left\{ X \in A_\infty \middle| \text{ conditions } (158) \text{ and } (159) \text{ hold } \right\} . \tag{160}$$

Hence [9, 19],

Proposition 4.2. *The ℓ-reduced KP hierarchy possesses $A^{(1)}_{\ell-1}$ symmetry.*

This does not mean that the equations in the ℓ-reduced hierarchies do not possess other symmetries as well. For example, the bilinear identity (155) is also obviously translation-invariant as well as gauge-invariant and thus its symmetries also include the Heisenberg algebra generated by the Bosons H_n (or a sub-algebra thereof, if one restricts himself to the reduced set of coordinates). It is however the $A^{(1)}_{\ell-1}$ algebra that provides the more fundamental symmetries, since these will correspond to Bäcklund transformations which can be used to generate solutions for the ℓ-reduced hierarchies, notably through the action of Darboux transformations, i.e., using solutions of linear equations (156). Moreover, besides these two sets of symmetries, there exist other symmetries as well, as we shall see in the next and final paragraph.

Dimensional reductions of the above type for B, C and D-type Kac-Moody algebras are discussed in [19] for the KP hierarchy, and in [31] for the 2D Toda lattice. There is a vast literature (see the refs. cited in [46]) on what are called constrained hierarchies or symmetry reductions, which generalize condition (153) so as to make a wider class of equations susceptible to the reduction process. The reader is referred to [46] for a discussion of the connections between those reductions and the ones described above.

Reduction of an $A^{(1)}_2$-Type System to the Painlevé IV Equation.

A symmetry which is omnipresent when it comes to discussing the reduction of (1+1)-dimensional partial differential equations to ordinary differential equations (again by a dimensional reduction), is the *scaling symmetry*. For example, for τ-functions one can define the following property,

Definition 4.1. *A KP τ-function is called* self-similar *if it possesses the following scaling property,*

$$\tau(\eta x_1, \eta^2 x_2, \dots) = \mathcal{K}(\eta)\, \tau(\boldsymbol{x})\,, \qquad \forall \eta \in \mathbb{C}\,, \tag{161}$$

or equivalently, if it is an eigenvector for the operator L,

$$L := \sum_{k=1}^{\infty} k x_k \partial_{x_k}\,, \qquad \exists c \in \mathbb{C} : L\tau = c\,\tau\,, \tag{162}$$

(compared to (161) we have that $c = \mathcal{K}'|_{\eta=1}$).

The L operator is part of the Virasoro algebra, ($n \in \mathbb{Z}$),

$$L_n := \sum_{i+j=n} \partial_{x_i} \partial_{x_j} + 2 \sum_{i-j=n} i x_i \partial_{x_j} + \sum_{i+j=-n} ij x_i x_j\,, \tag{163}$$

$$[L_m, L_n]_- = (m-n)L_{m+n} + \delta_{m+n,0} \frac{m^3 - m}{12}\,, \tag{164}$$

and $L = 1/2\, L_0$. L_0 is commonly known as the Virasoro energy operator and hence the eigenvalue c should be called the conformal weight of the function

τ. It can be shown that the Virasoro algebra can be realized within $\mathfrak{gl}(\infty)$ (see [25,7] for more details). In particular, $L \sim \sum_{j \in \mathbb{Z}+1/2}(j+1/2) : \psi_{-j}\psi_j^* :$. Furthermore, operator L, so to speak, survives the reduction from $\mathfrak{gl}(\infty)$ to the affine Lie algebras $A_{\ell-1}^{(1)}$ described in the previous paragraph. In fact, it turns out that such a reduced L can be realized as an element of the Cartan sub-algebra of $A_{\ell-1}^{(1)}$. For notational simplicity we shall use the same symbol L throughout, even in the context of reduced hierarchies where the functions upon which it acts no longer depend on all the coordinates.

Since Adler's seminal paper [1], it has been known that the Painlevé equations are somehow connected to periodic chains of Darboux transformations. This idea can be translated to the level of the KP τ-functions. As we did in Prop. 3.5, we consider a chain of τ-functions, generated by successive application of Darboux transformations. Starting from $\tau_{[0]} = \tau$, we construct $\tau_{[1]} = \tau_{[0]} \times \Phi_0$, where Φ_0 is a KP eigenfunction corresponding to $\tau_{[0]}$. Next we construct $\tau_{[2]} = \tau_{[1]} \times \Phi_1$ in terms of Φ_1 (corresponding to $\tau_{[1]}$), and so on ...

Please observe that Prop. 3.5 implies that any pair $(\tau_{[n]}, \tau_{[n+k]})$ satisfies the kth modified KP hierarchy.

Proposition 4.3.
i) If a Darboux τ-chain, $\tau_{[0]}, \ldots, \tau_{[n]}, \ldots$, is ℓ-periodic, i.e., if there exists a number ℓ such that $\tau_{[n+\ell]} = \tau_{[n]}$ ($\forall n = 0, 1, \ldots, \ell - 1$),then all $\tau_{[n]}$ are ℓ-reduced τ-functions.
ii) Conversely, from any ℓ-reduced τ-function, $\tau_{[0]}$, it is possible to construct an ℓ-periodic Darboux τ chain.

Let us now consider the case of a 3-periodic Darboux chain of self-similar τ-functions,

$$\{\tau_{[0]}, \tau_{[1]}, \tau_{[2]}, \tau_{[3]} = \tau_{[0]}\} , \quad \exists c_n \in \mathbb{C} : L\tau_{[n]} = c_n \tau_{[n]} . \tag{165}$$

Requiring the τ-functions in the chain to be self-similar effectively eliminates all symmetries generated by the Heisenberg algebra we discussed before. Instead, we focus our attention on the symmetries that remain as part of $A_2^{(1)}$.

As every pair $(\tau_{[n]}, \tau_{[n+1]})$ in the chain satisfies the 1st-modified KP hierarchy and especially its lowest-order member (110), we can represent such a chain by the following set of bilinear equations ($n = 0, 1, 2 ; \tau_{[3]} = \tau_{[0]}$),

$$\left(D_{x_2} - D_{x_1}^2\right) \tau_{[n+1]} \cdot \tau_{[n]} = 0 . \tag{166}$$

Now, the self-similarity of the τ-functions tells us that,

$$x_1(\tau_{[n]})_{x_1} + 2x_2(\tau_{[n]})_{x_2} + \cdots = c_n \tau_{[n]} , \tag{167}$$

a relation which can be used to eliminate the x_2-derivatives in (166) by setting

$$x_1 \equiv x , \quad x_2 \equiv -\frac{3}{2\varepsilon} , \quad x_n = 0 \ (\forall n \geq 3) , \tag{168}$$

in the τ-functions (for some $\varepsilon \in \mathbb{C}, \varepsilon \neq 0$). We then obtain,

$$\left(D_x^2 - \frac{\varepsilon x}{3} D_x - \kappa_n\right) \tau_{[n+1]} \cdot \tau_{[n]} = 0 , \tag{169}$$

where : $\kappa_n := \frac{\varepsilon}{3}(c_n - c_{n+1})$. This system provides a bilinear description of the well-known Painlevé IV (P_{IV}) equation [33,51], the transformation

$$g_n := \left(\log \frac{\tau_{[n+1]}}{\tau_{[n-1]}} \exp(-\frac{\varepsilon x^2}{6})\right)_x , \quad \alpha_n := \frac{\varepsilon}{3}(2c_n - c_{n+1} - c_{n-1} - 1) , \tag{170}$$

yielding the so-called *symmetric* form of the P_{IV} equation [6],

$$\begin{cases} (g_1)_x = g_1(g_3 - g_2) + \alpha_1 \\ (g_2)_x = g_2(g_1 - g_3) + \alpha_2 \\ (g_3)_x = g_3(g_2 - g_1) + \alpha_3 \end{cases} . \tag{171}$$

Expressing this system in terms of, e.g., $y(z) = \gamma g_1(x), x = \gamma z, \gamma^2 = 2/\varepsilon$, one obtains the P_{IV} equation in its standard form,

$$\frac{d^2 y}{dz^2} = \frac{1}{2y}\left(\frac{dy}{dz}\right)^2 + \frac{3}{2}y^3 + 4zy^2 + 2(z^2 - a)y + \frac{b}{y} , \tag{172}$$

$$a = \frac{\alpha_2 - \alpha_3}{\varepsilon} , \qquad b = -2\left(\frac{\alpha_1}{\varepsilon}\right)^2 . \tag{173}$$

Observe that (171) possesses one first integral $g_1 + g_2 + g_3 = -\varepsilon x$ which can be used in the elimination of g_2 and g_3. We can thus conclude that the P_{IV} equation is described by a 3-periodic, self-similar Darboux τ chain.

As we also know (by construction) that such τ's can be generated by elements of $A_2^{(1)}$, it is interesting to find out how much of this symmetry-algebra actually remains in such a chain. In other words, we want to study the Bäcklund transformations for this chain that are generated by Darboux transformations or binary Darboux transformations. Let us first point out that the eigenfunctions $\Phi_n = \tau_{[n+1]}/\tau_{[n]}$ ($n = 0, 1, 2$) used in the construction of the chain satisfy the linear equations,

$$(\Phi_n)_{2x} - \frac{\varepsilon x}{3}(\Phi_n)_x + 2\left(\log \tau_{[n]}\right)_{2x} \Phi_n = \kappa_n \Phi_n . \tag{174}$$

On the other hand, according to Lemma 3.1 (cf. Sec. 3.3), adjoint eigen-functions can be obtained by taking the inverses of these eigenfunctions, $\Phi_n^* := \Phi_{n-1}^{-1} = \tau_{[n-1]}/\tau_{[n]}$. These then satisfy

$$(\Phi_n^*)_{2x} + \frac{\varepsilon x}{3}(\Phi_n^*)_x + 2\left(\log \tau_{[n]}\right)_{2x} \Phi_n^* = \kappa_{n-1}\Phi_n^* . \tag{175}$$

It is easy to see [50] that the only Darboux transformations that give rise to Bäcklund transformations for the P_{IV} equation are the trivial cyclic permutations of the τ's in the chain,

$$S : \qquad S(\tau_{[n]}) := \tau_{[n+1]} \quad (= \tau_{[n]} \times \varPhi_n) \,, \qquad S(\kappa_n) := \kappa_{n+1} \,. \qquad (176)$$

Obviously, $S^3 = 1$.

Hence we are only left with the binary Darboux transformations to produce non-trivial Bäcklund transformations. Since a binary Darboux transformation is always accompanied by two commuting sets of Darboux transformations (Prop. 3.7) one can see [50] that only a binary Darboux transformation $\tau_{[n]} \to \widehat{\tau}_{[n]}$ of the following type will result in a Bäcklund transformation for the P_{IV} equation,

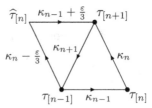

i.e., essentially, the diagram of Prop. 3.7 but with $\tau = \tau_{[n]}, \tau' = \tau_{[n-1]}, \tilde{\tau} = \tau_{[n+1]}$ and $\widehat{\tau} = \tau_{[n]} \times \Omega(\tau_{[n+1]}/\tau_{[n]}, \tau_{[n-1]}/\tau_{[n]})$, and with an additional arrow since the chains $\{\tau_{[n-1]}, \tau_{[n]}, \tau_{[n+1]}\}$ and $\{\tau_{[n-1]}, \widehat{\tau}_{[n]}, \tau_{[n+1]}\}$ are periodically closed.

Hence, there exist exactly three Bäcklund transformations that result from binary Darboux transformations, $(B_n(\tau_{[n]}) := \widehat{\tau}_{[n]})$ $(n = 0, 1, 2)$,

$$B_n(\tau_{[n]}) = \tau_{[n]} \times \partial^{-1} \frac{\tau_{[n+1]} \tau_{[n-1]}}{\tau_{[n]}^2}, \qquad B_n(\tau_{[n\pm 1]}) = \tau_{[n\pm 1]} \,. \qquad (177)$$

The resulting $\widehat{\tau}$'s are still self-similar, although now for an operator $L = x\partial_x - 2\varepsilon\partial_\varepsilon$, as obtained from (168),

$$L\widehat{\tau}_{[n]} = \widehat{c}_n \widehat{\tau}_{[n]} \,, \qquad \widehat{c}_n = c_{n+1} + c_{n-1} - c_n + 1 \,, \qquad (178)$$

which allows use to write the action of the B_n's induced on the weights c_m,

$$B_n(c_n) = c_{n+1} + c_{n-1} - c_n + 1 \,, \qquad B_n(c_{n\pm 1}) = c_{n\pm 1} \,. \qquad (179)$$

Furthermore, since the binary Darboux transformation which maps $\tau_{[n]}$ to $\widehat{\tau}_{[n]}$ can be shown to correspond to an element of $A_2^{(1)}$, essentially by the Boson-Fermion correspondence, we can represent $\widehat{\tau}_{[n]}$ as

$$\widehat{\tau}_{[n]} = \Gamma_{[n]}(\tau_{[n]}) \,, \qquad (180)$$

for some element $\Gamma_{[n]}$ in the vertex representation of $A_2^{(1)}$. In terms of this operator, (178) can be rewritten as

$$\left([L, \Gamma_{[n]}]_- + \frac{3\alpha_n}{\varepsilon} \Gamma_{[n]}\right) \tau_{[n]} = 0 \,, \qquad (181)$$

for α_n's as defined in (170). Since L, in this vertex representation, is part of the Cartan sub-algebra of $A_2^{(1)}$, this last relation shows that the α_n are directly related to the roots of $A_2^{(1)}$. Indeed, if one calculates the induced action of the $\boldsymbol{B}_n$'s on the α_n's by means of (179) and (170),

$$\boldsymbol{B}_n(\alpha_n) = -\alpha_n \ , \qquad \boldsymbol{B}_n(\alpha_{n\pm1}) = \alpha_{n\pm1} + \alpha_n \ , \tag{182}$$

one sees that this is exactly the action of the affine Weyl group $W(A_2^{(1)})$ on the roots of $A_2^{(1)}$, see, e.g., [34] for an exposé on the role such Weyl groups play in the case of the P_{IV} and P_{II} equations. It is easily verified that the action of the binary Darboux transformations $\boldsymbol{B}_n$ on the τ-functions indeed satisfies,

$$\boldsymbol{B}_n^2 = 1 \ , \qquad (\boldsymbol{B}_{n+1}\boldsymbol{B}_n)^3 = 1 \ . \tag{183}$$

This implies that, taken together with the symmetry $\boldsymbol{S}$, these $\boldsymbol{B}_n$ form an extended affine Weyl group $\widehat{W}(A_2^{(1)})$, i.e., in addition to (183), $\boldsymbol{S}\boldsymbol{B}_n = \boldsymbol{B}_{n+1}\boldsymbol{S}$.

Similar results can be obtained for periodic Darboux chains with general periods $\ell \geq 2$. The resulting systems will always allow for Bäcklund transformations (Darboux and binary Darboux) that form extended Weyl groups $\widehat{W}(A_{\ell-1}^{(1)})$. These systems are in fact equivalent to the systems described in [32]. In particular, since the P_{II} and P_V equations are obtained at $\ell = 2$ and $\ell = 4$, this result explains why these Painlevé equations are always accompanied by such extended Weyl groups.

References

1. V.E. Adler: Physica D **73**, 335 (1994)
2. M. Adler, E. Horozov, P. van Moerbeke: Int. Math. Res. Notices **11**, 569 (1999)
3. H. Aratyn, E. Nissimov, S. Pacheva: Comm. Math. Phys. **193**, 493 (1998)
4. L.V. Bogdanov, B.G. Konopelchenko: J. Math. Phys. **39**, 4683 (1998)
5. L.V. Bogdanov, B.G. Konopelchenko: J. Math. Phys. **39**, 4701 (1998)
6. F.J. Bureau: Bull. Ac. R. Belg. **66**, 280 (1980) [in French]
7. R. Carroll: Appl. Anal. **49**, 1 (1993)
8. E. Date, M. Jimbo, T. Miwa: J. Phys. Soc. Jpn. **51**, 4125 (1982)
9. E. Date, M. Jimbo, M. Kashiwara, T. Miwa: Publ. RIMS Kyoto Univ. **18**, 1077 (1982)
10. E. Date, M. Jimbo, M. Kashiwara, T. Miwa: Physica D **4**, 343 (1982)
11. E. Date, M. Kashiwara, M. Jimbo, T. Miwa: 'Transformation groups for soliton equations'. In: *Proceedings of RIMS symposium on Non-linear Integrable Systems-Classical Theory and Quantum Theory, Jyoto, Japan May 13 – ay 16, 1981*, ed. by M. Jimbo, T. Miwa (World Scientific Publ. Co., Singapore 1983) pp. 39–119
12. L.A. Dickey: *Soliton Equations and Hamiltonian Systems* (World Scientific, Singapore 1991)

13. A. Doliwa, M. Mañas, L. Martínez-Alonso, E. Medina, P.M. Santini: J. Phys. A **32**, 1197 (1999)
14. H.M. Farkas: Journal d'Analyse Math. **44**, 205 (1984)
15. R. Hirota: J. Phys. Soc. Jpn. **50**, 3785 (1981)
16. R. Hirota: 'Direct methods in soliton theory'. In: *Solitons*, ed. R.K. Bullough and P.J. Caudrey (Springer-Verlag, Berlin 1988) pp. 157–175
17. R. Hirota, Y. Ohta: J. Phys. Soc. Jpn **60**, 798 (1991)
18. S. Isojima, R. Willox, J. Satsuma: J. Phys. A **35**, 6893 (2002)
19. M. Jimbo, T. Miwa: Publ. RIMS Kyoto Univ. **19**, 943 (1983)
20. S. Kakei: RIMS kōkyūroku **1221**, 199 (2001) [in Japanese]
21. V.G. Kac, A.K. Raina: *Highest Weight Representations of Infinite Dimensional Lie Algebras*, Adv. Series in Math. Phys. Vol. **2** (World Scientific, Singapore 1987)
22. V.G. Kac: *Infinite Dimensinal Lie Algebras*, third edition (Cambridge Univ. Press, New York 1995)
23. V.G. Kac, J. van de Leur: 'The n-component KP Hierarchy and Representation Theory'. In: *Important Developments in Soliton Theory*, ed. by A.S. Fokas, V.E. Zakharov (Springer, Berlin 1993) pp. 302–342
24. V. Kac, J. van de Leur: CRM Proceedings and Lecture Notes **14**, 159 (1998)
25. N. Kawamoto, Y. Namikawa, A. Tsuchiya, Y. Yamada: Comm. Math. Phys. **116**, 247 (1988)
26. B. Konopelchenko: *Solitons in Multidimensions* (World Scientific, Singapore 1993)
27. I. Loris: Symmetry reductions in the Tau-Function Approach to Integrability. Doctoral Thesis, Free University of Brussels (V.U.B.), Brussels (1998)
28. V.B. Matveev, M.A. Salle: *Darboux Transformations and Solitons* (Spinger Verlag, Berlin 1991)
29. T. Miwa: Proc. Japan Acad. Ser. A **58**, 9 (1982)
30. T. Miwa, M. Jimbo, E. Date: *Solitons – Differential Equations, Symmetries and Infinite-Dimensional Algebras* (Cambridge Univ. Press, Cambridge, 2000)
31. J. Nimmo, R. Willox: Proc. R. Soc. London A **453**, 2497 (1997)
32. M. Noumi, Y. Yamada: Funkcialaj Ekvacioj **41**, 483 (1998)
33. M. Noumi, Y. Yamada: Nagoya Math. J. **153**, 53 (1999)
34. M. Noumi: *Painlevé Equations – An Introduction via Symmetry* (Asakura Shoten, Tokyo 2000) [in Japanese]
35. W. Oevel, Physica A **195**, 533 (1993)
36. Y. Ohta, J. Satsuma, D. Takahashi, T. Tokihiro: Prog. Theor. Phys. Suppl. **94**, 210 (1988)
37. C. Rogers, W.F. Shadwick: *Bäcklund Transformations and their Applications*, Mathematics in Science and Engineering Vol.**161** (Academic Press, New York 1982)
38. M. Sato: RIMS kōkyūroku **439**, 30 (1981)
39. M. Sato, Y. Sato: 'Soliton equations as dynamical systems on infinite dimensional Grassmann manifold'. In: *Nonlinear PDE in Applied Science. U.S.-Japan Seminar, Tokyo, 1982*, Lecture Notes in Num. Appl. Anal. **5** (North-Holland, New York 1983) pp. 259–271
40. J. Satsuma: Hirota bilinear method for nonlinear evolution equations, Lect. Notes Phys. **632**, 171–222 (2003)
41. K. Smith, L. Kahanpää, P. Kekäläinen, W. Traves: *An Invitation to Algebraic Geometry* (Springer, New York 2000)

42. T. Tokihiro, R. Willox, J. Satsuma: Phys. Lett. A **236**, 23 (1997)
43. K. Ueno, K. Takasaki: 'Toda lattice hierarchy'. In: *Group Representations and Systems of Differential Equations*, ed. by K. Okamoto, Adv. Stud. Pure Math. Vol. **4** (North-Holland, Amsterdam 1984) pp. 1–95
44. R. Willox, T. Tokihiro, J. Satsuma: J. Math. Phys. **38**, 6455 (1997)
45. R. Willox, T. Tokihiro, I. Loris, J. Satsuma: Inverse Problems **14**, 745 (1998)
46. R. Willox, I. Loris: J. Math. Phys. **40**, 6501 (1999)
47. R. Willox, Y. Ohta, C.R. Gilson, T. Tokihiro, J. Satsuma : Phys. Lett. A **252**, 163 (1999)
48. R. Willox: RIMS kōkyūroku **1170**, 111 (2000)
49. R. Willox: RIMS kōkyūroku **1280**, 1 (2002) [in Japanese]
50. R. Willox: RIMS kōkyūroku **1302**, 21 (2003) [in Japanese]
51. R. Willox, J. Hietarinta: J. Phys. A **36**, 10615 (2003)

Special Solutions
of Discrete Integrable Systems

Y. Ohta

Department of Mathematics, Kobe University, Rokko, Kobe 657-8501, Japan,
ohta@math.kobe-u.ac.jp

Abstract. Hierarchies of discrete soliton equations are constructed in bilinear form as a consequence of the algebraic identities satisfied by determinants and Pfaffians. Difference formulas for determinants and Pfaffians are derived from the discrete linear dispersion relations satisfied by their elements. For completeness, we first summarize the main algebraic properties of determinants and Pfaffians.

1 Introduction

Integrable systems have attracted a great deal of interest and have been studied for many years from various view points. One of those is the characterization of the space of solutions. In the case of soliton equations, there is a wide class of solutions which possess a simple representation in terms of determinants or Pfaffians. Even though it has not yet been rigorously proved that the general solution is given by determinants or Pfaffians, it is known that the spaces of determinant and Pfaffian solutions are large enough to determine the algebraic properties of soliton equations. This is because those spaces suffice to construct the corresponding hierarchies of soliton equations.

The purpose of this article is to provide a simple description of the bilinear theory of discrete soliton equations and their solutions in a self-contained way. Starting from a space of solutions, that is, a space of determinants or Pfaffians, hierarchies of bilinear equations satisfied by all of the elements of the space are constructed from the algebraic identities satisfied by determinants and Pfaffians. Because of the difficulty of the argument of convergence for the series expansion of the solution, it has not yet been clarified whether these hierarchies characterize the space or not. However, the spaces of determinants and Pfaffians of finite size contain many solutions, including solitons, dromions, rational exponential solitons, special function solutions, and others.

Determinants correspond to the continuous and discrete KP hierarchy and Pfaffians correspond to the continuous and discrete coupled KP hierarchies. Both for the determinants and for the Pfaffians, there are two ways to impose the differential and difference structures. One is the Wronski type and the other is the Gram type. The continuous KP hierarchies and their Wronski and Gram determinant solutions, and the continuous coupled KP

Y. Ohta, Special Solutions of Discrete Integrable Systems, Lect. Notes Phys. **644**, 57–83 (2004)
http://www.springerlink.com/   © Springer-Verlag Berlin Heidelberg 2004

hierarchies and their Wronski and Gram Pfaffian solutions are well known and have been studied for many years. The discrete KP hierarchies and their discrete Wronski and Gram determinant solutions, and the discrete coupled KP hierarchies and their discrete Wronski and Gram Pfaffian solutions have also appeared in the literature some years ago. In this article, the case of discrete coupled KP equations and their discrete Wronski and Gram Pfaffian solutions is described. The other cases can be derived from this in a direct way. The discrete KP and its determinant solutions are a special case, and the continuous cases can be recovered by the continuous limit.

In Sect. 2, we give a summary of the properties of determinants and Pfaffians for later use. The most important properties are the bilinear algebraic identities which turn out to be the bilinear form of soliton equations. The difference formulas for the discrete Wronski and Gram Pfaffians are given in Sect. 3. Using these formulas, we rewrite the algebraic identities as discrete bilinear equations in Sect. 4. Concluding remarks are made in Sect. 5.

2 Determinant and Pfaffian

In this section, we list the definitions of determinants and Pfaffians and those properties which will be used in the construction of hierarchies of soliton equations in the following sections.

2.1 Definition

Definition 2.1. *For a given square array of indeterminates,*

$$
\begin{array}{cccc}
a_{11} & a_{12} & \cdots & a_{1N} \\
a_{21} & a_{22} & \cdots & a_{2N} \\
\vdots & \vdots & & \vdots \\
a_{N1} & a_{N2} & \cdots & a_{NN}
\end{array}
$$

the determinant, $\det(a_{ij})_{1\le i,j\le N}$ *is defined as the polynomial in these variables,*

$$
\det(a_{ij})_{1\le i,j\le N} = \sum \mathrm{sign}\begin{matrix} j_1 & j_2 & \cdots & j_N \\ i_1 & i_2 & \cdots & i_N \end{matrix} a_{i_1 j_1} a_{i_2 j_2} \cdots a_{i_N j_N} \tag{1}
$$

where the summation is taken over all choices of N elements, $a_{i_1 j_1}$, $a_{i_2 j_2}$, $\cdots$, $a_{i_N j_N}$, from the square array a_{ij} $(1 \le i,j \le N)$ such that $i_\mu \ne i_\nu$ and $j_\mu \ne j_\nu$ for $\mu \ne \nu$, and sign denotes the signature of permutation of the indices. The number of terms is $N!$.

This quantity is well-defined because the sign does not depend on the order of the pairs, $\{i_1,j_1\}$, $\{i_2,j_2\}$, $\cdots$, $\{i_N,j_N\}$. We also use the following notation for a determinant,

$$\det(a_{i_\mu j_\nu})_{1\leq\mu,\nu\leq N} = \begin{pmatrix} j_1, j_2, \cdots, j_N \\ i_1, i_2, \cdots, i_N \end{pmatrix} = \begin{vmatrix} a_{i_1 j_1} & a_{i_1 j_2} & \cdots & a_{i_1 j_N} \\ a_{i_2 j_1} & a_{i_2 j_2} & \cdots & a_{i_2 j_N} \\ \vdots & \vdots & & \vdots \\ a_{i_N j_1} & a_{i_N j_2} & \cdots & a_{i_N j_N} \end{vmatrix}$$

Hence $\begin{pmatrix} j \\ i \end{pmatrix}$ means a 1×1 determinant, which is a monomial a_{ij}.

Definition 2.2. *For a given triangular array of indeterminates,*

$$\begin{matrix} a_{12} & a_{13} & \cdots & & a_{1,2N} \\ & a_{23} & \cdots & & a_{2,2N} \\ & & \ddots & & \vdots \\ & & & & a_{2N-1,2N} \end{matrix}$$

the Pfaffian, $\mathrm{Pf}(a_{ij})_{1\leq i<j\leq 2N}$ *is defined as the polynomial in these variables,*

$$\mathrm{Pf}(a_{ij})_{1\leq i<j\leq 2N} = \sum \mathrm{sign} \begin{matrix} 1 & 2 & \cdots & 2N \\ i_1 & i_2 & \cdots & i_{2N} \end{matrix} a_{i_1 i_2} a_{i_3 i_4} \cdots a_{i_{2N-1} i_{2N}} \qquad (2)$$

where the summation is taken over all choices of N elements, $a_{i_1 i_2}, a_{i_3 i_4}, \cdots,$ $a_{i_{2N-1} i_{2N}}$, *from the triangular array a_{ij} $(1 \leq i < j \leq N)$ such that $i_\mu \neq i_\nu$ for $\mu \neq \nu$, and* sign *denotes the signature of the permutation of the indices. The number of terms is $(2N-1)!!$.*

This quantity is well-defined because the signature does not depend on the order of the pairs, $\{i_1, i_2\}, \{i_3, i_4\}, \cdots, \{i_{2N-1}, i_{2N}\}$. We also use the following notation for a Pfaffian,

$$\mathrm{Pf}(a_{i_\mu i_\nu})_{1\leq\mu<\nu\leq 2N} = (i_1, i_2, \cdots, i_{2N}) = \begin{vmatrix} a_{i_1 i_2} & a_{i_1 i_3} & \cdots & a_{i_1 i_{2N}} \\ & a_{i_2 i_3} & \cdots & a_{i_2 i_{2N}} \\ & & \ddots & \vdots \\ & & & a_{i_{2N-1} i_{2N}} \end{vmatrix}$$

Hence (i, j) means a Pfaffian with one component, which is a monomial a_{ij}. For convenience, we define the all Pfaffians of odd size to be 0,

$$\mathrm{Pf}(a_{ij})_{1\leq i<j\leq 2N+1} = 0$$

and we define (i, j) for $i \geq j$ by the antisymmetry condition,

$$(i, j) = -(j, i), \qquad (i, i) = 0.$$

Lemma 2.1. *If $(i, j) = 0$ for $1 \leq i < j \leq N$, then*

$$(1, 2, \cdots, N, N', \cdots, 2', 1') = \begin{pmatrix} 1', 2', \cdots, N' \\ 1, 2, \cdots, N \end{pmatrix}$$

where $\begin{pmatrix} j' \\ i \end{pmatrix} = (i, j')$. In another notation,

$$
\begin{vmatrix}
0\,0 \cdots 0 & (1, N') & (1, N-1') & \cdots & (1, 2') & (1, 1') \\
0 \cdots 0 & (2, N') & (2, N-1') & \cdots & (2, 2') & (2, 1') \\
\ddots\;\; \vdots & \vdots & \vdots & & \vdots & \vdots \\
0\;(N-1, N') & (N-1, N-1') & \cdots & (N-1, 2') & (N-1, 1') \\
(N, N') & (N, N-1') & \cdots & (N, 2') & (N, 1') \\
 & (N', N-1') & \cdots & (N', 2') & (N', 1') \\
 & & \ddots & \vdots & \vdots \\
 & & & (3', 2') & (3', 1') \\
 & & & & (2', 1')
\end{vmatrix}
$$

$$
= \begin{vmatrix}
(1, 1') & (1, 2') & \cdots & (1, N') \\
(2, 1') & (2, 2') & \cdots & (2, N') \\
\vdots & \vdots & & \vdots \\
(N, 1') & (N, 2') & \cdots & (N, N')
\end{vmatrix}
$$

This lemma means that determinants are special cases of the Pfaffians. Therefore properties of determinants can be derived from those of the Pfaffians by specialization.

Proof. From the definition of Pfaffians (2), we obtain

$$
(1, 2, \cdots, N, N', \cdots, 2', 1')
$$
$$
= \sum \text{sign} \begin{matrix} 1 & 2 & \cdots & N & N' & \cdots & 2' & 1' \\ i_1 & i_2 & \cdots & i_N & i_{N+1} & \cdots & i_{2N-1} & i_{2N} \end{matrix}
$$
$$
\times (i_1, i_2)(i_3, i_4) \cdots (i_{2N-1}, i_{2N})
$$
$$
= \sum \text{sign} \begin{matrix} 1 & 2 & \cdots & N & N' & \cdots & 2' & 1' \\ i_1 & i_3 & \cdots & i_{2N-1} & i_{2N} & \cdots & i_4 & i_2 \end{matrix}
$$
$$
\times (i_1, i_2)(i_3, i_4) \cdots (i_{2N-1}, i_{2N})
$$

where the summation is taken over all choices of N elements, (i_1, i_2), (i_3, i_4), $\cdots$, (i_{2N-1}, i_{2N}), from (i, j) $(1 \leq i < j \leq N)$, (i, j') $(1 \leq i, j \leq N)$ and (i', j') $(N \geq i > j \geq 1)$ such that $i_\mu \neq i_\nu$ for $\mu \neq \nu$. If $(i, j) = 0$ for $1 \leq i < j \leq N$, the nonvanishing terms in the summation consist of the elements of the form (i, j') $(1 \leq i, j \leq N)$ only. By arranging the order of the elements in each term in such a way that $i_2 = 1'$, $i_4 = 2'$, $\cdots$, $i_{2N} = N'$, we obtain

$$
(1, 2, \cdots, N, N', \cdots, 2', 1')
$$
$$
= \sum \text{sign} \begin{matrix} 1 & 2 & \cdots & N \\ i_1 & i_3 & \cdots & i_{2N-1} \end{matrix} (i_1, 1')(i_3, 2') \cdots (i_{2N-1}, N')
$$

where the summation is taken over all choices of $i_1, i_3, \cdots, i_{2N-1}$ from 1, 2, $\cdots$, N such that $i_\mu \neq i_\nu$ for $\mu \neq \nu$. Thus the right-hand side coincides with the $N \times N$ determinant whose $\binom{j}{i}$-element is (i, j').

Proposition 2.1. *If* $(i, j') = 0$ *for* $1 \leq i, j \leq N$, *then*

$$(1, 2, \cdots, N, N', \cdots, 2', 1') = (1, 2, \cdots, N)(N', \cdots, 2', 1').$$

In another notation,

$$\begin{vmatrix} (1,2) & (1,3) & \cdots & (1,N) & 0 & 0 & \cdots & 0 & 0 \\ & (2,3) & \cdots & (2,N) & 0 & 0 & \cdots & 0 & 0 \\ & & \ddots & \vdots & \vdots & \vdots & & \vdots & \vdots \\ & & & (N-1,N) & 0 & 0 & \cdots & 0 & 0 \\ & & & & 0 & 0 & \cdots & 0 & 0 \\ & & & & (N',N-1') & \cdots & (N',2') & (N',1') \\ & & & & & \ddots & \vdots & \vdots \\ & & & & & & (3',2') & (3',1') \\ & & & & & & & (2',1') \end{vmatrix}$$

$$= \begin{vmatrix} (1,2) & (1,3) & \cdots & (1,N) \\ & (2,3) & \cdots & (2,N) \\ & & \ddots & \vdots \\ & & & (N-1,N) \end{vmatrix} \begin{vmatrix} (N',N-1') & \cdots & (N',2') & (N',1') \\ & \ddots & \vdots & \vdots \\ & & (3',2') & (3',1') \\ & & & (2',1') \end{vmatrix}.$$

Proof. From the definition of Pfaffians (2), we obtain

$$(1, 2, \cdots, N, N', \cdots, 2', 1')$$
$$= \sum \operatorname{sign} \begin{matrix} 1 & 2 & \cdots & N & N' & \cdots & 2' & 1' \\ i_1 & i_2 & \cdots & i_N & i_{N+1} & \cdots & i_{2N-1} & i_{2N} \end{matrix}$$
$$\times (i_1, i_2)(i_3, i_4) \cdots (i_{2N-1}, i_{2N}),$$

where the summation is taken over all choices of N elements, $(i_1, i_2), (i_3, i_4)$, $\cdots$, (i_{2N-1}, i_{2N}), from (i, j) $(1 \leq i < j \leq N)$, (i, j') $(1 \leq i, j \leq N)$ and (i', j') $(N \geq i > j \geq 1)$ such that $i_\mu \neq i_\nu$ for $\mu \neq \nu$. If $(i, j') = 0$ for $1 \leq i, j \leq N$, the nonvanishing terms on the right-hand side consist of the elements of the form (i, j) $(1 \leq i < j \leq N)$ and (i', j') $(N \geq i > j \geq 1)$ only. Thus when N is odd, all terms vanish and the proposition holds since a Pfaffian of odd size is defined to be 0. When N is even, by arranging the order of the elements in each term in such a way that $i_1, i_2, \cdots, i_N \in \{1, 2, \cdots, N\}$ and $i_{N+1}, i_{N+2}, \cdots, i_{2N} \in \{N', \cdots, 2', 1'\}$, we get

$$(1, 2, \cdots, N, N', \cdots, 2', 1')$$
$$= \sum \text{sign.}\begin{smallmatrix} 1 & 2 & \cdots & N \\ i_1 & i_2 & \cdots & i_N \end{smallmatrix} \text{sign.}\begin{smallmatrix} N' & \cdots & 2' & 1' \\ i_{N+1} & \cdots & i_{2N-1} & i_{2N} \end{smallmatrix}$$
$$\times (i_1, i_2)(i_3, i_4) \cdots (i_{2N-1}, i_{2N}),$$

where the summation is taken over all choices of (i_1, i_2), (i_3, i_4), $\cdots$, (i_{N-1}, i_N) from (i, j) $(1 \le i < j \le N)$, and (i_{N+1}, i_{N+2}), (i_{N+3}, i_{N+4}), $\cdots$, (i_{2N-1}, i_{2N}) from (i', j') $(N \ge i > j \ge 1)$. Thus the right-hand side coincides with the product of the two Pfaffians, $(1, 2, \cdots, N)$ and $(N', \cdots, 2', 1')$.

2.2 Linearity and Alternativity

For each term of the summation in the definition of a determinant (1), each number from 1, 2, $\cdots$, N appears once in the indices j_1, j_2, $\cdots$, j_N. Thus the determinant is linear with respect to its jth column for $j = 1, 2, \cdots, N$. Similarly each number from 1, 2, $\cdots$, N appears once in the indices i_1, i_2, $\cdots$, i_N, in definition (1). Thus the determinant is linear with respect to its ith row for $i = 1, 2, \cdots, N$. Therefore we have the next proposition.

Proposition 2.2.

$$\begin{vmatrix} \binom{1}{1} & \binom{2}{1} & \cdots & \binom{j}{1} + \binom{j}{1}' & \cdots & \binom{N}{1} \\ \binom{1}{2} & \binom{2}{2} & \cdots & \binom{j}{2} + \binom{j}{2}' & \cdots & \binom{N}{2} \\ \vdots & \vdots & & \vdots & & \vdots \\ \binom{1}{N} & \binom{2}{N} & \cdots & \binom{j}{N} + \binom{j}{N}' & \cdots & \binom{N}{N} \end{vmatrix}$$

$$= \begin{vmatrix} \binom{1}{1} & \binom{2}{1} & \cdots & \binom{j}{1} & \cdots & \binom{N}{1} \\ \binom{1}{2} & \binom{2}{2} & \cdots & \binom{j}{2} & \cdots & \binom{N}{2} \\ \vdots & \vdots & & \vdots & & \vdots \\ \binom{1}{N} & \binom{2}{N} & \cdots & \binom{j}{N} & \cdots & \binom{N}{N} \end{vmatrix} + \begin{vmatrix} \binom{1}{1} & \binom{2}{1} & \cdots & \binom{j}{1}' & \cdots & \binom{N}{1} \\ \binom{1}{2} & \binom{2}{2} & \cdots & \binom{j}{2}' & \cdots & \binom{N}{2} \\ \vdots & \vdots & & \vdots & & \vdots \\ \binom{1}{N} & \binom{2}{N} & \cdots & \binom{j}{N}' & \cdots & \binom{N}{N} \end{vmatrix}$$

and

$$\begin{vmatrix} \binom{1}{1} & \binom{2}{1} & \cdots & c\binom{j}{1} & \cdots & \binom{N}{1} \\ \binom{1}{2} & \binom{2}{2} & \cdots & c\binom{j}{2} & \cdots & \binom{N}{2} \\ \vdots & \vdots & & \vdots & & \vdots \\ \binom{1}{N} & \binom{2}{N} & \cdots & c\binom{j}{N} & \cdots & \binom{N}{N} \end{vmatrix} = c \begin{vmatrix} \binom{1}{1} & \binom{2}{1} & \cdots & \binom{j}{1} & \cdots & \binom{N}{1} \\ \binom{1}{2} & \binom{2}{2} & \cdots & \binom{j}{2} & \cdots & \binom{N}{2} \\ \vdots & \vdots & & \vdots & & \vdots \\ \binom{1}{N} & \binom{2}{N} & \cdots & \binom{j}{N} & \cdots & \binom{N}{N} \end{vmatrix}$$

Similarly

$$\left| \begin{array}{cccc} \binom{\substack{1\\1\\1\\2}} & \binom{\substack{2\\1\\2\\2}} & \cdots & \binom{\substack{N\\1\\N\\2}} \\ \vdots & \vdots & & \vdots \\ \binom{1}{i}+\binom{1}{i}' & \binom{2}{i}+\binom{2}{i}' & \cdots & \binom{N}{i}+\binom{N}{i}' \\ \vdots & \vdots & & \vdots \\ \binom{1}{N} & \binom{2}{N} & \cdots & \binom{N}{N} \end{array} \right|$$

$$= \left| \begin{array}{cccc} \binom{\substack{1\\1\\1\\2}} & \binom{\substack{2\\1\\2\\2}} & \cdots & \binom{\substack{N\\1\\N\\2}} \\ \vdots & \vdots & & \vdots \\ \binom{1}{i} & \binom{2}{i} & \cdots & \binom{N}{i} \\ \vdots & \vdots & & \vdots \\ \binom{1}{N} & \binom{2}{N} & \cdots & \binom{N}{N} \end{array} \right| + \left| \begin{array}{cccc} \binom{\substack{1\\1\\1\\2}} & \binom{\substack{2\\1\\2\\2}} & \cdots & \binom{\substack{N\\1\\N\\2}} \\ \vdots & \vdots & & \vdots \\ \binom{1}{i}' & \binom{2}{i}' & \cdots & \binom{N}{i}' \\ \vdots & \vdots & & \vdots \\ \binom{1}{N} & \binom{2}{N} & \cdots & \binom{N}{N} \end{array} \right|$$

and

$$\left| \begin{array}{cccc} \binom{\substack{1\\1\\1\\2}} & \binom{\substack{2\\1\\2\\2}} & \cdots & \binom{\substack{N\\1\\N\\2}} \\ \vdots & \vdots & & \vdots \\ c\binom{1}{i} & c\binom{2}{i} & \cdots & c\binom{N}{i} \\ \vdots & \vdots & & \vdots \\ \binom{1}{N} & \binom{2}{N} & \cdots & \binom{N}{N} \end{array} \right| = c \left| \begin{array}{cccc} \binom{\substack{1\\1\\1\\2}} & \binom{\substack{2\\1\\2\\2}} & \cdots & \binom{\substack{N\\1\\N\\2}} \\ \vdots & \vdots & & \vdots \\ \binom{1}{i} & \binom{2}{i} & \cdots & \binom{N}{i} \\ \vdots & \vdots & & \vdots \\ \binom{1}{N} & \binom{2}{N} & \cdots & \binom{N}{N} \end{array} \right|$$

In the case of a Pfaffian, there is no distinction between column and row. For each term of the summation in the definition of a Pfaffian (2), each number from $1, 2, \cdots, 2N$ appears once in the indices $i_1, i_2, \cdots, i_{2N}$. Thus the Pfaffian is linear with respect to its ith column for $i = 1, 2, \cdots, 2N$.

Proposition 2.3.

$$
\begin{vmatrix}
(1,2) & (1,3) & \cdots & (1,i)+(1,i)' & & \cdots & (1,2N) \\
 & (2,3) & \cdots & (2,i)+(2,i)' & & \cdots & (2,2N) \\
 & & \ddots & \vdots & & & \vdots \\
 & & & (i-1,i)+(i-1,i)' & & & \\
 & & & & (i,i+1)+(i,i+1)' & \cdots & (i,2N)+(i,2N)' \\
 & & & & & \ddots & \vdots \\
 & & & & & & (2N-1,2N)
\end{vmatrix}
$$

$$
=
\begin{vmatrix}
(1,2) & (1,3) & \cdots & (1,i) & & \cdots & (1,2N) \\
 & (2,3) & \cdots & (2,i) & & \cdots & (2,2N) \\
 & & \ddots & \vdots & & & \vdots \\
 & & & (i-1,i) & & & \\
 & & & & (i,i+1) & \cdots & (i,2N) \\
 & & & & & \ddots & \vdots \\
 & & & & & & (2N-1,2N)
\end{vmatrix}
$$

$$
+
\begin{vmatrix}
(1,2) & (1,3) & \cdots & (1,i)' & & \cdots & (1,2N) \\
 & (2,3) & \cdots & (2,i)' & & \cdots & (2,2N) \\
 & & \ddots & \vdots & & & \vdots \\
 & & & (i-1,i)' & & & \\
 & & & & (i,i+1)' & \cdots & (i,2N)' \\
 & & & & & \ddots & \vdots \\
 & & & & & & (2N-1,2N)
\end{vmatrix}
$$

and

$$
\begin{vmatrix}
(1,2) & (1,3) & \cdots & c(1,i) & & \cdots & (1,2N) \\
 & (2,3) & \cdots & c(2,i) & & \cdots & (2,2N) \\
 & & \ddots & \vdots & & & \vdots \\
 & & & c(i-1,i) & & & \\
 & & & & c(i,i+1) & \cdots & c(i,2N) \\
 & & & & & \ddots & \vdots \\
 & & & & & & (2N-1,2N)
\end{vmatrix}
$$

$$
= c
\begin{vmatrix}
(1,2) & (1,3) & \cdots & (1,i) & & \cdots & (1,2N) \\
 & (2,3) & \cdots & (2,i) & & \cdots & (2,2N) \\
 & & \ddots & \vdots & & & \vdots \\
 & & & (i-1,i) & & & \\
 & & & & (i,i+1) & \cdots & (i,2N) \\
 & & & & & \ddots & \vdots \\
 & & & & & & (2N-1,2N)
\end{vmatrix}
$$

The determinant is alternating with respect to its columns and to its rows, that is, the following propositions hold.

Proposition 2.4. *If two columns of a square array of variables are identical, then its determinant is 0,*

$$\begin{pmatrix} 1,2,\cdots,j,\cdots,j,\cdots,N \\ 1,2,\cdots\cdots\cdots\cdots,N \end{pmatrix} = 0$$

In another notation,

$$\begin{vmatrix} \begin{pmatrix}1\\1\end{pmatrix} & \begin{pmatrix}2\\1\end{pmatrix} & \cdots & \begin{pmatrix}j\\1\end{pmatrix} & \cdots & \begin{pmatrix}j\\1\end{pmatrix} & \cdots & \begin{pmatrix}N\\1\end{pmatrix} \\ \begin{pmatrix}1\\2\end{pmatrix} & \begin{pmatrix}2\\2\end{pmatrix} & \cdots & \begin{pmatrix}j\\2\end{pmatrix} & \cdots & \begin{pmatrix}j\\2\end{pmatrix} & \cdots & \begin{pmatrix}N\\2\end{pmatrix} \\ \vdots & \vdots & & \vdots & & \vdots & & \vdots \\ \begin{pmatrix}1\\N\end{pmatrix} & \begin{pmatrix}2\\N\end{pmatrix} & \cdots & \begin{pmatrix}j\\N\end{pmatrix} & \cdots & \begin{pmatrix}j\\N\end{pmatrix} & \cdots & \begin{pmatrix}N\\N\end{pmatrix} \end{vmatrix} = 0$$

Similarly if two rows of a square array of variables are identical, then its determinant is 0,

$$\begin{pmatrix} 1,2,\cdots\cdots\cdots\cdots,N \\ 1,2,\cdots,i,\cdots,i,\cdots,N \end{pmatrix} = 0$$

In another notation,

$$\begin{vmatrix} \begin{pmatrix}1\\1\end{pmatrix} & \begin{pmatrix}2\\1\end{pmatrix} & \cdots & \begin{pmatrix}N\\1\end{pmatrix} \\ \begin{pmatrix}1\\2\end{pmatrix} & \begin{pmatrix}2\\2\end{pmatrix} & \cdots & \begin{pmatrix}N\\2\end{pmatrix} \\ \vdots & \vdots & & \vdots \\ \begin{pmatrix}1\\i\end{pmatrix} & \begin{pmatrix}2\\i\end{pmatrix} & \cdots & \begin{pmatrix}N\\i\end{pmatrix} \\ \vdots & \vdots & & \vdots \\ \begin{pmatrix}1\\i\end{pmatrix} & \begin{pmatrix}2\\i\end{pmatrix} & \cdots & \begin{pmatrix}N\\i\end{pmatrix} \\ \vdots & \vdots & & \vdots \\ \begin{pmatrix}1\\N\end{pmatrix} & \begin{pmatrix}2\\N\end{pmatrix} & \cdots & \begin{pmatrix}N\\N\end{pmatrix} \end{vmatrix} = 0$$

Proof. First we prove the proposition in the case of identical columns. Let us assume that the jth column and the j'th column are identical ($j \neq j'$). In the summation of definition (1), a term including $\begin{pmatrix}j\\i\end{pmatrix}\begin{pmatrix}j'\\i'\end{pmatrix}$ cancels with the same term with j and j' exchanged, because $\binom{j}{i}\binom{j'}{i'} = \binom{j'}{i}\binom{j}{i'}$ and the signs of these terms are opposite. Therefore all terms in the summation cancel each other and the determinant vanishes. The case of rows is proved in the same way.

From this proposition and the linearity of determinant, the next one follows immediately.

Proposition 2.5. *By exchanging two columns, the determinant changes its sign, that is,*

$$\begin{pmatrix} 1,2,\cdots,j,\cdots,j',\cdots,N \\ 1,2,\cdots\cdots\cdots\cdots,N \end{pmatrix} = -\begin{pmatrix} 1,2,\cdots,j',\cdots,j,\cdots,N \\ 1,2,\cdots\cdots\cdots\cdots,N \end{pmatrix}$$

In another notation,

$$\begin{vmatrix} \binom{1}{1} & \binom{2}{1} & \cdots & \binom{j}{1} & \cdots & \binom{j'}{1} & \cdots & \binom{N}{1} \\ \binom{1}{2} & \binom{2}{2} & \cdots & \binom{j}{2} & \cdots & \binom{j'}{2} & \cdots & \binom{N}{2} \\ \vdots & \vdots & & \vdots & & \vdots & & \vdots \\ \binom{1}{N} & \binom{2}{N} & \cdots & \binom{j}{N} & \cdots & \binom{j'}{N} & \cdots & \binom{N}{N} \end{vmatrix}$$

$$= -\begin{vmatrix} \binom{1}{1} & \binom{2}{1} & \cdots & \binom{j'}{1} & \cdots & \binom{j}{1} & \cdots & \binom{N}{1} \\ \binom{1}{2} & \binom{2}{2} & \cdots & \binom{j'}{2} & \cdots & \binom{j}{2} & \cdots & \binom{N}{2} \\ \vdots & \vdots & & \vdots & & \vdots & & \vdots \\ \binom{1}{N} & \binom{2}{N} & \cdots & \binom{j'}{N} & \cdots & \binom{j}{N} & \cdots & \binom{N}{N} \end{vmatrix}$$

Similarly by exchanging two rows, the determinant changes its sign, that is,

$$\begin{pmatrix} 1,2,\cdots\cdots\cdots\cdots,N \\ 1,2,\cdots,i,\cdots,i',\cdots,N \end{pmatrix} = -\begin{pmatrix} 1,2,\cdots\cdots\cdots\cdots,N \\ 1,2,\cdots,i',\cdots,i,\cdots,N \end{pmatrix}$$

In another notation,

$$
\begin{vmatrix}
\binom{1}{1} & \binom{2}{1} & \cdots & \binom{N}{1} \\
\binom{1}{2} & \binom{2}{2} & \cdots & \binom{N}{2} \\
\vdots & \vdots & & \vdots \\
\binom{1}{i} & \binom{2}{i} & \cdots & \binom{N}{i} \\
\vdots & \vdots & & \vdots \\
\binom{1}{i'} & \binom{2}{i'} & \cdots & \binom{N}{i'} \\
\vdots & \vdots & & \vdots \\
\binom{1}{N} & \binom{2}{N} & \cdots & \binom{N}{N}
\end{vmatrix}
= -
\begin{vmatrix}
\binom{1}{1} & \binom{2}{1} & \cdots & \binom{N}{1} \\
\binom{1}{2} & \binom{2}{2} & \cdots & \binom{N}{2} \\
\vdots & \vdots & & \vdots \\
\binom{1}{i'} & \binom{2}{i'} & \cdots & \binom{N}{i'} \\
\vdots & \vdots & & \vdots \\
\binom{1}{i} & \binom{2}{i} & \cdots & \binom{N}{i} \\
\vdots & \vdots & & \vdots \\
\binom{1}{N} & \binom{2}{N} & \cdots & \binom{N}{N}
\end{vmatrix}
$$

By the linearity and alternativity of the determinant, we know that adding a column (resp., a row) multiplied by a constant to another column (resp., row) does not change the determinant.

The Pfaffian is also alternating with respect to its columns.

Proposition 2.6. *For $i \neq j$, if the ith and jth columns of a triangular array of variables are identical, that is, if $(i,k) = (j,k)$ for all k, then its Pfaffian is 0,*

$$(1, 2, \cdots, i, \cdots, i, \cdots, 2N) = 0 \tag{3}$$

In another notation,

$$
\begin{vmatrix}
(1,2) & (1,3) & \cdots & (1,i) & & (1,i) & \cdots & (1,2N) \\
 & (2,3) & \cdots & (2,i) & & (2,i) & \cdots & (2,2N) \\
 & & \ddots & \vdots & & \vdots & & \vdots \\
 & & & (i-1,i) & & (i-1,i) & & \\
 & & (i,i+1) & \cdots & (i,j-1) & 0 & (i,j+1) & \cdots & (i,2N) \\
 & & & \ddots & & -(i,i+1) & & & \\
 & & & & \ddots & \vdots & & \vdots \\
 & & & & & -(i,j-1) & & \\
 & & & & & (i,j+1) & \cdots & (i,2N) \\
 & & & & & & \ddots & \vdots \\
 & & & & & & & (2N-1,2N)
\end{vmatrix}
$$

$$= 0$$

Proof. Without loss of generality, we can assume $i < j$. In the summation of definition (2), for $k < l < i < j$, a term including $(k,i)(l,j)$ cancels with the same term with i and j exchanged, because $(k,i)(l,j) = (k,j)(l,i)$ and the signs of these terms are opposite. For $k < i < l < j$, a term including $(k,i)(l,j)$ cancels with the same term with $(k,i)(l,j)$ replaced by $(k,j)(i,l)$, because $(k,i)(l,j) = -(k,j)(i,l)$ and the signs of these terms are the same. For $k < i < j < l$, a term including $(k,i)(j,l)$ cancels with the same term with i and j exchanged, because $(k,i)(j,l) = (k,j)(i,l)$ and the signs of these terms are opposite. Similarly for $i < k < l < j$, $i < k < j < l$ and $i < j < k < l$, cancellation of the terms occurs. Finally a term including (i,j) vanishes, because $(i,j) = 0$. Therefore all terms in the summation cancel each other and the Pfaffian vanishes.

From this proposition and the linearity of the Pfaffian, the next one follows immediately.

Proposition 2.7. *By exchanging two columns, the Pfaffian changes sign, that is,*

$$(1, 2, \cdots, i, \cdots, j, \cdots, 2N) = -(1, 2, \cdots, j, \cdots, i, \cdots, 2N)$$

In another notation,

$$
\begin{vmatrix}
(1,2) & (1,3) \cdots & (1,i) & & \cdots & & (1,j) & & \cdots & (1,2N) \\
& (2,3) \cdots & (2,i) & & \cdots & & (2,j) & & \cdots & (2,2N) \\
& & \vdots & & & & \vdots & & & \vdots \\
& & (i-1,i) & & & & (i-1,j) & & & \\
& & & (i,i+1) \cdots & (i,j-1) & (i,j) & (i,j+1) \cdots & & (i,2N) \\
& & & & & (i+1,j) & & & \\
& & & & & \vdots & & & \vdots \\
& & & & & (j-1,j) & & & \\
& & & & & & (j,j+1) \cdots & & (j,2N) \\
& & & & & & & & \vdots \\
& & & & & & & & (2N-1,2N)
\end{vmatrix}
$$

$$
= -\begin{vmatrix}
(1,2) & (1,3) \cdots & (1,j) & & \cdots & & (1,i) & & \cdots & (1,2N) \\
& (2,3) \cdots & (2,j) & & \cdots & & (2,i) & & \cdots & (2,2N) \\
& & \vdots & & & & \vdots & & & \vdots \\
& & (i-1,j) & & & & (i-1,i) & & & \\
& & & -(i+1,j) \cdots & -(j-1,j) & -(i,j) & (j,j+1) \cdots & & (j,2N) \\
& & & & & -(i,i+1) & & & \\
& & & & & \vdots & & & \vdots \\
& & & & & -(i,j-1) & & & \\
& & & & & & (i,j+1) \cdots & & (i,2N) \\
& & & & & & & & \vdots \\
& & & & & & & & (2N-1,2N)
\end{vmatrix}
$$

By the linearity and alternativity of the Pfaffian, we know that adding a column multiplied by a constant to another column does not change the Pfaffian.

By using the above properties of determinants and Pfaffians, we can prove that the determinant of an antisymmetric array is the square of the Pfaffian.

Lemma 2.2. *If* $\begin{pmatrix} j \\ i \end{pmatrix}$ *is antisymmetric, that is,* $\begin{pmatrix} j \\ i \end{pmatrix} = -\begin{pmatrix} i \\ j \end{pmatrix}$ *and* $\begin{pmatrix} i \\ i \end{pmatrix} = 0$,

then

$$\begin{pmatrix} 1, 2, \cdots, N \\ 1, 2, \cdots, N \end{pmatrix} = (1, 2, \cdots, N)^2,$$

where $\begin{pmatrix} j \\ i \end{pmatrix} = (i, j)$. *In another notation,*

$$\begin{vmatrix} 0 & (1,2) & (1,3) & \cdots & & (1,N) \\ -(1,2) & 0 & (2,3) & \cdots & & (2,N) \\ -(1,3) & -(2,3) & \ddots & & \ddots & \vdots \\ \vdots & \vdots & \ddots & & \ddots & (N-1,N) \\ -(1,N) & -(2,N) & \cdots & & -(N-1,N) & 0 \end{vmatrix}$$

$$= \begin{vmatrix} (1,2) & (1,3) & \cdots & (1,N) \\ & (2,3) & \cdots & (2,N) \\ & & \ddots & \vdots \\ & & & (N-1,N) \end{vmatrix}^2$$

This lemma means that the Pfaffian can be regarded as a square root of the determinant of an antisymmetric array. Therefore the properties of Pfaffians can be derived from those of determinants by specialization.

Proof.

$$\begin{vmatrix} 0 & (1,2) & (1,3) & \cdots & & (1,N) \\ -(1,2) & 0 & (2,3) & \cdots & & (2,N) \\ -(1,3) & -(2,3) & 0 & & \ddots & \vdots \\ \vdots & \vdots & \ddots & & \ddots & (N-1,N) \\ -(1,N) & -(2,N) & \cdots & & -(N-1,N) & 0 \end{vmatrix}$$

$$= \begin{vmatrix} 0\,0\,0\cdots 0 & (1,N) & & \cdots & (1,3) & (1,2) & 0 \\ 0\,0\cdots 0 & (2,N) & & \cdots & (2,3) & 0 & -(1,2) \\ 0\cdots 0 & (3,N) & & \cdots & 0 & -(2,3) & -(1,3) \\ \ddots\ \vdots & \vdots & & & \vdots & \vdots & \vdots \\ 0 & (N-1,N) & 0 & & & & \\ 0 & -(N-1,N) & \cdots & -(3,N) & -(2,N) & -(1,N) \\ & (N-1,N) & \cdots & (3,N) & (2,N) & (1,N) \\ & & \ddots & \vdots & \vdots & \vdots \\ & & & (3,4) & (2,4) & (1,4) \\ & & & (2,3) & (1,3) \\ & & & & & (1,2) \end{vmatrix}$$

where we used Lemma 2.1,

$$
=
\begin{vmatrix}
(1,2) & (1,3) & (1,4) & \cdots & (1,N) & 0 & & \cdots & 0 & 0 & 0 \\
 & 0 & 0 & \cdots & 0 & (2,N) & & \cdots & (2,3) & 0 & -(1,2) \\
 & & 0 & \cdots & 0 & (3,N) & & \cdots & 0 & -(2,3) & -(1,3) \\
 & & & \ddots & \vdots & \vdots & & & \vdots & \vdots & \vdots \\
 & & & & 0 & (N-1,N) & 0 & & & & \\
 & & & & 0 & -(N-1,N) & \cdots & -(3,N) & -(2,N) & -(1,N) \\
 & & & & & (N-1,N) & \cdots & (3,N) & (2,N) & (1,N) \\
 & & & & & & \ddots & \vdots & \vdots & \vdots \\
 & & & & & & & (3,4) & (2,4) & (1,4) \\
 & & & & & & & & (2,3) & (1,3) \\
 & & & & & & & & & (1,2)
\end{vmatrix}
$$

where we added the $2N$th column to the 1st one,

$$
=
\begin{vmatrix}
(1,2) & (1,3) & (1,4) & \cdots & (1,N) & 0 & & \cdots & 0 & 0 & 0 \\
(2,3) & (2,4) & \cdots & (2,N) & 0 & & & \cdots & 0 & 0 & 0 \\
 & & 0 & \cdots & 0 & (3,N) & & \cdots & 0 & -(2,3) & -(1,3) \\
 & & & \ddots & \vdots & \vdots & & & \vdots & \vdots & \vdots \\
 & & & & 0 & (N-1,N) & 0 & & & & \\
 & & & & 0 & -(N-1,N) & \cdots & -(3,N) & -(2,N) & -(1,N) \\
 & & & & & (N-1,N) & \cdots & (3,N) & (2,N) & (1,N) \\
 & & & & & & \ddots & \vdots & \vdots & \vdots \\
 & & & & & & & (3,4) & (2,4) & (1,4) \\
 & & & & & & & & (2,3) & (1,3) \\
 & & & & & & & & & (1,2)
\end{vmatrix}
$$

where we added the $(2N-1)$th column to the 2nd one,

$$
=
\begin{vmatrix}
(1,2) & (1,3) & (1,4) & \cdots & (1,N) & 0 & & \cdots & 0 & 0 & 0 \\
(2,3) & (2,4) & \cdots & (2,N) & 0 & & \cdots & 0 & 0 & 0 \\
(3,4) & \cdots & (3,N) & 0 & & \cdots & 0 & 0 & 0 \\
 & \ddots & \vdots & \vdots & & & & \vdots & \vdots & \vdots \\
 & & (N-1,N) & 0 & 0 & & & & & \\
 & & & 0 & 0 & \cdots & 0 & 0 & 0 \\
 & & & (N-1,N) & \cdots & (3,N) & (2,N) & (1,N) \\
 & & & & \ddots & \vdots & \vdots & \vdots \\
 & & & & & (3,4) & (2,4) & (1,4) \\
 & & & & & & (2,3) & (1,3) \\
 & & & & & & & (1,2)
\end{vmatrix}
$$

where we repeatedly added the $(2N+1-k)$th column to the kth one for $k = 3, 4, \cdots, N$,

$$
=
\begin{vmatrix}
(1,2) & (1,3) & \cdots & (1,N) \\
 & (2,3) & \cdots & (2,N) \\
 & & \ddots & \vdots \\
 & & & (N-1,N)
\end{vmatrix}
\begin{vmatrix}
(N-1,N) & \cdots & (2,N) & (1,N) \\
 & \ddots & \vdots & \vdots \\
 & & (2,3) & (1,3) \\
 & & & (1,2)
\end{vmatrix}
$$

where we used Proposition 2.1. When N is odd, the determinant of an anti-symmetric array vanishes.

2.3 Cofactor and Expansion Formula

A cofactor of a determinant or Pfaffian is defined to be the coefficient of a certain term in the polynomial. We note that for any i and j, both determinant and Pfaffian are polynomials in $\binom{j}{i}$ or (i,j) of at most degree 1, thus the following concepts are well-defined.

Definition 2.3. *We define the* $\binom{j_1, j_2, \cdots, j_n}{i_1, i_2, \cdots, i_n}$*-cofactor of the determinant* $\binom{1,2,\cdots,N}{1,2,\cdots,N}$ *to be*

$$\binom{1,2,\cdots,N}{1,2,\cdots,N}_{i_1 i_2 \cdots i_n}^{j_1 j_2 \cdots j_n} = \left(\text{coefficient of } \binom{j_1}{i_1}\binom{j_2}{i_2}\cdots\binom{j_n}{i_n} \text{ in } \binom{1,2,\cdots,N}{1,2,\cdots,N} \right)$$

Definition 2.4. *We define the* $(i_1, i_2, \cdots, i_{2n})$*-cofactor of the Pfaffian* $(1,2, \cdots, 2N)$ *to be*

$$(1,2,\cdots,2N)_{i_1 i_2 \cdots i_{2n}} = \big(\text{coefficient of } (i_1, i_2)(i_3, i_4)\cdots(i_{2n-1}, i_{2n}) \\ \text{in } (1,2,\cdots,2N) \big)$$

From the definitions of determinants and Pfaffians, (1) and (2), it is clear that each cofactor is also expressed as a determinant or Pfaffian with an all over sign.

Proposition 2.8. *For* $1 \le i_1 < i_2 < \cdots < i_n \le N$ *and* $1 \le j_1 < j_2 < \cdots < j_n \le N$,

$$\binom{1,2,\cdots,N}{1,2,\cdots,N}_{i_1 i_2 \cdots i_n}^{j_1 j_2 \cdots j_n}$$
$$= (-1)^{i_1+j_1+i_2+j_2+\cdots+i_n+j_n} \binom{1,2,\cdots,\widehat{j_1},\cdots,\widehat{j_2},\cdots,\widehat{j_n},\cdots,N}{1,2,\cdots,\widehat{i_1},\cdots,\widehat{i_2},\cdots,\widehat{i_n},\cdots,N}$$

where ^ *means deletion.*

Proposition 2.9. *For permutations* σ *and* τ,

$$\binom{1,2,\cdots,N}{1,2,\cdots,N}_{i_{\tau(1)} i_{\tau(2)} \cdots i_{\tau(n)}}^{j_{\sigma(1)} j_{\sigma(2)} \cdots j_{\sigma(n)}} = \text{sign}\,\sigma \; \text{sign}\,\tau \binom{1,2,\cdots,N}{1,2,\cdots,N}_{i_1 i_2 \cdots i_n}^{j_1 j_2 \cdots j_n}.$$

Proposition 2.10. *For* $1 \le i_1 < i_2 < \cdots < i_{2n} \le 2N$,

$$(1,2,\cdots,2N)_{i_1 i_2 \cdots i_{2n}}$$
$$= (-1)^{i_1+i_2+\cdots+i_{2n}+n}(1,2,\cdots,\widehat{i_1},\cdots,\widehat{i_2},\cdots,\widehat{i_{2n}},\cdots,2N),$$

where ^ *means deletion.*

Proposition 2.11. *For a permutation σ,*

$$(1, 2, \cdots, 2N)_{i_{\sigma(1)} i_{\sigma(2)} \cdots i_{\sigma(2n)}} = \operatorname{sign} \sigma \ (1, 2, \cdots, 2N)_{i_1 i_2 \cdots i_{2n}}.$$

Since in the definition of determinants (1), the index j ($1 \leq j \leq N$) appears once in the column indices $j_1, j_2, \cdots, j_N$ in each term of the summation and the index i ($1 \leq i \leq N$) appears once in the row indices $i_1, i_2, \cdots, i_N$, we get the following expansion formula for determinants.

Proposition 2.12.

$$\begin{pmatrix} 1, 2, \cdots, N \\ 1, 2, \cdots, N \end{pmatrix} = \sum_{i=1}^{N} \begin{pmatrix} 1, 2, \cdots, N \\ 1, 2, \cdots, N \end{pmatrix}_i^j \begin{pmatrix} j \\ i \end{pmatrix} \qquad 1 \leq j \leq N$$

$$= \sum_{j=1}^{N} \begin{pmatrix} 1, 2, \cdots, N \\ 1, 2, \cdots, N \end{pmatrix}_i^j \begin{pmatrix} j \\ i \end{pmatrix} \qquad 1 \leq i \leq N.$$

Similarly since in the definition of Pfaffians (2), the index j ($1 \leq j \leq 2N$) appears once in each term of the summation, we get the following expansion formula for Pfaffians.

Proposition 2.13.

$$(1, 2, \cdots, 2N) = \sum_{i=1}^{2N} (1, 2, \cdots, 2N)_{ij} (i, j) \qquad 1 \leq j \leq 2N.$$

By taking $j = 2N$ in this proposition, we get

$$(1, 2, \cdots, 2N) = \sum_{i=1}^{2N-1} (-1)^{i-1} (1, \cdots, \widehat{i}, \cdots, 2N - 1)(i, 2N) \qquad (4)$$

where $\widehat{}$ means deletion. We note that this can be regarded as a recursive definition of Pfaffians, that is, Pfaffians of size $2N$ are inductively defined by using Pfaffians of size $2N - 2$.

2.4 Algebraic Identities

We now formulate the bilinear algebraic identity for Pfaffians.

Theorem 2.1. *The Pfaffians,*

$$\xi^{i_1 i_2 \cdots i_\nu} = (1, 2, \cdots, N, i_1, i_2, \cdots, i_\nu) \qquad (5)$$

satisfy the identity,

$$\sum_{l=1}^{n} (-1)^l \xi^{i_1 i_2 \cdots i_m j_l} \xi^{j_1 j_2 \cdots \widehat{j_l} \cdots j_n} + \sum_{k=1}^{m} (-1)^k \xi^{i_1 i_2 \cdots \widehat{i_k} \cdots i_m} \xi^{j_1 j_2 \cdots j_n i_k} = 0 \qquad (6)$$

where $\widehat{}$ means deletion.

Proof. Using the expansion formula of Pfaffians (4), we obtain

$$\sum_{l=1}^{L}(-1)^l(a_1, a_2, \cdots, a_K, b_l)(b_1, b_2, \cdots, \widehat{b_l}, \cdots, b_L)$$

$$+ \sum_{k=1}^{K}(-1)^k(a_1, a_2, \cdots, \widehat{a_k}, \cdots, a_K)(b_1, b_2, \cdots, b_L, a_k)$$

$$= \sum_{l=1}^{L}\sum_{k=1}^{K}(-1)^{k+l-1}(a_k, b_l)(a_1, a_2, \cdots, \widehat{a_k}, \cdots, a_K)(b_1, b_2, \cdots, \widehat{b_l}, \cdots, b_L)$$

$$+ \sum_{k=1}^{K}\sum_{l=1}^{L}(-1)^{k+l-1}(a_1, a_2, \cdots, \widehat{a_k}, \cdots, a_K)(b_l, a_k)(b_1, b_2, \cdots, \widehat{b_l}, \cdots, b_L)$$

$$= 0 \qquad\qquad K, L : \text{ odd}$$

Taking $K = N + m$, $L = N + n$, $a_k = b_k = k$ $(1 \le k \le N)$, $a_{N+k} = i_k$ $(1 \le k \le m)$ and $b_{N+k} = j_k$ $(1 \le k \le n)$ in the above equation and using (3), we recover the Pfaffian identity.

Using Lemma 2.1, we can derive a bilinear algebraic identity for determinants by specialization.

Theorem 2.2. *The determinants,*

$$\xi^{i_1 i_2 \cdots i_m} = \begin{pmatrix} 1, 2, \cdots, N - m, i_1, i_2, \cdots, i_m \\ 1, 2, \cdots\cdots\cdots\cdots\cdots\cdots\cdots\cdots, N \end{pmatrix} \tag{7}$$

satisfy the identity,

$$\sum_{l=1}^{m+1}(-1)^l \xi^{i_1 i_2 \cdots i_{m-1} j_l} \xi^{j_1 j_2 \cdots \widehat{j_l} \cdots j_{m+1}} = 0 \tag{8}$$

where $\widehat{}$ *means deletion.*

Proof. In the identity for Pfaffians (6), we replace m by $m - 1$, n by $m + 1$ and N by $2N - m$, and rewrite the indices $1, 2, \cdots, 2N - m$ as $N', \cdots, 2', 1', 1, 2, \cdots, N - m$. By taking $(i', j') = 0$ $(N \ge i > j \ge 1)$ and

$$(i', j) = \begin{pmatrix} j \\ i \end{pmatrix}$$

$$1 \le i \le N, \ j \in \{1, 2, \cdots, N - m, i_1, i_2, \cdots, i_{m-1}, j_1, j_2, \cdots, j_{m+1}\},$$

the Pfaffians in the first term in (6) are specialized to determinants given in (7) and the second term in (6) vanishes. Then the Pfaffian identity (6) reduces to the determinant one (8).

The determinant identity (8) is called the Plücker relation and the ξ's in (7) are called the Plücker coordinates. The algebraic identity (6) is the Pfaffian counterpart of the Plücker relation, and the ξ's in (5) are the Pfaffian counterparts of the Plücker coordinates.

2.5 Golden Theorem

The following golden theorem is useful in the construction of new identities for Pfaffians from simple identities.

Theorem 2.3. *For a given identity for Pfaffians, the equation which is obtained by replacing all Pfaffians $(i_1, i_2, \cdots, i_{2n})$ appearing in the identity by*
$$\frac{(1, 2, \cdots, 2N, i_1, i_2, \cdots, i_{2n})}{(1, 2, \cdots, 2N)} \text{ is also an identity.}$$

Example 2.1. We have

$$(a_1, a_2, a_3, a_4) = (a_1, a_2)(a_3, a_4) - (a_1, a_3)(a_2, a_4) + (a_1, a_4)(a_2, a_3)$$

By the above theorem, we get

$$\frac{(1, 2, \cdots, 2N, a_1, a_2, a_3, a_4)}{(1, 2, \cdots, 2N)}$$
$$= \frac{(1, 2, \cdots, 2N, a_1, a_2)}{(1, 2, \cdots, 2N)} \frac{(1, 2, \cdots, 2N, a_3, a_4)}{(1, 2, \cdots, 2N)}$$
$$- \frac{(1, 2, \cdots, 2N, a_1, a_3)}{(1, 2, \cdots, 2N)} \frac{(1, 2, \cdots, 2N, a_2, a_4)}{(1, 2, \cdots, 2N)}$$
$$+ \frac{(1, 2, \cdots, 2N, a_1, a_4)}{(1, 2, \cdots, 2N)} \frac{(1, 2, \cdots, 2N, a_2, a_3)}{(1, 2, \cdots, 2N)}$$

thus we obtain

$$(1, 2, \cdots, 2N, a_1, a_2, a_3, a_4)(1, 2, \cdots, 2N)$$
$$= (1, 2, \cdots, 2N, a_1, a_2)(1, 2, \cdots, 2N, a_3, a_4)$$
$$- (1, 2, \cdots, 2N, a_1, a_3)(1, 2, \cdots, 2N, a_2, a_4)$$
$$+ (1, 2, \cdots, 2N, a_1, a_4)(1, 2, \cdots, 2N, a_2, a_3) \tag{9}$$

Example 2.2. Applying the theorem to the expansion,

$$0 = (a_0, a_0, a_1, a_2, a_3, a_4)$$
$$= (a_0, a_1, a_2, a_3)(a_0, a_4) - (a_0, a_1, a_2, a_4)(a_0, a_3)$$
$$+ (a_0, a_1, a_3, a_4)(a_0, a_2) - (a_0, a_2, a_3, a_4)(a_0, a_1)$$

we get

$$(1, 2, \cdots, 2N, a_0, a_1, a_2, a_3)(1, 2, \cdots, 2N, a_0, a_4)$$
$$- (1, 2, \cdots, 2N, a_0, a_1, a_2, a_4)(1, 2, \cdots, 2N, a_0, a_3)$$
$$+ (1, 2, \cdots, 2N, a_0, a_1, a_3, a_4)(1, 2, \cdots, 2N, a_0, a_2)$$
$$- (1, 2, \cdots, 2N, a_0, a_2, a_3, a_4)(1, 2, \cdots, 2N, a_0, a_1) = 0. \tag{10}$$

Proof. In the algebraic identity for Pfaffians (6), by rewriting m to $2n-1$, n to 1, N to $2N$ and j_1 to i_{2n}, we obtain

$$(1, 2, \cdots, 2N, i_1, \cdots, i_{2n})(1, 2, \cdots, 2N)$$

$$= \sum_{k=1}^{2n-1} (-1)^{k-1}(1, 2, \cdots, 2N, i_1, \cdots, \widehat{i_k}, \cdots, i_{2n-1})(1, 2, \cdots, 2N, i_k, i_{2n})$$

thus

$$\frac{(1, 2, \cdots, 2N, i_1, \cdots, i_{2n})}{(1, 2, \cdots, 2N)}$$

$$= \sum_{k=1}^{2n-1} (-1)^{k-1}\frac{(1, 2, \cdots, 2N, i_1, \cdots, \widehat{i_k}, \cdots, i_{2n-1})}{(1, 2, \cdots, 2N)}\frac{(1, 2, \cdots, 2N, i_k, i_{2n})}{(1, 2, \cdots, 2N)}$$

This has the same form as the recursive definition of Pfaffians (4). From this fact, the golden theorem is immediately proved, because the replacement of $(i_1, i_2, \cdots, i_{2n})$ with $\dfrac{(1, 2, \cdots, 2N, i_1, i_2, \cdots, i_{2n})}{(1, 2, \cdots, 2N)}$ is valid in the recursive definition of Pfaffians.

In the case of determinants, we also have a golden theorem, which is again useful for the construction of new identities from simple ones.

Theorem 2.4. *For a given identity of determinants, the equation which is obtained by replacing all determinants* $\begin{pmatrix} j_1, j_2, \cdots, j_n \\ i_1, i_2, \cdots, i_n \end{pmatrix}$ *appearing in the identity by* $\dfrac{\begin{pmatrix} 1, 2, \cdots, N, j_1, j_2, \cdots, j_n \\ 1, 2, \cdots, N, i_1, i_2, \cdots, i_n \end{pmatrix}}{\begin{pmatrix} 1, 2, \cdots, N \\ 1, 2, \cdots, N \end{pmatrix}}$ *is also an identity.*

Example 2.3. We have

$$\begin{pmatrix} b_1, b_2 \\ a_1, a_2 \end{pmatrix} = \begin{pmatrix} b_1 \\ a_1 \end{pmatrix}\begin{pmatrix} b_2 \\ a_2 \end{pmatrix} - \begin{pmatrix} b_2 \\ a_1 \end{pmatrix}\begin{pmatrix} b_1 \\ a_2 \end{pmatrix}$$

By the above theorem, we get

$$\frac{\begin{pmatrix} 1, 2, \cdots, N, b_1, b_2 \\ 1, 2, \cdots, N, a_1, a_2 \end{pmatrix}}{\begin{pmatrix} 1, 2, \cdots, N \\ 1, 2, \cdots, N \end{pmatrix}}$$

$$= \frac{\begin{pmatrix} 1, 2, \cdots, N, b_1 \\ 1, 2, \cdots, N, a_1 \end{pmatrix}\begin{pmatrix} 1, 2, \cdots, N, b_2 \\ 1, 2, \cdots, N, a_2 \end{pmatrix}}{\begin{pmatrix} 1, 2, \cdots, N \\ 1, 2, \cdots, N \end{pmatrix}\begin{pmatrix} 1, 2, \cdots, N \\ 1, 2, \cdots, N \end{pmatrix}} - \frac{\begin{pmatrix} 1, 2, \cdots, N, b_2 \\ 1, 2, \cdots, N, a_1 \end{pmatrix}\begin{pmatrix} 1, 2, \cdots, N, b_1 \\ 1, 2, \cdots, N, a_2 \end{pmatrix}}{\begin{pmatrix} 1, 2, \cdots, N \\ 1, 2, \cdots, N \end{pmatrix}\begin{pmatrix} 1, 2, \cdots, N \\ 1, 2, \cdots, N \end{pmatrix}}$$

thus we obtain

$$\begin{pmatrix} 1,2,\cdots,N,b_1,b_2 \\ 1,2,\cdots,N,a_1,a_2 \end{pmatrix} \begin{pmatrix} 1,2,\cdots,N \\ 1,2,\cdots,N \end{pmatrix}$$
$$= \begin{pmatrix} 1,2,\cdots,N,b_1 \\ 1,2,\cdots,N,a_1 \end{pmatrix} \begin{pmatrix} 1,2,\cdots,N,b_2 \\ 1,2,\cdots,N,a_2 \end{pmatrix} - \begin{pmatrix} 1,2,\cdots,N,b_2 \\ 1,2,\cdots,N,a_1 \end{pmatrix} \begin{pmatrix} 1,2,\cdots,N,b_1 \\ 1,2,\cdots,N,a_2 \end{pmatrix}$$

This is called the Jacobi formula for determinants.

Proof. From the golden theorem for Pfaffians (Th. 2.3), we obtain

$$\frac{(1,2,\cdots,N,i_1,i_2,\cdots,i_{2n},N',\cdots,2',1')}{(1,2,\cdots,N,N',\cdots,2',1')}$$
$$= \mathrm{Pf}\left(\frac{(1,2,\cdots,N,i_\mu,i_\nu,N',\cdots,2',1')}{(1,2,\cdots,N,N',\cdots,2',1')} \right)_{1\le\mu<\nu\le2n}$$

By taking $(i,j)=0$ $(1\le i<j\le N)$, $(i,i_\mu)=0$ $(1\le i\le N,1\le\mu\le n)$, $(i_\mu,i_\nu)=0$ $(1\le\mu<\nu\le n)$, we get

$$\frac{\begin{pmatrix} 1,2,\cdots,N,j_1,j_2,\cdots,j_n \\ 1,2,\cdots,N,i_1,i_2,\cdots,i_n \end{pmatrix}}{\begin{pmatrix} 1,2,\cdots,N \\ 1,2,\cdots,N \end{pmatrix}} = \det\left(\frac{\begin{pmatrix} 1,2,\cdots,N,j_\nu \\ 1,2,\cdots,N,i_\mu \end{pmatrix}}{\begin{pmatrix} 1,2,\cdots,N \\ 1,2,\cdots,N \end{pmatrix}} \right)_{1\le\mu,\nu\le n}$$

where $\begin{pmatrix} j \\ i \end{pmatrix} = (i,j')$ $(1\le i,j\le N)$, $\begin{pmatrix} j_\nu \\ i \end{pmatrix} = (i,i_{2n+1-\nu})$ $(1\le i\le N, 1\le \nu\le n)$, $\begin{pmatrix} j \\ i_\mu \end{pmatrix} = (i_\mu,j')$ $(1\le\mu\le n,1\le j\le N)$ and $\begin{pmatrix} j_\nu \\ i_\mu \end{pmatrix} = (i_\mu,i_{2n+1-\nu})$ $(1\le\mu,\nu\le n)$. This has the same form as the definition of determinants (1). From this fact, the golden theorem is immediately proved, because the replacement of $\begin{pmatrix} j_1,j_2,\cdots,j_n \\ i_1,i_2,\cdots,i_n \end{pmatrix}$ with $\frac{\begin{pmatrix} 1,2,\cdots,N,j_1,j_2,\cdots,j_n \\ 1,2,\cdots,N,i_1,i_2,\cdots,i_n \end{pmatrix}}{\begin{pmatrix} 1,2,\cdots,N \\ 1,2,\cdots,N \end{pmatrix}}$ is valid in the definition of determinants.

Theorems 2.3 and 2.4 are called the Wick theorems.

2.6 Differential Formula

Assume that $\begin{pmatrix} j \\ i \end{pmatrix}$ and (i,j) are differentiable functions of a variable x.

Theorem 2.5.

$$\partial_x \begin{pmatrix} 1,2,\cdots,N \\ 1,2,\cdots,N \end{pmatrix} = \sum_{1\le i,j\le N} \begin{pmatrix} 1,2,\cdots,N \\ 1,2,\cdots,N \end{pmatrix}^j_i \partial_x \begin{pmatrix} j \\ i \end{pmatrix}$$

Proof. Since the determinant $\begin{pmatrix} 1,2,\cdots,N \\ 1,2,\cdots,N \end{pmatrix}$ is a polynomial in $\begin{pmatrix} j \\ i \end{pmatrix}$ of at most degree 1 for all i and j, we have

$$\partial_x \begin{pmatrix} 1,2,\cdots,N \\ 1,2,\cdots,N \end{pmatrix} = \sum_{1\leq i,j\leq N} \left(\text{coefficient of } \partial_x \begin{pmatrix} j \\ i \end{pmatrix} \right) \partial_x \begin{pmatrix} j \\ i \end{pmatrix}$$

and the above coefficient of $\partial_x \begin{pmatrix} j \\ i \end{pmatrix}$ is equal to the coefficient of $\begin{pmatrix} j \\ i \end{pmatrix}$ in the

determinant $\begin{pmatrix} 1,2,\cdots,N \\ 1,2,\cdots,N \end{pmatrix}$, which is $\begin{pmatrix} 1,2,\cdots,N \\ 1,2,\cdots,N \end{pmatrix}_i^j$ by definition.

Theorem 2.6.

$$\partial_x(1,2,\cdots,2N) = \sum_{1\leq i<j\leq 2N} (1,2,\cdots,2N)_{ij}\partial_x(i,j)$$

Proof. Since the Pfaffian $(1,2,\cdots,2N)$ is a polynomial in (i,j) of at most degree 1 for all i and j, we have

$$\partial_x(1,2,\cdots,2N) = \sum_{1\leq i<j\leq 2N} (\text{coefficient of } \partial_x(i,j))\partial_x(i,j)$$

and the above coefficient of $\partial_x(i,j)$ is equal to the coefficient of (i,j) in the Pfaffian $(1,2,\cdots,2N)$, which is $(1,2,\cdots,2N)_{ij}$ by definition.

3 Difference Formulas

In this section, we give the difference formulas for Pfaffians with discrete Wronski or Gram structure. Let the elements of a Pfaffian be functions of discrete independent variables, k_1, k_2, k_3 ,$\cdots$. In order to specify the values of the independent variables, we write them as a suffix, for example, $(i,j)_{k_1,k_2,k_3,\cdots}$, and we may omit unshifted independent variables for simplicity. We denote the difference interval for k_ν by a_ν.

3.1 Discrete Wronski Pfaffians

Definition 3.1. *If the (i,j)-element of a Pfaffian satisfies the discrete linear dispersion relation,*

$$\frac{1}{a_\nu}((i,j)_{k_\nu} - (i,j)_{k_\nu-1}) = (i+1,j)_{k_\nu} + (i,j+1)_{k_\nu} - a_\nu(i+1,j+1)_{k_\nu}$$

for all ν, then this Pfaffian is called a discrete Wronski Pfaffian.

Example 3.1. Let

$$(i,j) = \sum_n (\varphi_n^{(i)} \psi_n^{(j)} - \varphi_n^{(j)} \psi_n^{(i)}).$$

These are elements of a discrete Wronski Pfaffian. Here $\varphi_n^{(i)}$ and $\psi_n^{(j)}$ are arbitrary functions satisfying the discrete linear dispersion relations,

$$\frac{1}{a_\nu}\left(\varphi_n^{(i)}(k_\nu) - \varphi_n^{(i)}(k_\nu - 1)\right) = \varphi_n^{(i+1)}(k_\nu)$$

$$\frac{1}{a_\nu}\left(\psi_n^{(j)}(k_\nu) - \psi_n^{(j)}(k_\nu - 1)\right) = \psi_n^{(j+1)}(k_\nu),$$

for all ν.

Using the linearity and alternativity of Pfaffians proved in Sect. 2 and the above discrete linear dispersion relation for the (i,j)-element, the next theorem can be easily proved.

Theorem 3.1. *The discrete Wronski Pfaffian* $(1,2,\cdots,2N)$ *satisfies the difference formulas,*

$$(1,2,\cdots,2N)_{k_\nu+1} = (1,2,\cdots,2N-1,\overline{2N}_\nu)$$
$$a_\nu(1,2,\cdots,2N)_{k_\nu+1} = (1,2,\cdots,2N-1,\overline{2N-1}_\nu)$$
$$(a_\nu - a_\mu)(1,2,\cdots,2N)_{k_\nu+1,k_\mu+1} = (1,2,\cdots,2N-2,\overline{2N-1}_\mu,\overline{2N-1}_\nu)$$
$$a_\nu a_\mu(a_\nu - a_\mu)(1,2,\cdots,2N)_{k_\nu+1,k_\mu+1} = (1,2,\cdots,2N-2,\overline{2N-2}_\mu,\overline{2N-2}_\nu)$$
$$\prod_{1\le\nu<\mu\le n}(a_\nu - a_\mu)(1,2,\cdots,2N)_{k_1+1,k_2+1,\cdots,k_n+1}$$
$$= (1,2,\cdots,2N-n,\overline{2N-n+1}_n,\cdots,\overline{2N-n+1}_2,\overline{2N-n+1}_1)$$
$$\prod_{\nu=1}^n a_\nu \prod_{1\le\nu<\mu\le n}(a_\nu - a_\mu)(1,2,\cdots,2N)_{k_1+1,k_2+1,\cdots,k_n+1}$$
$$= (1,2,\cdots,2N-n,\overline{2N-n}_n,\cdots,\overline{2N-n}_2,\overline{2N-n}_1),$$

where

$$(i,\overline{j}_\nu) = (i,j)_{k_\nu+1} - a_\nu(i+1,j)_{k_\nu+1}$$

$$(\overline{j}_\mu,\overline{j}_\nu) = (a_\nu - a_\mu)(j,j+1)_{k_\nu+1,k_\mu+1}.$$

3.2 Discrete Gram Pfaffians

Definition 3.2. *If the (i,j)-element of a Pfaffian satisfies the discrete linear dispersion relation,*

$$\frac{1}{a_\nu}\left((i,j)_{k_\nu} - (i,j)_{k_\nu-1}\right) = \varphi_i^{(0)}(k_\nu)\bar{\varphi}_j^{(0)}(k_\nu - 1) - \varphi_j^{(0)}(k_\nu)\bar{\varphi}_i^{(0)}(k_\nu - 1),$$

for all ν, and if $\varphi_i^{(n)}$ and $\bar{\varphi}_j^{(n)}$ satisfy the discrete linear dispersion relations,

$$\frac{1}{a_\nu}\left(\varphi_i^{(n)}(k_\nu) - \varphi_i^{(n)}(k_\nu - 1)\right) = \varphi_i^{(n+1)}(k_\nu)$$

$$\frac{1}{a_\nu}\left(\bar{\varphi}_j^{(n)}(k_\nu + 1) - \bar{\varphi}_j^{(n)}(k_\nu)\right) = \bar{\varphi}_j^{(n+1)}(k_\nu),$$

for all ν, then this Pfaffian is called a discrete Gram Pfaffian.

Example 3.2. Let

$$(i,j) = c_{ij} + \sum_{n=0}^{\infty}(-1)^n\left(\varphi_i^{(n)}\bar{\varphi}_j^{(-n-1)} - \varphi_j^{(n)}\bar{\varphi}_i^{(-n-1)}\right).$$

They are elements of a discrete Gram Pfaffian. Here the c_{ij}'s are constants satisfying the antisymmetry, $c_{ij} = -c_{ji}$ and $c_{ii} = 0$.

Using the linearity and alternativity of Pfaffians proved in Sect. 2 and the above discrete linear dispersion relation for the (i,j)-element, the next theorem can be easily proved.

Theorem 3.2. *The discrete Gram Pfaffian* $(1, 2, \cdots, 2N)$ *satisfies the following difference formulas,*

$$\frac{1}{a_\nu}(1, 2, \cdots, 2N)_{k_\nu+1} = (\bar{d}^0, 1, 2, \cdots, 2N, d_\nu^0)$$

$$\frac{a_\nu - a_\mu}{(a_\nu a_\mu)^2}(1, 2, \cdots, 2N)_{k_\nu+1,k_\mu+1} = (\bar{d}^0, \bar{d}^1, 1, 2, \cdots, 2N, d_\nu^0, d_\mu^0)$$

$$\frac{\displaystyle\prod_{1\le\nu<\mu\le n}(a_\nu - a_\mu)}{\left(\displaystyle\prod_{\nu=1}^n a_\nu\right)^n}(1, 2, \cdots, 2N)_{k_1+1,k_2+1,\cdots,k_n+1}$$

$$= (\bar{d}^0, \bar{d}^1, \cdots, \bar{d}^{n-1}, 1, 2, \cdots, 2N, d_1^0, d_2^0, \cdots, d_n^0),$$

where

$$(i, d_\nu^0) = \varphi_i^{(0)}(k_\nu + 1) \qquad (d_\mu^0, d_\nu^0) = 0$$

$$(\bar{d}^n, j) = -\bar{\varphi}_j^{(n)} \qquad (\bar{d}^n, \bar{d}^m) = 0 \qquad (\bar{d}^n, d_\nu^0) = (-1)^n\frac{1}{a_\nu^{n+1}}.$$

4 Discrete Bilinear Equations

Both the discrete Wronski Pfaffians and discrete Gram Pfaffians satisfy the same bilinear equations, that is, the discrete coupled KP hierarchy. Below we give the bilinear form of discrete coupled KP equations for both the Wronski case and Gram case.

4.1 Discrete Wronski Pfaffian

For discrete Wronski Pfaffians, let us define

$$\tau_N = (1, 2, \cdots, 2N)$$

Then the difference formulas for τ_N are given by Theorem 3.1. The following bilinear equations for τ_N are a direct consequence of bilinear algebraic identities (9) and (10),

$$
\begin{aligned}
& a_\nu (a_\mu - a_\lambda)\tau_N(k_\mu + 1, k_\lambda + 1)\tau_N(k_\nu + 1) \\
+ & a_\mu (a_\lambda - a_\nu)\tau_N(k_\nu + 1, k_\lambda + 1)\tau_N(k_\mu + 1) \\
+ & a_\lambda (a_\nu - a_\mu)\tau_N(k_\nu + 1, k_\mu + 1)\tau_N(k_\lambda + 1) \\
= & a_\nu a_\mu a_\lambda (a_\mu - a_\lambda)(a_\lambda - a_\nu)(a_\nu - a_\mu)\tau_{N+1}(k_\nu + 1, k_\mu + 1, k_\lambda + 1)\tau_{N-1},
\end{aligned}
$$

$$
\begin{aligned}
& a_\mu a_\lambda (a_\mu - a_\lambda)\tau_{N+1}(k_\mu + 1, k_\lambda + 1)\tau_N(k_\nu + 1) \\
+ & a_\lambda a_\nu (a_\lambda - a_\nu)\tau_{N+1}(k_\nu + 1, k_\lambda + 1)\tau_N(k_\mu + 1) \\
+ & a_\nu a_\mu (a_\nu - a_\mu)\tau_{N+1}(k_\nu + 1, k_\mu + 1)\tau_N(k_\lambda + 1) \\
+ & (a_\mu - a_\lambda)(a_\lambda - a_\nu)(a_\nu - a_\mu)\tau_{N+1}(k_\nu + 1, k_\mu + 1, k_\lambda + 1)\tau_N = 0.
\end{aligned}
$$

These equations coincide with the bilinear equations of the discrete coupled KP hierarchy.

4.2 Discrete Gram Pfaffian

For discrete Gram Pfaffians, let us define

$$
\begin{aligned}
\tau &= (1, 2, \cdots, 2N) \\
\sigma &= (1, 2, \cdots, 2N, d^0, d^1) \\
\bar\sigma &= (\bar{d}^0, \bar{d}^1, 1, 2, \cdots, 2N),
\end{aligned}
$$

where

$$(i, d^n) = \varphi_i^{(n)} \qquad (d^m, d^n) = 0 \qquad (d^n, d_\nu^0) = 0.$$

Then the difference formulas for τ are given by Theorem 3.2. In the same manner, we can obtain the following difference formulas for σ and $\bar\sigma$,

$$a_\nu \sigma(k_\nu + 1) = (1, 2, \cdots, 2N, d^0, d_\nu^0)$$

$$(a_\nu - a_\mu)\sigma(k_\nu + 1, k_\mu + 1) = (1, 2, \cdots, 2N, d_\mu^0, d_\nu^0)$$

$$\frac{(a_\nu - a_\mu)(a_\nu - a_\lambda)(a_\mu - a_\lambda)}{a_\nu a_\mu a_\lambda}\sigma(k_\nu + 1, k_\mu + 1, k_\lambda + 1)$$

$$= (d^0, 1, 2, \cdots, 2N, d_\lambda^0, d_\mu^0, d_\nu^0)$$

$$\frac{1}{a_\nu^3}\bar\sigma(k_\nu + 1) = (d^0, d^1, d^2, 1, 2, \cdots, 2N, d_\nu^0).$$

The following bilinear equations for τ, σ and $\bar\sigma$ are direct consequences of bilinear algebraic identities (9) and (10),

$$a_\nu(a_\mu - a_\lambda)\tau(k_\mu + 1, k_\lambda + 1)\tau(k_\nu + 1)$$
$$+ a_\mu(a_\lambda - a_\nu)\tau(k_\nu + 1, k_\lambda + 1)\tau(k_\mu + 1)$$
$$+ a_\lambda(a_\nu - a_\mu)\tau(k_\nu + 1, k_\mu + 1)\tau(k_\lambda + 1)$$
$$= a_\nu a_\mu a_\lambda (a_\mu - a_\lambda)(a_\lambda - a_\nu)(a_\nu - a_\mu)\sigma(k_\nu + 1, k_\mu + 1, k_\lambda + 1)\bar\sigma$$

$$a_\mu a_\lambda(a_\mu - a_\lambda)\sigma(k_\mu + 1, k_\lambda + 1)\tau(k_\nu + 1)$$
$$+ a_\lambda a_\nu(a_\lambda - a_\nu)\sigma(k_\nu + 1, k_\lambda + 1)\tau(k_\mu + 1)$$
$$+ a_\nu a_\mu(a_\nu - a_\mu)\sigma(k_\nu + 1, k_\mu + 1)\tau(k_\lambda + 1)$$
$$+ (a_\mu - a_\lambda)(a_\lambda - a_\nu)(a_\nu - a_\mu)\sigma(k_\nu + 1, k_\mu + 1, k_\lambda + 1)\tau = 0$$

$$a_\mu a_\lambda(a_\mu - a_\lambda)\tau(k_\mu + 1, k_\lambda + 1)\bar\sigma(k_\nu + 1)$$
$$+ a_\lambda a_\nu(a_\lambda - a_\nu)\tau(k_\nu + 1, k_\lambda + 1)\bar\sigma(k_\mu + 1)$$
$$+ a_\nu a_\mu(a_\nu - a_\mu)\tau(k_\nu + 1, k_\mu + 1)\bar\sigma(k_\lambda + 1)$$
$$+ (a_\mu - a_\lambda)(a_\lambda - a_\nu)(a_\nu - a_\mu)\tau(k_\nu + 1, k_\mu + 1, k_\lambda + 1)\bar\sigma = 0.$$

These equations coincide with the bilinear equations for discrete Wronski Pfaffians and they also coincide with the bilinear equations of the discrete coupled KP hierarchy.

5 Concluding Remarks

We have obtained bilinear equations for the hierarchies of discrete soliton equations from the algebraic identities and the difference formulas for determinants and Pfaffians. Whereas the identities for determinants correspond to the KP hierarchy, the identities for Pfaffians correspond to the coupled KP hierarchy both in the continuous and in the discrete case. Various properties of determinants and Pfaffians can be found in, for example, [1]. Both determinants and Pfaffians are important in the solution of the hierarchies of soliton equations.

There are two types of differential or difference structure, the Wronski type and the Gram type, both for determinants and for Pfaffians and both for the continuous and for the discrete case. Continuous Wronski determinants have been intensively studied [2–6]. The continuous Gram determinants have the same algebraic properties as the Wronski ones [7,8]. The case of discrete Wronski and Gram determinants was analyzed in [9]. It was shown that Pfaffians play a role similar to that of determinants, both in the continuous case [10, 11] and in the discrete case [12]. The methods for generating discrete soliton equations were studied many years ago [13, 14]. Here they were interpreted in the language of determinant and Pfaffian identities.

So far it is not known whether all solutions of the hierarchies of soliton equations are determinants or Pfaffians, or whether these are only special solutions. The Sato theory showed that the solutions of the KP hierarchies can be characterized as orbits in the universal Grassmann manifold [15]- [19]. Date, Kashiwara, Jimbo and Miwa developed the theory of transformation groups for soliton equations and showed that the Lie algebra of symmetries of the space of solutions is an affine Lie algebra [20–22]. The relation between the continuous soliton equations and the discrete ones is described by the Miwa transformation [23].

We observe that the techniques used to construct subhierarchies of soliton equations such as reduction, truncation [24], restriction to BKP, CKP, $\cdots$, multicomponentization, and so on, are applicable to the Pfaffian case. These subhierarchies have not been fully studied yet.

The formulations in terms of determinants and Pfaffians, which are obtained by means of Darboux transformations of a trivial vacuum solution, can be generalized to the case where the initial solution is a nontrivial vacuum. The Darboux transformations then yield determinant and Pfaffian solutions depending implicitly on the chosen vacuum solution, and the nontrivial vacuum appears in the coefficients of the associated bilinear equations [25]. This result could be applied to similarity reductions and nonautonomous systems.

It is also known that the determinant and Pfaffian solutions have deep relations with matrix integrals and orthogonal polynomials [26, 27]. The soliton equations and their solutions can be derived from the theory of random matrices. There are so many related topics that we had to keep the number of references to a minimum, and we can not predict in what directions the theory will develop.

Acknowledgements

The author would like to express his sincere thanks to Professor R. Hirota and Dr. S. Tsujimoto for valuable discussions and useful comments.

References

1. T. Muir: *A Treatise on the Theory of Determinants* (Macmillan and Co., London 1882).
2. J. Satsuma: J. Phys. Soc. Jpn. **46**, 359 (1979).
3. N. C. Freeman, J. J. C. Nimmo: Phys. Lett. **95A**, 1 (1983).
4. N. C. Freeman: IMA J. Appl. Math. **32**, 125 (1984).
5. R. Hirota: J. Phys. Soc. Jpn. **55**, 2137 (1986).
6. J. J. C. Nimmo: Symmetric functions and the KP hierarchy. In: *Nonlinear Evolutions*, ed. by J. J. P. Leon (World Scientific, Singapore 1988) pp 245–261.
7. A. Nakamura: J. Phys. Soc. Jpn. **58**, 412 (1989).
8. S. Miyake, Y. Ohta, J. Satsuma: J. Phys. Soc. Jpn. **59**, 48 (1990).
9. Y. Ohta, R. Hirota, S. Tsujimoto, T. Imai: J. Phys. Soc. Jpn. **62**, 1872 (1993).
10. R. Hirota, Y. Ohta: J. Phys. Soc. Jpn. **60**, 798 (1991).
11. C. R. Gilson, J. J. C. Nimmo: Theo. Math. Phys. **128**, 870 (2001).
12. C. R. Gilson, J. J. C. Nimmo, S. Tsujimoto: J. Phys. A **34**, 10569 (2001).
13. R. Hirota: J. Phys. Soc. Jpn. **43**, 1424, 2074, 2079 (1977); ibid **45**, 321 (1978).
14. E. Date, M. Jimbo, T. Miwa: J. Phys. Soc. Jpn. **51**, 4116, 4125 (1982); ibid **52**, 388, 761, 766 (1983).
15. M. Sato: RIMS Kokyuroku **439**, 30 (1981).
16. M. Sato, Y. Sato: Soliton equations as dynamical systems on infinite-dimensional Grassmann manifold, In: *Nonlinear Partial Differential Equations in Applied Science*, ed. by H. Fujita, P. D. Lax, G. Strang (Kinokuniya, North-Holland, Tokyo 1983) pp 259–271.
17. M. Sato: *Soliton Equations and the Universal Grassmann Manifolds*, notes by M. Noumi (Sophia Kokyuroku in Mathematics 18, Dept. Math. Sophia Univ., Tokyo 1984).
18. Y. Ohta, J. Satsuma, D. Takahashi, T. Tokihiro: Prog. Theo. Phys. Suppl. **94**, 210 (1988).
19. J. Satsuma, R. Willox: Sato theory and transformation group theory approach to integrable systems. in this volume.
20. E. Date, M. Kashiwara, M. Jimbo, T. Miwa: Transformation groups for soliton equations. In: *Non-linear Integrable Systems — Classical Theory and Quantum Theory*, ed. by M. Jimbo, T. Miwa (World Scientific, Singapore 1983) pp 39-119.
21. M. Jimbo, T. Miwa: Publ. RIMS, Kyoto Univ. **19**, 943 (1983).
22. T. Miwa, M. Jimbo, E. Date: *Solitons. Differential Equations, Symmetries and Infinite-Dimensional Algebras*, translated by M. Reid (Cambridge Tracts in Mathematics 135, Cambridge University Press, Cambridge 2000).
23. T. Miwa: Proc. Jpn. Acad. **58A**, 9 (1982).
24. H. Harada: J. Phys. Soc. Jpn. **54**, 4507 (1985); ibid **56**, 3847 (1987).
25. J. J. C. Nimmo: Inverse Problems **8**, 219 (1992); J. Phys. A **30**, 8693 (1997).
26. S. Kakei: J. Phys. Soc. Jpn. **68**, 2875 (1999).
27. M. Adler, T. Shiota, P. van Moerbeke: Math. Ann. **322**, 423 (2002).

Discrete Differential Geometry. Integrability as Consistency

Alexander I. Bobenko*

Institut für Mathematik, Fakultät 2, Technische Universität Berlin, Strasse des 17. Juni 136, 10623 Berlin, Germany, bobenko@math.tu-berlin.de

Abstract. We discuss a new geometric approach to discrete integrability coming from discrete differential geometry. A d–dimensional equation is called consistent if it is valid for all d–dimensional sublattices of a $(d+1)$–dimensional lattice. This algorithmically verifiable property implies analytical structures characteristic of integrability, such as the zero-curvature representation, and allows one to classify discrete integrable equations within certain natural classes. These ideas also apply to the noncommutative case. Theorems about the smooth limit of the theory are also presented.

1 Introduction

The original results presented in these lectures were proved in the recent series of papers [3, 4, 6, 7]. We refer to these papers for more details, further references and complete proofs. For the geometric background in discrete differential geometry see in particular [2].

2 Origin and Motivation: Differential Geometry

Long before the theory of solitons, geometers used integrable equations to describe various special curves, surfaces etc. At that time no relation to mathematical physics was known, and quite different geometries which appeared in this context (called integrable nowadays) were unified by their common geometric features:

- Integrable surfaces, curves etc. have nice geometric properties,
- Integrable geometries come with their interesting transformations acting within the class,
- These transformations are permutable (Bianchi permutability).

Since 'nice' and 'interesting' can hardly be treated as mathematically formulated features, let us discuss the permutability property. We shall explain

* Partially supported by the SFB 288 "Differential geometry and quantum physics" and by the DFG research center "Mathematics for key technologies" (FZT 86) in Berlin.

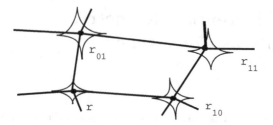

Fig. 1. Permutability of the Bäcklund transformations

it in more detail for the classical example of surfaces with constant negative Gaussian curvature (K-surface) with their Bäcklund transformations.

Let $r : \mathbb{R}^2 \to \mathbb{R}^3$ be a K-surface, and r_{10} and r_{01} two K-surfaces obtained by Bäcklund transformations of r. The classical Bianchi permutability theorem claims that there exists a unique K-surface r_{11} which is a Bäcklund transform of r_{10} and r_{01} (Fig. 1). Moreover,

(i) the straight line connecting the points $r(x,y)$ and $r_{10}(x,y)$ lies in the tangent planes of the surfaces r and r_{10} at these points,

(ii) the opposite edges of the quadrilateral $(r, r_{10}, r_{01}, r_{11})$ have equal lengths,

$$\|r_{10} - r\| = \|r_{11} - r_{01}\|, \qquad \|r_{01} - r\| = \|r_{11} - r_{10}\|.$$

This way a $\mathbb{Z}^2$ lattice $r_{k,\ell}$ obtained by permutable Bäcklund transformations gives rise to discrete K-surfaces. Indeed, fixing the smooth parameters (x,y) one observes that

(i) the points $r_{k,\ell}, r_{k,\ell\pm1}, r_{k\pm1,\ell}$ lie in one plane (the 'tangent' plane of the discrete K-surface at the vertex $r_{k,\ell}$),

(ii) the opposite edges of the quadrilateral $r_{k,\ell}, r_{k+1,\ell}, r_{k+1,\ell+1}, r_{k,\ell+1}$ have equal lengths.

These are exactly the characteristic properties [1] of the *discrete K-surfaces*, $r : \mathbb{Z}^2 \to \mathbb{R}^3$.

One immediately observes that the discrete K-surfaces have the same properties as their smooth counterparts. There exist deep reasons for that. The classical differential geometry of integrable surfaces may be obtained from a unifying multi-dimensional discrete theory by a refinement of the coordinate mesh-size in some of the directions.

Indeed, by refining of the coordinate mesh-size,

$$r : (\epsilon\mathbb{Z})^2 \to \mathbb{R}^3 \quad \longrightarrow \quad r : \mathbb{R}^2 \to \mathbb{R}^3,$$

discrete surface $\epsilon \to 0$ smooth surface,

in the limit one obtains classical smooth K-surfaces from discrete K-surfaces. This statement is visualized in Fig. 2 which shows an example of a continuous

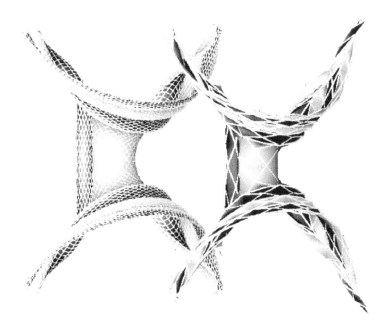

Fig. 2. A continuous and a discrete Amsler surfaces

Amsler surface and its discrete analogue. The subclass of Amsler surfaces is characterized by the condition that the K-surface (smooth or discrete) should contain two straight lines.

Moreover, the classical Bianchi permutability implies n–dimensional permutability of the Bäcklund transformations. This means that the set of a given K-surface, $r : \mathbb{R}^2 \to \mathbb{R}^3$, with its n Bäcklund transforms, $r_{10\ldots0}$, $r_{010\ldots0}, \ldots, r_{0\ldots01}$, can be completed to 2^n different K-surfaces, $r_{i_1\ldots i_n}, i_k \in \{0,1\}$, associated to the vertices of the n–dimensional cube $C = \{0,1\}^n$. The surfaces associated to vertices of C connected by edges are Bäcklund transforms of each other.

Similar to the 2–dimensional case, this description can be extended to an n–dimensional lattice. Fixing the smooth parameters (x, y), one obtains a map,

$$r : (\epsilon_1 \mathbb{Z}) \times \ldots \times (\epsilon_n \mathbb{Z}) \to \mathbb{R}^3,$$

which is an n–dimensional net obtained from one point of a K-surface by permutable Bäcklund transformations. It turns out that the whole smooth theory can be recovered from this description. Indeed, completely changing the point of view, in the limit $\epsilon_1 \to 0, \epsilon_2 \to 0, \epsilon_3 = \ldots = \epsilon_n = 1$, one arrives at a smooth K-surface with an $(n-2)$–dimensional discrete family of permutable Bäcklund transforms (Fig. 3),

$$r : \mathbb{R}^2 \times \mathbb{Z}^{n-2} \to \mathbb{R}^3.$$

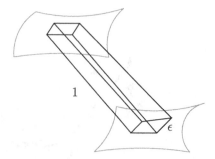

Fig. 3. Surfaces and their transformations as a limit of multidimensional lattices

This simple idea is quite fruitful. In the discrete case all directions of the multi–dimensional lattices appear in a quite symmetric way. It leads to

- A unification of surfaces and their transformations. Discrete surfaces and their transformations are indistinguishable.
- A fundamental consistency principle. Due to the symmetry of the discrete setup the same equations hold on all elementary faces of the lattice. This leads us beyond the pure differential geometry to a new understanding of the integrability, classification of integrable equations and derivation of the zero curvature (Lax) representation from the first principles.
- Interesting generalizations for $d > 2$–dimensional systems, quantum systems, discrete systems with the fields on various lattice elements (vertices, edges, faces, etc.).

As it was mentioned above, all this suggests that it might be possible to develop the classical differential geometry, including both the theory of surfaces and of their transformations, as a mesh-refining limit of the discrete constructions. On the other hand, the good quantitative properties of approximations provided by the discrete differential geometry suggest that they might be put at the basis of the practical numerical algorithms for computations in differential geometry. However until recently there were no rigorous mathematical results supporting this observation.

The first step in closing this gap was made in the paper [7] where a geometric numerical scheme for a class of nonlinear hyperbolic equations was developed and general convergence results were proved. We return to this problem in Sect. 6, considering in particular the sine–Gordon equation and discrete and smooth K-surfaces.

3 Equations on Quad-Graphs. Integrability as Consistency

Traditionally, discrete integrable systems were considered for fields defined on the $\mathbb{Z}^2$ lattice. Having in mind geometric applications, it is natural to generalize this setup to include distinguished vertices with different combinatorics, and moreover to consider graphs with various global properties. A direct generalization of the Lax representation from the $\mathbb{Z}^2$ lattice to more general lattices leads to a concept of

3.1 Discrete Flat Connections on Graphs

Integrable systems on graphs can be defined as flat connections whose values are in loop groups. More precisely, this notion includes the following component elements:

- A *cellular decomposition* $\mathcal{G}$ of an oriented surface. The set of its vertices will be denoted by $V(\mathcal{G})$, the set of its edges will be denoted by $E(\mathcal{G})$, and the set of its faces will be denoted by $F(\mathcal{G})$. For each edge, one of its possible orientations is fixed.
- A *loop group* $G[\lambda]$, whose elements are functions from $\mathbb{C}$ into some group G. The complex argument, λ, of these functions is known in the theory of integrable systems as the *spectral parameter*.
- A *wave function* $\Psi : V(\mathcal{G}) \to G[\lambda]$, defined on the vertices of $\mathcal{G}$.
- A collection of *transition matrices*, $L : E(\mathcal{G}) \to G[\lambda]$, defined on the edges of $\mathcal{G}$.

It is supposed that for any oriented edge, $e = (v_1, v_2) \in E(\mathcal{G})$, the values of the wave functions at its ends are connected by

$$\Psi(v_2, \lambda) = L(e, \lambda)\Psi(v_1, \lambda). \tag{1}$$

Therefore the following *discrete zero-curvature condition* is supposed to be satisfied. Consider any closed contour consisting of a finite number of edges of $\mathcal{G}$,

$$e_1 = (v_1, v_2), \quad e_2 = (v_2, v_3), \quad \ldots, \quad e_n = (v_n, v_1).$$

Then

$$L(e_n, \lambda) \cdots L(e_2, \lambda)L(e_1, \lambda) = I. \tag{2}$$

In particular, for any edge $e = (v_1, v_2)$, if $e^{-1} = (v_2, v_1)$, then

$$L(e^{-1}, \lambda) = \left(L(e, \lambda)\right)^{-1}. \tag{3}$$

Actually, in applications the matrices $L(e, \lambda)$ also depend on a point of some set X (the phase-space of an integrable system), so that some elements $x(e) \in X$ are attached to the edges e of $\mathcal{G}$. In this case the discrete zero-curvature condition (2) becomes equivalent to the collection of equations relating the fields $x(e_1), \ldots, x(e_n)$ attached to the edges of each closed contour. We say that this collection of equations admits a *zero-curvature representation*.

3.2 Quad-Graphs

Although one can, in principle, consider integrable systems in the sense of the traditional definition of Sect. 3.1 on very different kinds of graph, one should not go that far with the generalization.

As we have shown in [3], there is a special class of graph, called *quad-graphs*, supporting the most fundamental properties of integrability theory. This notion turns out to be a proper generalization of the $\mathbb{Z}^2$ lattice as far as integrability theory is concerned.

Definition 3.1. *A cellular decomposition, $\mathcal{G}$, of an oriented surface is called a quad-graph, if all its faces are quadrilateral.*

Here we mainly consider the local theory of integrable systems on quad-graphs. Therefore, in order to avoid the discussion of some subtle boundary and topological effects, we shall always suppose that the surface carrying the quad-graph is a topological disk; no boundary effects will be considered.

Before we proceed to integrable systems, we would like to propose a construction which, from an arbitrary cellular decomposition, produces a certain quad-graph. Towards this aim, we first recall the notion of the *dual graph*, or, more precisely, of the *dual cellular decomposition* $\mathcal{G}^*$. The vertices in $V(\mathcal{G}^*)$ are in one-to-one correspondence with the faces in $F(\mathcal{G})$ (actually, they can be chosen to be certain points inside the corresponding faces, cf. Fig. 4). Each $\mathfrak{e} \in E(\mathcal{G})$ separates two faces in $F(\mathcal{G})$, which in turn correspond to two vertices in $V(\mathcal{G}^*)$. A path between these two vertices is then declared to be an edge $\mathfrak{e}^* \in E(\mathcal{G}^*)$ dual to $\mathfrak{e}$. Finally, the faces in $F(\mathcal{G}^*)$ are in a one-to-one correspondence with the vertices in $V(\mathcal{G})$. If $v_0 \in V(\mathcal{G})$, and $v_1, \ldots, v_n \in V(\mathcal{G})$ are its neighbors connected with v_0 by the edges $\mathfrak{e}_1 = (v_0, v_1), \ldots, \mathfrak{e}_n = (v_0, v_n) \in E(\mathcal{G})$, then the face in $F(\mathcal{G}^*)$ corresponding to v_0 is defined by its boundary, $\mathfrak{e}_1^* \cup \ldots \cup \mathfrak{e}_n^*$ (cf. Fig. 5).

   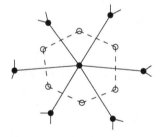

Fig. 4. The vertex in $V(\mathcal{G}^*)$ dual to the face in $F(\mathcal{G})$.

Fig. 5. The face in $F(\mathcal{G}^*)$ dual to the vertex in $V(\mathcal{G})$.

Now we introduce a new complex, the *double* $\mathcal{D}$, constructed from $\mathcal{G}$, $\mathcal{G}^*$. The set of vertices of the double ,$\mathcal{D}$, is $V(\mathcal{D}) = V(\mathcal{G}) \cup V(\mathcal{G}^*)$. Each

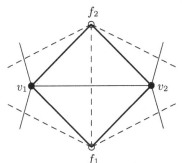

Fig. 6. A face of the double

pair of dual edges, say $\mathfrak{e} = (v_1, v_2)$ and $\mathfrak{e}^* = (f_1, f_2)$, as in Fig. 6, defines a quadrilateral (v_1, f_1, v_2, f_2), and all these quadrilaterals constitute the faces of the cell decomposition (quad-graph) $\mathcal{D}$. Let us stress that the edges of $\mathcal{D}$ belong neither to $E(\mathcal{G})$ nor to $E(\mathcal{G}^*)$. See Fig. 6.

Quad-graphs $\mathcal{D}$ arising as doubles have the following property, the set $V(\mathcal{D})$ may be decomposed into two complementary halves, $V(\mathcal{D}) = V(\mathcal{G}) \cup V(\mathcal{G}^*)$ ("black" and "white" vertices), such that the endpoints of each edge of $E(\mathcal{D})$ are of different colors. One can always color a quad-graph this way if it has no non-trivial periods, $i.e.$, it comes from the cellular decomposition $\mathcal{G}$ of a disk.

Conversely, any such quad-graph $\mathcal{D}$ may be considered to be the double of some cellular decomposition $\mathcal{G}$. The edges in $E(\mathcal{G})$, say, are defined then as paths joining two "black" vertices of each face in $F(\mathcal{D})$. (This decomposition of $V(\mathcal{D})$ into $V(\mathcal{G})$ and $V(\mathcal{G}^*)$ is unique, up to interchanging the roles of $\mathcal{G}$ and $\mathcal{G}^*$.)

Again, since we are mainly interested in the local theory, we avoid global considerations. Therefore we always assume (without mentioning it explicitly) that our quad-graphs are cellular decompositions of a disk, thus $\mathcal{G}$ and $\mathcal{G}^*$ may be well-defined.

For the integrable systems on quad-graphs we consider here the fields z attached to the vertices of the graph[1]. They are subject to an equation

$$Q(z_1, z_2, z_3, z_4) = 0, \tag{4}$$

relating four fields residing on the four vertices of an arbitrary face in $F(\mathcal{D})$. Moreover, in all our examples it will be possible to solve equation (4) uniquely for any field $z_1, \dots, z_4$ in terms of the other three.

---

[1] The systems with the fields on the edges are also very interesting, and are related to the Yang-Baxter maps (see Sect. 5.1).

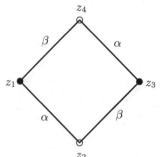

Fig. 7. A face of the labelled quad-graph

The Hirota equation,

$$\frac{z_4}{z_2} = \frac{\alpha z_3 - \beta z_1}{\beta z_3 - \alpha z_1}, \tag{5}$$

is such an example. We observe that the equation carries parameters α and β which can be naturally associated to the edges, and the opposite edges of an elementary quadrilateral carry equal parameters (see Fig. 7). At this point we specify the setup further. The example illustrated in Fig. 7 can be naturally generalized. An integrable system on a quad-graph,

$$Q(z_1, z_2, z_3, z_4; \alpha, \beta) = 0 \tag{6}$$

is parametrized by a function on the set of edges, $E(\mathcal{D})$, of the quad-graph which takes equal values on the opposite edges of any elementary quadrilateral. We call such a function a *labelling* of the quad-graph. Obviously, there exist infinitely many labellings, all of which may be constructed as follows: choose some value of α for an arbitrary edge of $\mathcal{D}$, and assign consecutively the same value to all "parallel" edges along a strip of quadrilaterals, according to the definition of labelling. After that, take an arbitrary edge still without a label, choose some value of α for it, and extend the same value along the corresponding strip of quadrilaterals. Proceed similarly, till all edges of $\mathcal{D}$ are exhausted.

An elementary quadrilateral of a quad-graph can be viewed from various directions. This implies that system (6) is well defined on a general quad-graph only if it possesses the rhombic symmetry, *i.e.*, each of the equations

$$Q(z_1, z_4, z_3, z_2; \beta, \alpha) = 0, \quad Q(z_3, z_2, z_1, z_4; \beta, \alpha) = 0$$

is equivalent to (6).

3.3 3D-Consistency

Now we introduce a crucial property of discrete integrable systems which will be taken to be characteristic.

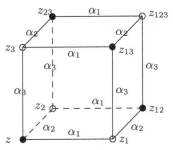

Fig. 8. Elementary cube

Let us extend the planar quad-graph $\mathcal{D}$ into the third dimension. Formally speaking, we consider a second copy $\mathcal{D}'$ of $\mathcal{D}$, and add edges connecting each vertex $v \in V(\mathcal{D})$ with its copy $v' \in V(\mathcal{D}')$. In this way we obtain a "3–dimensional quad-graph", **D**, whose set of vertices is

$$V(\mathbf{D}) = V(\mathcal{D}) \cup V(\mathcal{D}'),$$

and whose set of edges is

$$E(\mathbf{D}) = E(\mathcal{D}) \cup E(\mathcal{D}') \cup \{(v, v') : v \in V(\mathcal{D})\}.$$

Elementary building blocks of **D** are "cubes" as shown in Fig. 8. Clearly, we can still consistently subdivide the vertices of **D** into "black" and "white" vertices, so that the vertices connected by an edge have opposite colors. In the same way the labelling on $E(\mathcal{D})$ is extended to a labelling of $E(\mathbf{D})$. The opposite edges of all elementary faces (including the "vertical" ones) carry equal parameters (see Fig. 8).

Now, the fundamental property of discrete integrable systems mentioned above is the *three–dimensional consistency*.

Definition 3.2. *Consider an elementary cube, as in Fig. 8. Suppose that the values of the field, z, z_1, z_2, and z_3, are given at a vertex and at its three neighbors . Then equation (6) uniquely determines the values z_{12}, z_{23}, and z_{13}. After that the same equation (6) produces three a priori different values for the value of the field z_{123} at the eighth vertex of the cube, coming from the faces $[z_1, z_{12}, z_{123}, z_{13}]$, $[z_2, z_{12}, z_{123}, z_{23}]$ and $[z_3, z_{13}, z_{123}, z_{23}]$, respectively. Equation (6) is called 3D-consistent if these three values for z_{123} coincide for any choice of the initial data z, z_1, z_2, z_3.*

Proposition 3.1. *The Hirota equation,*

$$\frac{z_{12}}{z} = \frac{\alpha_2 z_1 - \alpha_1 z_2}{\alpha_1 z_1 - \alpha_2 z_2},$$

is 3D-consistent.

This can be verified by a straightforward computation. For the field at the eighth vertex of the cube one obtains

$$z_{123} = \frac{(l_{21} - l_{12})z_1 z_2 + (l_{32} - l_{23})z_2 z_3 + (l_{13} - l_{31})z_1 z_3}{(l_{23} - l_{32})z_1 + (l_{31} - l_{13})z_2 + (l_{12} - l_{21})z_3}, \tag{7}$$

where $l_{ij} = \dfrac{\alpha_i}{\alpha_j}$.

In [3] and [4] we suggested treating the consistency property (in the sense of Definition 3.2) as the characteristic one for discrete integrable systems. Thus we come to the central definition of these lectures.

Definition 3.3. *A discrete equation is called* integrable *if it is* consistent.

Note that this definition of the integrability is conceptually transparent and algorithmic: the integrability of any equation can be easily verified.

3.4 Zero-Curvature Representation from the 3D-Consistency

We show that our Definition 3.3 of discrete integrable systems is more fundamental then the traditional one discussed in Sect. 3.1. Recall that normally the problem of finding a zero-curvature representation for a given system is a difficult task whose successful solution is only possible with a large amount of luck in the guess-work. We show that finding the zero-curvature representation for a given discrete system with the consistency property becomes an algorithmically solvable problem, and we demonstrate how the corresponding flat connection in a loop group can be derived from the equation.

We get rid of our symmetric notations and consider the system

$$Q(z_1, z_2, z_3, z_4; \alpha, \beta) = 0 \tag{8}$$

on the base face of the cube, and choose the vertical direction to carry an additional (spectral) parameter λ (see Fig. 9).

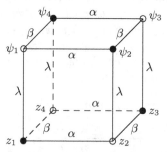

Fig. 9. Zero-curvature representation from the consistency

Assume that the left-hand-side of (8) is affine in each z_k. This gives z_4 as a fractional–linear (Möbius) transformation of z_2 with the coefficients depending on z_1 and z_3 and on α and β. One can of course freely interchange $z_1, \ldots, z_4$ in this statement. Now consider the equations on the vertical faces of the cube in Fig. 9. One obtains ψ_2 as a Möbius transformation of ψ_1

$$\psi_2 = L(z_2, z_1; \alpha, \lambda)[\psi_1],$$

with coefficients depending on the fields z_2 and z_1, on the parameter α in system (8) and on the additional parameter λ, which is to be treated as the spectral parameter. Mapping $L(z_2, z_1; \alpha, \lambda)$ is associated to the oriented edge, (z_1, z_2). For the reverse edge, (z_2, z_1), one obviously obtains the inverse transformation

$$L(z_1, z_2; \alpha, \lambda) = L(z_2, z_1; \alpha, \lambda)^{-1}.$$

Going once around the horizontal face of the cube one obtains

$$\psi_1 = L(z_1, z_4; \beta, \lambda)L(z_4, z_3; \alpha, \lambda)L(z_3, z_2; \beta, \lambda)L(z_2, z_1; \alpha, \lambda)[\psi_1].$$

The composed Möbius transformation in the right-hand-side is the identity because of the arbitrariness of ψ_1.

Using the matrix notation for the action of the Möbius transformations,

$$\frac{az + b}{cz + d} = L[z], \quad \text{where} \quad L = \begin{pmatrix} a & b \\ c & d \end{pmatrix},$$

and normalizing these matrices (for example by the condition $\det L = 1$), we derive the zero-curvature representation,

$$L(z_1, z_4; \beta, \lambda)L(z_4, z_3; \alpha, \lambda)L(z_3, z_2; \beta, \lambda)L(z_2, z_1; \alpha, \lambda) = I, \qquad (9)$$

for (8), where the L's are elements of the corresponding loop group. Equivalently, (9) can be written as

$$L(z_3, z_2; \beta, \lambda)L(z_2, z_1; \alpha, \lambda) = L(z_3, z_4; \alpha, \lambda)L(z_4, z_1; \beta, \lambda), \qquad (10)$$

where one has a little more freedom in normalizations.

Let us apply this derivation method to the Hirota equation. Equation (5) can be written as $Q = 0$ with

$$Q(z_1, z_2, z_3, z_4; \alpha, \beta) = \alpha(z_2 z_3 + z_1 z_4) - \beta(z_3 z_4 + z_1 z_2).$$

Performing the computations as above in this case we derive the zero-curvature representation for the Hirota equation (10) with the matrices

$$L(z_2, z_1, \alpha, \lambda) = \begin{pmatrix} \alpha & -\lambda z_2 \\ \dfrac{\lambda}{z_1} & -\alpha \dfrac{z_2}{z_1} \end{pmatrix}. \qquad (11)$$

Another important example is the cross-ratio equation,

$$\frac{(z_1 - z_2)(z_3 - z_4)}{(z_2 - z_3)(z_4 - z_1)} = \frac{\alpha}{\beta}. \tag{12}$$

It is easy to show that it is 3D-consistent. Written in the form

$$\alpha(z_2 - z_3)(z_4 - z_1) - \beta(z_1 - z_2)(z_3 - z_4) = 0,$$

it obviously belongs to the class discussed in this section. By direct computation we obtain the L-matrices,

$$\tilde{L}(z_2, z_1, \alpha, \lambda) = \begin{pmatrix} 1 + \dfrac{\lambda\alpha z_2}{(z_1 - z_2)} & -\dfrac{\lambda\alpha z_1 z_2}{(z_1 - z_2)} \\ \dfrac{\lambda\alpha}{(z_1 - z_2)} & 1 - \dfrac{\lambda\alpha z_1}{(z_1 - z_2)} \end{pmatrix},$$

which are gauge-equivalent to

$$L(z_2, z_1, \alpha, \lambda) = \begin{pmatrix} 1 & z_1 - z_2 \\ \dfrac{\alpha}{\lambda(z_1 - z_2)} & 1 \end{pmatrix}. \tag{13}$$

4 Classification

We have seen that the idea of consistency is at the core of the integrability theory and may be even suggested as a definition of integrability.

Here we give a further application of the consistency approach. We show that it provides an effective tool for finding and classifying all integrable systems in certain classes of equations. In the previous section we presented two important systems which belong to our theory. Here we complete the list of the examples, classifying all integrable (in the sense of Definition 3.3) one-field equations on quad-graphs satisfying some natural symmetry conditions.

We consider equations

$$Q(x, u, v, y; \alpha, \beta) = 0, \tag{14}$$

on quad-graphs. Equations are associated to elementary quadrilaterals, the fields $x, u, v, y \in \mathbb{C}$ are assigned to the four vertices of the quadrilateral, and the parameters $\alpha, \beta \in \mathbb{C}$ are assigned to its edges, as shown in Fig. 10.

We now list more precisely the assumptions under which we classify the equations.

1) Consistency. Equation (14) is integrable (in the sense that it is 3D-consistent). As explained in the previous section, this property means that this equation may be consistently embedded in a three–dimensional lattice, so that the same equations hold for all six faces of any elementary cube, as in Fig. 8.

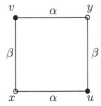

Fig. 10. An elementary quadrilateral; fields are assigned to vertices

Further, we assume that equations (14) can be uniquely solved for any one of their arguments, $x, u, v, y \in \widehat{\mathbb{CP}}^1$. Therefore, the solutions have to be fractional–linear in each of their arguments. This naturally leads to the following condition.

2) Linearity. The function $Q(x, u, v, y; \alpha, \beta)$ is linear in each argument (affine linear):

$$Q(x, u, v, y; \alpha, \beta) = a_1 xuvy + \cdots + a_{16}, \tag{15}$$

where coefficients a_i depend on α and β.

Third, we are interested in equations on quad-graphs of arbitrary combinatorics, hence it will be natural to assume that all variables involved in equations (14) are on equal footing. Therefore, our next assumption reads as follows.

3) Symmetry. Equation (14) is invariant under the group D_4 of the symmetries of the square (Fig. 11), that is, function Q satisfies the symmetry properties

$$Q(x, u, v, y; \alpha, \beta) = \varepsilon Q(x, v, u, y; \beta, \alpha) = \sigma Q(u, x, y, v; \alpha, \beta) \tag{16}$$

with $\varepsilon, \sigma = \pm 1$. Of course, due to symmetries (16), not all coefficients a_i in (15) are independent.

Finally, it is worth looking more attentively at expression (7) for the eighth point in the cube for the Hirota equation and at the similar formula

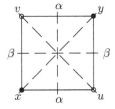

Fig. 11. D_4 symmetry

for the cross-ratio equation

$$z_{123} = \frac{(\alpha_1 - \alpha_2)z_1 z_2 + (\alpha_3 - \alpha_1)z_3 z_1 + (\alpha_2 - \alpha_3)z_2 z_3}{(\alpha_3 - \alpha_2)z_1 + (\alpha_1 - \alpha_3)z_2 + (\alpha_2 - \alpha_1)z_3}. \tag{17}$$

Looking ahead, we mention a very amazing and unexpected feature of these expressions: value z_{123} actually depends on z_1, z_2, z_3 only, and does not depend on z. In other words, four black points in Fig. 8 (the vertices of a tetrahedron) are related by a well-defined equation. This property, being rather strange at first glance, actually is valid not only in this but in all known nontrivial examples. We take it as an additional assumption in our solution of the classification problem.

4) Tetrahedron Property. Function $z_{123} = f(z, z_1, z_2, z_3; \alpha_1, \alpha_2, \alpha_3)$, existing due to the 3D-consistency, actually does not depend on variable z, that is, $f_z = 0$.

Under the tetrahedron property we can paint the vertices of the cube black and white, as in Fig. 8, and the vertices of each of two tetrahedrons satisfy an equation of the form,

$$\widehat{Q}(z_1, z_2, z_3, z_{123}; \alpha_1, \alpha_2, \alpha_3) = 0. \tag{18}$$

It is easy to see that under assumption 2) (linearity) function $\widehat{Q}$ may be also taken to be linear in each argument. (Clearly, formulas (7) and (17) may also be written in such a form.)

We identify equations related by certain natural transformations. First, acting simultaneously on all variables z by one and the same Möbius transformation does not violate our three assumptions. Second, the same holds for the simultaneous point change of all parameters, $\alpha \mapsto \varphi(\alpha)$.

Theorem 4.1. *[4] Up to common Möbius transformations of variables z and point transformations of the parameters α, the 3D-consistent quad-graph equations (14) with the properties 2), 3), 4) (linearity, symmetry and the tetrahedron property) are exhausted by the following three lists Q, H, and A, where $x = z$, $u = z_1$, $v = z_2$, $y = z_{12}$, $\alpha = \alpha_1$, $\beta = \alpha_2$:*
    List Q

(Q1) $\alpha(x - v)(u - y) - \beta(x - u)(v - y) + \delta^2 \alpha\beta(\alpha - \beta) = 0,$

(Q2) $\alpha(x - v)(u - y) - \beta(x - u)(v - y) + \alpha\beta(\alpha - \beta)(x + u + v + y)$
$$-\alpha\beta(\alpha - \beta)(\alpha^2 - \alpha\beta + \beta^2) = 0,$$

(Q3) $(\beta^2 - \alpha^2)(xy + uv) + \beta(\alpha^2 - 1)(xu + vy) - \alpha(\beta^2 - 1)(xv + uy)$
$$-\delta^2(\alpha^2 - \beta^2)(\alpha^2 - 1)(\beta^2 - 1)/(4\alpha\beta) = 0,$$

(Q4) $a_0 xuvy + a_1(xuv + uvy + vyx + yxu) + a_2(xy + uv) + \bar{a}_2(xu + vy)$
$$+\tilde{a}_2(xv + uy) + a_3(x + u + v + y) + a_4 = 0,$$

where the coefficients a_i are expressed in terms of (α, a) and (β, b) with $a^2 = r(\alpha)$, $b^2 = r(\beta)$, $r(x) = 4x^3 - g_2 x - g_3$, by the following formulas:

$$a_0 = a + b, \quad a_1 = -\beta a - \alpha b, \quad a_2 = \beta^2 a + \alpha^2 b,$$

$$\bar{a}_2 = \frac{ab(a+b)}{2(\alpha - \beta)} + \beta^2 a - (2\alpha^2 - \frac{g_2}{4})b,$$

$$\tilde{a}_2 = \frac{ab(a+b)}{2(\beta - \alpha)} + \alpha^2 b - (2\beta^2 - \frac{g_2}{4})a,$$

$$a_3 = \frac{g_3}{2} a_0 - \frac{g_2}{4} a_1, \quad a_4 = \frac{g_2^2}{16} a_0 - g_3 a_1.$$

List H:

(H1)$(x - y)(u - v) + \beta - \alpha = 0,$
(H2)$(x - y)(u - v) + (\beta - \alpha)(x + u + v + y) + \beta^2 - \alpha^2 = 0,$
(H3)$\alpha(xu + vy) - \beta(xv + uy) + \delta(\alpha^2 - \beta^2) = 0.$

List A:

(A1)$\alpha(x + v)(u + y) - \beta(x + u)(v + y) - \delta^2 \alpha\beta(\alpha - \beta) = 0,$
(A2)$(\beta^2 - \alpha^2)(xuvy + 1) + \beta(\alpha^2 - 1)(xv + uy) - \alpha(\beta^2 - 1)(xu + vy) = 0.$

The proof of this theorem is rather involved and is given in [4].

Remarks 1) List A can be omitted by allowing an extended group of Möbius transformations, which act on the variables x, y differently than on u, v, white and black sublattices on Figs. 10 and 8. In this manner (A1) is related to (Q1) by the change $u \to -u$, $v \to -v$, and (A2) is related to (Q3) with $\delta = 0$ by the change $u \to 1/u$, $v \to 1/v$. So, really independent equations are given by the lists Q and H.

2) In both lists, Q and H, the last equations are the most general ones. This means that (Q1)–(Q3) and (H1), (H2) may be obtained from (Q4) and (H3), respectively, by certain degenerations and/or limit procedures. So, one might be tempted to shorten these lists to one item each. However, on the one hand, these limit procedures are outside our group of admissible (Möbius) transformations, and, on the other, in many situations the "degenerate" equations (Q1)–(Q3) and (H1), (H2) are of interest in themselves. This resembles the situation with the six Painlevé equations and the coalescences connecting them.

3) Parameter δ in (Q1), (Q3), (H3) can be scaled away, so that one can assume without loss of generality that $\delta = 0$ or $\delta = 1$.

4) It is natural to set in (Q4) $(\alpha, a) = (\wp(A), \wp'(A))$ and, similarly, $(\beta, b) = (\wp(B), \wp'(B))$. So, this equation is actually parametrized by two points of the elliptic curve $\mu^2 = r(\lambda)$. The appearance of an elliptic curve in our classification problem is by no means obvious from the beginning. If r has multiple roots, the elliptic curve degenerates into a rational one, and (Q4)

degenerates to one of the previous equations of list Q; for example, if $g_2 = g_3 = 0$, then inversion $x \to 1/x$ turns (Q4) into (Q2).

5) Note that the list contains the fundamental equations only. A discrete equation which is derived from an equation with the consistency property will usually lose this property.

5 Generalizations: Multidimensional and Non-commutative (Quantum) Cases

5.1 Yang-Baxter Maps

As we mentioned, however, to assign fields to the vertices is not the only possibility. Another large class of 2–dimensional systems on quad-graphs consists of those where the fields are assigned to the *edges*, see Fig. 12. In this situation it is natural to assume that each elementary quadrilateral carries a map $R : \mathcal{X}^2 \to \mathcal{X}^2$, where $\mathcal{X}$ is the space where the fields a and b take values, so that $(a_2, b_1) = R(a, b; \alpha, \beta)$. The question of the three–dimensional consistency of such maps is also legitimate and, moreover, has recently begun to be studied. The corresponding property can be encoded in the formula

$$R_{23} \circ R_{13} \circ R_{12} = R_{12} \circ R_{13} \circ R_{23}, \tag{19}$$

where each $R_{ij} : \mathcal{X}^3 \to \mathcal{X}^3$ acts as the map R on the factors i and j of the cartesian product $\mathcal{X}^3$, and acts identically on the third factor. This equation should be understood as follows. The fields a and b are supposed to be attached to the edges parallel to the 1st and the 2nd coordinate axes, respectively. Additionally, consider the fields c attached to the edges parallel to the 3rd coordinate axis. Then the left-hand side of (19) corresponds to the chain of maps along the three rear faces of the cube in Fig. 13,

$$(a, b) \mapsto (a_2, b_1), \quad (a_2, c) \mapsto (a_{23}, c_1), \quad (b_1, c_1) \mapsto (b_{13}, c_{12}),$$

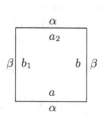

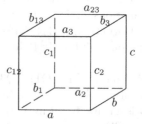

Fig. 12. An elementary quadrilateral; both fields and labels are assigned to edges

Fig. 13. Three–dimensional consistency; fields assigned to edges

while its right-hand side corresponds to the chain of maps along the three front faces of the cube,

$$(b, c) \mapsto (b_3, c_2), \quad (a, c_2) \mapsto (a_3, c_{12}), \quad (a_3, b_3) \mapsto (a_{23}, b_{13})$$

So (19) assures that the two ways of obtaining (a_{23}, b_{13}, c_{12}) from the initial data (a, b, c) lead to the same results. The maps with this property were introduced by Drinfeld under the name of "set–theoretical solutions of the Yang–Baxter equation", an alternative name is "Yang–Baxter maps" used by Veselov. Under some circumstances, systems with fields on vertices can be regarded as systems with fields on edges or vice versa (this is the case, *e.g.*, for systems (Q1), (Q3)$_{\delta=0}$, (H1), (H3)$_{\delta=0}$ of our list, for which the variables X enter only in combinations like $X - U$ for edges (x, u)), but in general the two classes of systems should be considered to be different. The problem of classifying Yang–Baxter maps, like the one achieved in the previous section, has been recently solved in [5].

5.2 Four-Dimensional Consistency of Three-Dimensional Systems

The consistency principle can be obviously generalized to an arbitrary dimension. We say that

a d–dimensional discrete equation possesses the consistency property, if it may be imposed in a consistent way on all d–dimensional sublattices of a $(d + 1)$–dimensional lattice.

In the three–dimensional context there are also *a priori* many kinds of systems, according to where the fields are defined: on the vertices, on the edges, or on the elementary squares of the cubic lattice.

Consider 3–dimensional systems with the fields at the vertices. In this case each elementary cube carries just one equation,

$$Q(z, z_1, z_2, z_3, z_{12}, z_{23}, z_{13}, z_{123}) = 0, \tag{20}$$

relating the fields in all its vertices. Such an equation should be solvable for any of its arguments in terms of the other seven arguments. The four–dimensional consistency of such equations is defined in the obvious way.

- Starting with initial data z, z_i $(1 \leq i \leq 4)$, z_{ij} $(1 \leq i < j \leq 4)$, equation (20) allows us to determine all fields z_{ijk} $(1 \leq i < j < k \leq 4)$ uniquely. Then we have *four* different ways of finding z_{1234} corresponding to four 3–dimensional cubic faces adjacent to the vertex z_{1234} of the four–dimensional hypercube (Fig. 14). All four values actually coincide.

So, one can consistently impose equations (20) on all elementary cubes of the three–dimensional cubical complex, which is a three–dimensional generalization of the quad-graph. It is tempting to accept the four–dimensional

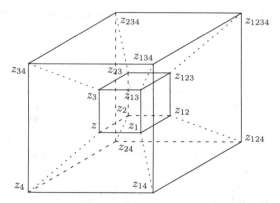

Fig. 14. Hypercube

consistency of equations of type (20) as the constructive definition of their integrability. It is important to solve the correspondent classification problem.

Let us give here some examples. Consider the equation

$$\frac{(z_1 - z_3)(z_2 - z_{123})}{(z_3 - z_2)(z_{123} - z_1)} = \frac{(z - z_{13})(z_{12} - z_{23})}{(z_{13} - z_{12})(z_{23} - z)}. \tag{21}$$

It is not difficult to see that (21) admits as its symmetry group the group D_8 of the cube. This equation can be uniquely solved for a field at an arbitrary vertex of a 3–dimensional cube, provided the fields at the seven other vertices are known.

The fundamental fact is:

Proposition 5.1. *Equation (21) is four–dimensionally consistent.*

A different factorization of the face variables into the vertex ones leads to another remarkable three–dimensional system known as the discrete BKP equation. For any solution $x : \mathbb{Z}^4 \to \mathbb{C}$ of (21), define a function $\tau : \mathbb{Z}^4 \to \mathbb{C}$ by the equations

$$\frac{\tau_i \tau_j}{\tau \tau_{ij}} = \frac{x_{ij} - x}{x_i - x_j}, \quad i < j. \tag{22}$$

Equation (21) assures that this can be done in an essentially unique way (up to initial data on the coordinate axes, whose influence is a trivial scaling of the solution). On any 3–dimensional cube the function τ satisfies the discrete BKP equation,

$$\tau \tau_{ijk} - \tau_i \tau_{jk} + \tau_j \tau_{ik} - \tau_k \tau_{ij} = 0, \quad i < j < k. \tag{23}$$

Proposition 5.2. *Equation (23) is four–dimensionally consistent.*

Moreover, for the value τ_{1234} one finds a remarkable equation,

$$\tau\tau_{1234} - \tau_{12}\tau_{34} + \tau_{13}\tau_{24} - \tau_{23}\tau_{34} = 0, \tag{24}$$

which essentially reproduces the discrete BKP equation. So τ_{1234} does not actually depend on the values τ_i, $1 \le i \le 4$. This can be considered to be an analogue of the tetrahedron property of Sect. 4.

5.3 Noncommutative (Quantum) Cases

As it was shown in [6], the consistency approach works also in the noncommutative case, where the participating fields live in an arbitrary associative (not necessary commutative) algebra $\mathcal{A}$ (over the field $\mathcal{K}$). It turns out that finding the zero-curvature representation does not hinge on the particular algebra $\mathcal{A}$ nor on prescribing some particular commutation rules for fields in the neighboring vertices. The fact that some commutation relations are preserved by the evolution is thus conceptually separate from the integrability.

As before, we deal with equations on quadrilaterals,

$$Q(x, u, v, y; \alpha, \beta) = 0.$$

Now $x, u, v, y \in \mathcal{A}$ are the fields assigned to the four vertices of the quadrilateral, and $\alpha, \beta \in \mathcal{K}$ are the parameters assigned to its edges.

We start our considerations with the following, more special equation,

$$yx^{-1} = f_{\alpha\beta}(uv^{-1}). \tag{25}$$

(Here and below, any time we encounter the inverse x^{-1} of a non-zero element, $x \in \mathcal{A}$, its existence is assumed.) We require that this equation do not depend on how we regard the elementary quadrilateral (recall that we consider equations on the quad-graphs). It is not difficult to see that this implies the following symmetries:

$$f_{\alpha\beta}(A) = f_{\beta\alpha}(A^{-1}), \tag{26}$$

$$f_{\alpha\beta}(A^{-1}) = (f_{\alpha\beta}(A))^{-1}, \tag{27}$$

$$f_{\beta\alpha}(A) = f_{\alpha\beta}^{-1}(A^{-1}). \tag{28}$$

In (28) $f_{\alpha\beta}^{-1}$ stands for the inverse function to $f_{\alpha\beta}$, which has to be distinguished from the inversion in the algebra $\mathcal{A}$ in the formula (27).

All the conditions (26)–(28) are satisfied for the function which characterizes the Hirota equation,

$$f_{\alpha\beta}(A) = \frac{1 - (\beta/\alpha)A}{(\beta/\alpha) - A}. \tag{29}$$

The 3D-consistency condition for equation (25) is

$$f_{\alpha_j \alpha_k}\left(f_{\alpha_i \alpha_j}(z_i z_j^{-1})(f_{\alpha_i \alpha_k}(z_i z_k^{-1}))^{-1}\right) =$$
$$f_{\alpha_i \alpha_k}\left(f_{\alpha_i \alpha_j}(z_i z_j^{-1})(f_{\alpha_j \alpha_k}(z_j z_k^{-1}))^{-1}\right) z_j z_i^{-1}.$$

Taking into account that $f_{\alpha\beta}$ actually depends only on β/α, we slightly abuse the notations and write $f_{\alpha\beta} = f_{\beta/\alpha}$. Setting $\lambda = \alpha_j/\alpha_i$, $\mu = \alpha_k/\alpha_j$, and $A = z_i z_j^{-1}$, $B^{-1} = z_j z_k^{-1}$, and taking into account property (27), we rewrite the above equation as

$$f_\mu\left(f_\lambda(A)f_{\lambda\mu}(BA^{-1})\right) = f_{\lambda\mu}\left(f_\lambda(A)f_\mu(B)\right) A^{-1}. \tag{30}$$

Proposition 5.3. *The non–commutative Hirota equation is 3D-consistent.*

To prove this theorem, one proves that function (29) satisfies this functional equation for any $\lambda, \mu \in \mathcal{K}$ and for any $A, B \in \mathcal{A}$.

Alternatively, one proves the consistency by deriving the zero-curvature representation. We show that the following two schemes for computing z_{123} lead to one and the same result:

- $(z, z_1, z_2) \mapsto z_{12}$, $(z, z_1, z_3) \mapsto z_{13}$, $(z_1, z_{12}, z_{13}) \mapsto z_{123}$.
- $(z, z_1, z_2) \mapsto z_{12}$, $(z, z_2, z_3) \mapsto z_{23}$, $(z_2, z_{12}, z_{23}) \mapsto z_{123}$.

The Hirota equation on face (z, z_1, z_{13}, z_3),

$$z_{13} z^{-1} = f_{\alpha_3 \alpha_1}(z_3 z_1^{-1}),$$

can be written as a formula which gives z_{13} as a fractional–linear transformation of z_3,

$$z_{13} = (\alpha_1 z_3 - \alpha_3 z_1)(\alpha_3 z_3 - \alpha_1 z_1)^{-1} z = L(z_1, z, \alpha_1, \alpha_3)[z_3], \tag{31}$$

where

$$L(z_1, z, \alpha_1, \alpha_3) = \begin{pmatrix} \alpha_1 & -\alpha_3 z_1 \\ \alpha_3 z^{-1} & -\alpha_1 z^{-1} z_1 \end{pmatrix}. \tag{32}$$

We use here the notation which is common for Möbius transformations on $\mathbb{C}$ represented as a linear action of the group $GL(2, \mathbb{C})$. In the present case we define the action of the group $GL(2, \mathcal{A})$ on $\mathcal{A}$ by the formula

$$\begin{pmatrix} a & b \\ c & d \end{pmatrix} [z] = (az + b)(cz + d)^{-1}, \qquad a, b, c, d, z \in \mathcal{A}.$$

It is easy to see that this is indeed the left action of the group, provided that the multiplication in $GL(2, \mathcal{A})$ is defined by the natural formula

$$\begin{pmatrix} a' & b' \\ c' & d' \end{pmatrix} \begin{pmatrix} a & b \\ c & d \end{pmatrix} = \begin{pmatrix} a'a + b'c & a'b + b'd \\ c'a + d'c & c'b + d'd \end{pmatrix}.$$

Absolutely similarly to (31), we find that

$$z_{23} = L(z_2, z, \alpha_2, \alpha_3)[z_3]. \tag{33}$$

From (33) we derive, by a shift in the direction of the first coordinate axis, the expression for z_{123} obtained by the first scheme above,

$$z_{123} = L(z_{12}, z_1, \alpha_2, \alpha_3)[z_{13}], \tag{34}$$

while from (31) we find the expression for z_{123} corresponding to the second scheme,

$$z_{123} = L(z_{12}, z_2, \alpha_1, \alpha_3)[z_{23}]. \tag{35}$$

Substituting (31) and (33) on the right-hand sides of (34) and (35), respectively, we represent the equality we want to demonstrate in the following form,

$$L(z_{12}, z_1, \alpha_2, \alpha_3)L(z_1, z, \alpha_1, \alpha_3)[z_3]$$
$$= L(z_{12}, z_2, \alpha_1, \alpha_3)L(z_2, z, \alpha_2, \alpha_3)[z_3]. \tag{36}$$

It is not difficult to prove that the stronger claim holds, namely that

$$L(z_{12}, z_1, \alpha_2, \alpha_3)L(z_1, z, \alpha_1, \alpha_3) = L(z_{12}, z_2, \alpha_1, \alpha_3)L(z_2, z, \alpha_2, \alpha_3). \tag{37}$$

The last equation is nothing else but the zero-curvature condition for the noncommutative Hirota equation.

Proposition 5.4. *The Hirota equation admits a zero-curvature representation with matrices from the loop group* $GL(2, \mathcal{A})[\lambda]$. *The transition matrix along the (oriented) edge* (x, u) *carrying the label* α *is determined by*

$$L(u, x, \alpha; \lambda) = \begin{pmatrix} \alpha & -\lambda u \\ \lambda x^{-1} & -\alpha x^{-1} u \end{pmatrix}. \tag{38}$$

Quite similar claims (3D-consistency, derivation of the zero-curvature representation) hold for the non–commutative cross-ratio equation,

$$(x - u)(u - y)^{-1}(y - v)(v - x)^{-1} = \frac{\alpha}{\beta}.$$

6 Smooth Theory from the Discrete One

Let us return to smooth and discrete surfaces with constant negative Gaussian curvature. The philosophy of discrete differential geometry was explained in Sect. 2. Surfaces and their transformations are obtained as a special limit of a discrete master-theory. The latter treats the corresponding discrete surfaces and their transformations in an absolutely symmetric way. This is possible because they are merged into multidimensional nets such that all their sublattices have the same geometric properties. The possibility of this multidimensional extension results in the permutability of the corresponding difference equations characterizing the geometry.

Let us recall the analytic description of smooth and discrete K-surfaces. Let F be a K-surface parametrized by its asymtotic lines,

$$F : \Omega(r) = [0, r] \times [0, r] \to \mathbb{R}^3.$$

This means that the vectors $\partial_x F$, $\partial_y F$, $\partial_x^2 F$ and $\partial_y^2 F$ are orthogonal to the normal vector $N : \Omega(r) \to S^2$. Reparametrizing the asymptotic lines, if necessary, we assume that $|\partial_x F| = 1$ and $|\partial_y F| = 1$. Angle $\phi = \phi(x, y)$ between the vectors $\partial_x F$, and $\partial_y F$ satisfies the sine–Gordon equation,

$$\partial_x \partial_y \phi = \sin \phi. \tag{39}$$

Moreover, a K-surface is determined by a solution to (39) essentially uniquely. The corresponding construction is as follows. Consider the matrices U and V defined by the formulas

$$U(a; \lambda) = \frac{i}{2} \begin{pmatrix} a & -\lambda \\ -\lambda & -a \end{pmatrix}, \tag{40}$$

$$V(b; \lambda) = \frac{i}{2} \begin{pmatrix} 0 & \lambda^{-1} \exp(ib) \\ \lambda^{-1} \exp(-ib) & 0 \end{pmatrix}, \tag{41}$$

taking values in the twisted loop algebra,

$$g[\lambda] = \{\xi : \mathbb{R}_* \to \mathrm{su}(2) : \xi(-\lambda) = \sigma_3 \xi(\lambda) \sigma_3\}, \quad \sigma_3 = \begin{pmatrix} 1 & 0 \\ 0 & -1 \end{pmatrix}.$$

Suppose now that a and b are real-valued functions on $\Omega(r)$. Then the zero-curvature condition,

$$\partial_y U - \partial_x V + [U, V] = 0, \tag{42}$$

is satisfied identically in λ, if and only if (a, b) satisfy the system

$$\partial_y a = \sin b, \qquad \partial_x b = a, \tag{43}$$

or, in other words, if $a = \partial_x \phi$ and $b = \phi$, where ϕ is a solution of (39). Given a solution ϕ, that is, a pair of matrices (40), (41) satisfying (42), the following system of linear differential equations is uniquely solvable,

$$\partial_x \Phi = U\Phi, \quad \partial_y \Phi = V\Phi, \quad \Phi(0, 0; \lambda) = 1. \tag{44}$$

Here $\Phi : \Omega(r) \mapsto G[\lambda]$ takes values in the twisted loop group,

$$G[\lambda] = \{\Xi : \mathbb{R}_* \to \mathrm{SU}(2) : \Xi(-\lambda) = \sigma_3 \Xi(\lambda)\sigma_3\}.$$

The solution $\Phi(x, y; \lambda)$ yields the immersion $F(x, y)$ by the Sym formula,

$$F(x, y) = \left. \left(2\lambda \Phi(x, y; \lambda)^{-1} \partial_\lambda \Phi(x, y; \lambda) \right) \right|_{\lambda=1}. \tag{45}$$

(Here the canonical identification of su(2) with $\mathbb{R}^3$ is used.) Moreover, the right-hand side of (45) at values of λ different from $\lambda = 1$ determines a family of immersions, $F_\lambda : \Omega(r) \to \mathbb{R}^3$, all of which are K-surfaces parametrized by asymptotic lines. These surfaces F_λ constitute the so-called *associated family* of F.

Now we turn to the analytic description of discrete K-surfaces. Let ε be a discretization parameter, and we introduce discrete domains,

$$\Omega^\varepsilon(r) = [0, r]^\varepsilon \times [0, r]^\varepsilon \subset (\varepsilon \mathbb{Z})^2,$$

where $[0, r]^\varepsilon = [0, r] \cap (\varepsilon \mathbb{Z})$. Each $\Omega^\varepsilon(r)$ contains $O(\varepsilon^{-2})$ grid points. Let F^ε be a discrete surface parametrized by asymptotic lines, *i.e.*, an immersion,

$$F^\varepsilon : \Omega^\varepsilon(r) \to \mathbb{R}^3, \tag{46}$$

such that for each $(x, y) \in \Omega^\varepsilon(r)$ the five points $F^\varepsilon(x, y)$, $F^\varepsilon(x \pm \varepsilon, y)$, and $F^\varepsilon(x, y \pm \varepsilon)$ lie in a single plane, $\mathcal{P}(x, y)$. Let us introduce the difference analogues of the partial derivatives,

$$\delta_x^\varepsilon p(x, y) = \frac{1}{\varepsilon}\Big(p(x + \varepsilon, y) - p(x, y) \Big), \quad \delta_y^\varepsilon p(x, y) = \frac{1}{\varepsilon}\Big(p(x, y + \varepsilon) - p(x, y) \Big). \tag{47}$$

It is required that all edges of the discrete surface F^ε have the same length, $\varepsilon \ell$, that is, $|\delta_x^\varepsilon F^\varepsilon| = |\delta_y^\varepsilon F^\varepsilon| = \ell$, and it turns out to be convenient to assume that $\ell = (1 + \varepsilon^2/4)^{-1}$. The same relation we presented between K-surfaces and solutions to the (classical) sine–Gordon equation (39) can be found between discrete K-surfaces and solutions to the sine–Gordon equation in Hirota's discretization,

$$\sin \frac{1}{4}\big(\phi(x + \varepsilon, y + \varepsilon) - \phi(x + \varepsilon, y) - \phi(x, y + \varepsilon) + \phi(x, y)\big)$$

$$= \frac{\varepsilon^2}{4} \sin \frac{1}{4}\big(\phi(x + \varepsilon, y + \varepsilon) + \phi(x + \varepsilon, y) + \phi(x, y + \varepsilon) + \phi(x, y)\big). \tag{48}$$

Consider the matrices $\mathcal{U}^\varepsilon$, $\mathcal{V}^\varepsilon$ defined by the formulas

$$\mathcal{U}^\varepsilon(a; \lambda) = (1 + \varepsilon^2 \lambda^2/4)^{-1/2} \begin{pmatrix} \exp(i\varepsilon a/2) & -i\varepsilon\lambda/2 \\ -i\varepsilon\lambda/2 & \exp(-i\varepsilon a/2) \end{pmatrix}, \tag{49}$$

$$\mathcal{V}^\varepsilon(b; \lambda) = (1 + \varepsilon^2 \lambda^{-2}/4)^{-1/2} \begin{pmatrix} 1 & (i\varepsilon\lambda^{-1}/2)\exp(ib) \\ (i\varepsilon\lambda^{-1}/2)\exp(-ib) & 1 \end{pmatrix}. \tag{50}$$

Let a and b be real-valued functions on $\Omega^\varepsilon(r)$, and consider the discrete zero-curvature condition,

$$\mathcal{U}^\varepsilon(x, y + \varepsilon; \lambda) \cdot \mathcal{V}^\varepsilon(x, y; \lambda) = \mathcal{V}^\varepsilon(x + \varepsilon, y; \lambda) \cdot \mathcal{U}^\varepsilon(x, y; \lambda), \qquad (51)$$

where $\mathcal{U}^\varepsilon$ and $\mathcal{V}^\varepsilon$ depend on $(x, y) \in \Omega^\varepsilon(r)$ by the dependence of a and b on (x, y), respectively. A direct calculation shows that (51) is equivalent to the system

$$\delta_y^\varepsilon a = \frac{2}{i\varepsilon^2} \log \frac{1 - (\varepsilon^2/4) \exp(-ib - i\varepsilon a/2)}{1 - (\varepsilon^2/4) \exp(ib + i\varepsilon a/2)}, \qquad \delta_x^\varepsilon b = a + \frac{\varepsilon}{2} \delta_y^\varepsilon a, \qquad (52)$$

or, in other words, to equation (48) for the function ϕ defined by

$$a = \delta_x^\varepsilon \phi, \qquad b = \phi + \frac{\varepsilon}{2} \delta_y^\varepsilon \phi. \qquad (53)$$

The formula (51) is the compatibility condition of the following system of linear difference equations:

$$\begin{aligned}
\Psi^\varepsilon(x + \varepsilon, y; \lambda) &= \mathcal{U}^\varepsilon(x, y; \lambda)\Psi^\varepsilon(x, y; \lambda), \\
\Psi^\varepsilon(x, y + \varepsilon; \lambda) &= \mathcal{V}^\varepsilon(x, y; \lambda)\Psi^\varepsilon(x, y; \lambda), \\
\Psi^\varepsilon(0, 0; \lambda) &= 1.
\end{aligned} \qquad (54)$$

So any solution of (48) uniquely defines a matrix, $\Psi^\varepsilon : \Omega^\varepsilon(r) \to G[\lambda]$, satisfying (54). This can be used to finally construct the immersion by an analogue of the Sym formula,

$$F^\varepsilon(x, y) = \left(2\lambda\Psi^\varepsilon(x, y; \lambda)^{-1}\partial_\lambda\Psi^\varepsilon(x, y; \lambda)\right)\Big|_{\lambda=1}. \qquad (55)$$

The geometric meaning of the function ϕ is the following. The angle between edges $F^\varepsilon(x + \varepsilon, y) - F^\varepsilon(x, y)$ and $F^\varepsilon(x, y + \varepsilon) - F^\varepsilon(x, y)$ is equal to $(\phi(x + \varepsilon, y) + \phi(x, y + \varepsilon))/2$; the angle between edges $F^\varepsilon(x, y + \varepsilon) - F^\varepsilon(x, y)$ and $F^\varepsilon(x - \varepsilon, y) - F^\varepsilon(x, y)$ is equal to $\pi - (\phi(x, y + \varepsilon) + \phi(x - \varepsilon, y))/2$; the angle between edges $F^\varepsilon(x - \varepsilon, y) - F^\varepsilon(x, y)$ and $F^\varepsilon(x, y - \varepsilon) - F^\varepsilon(x, y)$ is equal to $(\phi(x - \varepsilon, y) + \phi(x, y - \varepsilon))/2$; and the angle between edges $F^\varepsilon(x, y - \varepsilon) - F^\varepsilon(x, y)$ and $F^\varepsilon(x + \varepsilon, y) - F^\varepsilon(x, y)$ is equal to $\pi - (\phi(x, y - \varepsilon) + \phi(x + \varepsilon, y))/2$. In particular, the sum of these angles is 2π, so that the four neighboring vertices of $F^\varepsilon(x, y)$ lie in one plane, as they should. Again, the right-hand side of (55), at values of λ different from $\lambda = 1$ determines an associated family F_λ^ε of discrete K-surfaces parametrized by asymptotic lines.

Now we are prepared to state the approximation theorem for K-surfaces.

Theorem 6.1. *Let $a_0 : [0, r] \to \mathbb{R}$ and $b_0 : [0, r] \to S^1 = \mathbb{R}/(2\pi\mathbb{Z})$ be smooth functions. Then*

- *there exists a unique K-surface parametrized by asymptotic lines, $F : \Omega(r) \to \mathbb{R}^3$ such that its characteristic angle, $\phi : \Omega(r) \to S^1$, satisfies*

$$\partial_x\phi(x, 0) = a_0(x), \quad \phi(0, y) = b_0(y), \quad x, y \in [0, r], \qquad (56)$$

- *for any $\varepsilon > 0$ there exists a unique discrete K-surface $F^\varepsilon : \Omega^\varepsilon(r) \to \mathbb{R}^3$ such that its characteristic angle $\phi^\varepsilon : \Omega^\varepsilon(r) \to S^1$ satisfies*

$$\phi^\varepsilon(x+\varepsilon, 0) - \phi^\varepsilon(x, 0) = \varepsilon a_0(x), \quad \phi^\varepsilon(0, y+\varepsilon) + \phi^\varepsilon(0, y) = 2b_0(y), \quad (57)$$

for $x, y \in [0, r-\varepsilon]^\varepsilon$,
- *The inequality*

$$\sup_{\Omega^\varepsilon(r)} |F^\varepsilon - F| \le C\varepsilon, \quad (58)$$

where C does not depend on ε, is satisfied. Moreover, for a pair (m, n) of nonnegative integers

$$\sup_{\Omega^\varepsilon(r-k\varepsilon)} |(\delta_x^\varepsilon)^m (\delta_y^\varepsilon)^n F^\varepsilon - \partial_x^m \partial_y^n F| \to 0 \quad \text{as} \quad \varepsilon \to 0, \quad (59)$$

- *the estimates (58), (59) are satisfied, uniformly for $\lambda \in [\Lambda^{-1}, \Lambda]$ with any $\Lambda > 1$, if one replaces, in these estimates, the immersions F, F^ε by their associated families, F_λ, F_λ^ε, respectively.*

The complete proof of this theorem and its generalizations for nonlinear hyperbolic equations and their discretizations is presented in [7]. It is accomplished in two steps: first, the corresponding approximation results are proven for the Goursat problems for the hyperbolic systems (52) and (43), and then the approximation property is lifted to the frames Ψ^ε, Φ and finally to the surfaces F^ε, F. The proof of the C^∞-approximation goes along the same lines.

Moreover, a stronger approximation result follows from the consistency of the corresponding hyperbolic difference equations. As it was explained in Sect. 2, considering K-nets of higher dimensions and the corresponding consistent discrete hyperbolic systems, one obtains in the limit smooth K-surfaces with their Bäcklund transforms. The approximation results of Theorem 6.1 hold true also in this case. Permutability of the classical Bäcklund transformations then also easily follows.

References

1. A.I. Bobenko, U. Pinkall, Discrete surfaces with constant negative Gaussian curvature and the Hirota equation, J. Diff. Geom. **43** (1996) 527–611.
2. A.I. Bobenko, R. Seiler, eds., *Discrete Integrable Geometry and Physics*, Oxford, Clarendon Press 1999.
3. A.I. Bobenko, Yu.B. Suris, Integrable systems on quad-graphs, International Math. Research Notices **11** (2002) 573–612.
4. V.E. Adler, A.I. Bobenko, Yu.B. Suris, Classification of integrable equations on quad-graphs. The consistency approach, Comm. Math. Phys. **233** (2003) 513–543.

5. V.E. Adler, A.I. Bobenko, Yu.B. Suris, Geometry of Yang-Baxter maps: pencils of conics and quadrirational mappings, to appear in Comm. in Analysis and Geometry, arXiv:math.QA/0307009 (2003).
6. A.I. Bobenko, Yu.B. Suris, Integrable non-commutative equations on quadgraphs. The consistency approach, Lett. Math. Phys. **61** (2002) 241–254.
7. A.I. Bobenko, D. Matthes, Yu.B. Suris, Nonlinear hyperbolic equations in surface theory: integrable discretizations and approximation results, arXiv:math.NA/0208042 (2002).

Discrete Lagrangian Models

Yu. B. Suris

Institut für Mathematik, Technische Universität Berlin, Str. des 17. Juni 136,
10623 Berlin, Germany, `suris@sfb288.math.tu-berlin.de`

Abstract. These lectures are devoted to discrete integrable Lagrangian models.
A large collection of integrable models is presented in the Lagrangian fashion, along
with their integrable discretizations: the Neumann system, the Garnier system,
three systems from the rigid-body dynamics (multidimensional versions of the Euler top, the Lagrange top, and the top in a quadratic potential), the Clebsch case of
the Kirchhoff equations for a rigid body in an ideal fluid, and certain lattice systems
of the Toda type. The presentation of examples is preceded by the relevant theoretical background material on Hamiltonian mechanics on Poisson and symplectic
manifolds, complete integrability and Lax representations, Lagrangian mechanics
with continuous and discrete time on general manifolds and, in particular, on Lie
groups.

1 Introduction

These lectures are devoted to discrete integrable Lagrangian models. [1]
Though Hamiltonian mechanics on general Poisson manifolds is an extremely
powerful approach, it turns out that the majority of physically interesting
models may be better understood from the Lagrangian (variational) view-
point. Thus, we present here a large collection of integrable models in the
Lagrangian fashion, along with their integrable discretizations. The presentation of numerous examples is preceded by some theoretical background
material (Sects. 2–11) on Hamiltonian mechanics on Poisson and symplectic
manifolds, complete integrability and Lax representations, Lagrangian mechanics with continuous and discrete time on general manifolds and, in particular, on Lie groups. The list of concrete examples includes: the Neumann
system, the Garnier system, three systems from the rigid-body dynamics
(multidimensional versions of the Euler top, the Lagrange top, and the top
in a quadratic potential), the Clebsch case of the Kirchhoff equations for a
rigid body in an ideal fluid, and certain lattice systems of the Toda type, all
of them along with integrable discretizations. All bibliographical remarks are
collected in the concluding section.

---

[1] This is an updated and corrected version of selected sections from the book [50],
where all the proofs omitted here can be found. These sections are reproduced
by permission of Birkhäuser-Verlag, which is gratefully acknowledged.

Yu.B. Suris, Discrete Lagrangian Models, Lect. Notes Phys. **644**, 111–184 (2004)
`http://www.springerlink.com/`   © Springer-Verlag Berlin Heidelberg 2004

2 Poisson Brackets and Hamiltonian Flows

In what follows we denote by $\mathcal{F}(\mathcal{P})$ the set of smooth real–valued functions on a smooth manifold $\mathcal{P}$, and by $\mathfrak{X}(\mathcal{P})$ the set of vector fields on $\mathcal{P}$. More notational material is given in an Appendix to this section.

Definition 2.1. *A **Poisson bracket** (or **Poisson structure**) on a manifold $\mathcal{P}$ is a bilinear operation on the space $\mathcal{F}(\mathcal{P})$ of smooth functions on $\mathcal{P}$ denoted by $\{\cdot,\cdot\}$ and possessing the following properties:*

1. Skew–symmetry:

$$\{F,G\} = -\{G,F\} \qquad \forall F, G \in \mathcal{F}(\mathcal{P}) \, ;$$

2. Leibniz rule: $\{\cdot,\cdot\}$ is a derivation in each argument, i.e.

$$\{FG,H\} = \{F,H\}G + F\{G,H\} \qquad \forall F, G, H \in \mathcal{F}(\mathcal{P}) \, ;$$

3. Jacobi identity:

$$\{F,\{G,H\}\} + \{G,\{H,F\}\} + \{H,\{F,G\}\} = 0 \qquad \forall F, G, H \in \mathcal{F}(\mathcal{P}) \, .$$

*The pair $\left(\mathcal{P},\{\cdot,\cdot\}\right)$ is called a **Poisson manifold**.*

This definition is the most direct approach to the notion of a *Hamiltonian flow*. Since any derivation on $\mathcal{F}(\mathcal{P})$ is represented by a vector field, we accept the following definition.

Definition 2.2. *Let $\left(\mathcal{P},\{\cdot,\cdot\}\right)$ be a Poisson manifold. A **Hamiltonian vector field** X_H corresponding to the function $H \in \mathcal{F}(\mathcal{P})$ is the unique vector field on $\mathcal{P}$ satisfying*

$$X_H \cdot F = \{H,F\} \qquad \forall F \in \mathcal{F}(\mathcal{P}) \, .$$

*The function H is called a **Hamilton function** of X_H. The flow, $\varphi^t : \mathcal{P} \to \mathcal{P}$, of the Hamiltonian vector field X_H is called the **Hamiltonian flow** of the Hamilton function H. Another notation for X_H is $\{H,\cdot\}$.*

From the definition above the following statement follows immediately.

Proposition 2.1. *The map $H \to X_H$ from $\mathcal{F}(\mathcal{P})$ to $\mathfrak{X}(\mathcal{P})$ is a Lie algebra homomorphism, i.e.,*

$$[X_{H_1}, X_{H_2}] = X_{\{H_1,H_2\}} \qquad \forall H_1, H_2 \in \mathcal{F}(\mathcal{P}) \, .$$

Corollary. *Let φ^t be the Hamiltonian flow of X_H, and ψ^t be the Hamiltonian flow of X_F. If $\{H,F\} = 0$, then the flows φ^t, ψ^t commute,*

$$\varphi^t \cdot \psi^s = \psi^s \cdot \varphi^t \qquad \forall t, s \in \mathbb{R} \, .$$

One says that the functions H and F are *in involution* with respect to the bracket $\{\cdot, \cdot\}$ if $\{H, F\} = 0$. So, two Hamiltonian flows commute if and only if their Hamilton functions are in involution. Another property of involutive functions is expressed by the following proposition which is a direct consequence of the definitions.

Proposition 2.2. *Let φ^t be a Hamiltonian flow with the Hamilton function H. Then*

$$H \circ \varphi^t = H \, ,$$

and

$$\frac{d}{dt}(F \circ \varphi^t) = \{H, F \circ \varphi^t\} \, .$$

In particular, a function F is an integral of motion of the flow φ^t if and only if $\{H, F\} = 0$, that is, if H and F are *in involution*; the Hamilton function H itself is always an integral of motion of φ^t.

The most important property of the Hamiltonian flows is that each of the maps constituting such flow preserves Poisson brackets.

Definition 2.3. *Let $\left(\mathcal{P}, \{\cdot, \cdot\}_{\mathcal{P}}\right)$ and $\left(\mathcal{M}, \{\cdot, \cdot\}_{\mathcal{M}}\right)$ be two Poisson manifolds, and let $\varphi : \mathcal{P} \to \mathcal{M}$ be a smooth map. It is called a **Poisson map** if*

$$\{F, G\}_{\mathcal{M}} \circ \varphi = \{F \circ \varphi, G \circ \varphi\}_{\mathcal{P}} \qquad \forall F, G \in \mathcal{F}(\mathcal{M}) \, .$$

Theorem 2.1. *If $\varphi^t : \mathcal{P} \to \mathcal{P}$ is a Hamiltonian flow on $\mathcal{P}$, then for each $t \in \mathbb{R}$ the map φ^t is Poisson.*

Appendix: Gradients, Vector Fields, and Other Notations

Gradients of functions on vector spaces and on manifolds. If V is a vector space, and $f \in \mathcal{F}(V)$ is a smooth function, then the gradient $\nabla f : V \to V^*$ is defined via the formula

$$\langle \nabla f(x), y \rangle = \frac{d}{d\epsilon} f(x + \epsilon y)\Big|_{\epsilon=0} \, , \qquad \forall x, y \in V \, .$$

Similarly, for a function $f \in \mathcal{F}(\mathcal{M})$ on a smooth manifold $\mathcal{M}$, its gradient $\nabla f : \mathcal{M} \to T^*\mathcal{M}$ is defined in the following way. Let $Q \in \mathcal{M}$; then $\nabla f(Q)$ is an element of $T_Q^*\mathcal{M}$ satisfying

$$\langle \nabla f(Q), \dot{Q} \rangle = \frac{d}{d\epsilon} f(Q(\epsilon))\Big|_{\epsilon=0} \qquad \forall \dot{Q} \in T_Q\mathcal{M} \, ,$$

where $Q(\epsilon)$ stands for an arbitrary curve in $\mathcal{M}$ through $Q(0) = Q$ with the tangent vector $\dot{Q}(0) = \dot{Q}$.

Vector Fields. A vector field X on a manifold $\mathcal{M}$ is a map $X : \mathcal{M} \to T\mathcal{M}$ such that $X(Q) \in T_Q\mathcal{M}$ for any $Q \in \mathcal{M}$. The set of all vector fields on $\mathcal{M}$ is denoted by $\mathfrak{X}(\mathcal{M})$.

The flow of the vector field X is the one–parameter family of maps $\varphi^t : \mathcal{M} \to \mathcal{M}$ such that $t \to \varphi^t(Q)$ is the integral curve of X with the initial condition Q, i.e.,

$$\frac{d}{dt}\varphi^t(Q) = X(\varphi^t(Q)) , \qquad \varphi^0(Q) = Q .$$

The Lie derivative $X \cdot F$ of a function $F \in \mathcal{F}(\mathcal{M})$ along the vector field $X \in \mathfrak{X}(\mathcal{M})$ is defined as

$$(X \cdot F)(Q) = \frac{d}{dt} F(\varphi^t(Q))\Big|_{t=0} = \langle \nabla F(Q), X(Q) \rangle .$$

In local coordinates q_j,

$$X \cdot F = \sum_j X_j \frac{\partial F}{\partial q_j} .$$

The map $F \to X \cdot F$ is a derivation, and any derivation on $\mathcal{F}(\mathcal{M})$ is generated by some vector field. So, $X \in \mathfrak{X}(\mathcal{M})$ may be identified with a derivation written in local coordinates as

$$X = \sum_j X_j \frac{\partial}{\partial q_j} .$$

In particular, for any two vector fields $X, Y \in \mathfrak{X}(\mathcal{M})$, the following expression defines a derivation on $\mathcal{F}(\mathcal{M})$,

$$F \to X \cdot (Y \cdot F) - Y \cdot (X \cdot F) .$$

The corresponding vector field is denoted $[X, Y]$ and is called the Jacobi–Lie bracket of the vector fields X, Y. In local coordinates:

$$[X, Y]_j = \sum_i \left(X_i \frac{\partial Y_j}{\partial q_i} - Y_i \frac{\partial X_j}{\partial q_i} \right) .$$

If φ^t, ψ^t are the flows of the vector fields X, Y, respectively, then the necessary and sufficient condition for these flows to commute is the vanishing of the Jacobi–Lie bracket of X, Y, i.e.,

$$\varphi^t \circ \psi^s = \psi^s \circ \varphi^t \quad \forall t, s \in \mathbb{R} \quad \Leftrightarrow \quad [X, Y] = 0 .$$

3 Symplectic Manifolds

A somewhat more traditional approach to Hamiltonian mechanics is based on another choice of the fundamental structure, namely that of a symplectic manifold.

Definition 3.1. *A **symplectic structure** on a manifold $\mathcal{P}$ is a nondegenerate closed two–form Ω on $\mathcal{P}$. The pair $(\mathcal{P}, \Omega)$ is called a **symplectic manifold**.*

Actually, as we shall see, this structure is a particular case of the Poisson bracket structure. One can immediately define Hamiltonian vector fields with respect to a symplectic structure. The following definition is parallel to Definition 2.2.

Definition 3.2. *Let $(\mathcal{P}, \Omega)$ be a symplectic manifold. A **Hamiltonian vector field** X_H corresponding to the function $H \in \mathcal{F}(\mathcal{P})$, is the unique vector field on $\mathcal{P}$ satisfying*

$$\Omega(\xi, X_H(Q)) = \langle \nabla H(Q), \xi \rangle \qquad \forall \xi \in T_Q \mathcal{P} .$$

*H is called a **Hamilton function** of X_H. The flow $\varphi^t : \mathcal{P} \to \mathcal{P}$ of the Hamiltonian vector field X_H is called the **Hamiltonian flow** of the Hamilton function H.*

Actually, behind this definition the following construction is hidden. A symplectic structure on a manifold yields a vector bundle isomorphism between $T^*\mathcal{P}$ and $T\mathcal{P}$. Indeed, to any vector $\eta \in T_Q\mathcal{P}$ there corresponds a one–form, $\omega_\eta \in T_Q^*\mathcal{P}$, defined as

$$\omega_\eta(\xi) = \Omega(\xi, \eta) \qquad \forall \xi \in T_Q \mathcal{P} .$$

It is easy to see that the correspondence $\eta \to \omega_\eta$ is an isomorphism between $T_Q\mathcal{P}$ and $T_Q^*\mathcal{P}$. Denote the inverse isomorphism by $\mathcal{J} : T_Q^*\mathcal{P} \to T_Q\mathcal{P}$. Then Definition 3.2 of a Hamiltonian vector field may be represented as

$$X_H = \mathcal{J}(\nabla H) . \tag{1}$$

At any point $Q \in \mathcal{P}$, the tangent space $T_Q\mathcal{P}$ is spanned by the Hamiltonian vector fields at the point Q.

Definition 3.3. *Let $(\mathcal{P}_1, \Omega_1)$ and $(\mathcal{P}_2, \Omega_2)$ be two symplectic manifolds. A smooth map $\varphi : \mathcal{P}_1 \to \mathcal{P}_2$ is called **symplectic** if the pull–back of the form Ω_2 with respect to φ coincides with Ω_1, i.e., if*

$$\Omega_1(\xi, \eta) = \Omega_2(T_Q\varphi(\xi), T_Q\varphi(\eta)) ,$$

the form Ω_1 on the left-hand side being evaluated at in an arbitrary point $Q \in \mathcal{P}_1$, while the form Ω_2 on the right-hand side is evaluated at the corresponding point $\varphi(Q) \in \mathcal{P}_2$.

Hamiltonian flows on symplectic manifolds consist of symplectic maps. This is their characteristic property.

Theorem 3.1. *The flow φ^t of a vector field $X \in \mathfrak{X}(\mathcal{P})$ on a symplectic manifold $\left(\mathcal{P}, \Omega\right)$ consists of symplectic maps if and only if this field is locally Hamiltonian, i.e. if there exists locally a function $H \in \mathcal{F}(\mathcal{P})$ such that $X = X_H = \mathcal{J}(\nabla H)$.*

Finally, let us show how to include symplectic Hamiltonian mechanics into the Poisson bracket framework.

Theorem 3.2. *Let $(\mathcal{P}, \Omega)$ be a symplectic manifold. Then it is a Poisson manifold if one defines a Poisson bracket by the following formula,*

$$\{F, G\} = \Omega(X_F, X_G) = \Omega(\mathcal{J}(\nabla F), \mathcal{J}(\nabla G)) . \tag{2}$$

For an arbitrary function $F \in \mathcal{F}(\mathcal{P})$, by (1) and Definition 3.2, the following relation is satisfied,

$$X_H \cdot F = \langle \nabla F, \mathcal{J}(\nabla H) \rangle = \Omega(\mathcal{J}(\nabla H), X_F) .$$

Definition (2) allows us to rewrite the last formula as

$$X_H \cdot F = \{H, F\} ,$$

assuring the consistency of our present notations with those of Sect. 2.

Having given an intrinsic definition of symplectic manifolds, we can now characterize them as a subclass of Poisson manifolds. Let $\left(\mathcal{P}, \{\cdot, \cdot\}\right)$ be a d–dimensional Poisson manifold. Let $Q \in \mathcal{P}$, and consider local coordinates $x_1, \dots, x_d$ in the neighborhood of Q. The skew–symmetric $d \times d$ matrix

$$A_{kj} = \{x_k, x_j\} \tag{3}$$

is a coordinate representation of an intrinsic object called the **Poisson tensor**. We have

$$\{F, G\} = \sum_{k,j=1}^{d} A_{kj} \frac{\partial F}{\partial x_k} \frac{\partial G}{\partial x_j} \qquad \forall F, G \in \mathcal{F}(\mathcal{P}) .$$

Definition 3.4. *The rank of the matrix $(A_{kj})_{k,j=1}^{d}$ is called the **rank of the Poisson structure** at the point P.*

Proposition 3.1. *A Poisson manifold $\left(\mathcal{P}, \{\cdot, \cdot\}\right)$ is symplectic if the rank of the Poisson structure is everywhere equal to the dimension of $\mathcal{P}$.*

Since the matrix A is skew–symmetric, it can have a full rank only if d is even. Hence the dimension of a symplectic manifold is always an even number. We now give two important examples of symplectic manifolds.

Example 1: Constant symplectic structure on a vector space. Any non–degenerate, skew–symmetric bilinear form on an even–dimensional vector space V defines a symplectic structure on V. It can be shown that by a linear change of variables this structure may be transformed into the following *canonical* form. Let $V = \mathbb{R}^{2N}(\mathbf{x}, \mathbf{p})$, where $\mathbf{x} = (x_1, \dots, x_N)^{\mathrm{T}}$, $\mathbf{p} = (p_1, \dots, p_N)^{\mathrm{T}}$. The canonical skew–symmetric bilinear form on V is

$$\Omega((\mathbf{x}_1, \mathbf{p}_1), (\mathbf{x}_2, \mathbf{p}_2)) = \langle \mathbf{p}_1, \mathbf{x}_2 \rangle - \langle \mathbf{p}_2, \mathbf{x}_1 \rangle ,$$

where $\langle \cdot, \cdot \rangle$ is the usual Euclidean scalar product on $\mathbb{R}^N$. Another way of writing this is

$$\Omega = \sum_{k=1}^{N} dp_k \wedge dx_k .$$

The corresponding Poisson bracket is defined as

$$\{F, G\} = \sum_{k=1}^{N} \left(\frac{\partial F}{\partial p_k} \frac{\partial G}{\partial x_k} - \frac{\partial F}{\partial x_k} \frac{\partial G}{\partial p_k} \right) . \tag{4}$$

Another way to define this Poisson bracket is to give its values for all pairs of coordinate functions,

$$\{x_k, x_j\} = \{p_k, p_j\} = 0 , \qquad \{p_k, x_j\} = \delta_{kj} . \tag{5}$$

We call this symplectic manifold the *canonical phase space*, and the coordinates (x, p) on this space the *canonically conjugate coordinates*.

Example 2: The symplectic structure on a cotangent bundle. The cotangent bundle $T^*\mathcal{P}$ of any smooth manifold $\mathcal{P}$ carries a natural structure of a symplectic manifold. To define canonically conjugate coordinates on $T^*\mathcal{P}$, let $\{q_j\}_{j=1}^{N}$ be local coordinates on $\mathcal{P}$ in the neighborhood of a point $Q \in \mathcal{P}$. They define local coordinates $\{q_j, \dot{q}_j\}_{j=1}^{N}$ in the neighborhood of a point $(Q, \dot{Q}) \in T\mathcal{P}$. Now the canonical local coordinates $\{q_j, p_j\}_{j=1}^{N}$ in the neighborhood of a point $(Q, \Pi) \in T^*\mathcal{P}$ are defined by the relation

$$\langle \Pi, \dot{Q} \rangle = \sum_{k=1}^{N} p_k \dot{q}_k .$$

The Poisson bracket on $T^*\mathcal{P}$, corresponding to the standard symplectic structure, is given in these local coordinates by (4).

4 Poisson Reduction

We will discuss a general construction for producing new Poisson manifolds with the help of symmetry considerations. The simplest framework for the Poisson reduction is the following one. Let $\mathcal{P}$ be a smooth manifold, let G be a Lie group, and let $\Phi : G \times \mathcal{P} \to \mathcal{P}$ be a (left) **group action** of G on $\mathcal{P}$. We write it as $\Phi_g(Q)$ for $g \in G$ and $Q \in \mathcal{P}$. The axioms of a group action are the following:

- $\Phi_e(Q) = Q$ for all $Q \in \mathcal{P}$; here e is the unit element of the group G;
- $\Phi_{g_1}(\Phi_{g_2}(Q)) = \Phi_{g_1 g_2}(Q)$ for $g_1, g_2 \in G$ and $Q \in \mathcal{P}$.

Observe that in the infinitesimal limit the action of one–parameter subgroups of G define vector fields $\phi_\xi \in \mathfrak{X}(\mathcal{P})$ called the **infinitesimal generators** of the action Φ,

$$\phi_\xi(Q) = \left.\frac{d}{d\epsilon}\right|_{\epsilon=0} \Phi_{\exp(\epsilon\xi)}(Q) , \qquad \xi \in \mathfrak{g} . \tag{6}$$

Here $\mathfrak{g}$ stands for the Lie algebra of the Lie group G.

A group action defines an equivalence relation on $\mathcal{P}$. For $Q_1, Q_2 \in \mathcal{P}$ we write $Q_1 \simeq Q_2$, if there exists $g \in G$ such that $\Phi_g(Q_1) = Q_2$. The equivalence classes of this relation are the **orbits** of the action Φ, i.e., the sets

$$O_Q = \{\Phi_g(Q) : g \in G\} \subset \mathcal{P} .$$

The set of orbits, denoted by $\mathcal{P}/\simeq$, or else by $\mathcal{P}/G$, and called sometimes the **orbit space**, carries a natural topology. Namely, define $\pi : \mathcal{P} \to \mathcal{P}/G$ by $\pi(Q) = O_Q$, and declare $U \subset \mathcal{P}/G$ to be open if $\pi^{-1}(U)$ is an open set in $\mathcal{P}$. Under some additional conditions on the action Φ, the orbit space is a smooth manifold. For example, it is so if the action is free (has no fixed points) and proper. We shall always suppose that $\mathcal{P}/G$ is a smooth manifold.

Theorem 4.1. Let $\left(\mathcal{P}, \{\cdot, \cdot\}\right)$ be a Poisson manifold. Suppose that each map $\Phi_g : \mathcal{P} \to \mathcal{P}$ is Poisson. Then there exists a unique Poisson structure on $\mathcal{P}/G$ such that π is a Poisson map.

5 Complete Integrability

The key notion for these lectures is the *integrability* of a given Hamiltonian system. Among many existing definitions of integrability, our presentation will be based on the notion of the complete integrability *à la* Liouville–Arnold. The corresponding theorem tells how many integrals of motion assure integrability of a given Hamiltonian system, and describes the motion on the common level set of these integrals.

Theorem 5.1. (a) *Let* $\left(\mathcal{P}, \{\cdot, \cdot\}_{\mathcal{P}}\right)$ *be a* $2N$*-dimensional symplectic manifold. Suppose that there exist* N *functions* $F_1, \dots, F_N \in \mathcal{F}(\mathcal{P})$ *such that*

- $F_1, \dots, F_N$ *are functionally independent, i.e., the gradients* ∇F_k *are linearly independent everywhere on* $\mathcal{P}$;
- $F_1, \dots, F_N$ *are in involution:*

$$\{F_k, F_j\} = 0 \qquad 1 \le k, j \le N .$$

Let $\mathcal{T}$ *be a connected component of a common level set*

$$\left\{ Q \in \mathcal{P} : \ F_k(Q) = c_k, \ k = 1, \dots, N \right\} .$$

Then $\mathcal{T}$ *is diffeomorphic to* $\mathbb{T}^d \times \mathbb{R}^{N-d}$ *for some* $0 \le d \le N$. *In particular, if* $\mathcal{T}$ *is compact, it is necessarily diffeomorphic to* $\mathbb{T}^N$.

    (b) *If* $\mathcal{T}$ *is compact, then in some neighborhood* $\mathcal{T} \times \Omega$ *of* $\mathcal{T}$, *where* $\Omega \subset \mathbb{R}^N$ *is an open ball, there exist coordinates* $(I, \theta) = (I_k, \theta_k)_{k=1}^N$, *where* $I \in \Omega$ *and* $\theta \in \mathbb{T}^N$ (*action–angle coordinates*), *with the following properties:*

- *The actions* I_k *depend only on* F_j*'s,*

$$I_k = I_k(F_1, \dots, F_N), \quad k = 1, \dots, N .$$

- *The Poisson brackets of the coordinate functions are canonical,*

$$\{I_k, I_j\} = \{\theta_k, \theta_j\} = 0 , \quad \{I_k, \theta_j\} = \delta_{kj} , \quad 1 \le k, j \le N .$$

Hence

- *For an arbitrary Hamilton function,* $H = H(F_1, \dots, F_N)$, *depending only on* F_j*'s, the Hamiltonian equations of motion on* $\mathcal{P}$ *have the form*

$$\dot{I}_k = 0 , \quad \dot{\theta}_k = \omega_k(I_1, \dots, I_N) , \quad k = 1, \dots, N .$$

- *For an arbitrary symplectic map* $\Phi : \mathcal{P} \to \mathcal{P}$ *admitting* F_1, ..., F_N *as integrals of motion, the equations of motion in the coordinates* (I, θ) *take the form*

$$\widetilde{I}_k = I_k , \quad \widetilde{\theta}_k = \theta_k + \Omega_k(I_1, \dots, I_N) , \quad k = 1, \dots, N .$$

Hamiltonian flows and Poisson maps on $2N$–dimensional symplectic manifolds possessing N functionally independent integrals of motion, which are in involution, are called ***completely integrable*** (in the Liouville–Arnold sense).

6 Lax Representations

Almost all (perhaps, all) known integrable systems possess ***Lax representations***. In the situation of systems described by ordinary differential equations, a Lax representation for a given system means that there exist two maps, $L : \mathcal{P} \to \mathfrak{g}$ and $B : \mathcal{P} \to \mathfrak{g}$, from the system's phase space $\mathcal{P}$ into some Lie algebra $\mathfrak{g}$ such that the equations of motion are equivalent to

$$\dot{L} = [L, B] . \tag{7}$$

Matrix L, or, better, map $L : \mathcal{P} \to \mathfrak{g}$ is called the **Lax matrix**, while the matrix B is called the **auxiliary matrix** of the Lax representation. The pair (L, B) is called the **Lax pair** (and sometimes one uses, somewhat loosely, this term for the equation (7) itself). Finding a Lax representation for a given system usually implies its integrability, due to the fact that Ad–invariant functions on the Lie algebra $\mathfrak{g}$ are integrals of motion of the systems of the type (7), and therefore the values of such functions composed with the map L deliver functions on $\mathcal{P}$ serving as integrals of motion of the original system. One says that matrix L undergoes an **isospectral evolution**.

An important case often encountered in the theory of integrable systems occurs when the underlying algebra is a tensor product, $\mathfrak{g} = g^{\otimes N}$, of several copies of some algebra g. The corresponding Lax equations are of the form

$$\dot{L}_j = L_j B_{j-1} - B_j L_j , \qquad j \in \mathbb{Z}/N\mathbb{Z}. \tag{8}$$

Such equations yield an isospectral evolution of the so called **monodromy matrices**,

$$\dot{T}_j = [L_j, B_j] , \qquad T_j = L_j \cdot \ldots \cdot L_1 \cdot L_N \cdot \ldots \cdot L_{j+1} . \tag{9}$$

Equations of the type (8), also called sometimes **Lax triads**, are typical in the theory of integrable lattice systems.

Of course, in the Hamiltonian context, the isospectrality is not quite enough in order to establish the complete integrability. One has to show that the number of functionally independent integrals thus found is large enough, and that they are in involution. There exists an approach which incorporates an involutivity property in the very construction of Lax equations, namely the **r–matrix approach**. It uses a remarkable feature of equations (7), (8), namely that they can often be included into an abstract framework of Hamiltonian equations on $\mathfrak{g}$. More precisely, Poisson structures on $\mathfrak{g}$ can be defined, such that the corresponding Hamiltonian equations have the form (7), and the map $L : \mathcal{P} \to \mathfrak{g}$ is Poisson. In such a situation one says that the Lax representation admits a Hamiltonian interpretation.

Finally, we point out the natural discrete time analogues of the Lax equations. In formulating them, we adopt the following notations. All functions depend on $t \in h\mathbb{Z}$ with a small $h > 0$. The tilde denotes the discrete time shift by h: if $L = L(t)$, then $\widetilde{L} = L(t+h)$. The discrete time analogues of the Lax equations (7) and the Lax triads (8) are

$$\widetilde{L} = B^{-1}LB , \tag{10}$$

and

$$\widetilde{L}_j = B_j^{-1} L_j B_{j-1} \, , \tag{11}$$

respectively. Indeed, these equations yield the isospectrality of the discrete time evolution of L, resp., of T_j. In order for these discrete time equations to approximate the continuous time ones, it is required that

$$B_j = I + hB_j + O(h^2).$$

7 Lagrangian Mechanics on $\mathbb{R}^N$

Here we consider one of the basic constructions leading to Hamiltonian systems on symplectic manifolds, namely the variational principles of mechanics.

Consider **Newtonian equations of motion**, i.e., a system of second order differential equations,

$$\ddot{x}_k = F_k(x, \dot{x}) \, , \qquad x = (x_1, \dots, x_N) \in \mathbb{R}^N \, . \tag{12}$$

One says that they are variational equations if there exists a **Lagrange function** $\mathbf{L}(x, v)$ on $\mathbb{R}^{2N}(x, v)$ such that (12) is equivalent to the **Euler–Lagrange equations**

$$\frac{d}{dt} \frac{\partial \mathbf{L}(x, \dot{x})}{\partial \dot{x}_k} - \frac{\partial \mathbf{L}(x, \dot{x})}{\partial x_k} = 0 \, . \tag{13}$$

Of course, matrix $(\partial^2 \mathbf{L}(x, \dot{x}) / \partial \dot{x}_k \partial \dot{x}_j)_{k,j=1}^N$ has to be nondegenerate in order for (13) to be solved for $\ddot{x}_k$. Euler–Lagrange equations are necessary conditions for the corresponding integral curves $(x(t), \dot{x}(t))_{t=a}^{t=b}$ with fixed values of $x(a)$ and $x(b)$ to be critical points of the **action functional**

$$\mathbf{S} = \int_a^b \mathbf{L}(x(t), \dot{x}(t)) \, dt \, .$$

By the *Lagrangian formulation* of the Newtonian system (12) we understand its representation in the form

$$p_k = \partial \mathbf{L}(x, \dot{x}) / \partial \dot{x}_k \, , \qquad \dot{p}_k = \partial \mathbf{L}(x, \dot{x}) / \partial x_k \, . \tag{14}$$

Due to the above nondegeneracy condition for the matrix of second partial derivatives of $\mathbf{L}(x, \dot{x})$ with respect to $\dot{x}$, the set of equations

$$p_k = \partial \mathbf{L}(x, \dot{x}) / \partial \dot{x}_k \, , \quad 1 \le k \le N \, ,$$

may be solved for the quantities $\dot{x}_k$. It is well known in classical mechanics (and easily proven directly) that the resulting system may be written in the Hamiltonian form,

$$\dot{x}_k = \partial H(x, p) / \partial p_k \, , \qquad \dot{p}_k = -\partial H(x, p) / \partial x_k \, , \tag{15}$$

where the Hamilton function $H(x, p)$ is related to the Lagrange function by means of the famous **Legendre transformation**,

$$H(x, p) = \sum_{k=1}^{N} \dot{x}_k p_k - \mathbf{L}(x, \dot{x}) . \tag{16}$$

(Of course, in the last formula, $\dot{x}$ must be expressed in terms of p.) Equations (15) may be also written in the standard form,

$$\dot{x}_k = \{H, x_k\} , \qquad \dot{p}_k = \{H, p_k\} ,$$

with respect to the **canonical Poisson bracket** on $\mathbb{R}^{2N}(x, p)$ given by either of the formulas (4), (5). As we know, the most important property of the Hamiltonian equations of motion (15) is the preservation of bracket (5) by the corresponding flow. In other words, the Hamiltonian flow generated by (15) consists of symplectic maps.

The most natural discrete time analogue of the Newtonian equations of motion may be found, if one again starts from a variational principle. The **discrete–time action functional** is

$$\mathbb{S} = \sum_{n=a}^{b-1} \mathbb{L}\Big(x(n+1), x(n)\Big) , \tag{17}$$

where $(x(n))_{n=a}^{n=b}$ is a sequence of points in $\mathbb{R}^N$ with fixed values of $x(a)$ and $x(b)$. Here $\mathbb{L}(x, y)$ is the **discrete–time Lagrange function**. The necessary condition for a sequence $(x(n))_{n=a}^{n=b}$ to be a critical point of the above functional is given by the **discrete–time Euler–Lagrange equations**,

$$\frac{\partial}{\partial x_k}\Big(\mathbb{L}(\widetilde{x}, x) + \mathbb{L}(x, \underset{\sim}{x})\Big) = 0 , \tag{18}$$

where $x = x(n)$, $\widetilde{x} = x(n+1)$, $\underset{\sim}{x} = x(n-1)$. In order for these equations to be solvable for $\widetilde{x}$, the matrix of the second derivatives, $(\partial^2 \mathbb{L}(x, y)/\partial x_k \partial y_j)_{k,j=1}^N$, has to be nondegenerate.

By the *Lagrangian formulation* of such equations of motion we shall mean the system consisting of the following equations,

$$p_k = -\partial \mathbb{L}(\widetilde{x}, x)/\partial x_k , \tag{19}$$

$$\widetilde{p}_k = \partial \mathbb{L}(\widetilde{x}, x)/\partial \widetilde{x}_k . \tag{20}$$

The Hamiltonian formulation in the discrete time case is not defined. In particular, there exists no analogue of the Hamilton function which would be an integral of motion. However, the main qualitative feature of the Hamiltonian systems is inherited by the discrete–time Lagrangian systems. The map $(x, p) \to (\widetilde{x}, \widetilde{p})$ generated by equations (19), (20) is *symplectic* with respect to the standard symplectic bracket (5). In fact, these equations may be considered as one of the classical forms of generating functions for canonical transformations.

8 Lagrangian Mechanics on $T\mathcal{P}$ and on $\mathcal{P} \times \mathcal{P}$

An important generalization of the constructions of the previous section appears if we replace the Euclidean space, $\mathbb{R}^N(x)$, by an arbitrary smooth manifold $\mathcal{P}$. A continuous–time Lagrangian system is defined by a smooth function $\mathbf{L} \in \mathcal{F}(T\mathcal{P})$ on the tangent bundle of $\mathcal{P}$. The function $\mathbf{L}$ is called the **Lagrange function**. For an arbitrary function $Q : [a, b] \to \mathcal{P}$ one can consider the **action functional**

$$\mathbf{S} = \int_a^b \mathbf{L}(Q(t), \dot{Q}(t))dt . \tag{21}$$

A standard argument shows that the functions $Q(t)$ yielding extrema of this functional (in the class of variations preserving $Q(a)$ and $Q(b)$), satisfy necessarily the **Euler–Lagrange equations**: in local coordinates $\{q_j\}$ on $\mathcal{P}$,

$$\frac{d}{dt}\left(\frac{\partial \mathbf{L}}{\partial \dot{q}_j}\right) = \frac{\partial \mathbf{L}}{\partial q_j} . \tag{22}$$

The action functional $\mathbf{S}$ is independent of the choice of local coordinates, and thus the Euler–Lagrange equations are actually coordinate–independent as well.

Introducing the quantities

$$\Pi = \nabla_{\dot{Q}}\mathbf{L} \in T_Q^*\mathcal{P} , \tag{23}$$

one defines the **Legendre transformation**,

$$(Q, \dot{Q}) \in T\mathcal{P} \to (Q, \Pi) \in T^*\mathcal{P} . \tag{24}$$

If it is invertible, i.e., if $\dot{Q}$ can be expressed in terms of (Q, Π), then the Legendre transformation of the Euler–Lagrange equations (22) yields a Hamiltonian system on $T^*\mathcal{P}$ with respect to the standard symplectic structure on $T^*\mathcal{P}$, and with the Hamilton function

$$H(Q, \Pi) = \langle \Pi, \dot{Q} \rangle - \mathbf{L}(Q, \dot{Q}) , \tag{25}$$

where, of course, $\dot{Q}$ has to be expressed in terms of (Q, Π).

We will now discuss the famous **Noether theorem** which explains the existence of integrals of motion for Lagrangian systems with symmetries. We use the same notations as in Sect. 4.

Theorem 8.1. *Let $\Phi : G \times \mathcal{P} \to \mathcal{P}$ be an action of the Lie group G on $\mathcal{P}$, with infinitesimal generators $\phi_\xi \in \mathfrak{X}(\mathcal{P})$, $\xi \in \mathfrak{g}$. Let the Lagrange function $\mathbf{L} \in \mathcal{F}(T\mathcal{P})$ be invariant with respect to the action of G on $T\mathcal{P}$ induced by Φ:*

$$\mathbf{L}(\Phi_g(Q), \Phi_{g*}(\dot{Q})) = \mathbf{L}(Q, \dot{Q}) \qquad \forall g \in G . \tag{26}$$

Then the functions $I_\xi \in \mathcal{F}(T\mathcal{P})$,

$$I_\xi(Q, \dot{Q}) = \langle \nabla_{\dot{Q}} \mathbf{L}, \phi_\xi(Q) \rangle \,, \tag{27}$$

are integrals of motion of the Euler–Lagrange equations (22). Under the Legendre transformation these functions become integrals,

$$J_\xi(Q, \Pi) = \langle \Pi, \phi_\xi(Q) \rangle \,, \tag{28}$$

of the corresponding Hamiltonian system on $T^*\mathcal{P}$.

We now turn to the discrete time analogue of these constructions. The tangent bundle $T\mathcal{P}$ does not appear in the discrete time context at all, but the cotangent bundle $T^*\mathcal{P}$ does play an important role in the discrete time theory, as a phase space with the canonical invariant symplectic structure. Almost all constructions and results of the continuous time Lagrangian mechanics have their discrete time analogues. The only exception is the existence of the "energy" integral (25).

Let $\mathbb{L} \in \mathcal{F}(\mathcal{P} \times \mathcal{P})$ be a smooth function, called the **discrete–time Lagrange function**. For an arbitrary sequence, $\{Q(n) \in \mathcal{P},\ n = a, a + 1, \dots, b\}$, one can consider the **discrete–time action functional**,

$$\mathbb{S} = \sum_{n=a}^{b-1} \mathbb{L}(Q(n), Q(n+1)) \,. \tag{29}$$

Obviously, the sequences $\{Q(n)\}$ delivering extrema of this functional (in the class of variations preserving $Q(a)$ and $Q(b)$), necessarily satisfy the **discrete–time Euler–Lagrange equations**,

$$\nabla_1 \mathbb{L}(Q(n), Q(n+1)) + \nabla_2 \mathbb{L}(Q(n-1), Q(n)) = 0 \,. \tag{30}$$

Here $\nabla_{1,2}\mathbb{L}(Q_1, Q_2)$ denotes the gradients of $\mathbb{L}(Q_1, Q_2)$ with respect to the first argument, Q_1 (resp., the second argument, Q_2). This equation is written in intrinsic terms, i.e. independently of a choice of a coordinate chart. As pointed out above, an invariant formulation of the Euler–Lagrange equations in the continuous time case is more sophisticated, since the tangent bundle is not a direct product manifold. This seems to indicate that the discrete Euler–Lagrange equations are of a fundamental character.

Equation (30), which we shall also write as

$$\nabla_1 \mathbb{L}(Q, \widetilde{Q}) + \nabla_2 \mathbb{L}(\underset{\sim}{Q}, Q) = 0 \,, \tag{31}$$

is an implicit equation for $\widetilde{Q}$. In general, it has more than one solution, and therefore defines a correspondence (multi–valued map) $(\underset{\sim}{Q}, Q) \to (Q, \widetilde{Q})$. To discuss the symplectic properties of this correspondence, one defines the momentum,

$$\Pi = \nabla_2 \mathbb{L}(Q, \underset{\sim}{Q}) \in T^*_Q \mathcal{P} . \tag{32}$$

Then (31) may be rewritten as

$$\begin{cases} \Pi = -\nabla_1 \mathbb{L}(Q, \widetilde{Q}) , \\ \widetilde{\Pi} = \nabla_2 \mathbb{L}(Q, \widetilde{Q}) . \end{cases} \tag{33}$$

This system defines a (multi–valued) map $(Q, \Pi) \to (\widetilde{Q}, \widetilde{\Pi})$ on $T^*\mathcal{P}$. More precisely, the first equation in (33) is an implicit equation for $\widetilde{Q}$, while the second one allows for an explicit calculation of $\widetilde{\Pi}$ in terms of Q and $\widetilde{Q}$.

Theorem 8.2. *Each branch of the map $T^*\mathcal{P} \to T^*\mathcal{P}$ defined by (33) is symplectic with respect to the standard symplectic structure on $T^*\mathcal{P}$.*

Finally, we turn to the ***discrete-time Noether theorem***.

Theorem 8.3. *Let $\Phi : G \times \mathcal{P} \to \mathcal{P}$ be an action of the Lie group G on $\mathcal{P}$, with infinitesimal generators $\phi_\xi \in \mathfrak{X}(\mathcal{P})$, $\xi \in \mathfrak{g}$. Let the Lagrange function $\mathbb{L} \in \mathcal{F}(\mathcal{P} \times \mathcal{P})$ be invariant with respect to the action of G on $\mathcal{P} \times \mathcal{P}$ induced by Φ,*

$$\mathbb{L}(\Phi_g(Q_1), \Phi_g(Q_2)) = \mathbb{L}(Q_1, Q_2) \qquad \forall g \in G . \tag{34}$$

Then functions $I_\xi \in \mathcal{F}(\mathcal{P} \times \mathcal{P})$,

$$I_\xi(Q, \underset{\sim}{Q}) = \langle \nabla_2 \mathbb{L}(Q, \underset{\sim}{Q}), \phi_\xi(Q) \rangle , \tag{35}$$

are integrals of motion of the Euler–Lagrange equations (31). Under the discrete time Legendre transformation (32), these functions become integrals,

$$J_\xi(Q, \Pi) = \langle \Pi, \phi_\xi(Q) \rangle , \tag{36}$$

of the corresponding symplectic map on $T^*\mathcal{P}$.

It is important to observe that while functions (35) in $\mathcal{F}(\mathcal{P} \times \mathcal{P})$ clearly serve as *difference approximations* to functions (27) in $\mathcal{F}(T\mathcal{P})$, their expressions in terms of the cotangent bundle variables (Q, Π) coincide.

9 Lagrangian Mechanics on Lie Groups

Now we turn our attention to an important particular case of Lagrangian mechanics, the one where the basic manifold $\mathcal{P}$ carries the additional structure of a Lie group. We shall denote it by $\mathcal{P} = G$, and its typical element by $g \in G$, in order to identify the particular features of this case in its notation. The Lie algebra of G will be denoted by $\mathfrak{g}$. Some further important notations from Lie group theory are collected in the Appendix to this section. In particular, we shall use the notions of left Lie derivative, dF, and right Lie derivative, $d'F$, of a function $F \in \mathcal{F}(G)$. The functions $dF : G \to \mathfrak{g}^*$ and $d'F : G \to \mathfrak{g}^*$ are defined by the formulas

$$\langle dF(g), \eta \rangle = \frac{d}{d\epsilon} F(e^{\epsilon\eta}g) \Big|_{\epsilon=0} , \qquad \forall \eta \in \mathfrak{g} ,$$

$$\langle d'F(g), \eta \rangle = \frac{d}{d\epsilon} f(ge^{\epsilon\eta}) \Big|_{\epsilon=0} , \qquad \forall \eta \in \mathfrak{g} .$$

They are related to the gradient $\nabla F(g) \in T_g^* G$ by

$$\nabla F(g) = R_{g^{-1}}^* \, dF(g) = L_{g^{-1}}^* \, d'F(g) .$$

One of the most important features of the Lie group situation is the possibility of trivializing the tangent and the cotangent bundles. The **left trivialization of the tangent bundle**, $TG \to G \times \mathfrak{g}$, is achieved by the map

$$(g, \dot{g}) \in TG \to (g, \Omega) \in G \times \mathfrak{g} , \tag{37}$$

where

$$\Omega = L_{g^{-1}*}\dot{g} \quad \Leftrightarrow \quad \dot{g} = L_{g*}\Omega . \tag{38}$$

The corresponding **left trivialization of the cotangent bundle**, $T^*G \to G \times \mathfrak{g}^*$, is given by

$$(g, \Pi) \in T^*G \to (g, M) \in G \times \mathfrak{g}^* , \tag{39}$$

where

$$M = L_g^* \Pi \quad \Leftrightarrow \quad \Pi = L_{g^{-1}}^* M . \tag{40}$$

Similarly, the **right trivialization of the tangent bundle** is the map, $TG \to G \times \mathfrak{g}$,

$$(g, \dot{g}) \in TG \to (g, \omega) \in G \times \mathfrak{g} , \tag{41}$$

where

$$\omega = R_{g^{-1}*}\dot{g} \quad \Leftrightarrow \quad \dot{g} = R_{g*}\Omega . \tag{42}$$

The corresponding **right trivialization of the cotangent bundle**, $T^*G \to G \times \mathfrak{g}^*$, is given by

$$(g, \Pi) \in T^*G \to (g, m) \in G \times \mathfrak{g}^* , \tag{43}$$

where

$$m = R_g^* \Pi \quad \Leftrightarrow \quad \Pi = R_{g^{-1}}^* m . \tag{44}$$

Observe that the elements $\Omega, \omega \in \mathfrak{g}$ and $M, m \in \mathfrak{g}^*$ are related by the formulas

$$\Omega = \operatorname{Ad} g^{-1} \cdot \omega , \tag{45}$$

$$M = \operatorname{Ad}^* g \cdot m . \tag{46}$$

Our first task will be to push forward the standard symplectic Poisson bracket on T^*G with respect to both trivialization maps (39) and (43).

Proposition 9.1. *a) The Poisson structure $\{\cdot, \cdot\}^{(l)}$ on $G \times \mathfrak{g}^*$, which is the standard symplectic structure on T^*G pushed forward by the left trivialization map (39), is*

$$\{f_1, f_2\}^{(l)}(g, M) = -\langle d_g' f_1, \nabla_M f_2 \rangle + \langle d_g' f_2, \nabla_M f_1 \rangle + \langle M, [\nabla_M f_1, \nabla_M f_2] \rangle . \tag{47}$$

b) The Poisson structure $\{\cdot, \cdot\}^{(r)}$ on $G \times \mathfrak{g}^*$, which is the standard symplectic structure on T^*G pushed forward by the right trivialization map (43), is

$$\{f_1, f_2\}^{(r)}(g, m) = -\langle d_g f_1, \nabla_m f_2 \rangle + \langle d_g f_2, \nabla_m f_1 \rangle - \langle m, [\nabla_m f_1, \nabla_m f_2] \rangle . \tag{48}$$

Now we look at the Noether integrals of motion under the action of the trivialization maps. Suppose that there is a group action, $\Phi : K \times G \to G$, of some Lie group K on G. Denote by $\mathfrak{k}$ the Lie algebra of K, and, as usual, by $\phi_\xi \in \mathfrak{X}(G)$ the infinitesimal generators of the action Φ (here $\xi \in \mathfrak{k}$). Assuming invariance of the Lagrange functions under the action of K, we consider the Noether conserved quantities (28), (36) (observe that, in terms of the cotangent bundle, these quantities coincide). Under the left trivialization these conserved quantities are transformed into

$$J_\xi^{(l)}(g, M) = \langle M, L_{g^{-1}*} \phi_\xi(g) \rangle , \tag{49}$$

while under the right trivialization, we obtain

$$J_\xi^{(r)}(g, m) = \langle m, R_{g^{-1}*} \phi_\xi(g) \rangle . \tag{50}$$

Appendix: Notations from Lie Group Theory

Let G be a Lie group with Lie algebra $\mathfrak{g}$, and let $\mathfrak{g}^*$ be the dual vector space to $\mathfrak{g}$. We identify $\mathfrak{g}$ and $\mathfrak{g}^*$ with the tangent space and the cotangent space to G at the group unit, respectively,

$$\mathfrak{g} = T_e G , \qquad \mathfrak{g}^* = T_e^* G .$$

The pairing between the cotangent and the tangent spaces T_g^*G and T_gG at an arbitrary point, $g \in G$, is denoted by $\langle \cdot, \cdot \rangle$. The left and right translations in the group are the maps, L_g, $R_g : G \to G$, defined by

$$L_g h = gh, \qquad R_g h = hg \qquad \forall h \in G,$$

while L_{g*} and R_{g*} stand for the differentials of these maps,

$$L_{g*} : T_hG \to T_{gh}G, \qquad R_{g*} : T_hG \to T_{hg}G.$$

We denote by

$$\mathrm{Ad}\, g = L_{g*}R_{g^{-1}*} : \mathfrak{g} \to \mathfrak{g}$$

the adjoint action of the Lie group G on its Lie algebra, $\mathfrak{g} = T_eG$. The linear operators,

$$L_g^* : T_{gh}^*G \to T_h^*G, \qquad R_g^* : T_{hg}^*G \to T_h^*G$$

are adjoint to L_{g*}, R_{g*}, respectively, by the pairing $\langle \cdot, \cdot \rangle$,

$$\langle L_g^*\xi, \eta \rangle = \langle \xi, L_{g*}\eta \rangle \quad \text{for} \quad \xi \in T_{gh}^*G, \ \eta \in T_hG,$$

$$\langle R_g^*\xi, \eta \rangle = \langle \xi, R_{g*}\eta \rangle \quad \text{for} \quad \xi \in T_{hg}^*G, \ \eta \in T_hG.$$

The coadjoint action of the group,

$$\mathrm{Ad}^*\, g = L_g^*R_{g^{-1}}^* : \mathfrak{g}^* \to \mathfrak{g}^*,$$

is adjoint to $\mathrm{Ad}\, g$ by the pairing $\langle \cdot, \cdot \rangle$,

$$\langle \mathrm{Ad}^*\, g \cdot \xi, \eta \rangle = \langle \xi, \mathrm{Ad}\, g \cdot \eta \rangle \quad \text{for} \quad \xi \in \mathfrak{g}^*, \ \eta \in \mathfrak{g}.$$

The differentials of $\mathrm{Ad}\, g$ and $\mathrm{Ad}^*\, g$ with respect to g at the group unit e are the operators

$$\mathrm{ad}\,\eta : \mathfrak{g} \to \mathfrak{g} \qquad \text{and} \qquad \mathrm{ad}^*\,\eta : \mathfrak{g}^* \to \mathfrak{g}^*,$$

respectively, also adjoint by the pairing $\langle \cdot, \cdot \rangle$,

$$\langle \mathrm{ad}^*\, \eta \cdot \xi, \zeta \rangle = \langle \xi, \mathrm{ad}\, \eta \cdot \zeta \rangle \qquad \forall \xi \in \mathfrak{g}^*, \ \zeta \in \mathfrak{g}.$$

The action of ad is given by applying the Lie bracket in $\mathfrak{g}$,

$$\mathrm{ad}\,\eta \cdot \zeta = [\eta, \zeta], \quad \forall \zeta \in \mathfrak{g}.$$

10 Invariant Lagrangians and the Lie–Poisson Bracket

We now consider Lagrangian systems on Lie groups with invariant Lagrange functions.

10.1 Continuous–Time Case

Consider a Lagrange function $\mathbf{L} \in \mathcal{F}(TG)$, invariant with respect to left multiplications,

$$\mathbf{L}(g_0 g, L_{g_0 *} \dot{g}) = \mathbf{L}(g, \dot{g}) , \quad \forall g_0 \in G . \tag{51}$$

Obviously, this property is equivalent to the property that, under the left trivialization, the Lagrange function does not depend on g, i.e., it depends solely on $\Omega = L_{g^{-1} *} \dot{g}$,

$$\mathbf{L}(g, \dot{g}) = \mathbf{L}^{(l)}(\Omega) . \tag{52}$$

Now we want to reduce the Euler–Lagrange equations with respect to the action of G on TG induced by left multiplications. We realize the factor TG/G as $\mathfrak{g}$, the reduction map being

$$(g, \dot{g}) \in TG \;\rightarrow\; \Omega = L_{g^{-1} *} \dot{g} \in \mathfrak{g} .$$

It is easily seen that extremizing the functional $\mathbf{S}$ with a left–invariant Lagrange function $\mathbf{L}$ is equivalent to extremizing the functional

$$\mathbf{S}^{(l)} = \int_a^b \mathbf{L}^{(l)}(\Omega(t)) dt$$

with respect to variations $\Omega(t, \epsilon)$ of $\Omega(t)$ of the form

$$\Omega(t, \epsilon) = \Omega(t) + \epsilon \Big(\dot{\eta}(t) + [\Omega(t), \eta(t)] \Big) ,$$

where $\eta : [a, b] \to \mathfrak{g}$ is an arbitrary function vanishing at $t = a$ and $t = b$.

Theorem 10.1. *The differential equation for extremals of the functional* $\mathbf{S}^{(l)}$ *are*

$$\dot{M} = \mathrm{ad}^* \, \Omega \cdot M , \tag{53}$$

where

$$M = \nabla \mathbf{L}^{(l)}(\Omega) \in \mathfrak{g}^* . \tag{54}$$

If the Legendre transformation,

$$\Omega \in \mathfrak{g} \to M \in \mathfrak{g}^*, \tag{55}$$

is invertible, it transforms (53) *into a Hamiltonian system on* $\mathfrak{g}^*$ *with respect to the bracket,*

$$\{f_1, f_2\}^{(l)}(M) = \langle M, [\nabla_M f_1, \nabla_M f_2] \rangle , \tag{56}$$

with the Hamilton function,

$$H^{(l)}(M) = \langle M, \Omega \rangle - \mathbf{L}^{(l)}(\Omega) \,, \tag{57}$$

where Ω has to be expressed in terms of M. The motion in the Lie group is reconstructed by solving the linear differential equation

$$\dot{g} = L_{g*}\Omega \,. \tag{58}$$

The element

$$m = \mathrm{Ad}^* g^{-1} \cdot M \in \mathfrak{g}^* \tag{59}$$

is conserved in the evolution described by (53) and (58).

Similarly, suppose that the Lagrange function is invariant with respect to right multiplications,

$$\mathbf{L}(gg_0, R_{g_0*}\dot{g}) = \mathbf{L}(g, \dot{g}) \,, \quad \forall g_0 \in G \,. \tag{60}$$

This is equivalent to the property that under right trivialization the Lagrange function depends only on $\omega = R_{g^{-1}*}\dot{g}$, and not on g,

$$\mathbf{L}(g, \dot{g}) = \mathbf{L}^{(r)}(\omega) \,. \tag{61}$$

Reducing the Euler–Lagrange equations with respect to the action of G on TG induced by right multiplications, we still regard $\mathfrak{g}$ as TG/G, the reduction map this time being

$$(g, \dot{g}) \in TG \;\rightarrow\; \omega = R_{g^{-1}*}\dot{g} \in \mathfrak{g} \,.$$

The differential equations for extremals of the functional $\mathbf{S}$ with a right–invariant Lagrange function $\mathbf{L}$ are the same as for extremals of the functional

$$\mathbf{S}^{(r)} = \int_a^b \mathbf{L}^{(r)}(\omega(t))dt$$

with respect to variations $\omega(t, \epsilon)$ of $\omega(t)$ of the form

$$\omega(t, \epsilon) = \omega(t) + \epsilon\Big(\dot{\eta}(t) + [\eta(t), \omega(t)]\Big) \,,$$

where $\eta : [a, b] \to \mathfrak{g}$ is an arbitrary function vanishing at $t = a$ and $t = b$.

Theorem 10.2. *The differential equations for extremals of the functional* $\mathbf{S}^{(r)}$ *are*

$$\dot{m} = -\mathrm{ad}^* \omega \cdot m \,, \tag{62}$$

where

$$m = \nabla \mathbf{L}^{(r)}(\omega) \in \mathfrak{g}^* \, . \tag{63}$$

If the Legendre transformation

$$\omega \in \mathfrak{g} \to m \in \mathfrak{g}^* \tag{64}$$

is invertible, it transforms (62) *into a Hamiltonian system on* $\mathfrak{g}^*$ *with respect to the bracket*

$$\{f_1, f_2\}^{(r)}(m) = -\langle m, [\nabla_m f_1, \nabla_m f_2] \rangle \, , \tag{65}$$

with the Hamilton function

$$H^{(r)}(m) = \langle m, \omega \rangle - \mathbf{L}^{(r)}(\omega) \, , \tag{66}$$

where ω must be expressed in terms of m. The motion in the Lie group is reconstructed by solving the linear differential equation

$$\dot{g} = R_{g*}\omega \, . \tag{67}$$

The element

$$M = \mathrm{Ad}^* g \cdot m \in \mathfrak{g}^* \tag{68}$$

is conserved in the evolution described by (62) *and* (67).

The differential equations (53) and (62), when considered as equations on the Lie algebra $\mathfrak{g}$, i.e., the equations for $\Omega \in \mathfrak{g}$, resp. for $\omega \in \mathfrak{g}$, are called the **Euler–Poincaré equations**. However, it is easy to see that the following expressions are valid: $\Omega = \nabla_M H^{(l)}(M)$, resp. $\omega = \nabla_m H^{(r)}(m)$. This allows us to consider (53) and (62) also as equations on $\mathfrak{g}^*$, i.e., as equations for $M \in \mathfrak{g}^*$, resp., for $m \in \mathfrak{g}^*$, formulated completely in terms of the corresponding Hamilton function. As such, they are termed **Lie–Poisson equations** (sometimes also **Euler equations**). The Poisson brackets (56) and (65) are the simplest non–symplectic Poisson brackets. They are defined on the dual space to an arbitrary Lie algebra, and are known as the **Lie–Poisson brackets**.

10.2 Discrete–Time Case

Let the discrete–time Lagrange function $\mathbb{L}(g_1, g_2)$ be invariant with respect to left multiplications,

$$\mathbb{L}(g_0 g_1, g_0 g_2) = \mathbb{L}(g_1, g_2) \, , \quad \forall g_0 \in G \, . \tag{69}$$

This is equivalent to the fact that function $\mathbb{L}(g_n, g_{n+1})$ depends only on $W_n = g_n^{-1}g_{n+1}$,

$$\mathbb{L}(g_n, g_{n+1}) = \mathbb{L}^{(l)}(W_n) . \qquad (70)$$

Therefore we would like to reduce the discrete–time Euler–Lagrange equations with respect to the action of G on $G \times G$ by (componentwise) left multiplications, whereby the factor $G \times G/G$ will be realized as G by the reduction map

$$(g_1, g_2) \in G \times G \rightarrow W = g_1^{-1}g_2 \in G .$$

It is easy to see that extremizing the functional $\mathbb{S}$ with a left–invariant Lagrange function $\mathbb{L}$ is equivalent to extremizing the functional

$$\mathbb{S}^{(l)} = \sum_{n=a}^{b-1} \mathbb{L}^{(l)}(W_n) ,$$

with respect to variations $\{W_n(\epsilon)\}$ of the sequence $\{W_n\}$ of the form

$$W_n(\epsilon) = W_n e^{\epsilon \eta_{n+1} - \epsilon \mathrm{Ad}\, W_n^{-1} \cdot \eta_n} ,$$

where $\{\eta_n\}_{n=a}^{b}$ is an arbitrary sequence of elements of the Lie algebra $\mathfrak{g}$ such that $\eta_a = \eta_b = 0$.

Theorem 10.3. *The difference equations for extremals of the functional* $\mathbb{S}^{(l)}$ *are*

$$M_{n+1} = \mathrm{Ad}^* W_n \cdot M_n , \qquad (71)$$

where

$$M_n = d'\mathbb{L}^{(l)}(W_{n-1}) \in \mathfrak{g}^* . \qquad (72)$$

If the Legendre transformation,

$$W_{n-1} \in G \rightarrow M_n \in \mathfrak{g}^*, \qquad (73)$$

is invertible, then (71) define a map, $M_n \rightarrow M_{n+1}$, *which is Poisson with respect to the Lie–Poisson bracket (56). The motion in the Lie group is reconstructed by solving the linear difference equation*

$$g_{n+1} = g_n W_n . \qquad (74)$$

The element

$$m_n = \mathrm{Ad}^* g_n^{-1} \cdot M_n \in \mathfrak{g}^* \qquad (75)$$

is conserved in the evolution described by (71) and (74).

Next, consider the situation with the discrete–time Lagrange function $\mathbb{L}(g_1, g_2)$ invariant with respect to right multiplications,

$$\mathbb{L}(g_1 g_0, g_2 g_0) = \mathbb{L}(g_1, g_2) , \quad \forall g_0 \in G .\tag{76}$$

This is equivalent to the fact that function $\mathbb{L}(g_n, g_{n+1})$ depends only on $w_n = g_{n+1} g_n^{-1}$,

$$\mathbb{L}(g_n, g_{n+1}) = \mathbb{L}^{(r)}(w_n) .\tag{77}$$

This time we reduce the discrete–time Euler–Lagrange equations with respect to the action of G on $G \times G$ by right multiplications, the reduction map being

$$(g_1, g_2) \in G \times G \to w = g_2 g_1^{-1} \in G .$$

Extremizing the functional $\mathbb{S}$ with a right–invariant Lagrange function $\mathbb{L}$ is equivalent to extremizing the functional

$$\mathbb{S}^{(r)} = \sum_{n=a}^{b-1} \mathbb{L}^{(r)}(w_n) ,$$

with respect to variations $\{w_n(\epsilon)\}$ of the sequence $\{w_n\}$ satisfying the constraint

$$w_n(\epsilon) = e^{\epsilon \eta_{n+1} - \epsilon \mathrm{Ad}\, w_n \cdot \eta_n} w_n ,$$

where $\{\eta_n\}_{n=a}^b$ is an arbitrary sequence of elements of $\mathfrak{g}$ such that $\eta_a = \eta_b = 0$.

Theorem 10.4. *The difference equations for extremals of the functional $\mathbb{S}^{(r)}$ are*

$$m_{n+1} = \mathrm{Ad}^* w_n^{-1} \cdot m_n ,\tag{78}$$

where

$$m_n = d\mathbb{L}^{(r)}(w_{n-1}) \in \mathfrak{g}^* .\tag{79}$$

If the Legendre transformation,

$$w_{n-1} \in G \to m_n \in \mathfrak{g}^*,\tag{80}$$

is invertible, then (78) defines a map, $m_n \to m_{n+1}$, which is Poisson with respect to the Lie–Poisson bracket (65). The motion in the Lie group is reconstructed by solving the linear difference equation

$$g_{n+1} = w_n g_n .\tag{81}$$

The element

$$M_n = \mathrm{Ad}^* g_n \cdot m_n \in \mathfrak{g}^*\tag{82}$$

is conserved in the evolution described by (78) and (81).

Equations (71) and (78) are naturally considered to be *discrete–time Euler–Poincaré equations* on $\mathfrak{g}$, and may also be seen as *discrete–time Lie–Poisson equations* on $\mathfrak{g}^*$, provided the Legendre transformations are invertible. However, as opposed to their continuous time analogues, they do not necessarily possess an integral of motion analogous to the Hamilton function (although, of course, Casimir functions of the Lie–Poisson brackets on $\mathfrak{g}^*$ serve as integrals of motion also in the discrete time case).

11 Lagrangian Reduction and Euler–Poincaré Equations on Semidirect Products

A very important case of the Lagrangian mechanics on Lie groups is constituted by Lagrangian functions invariant with respect to the left or right multiplications by elements of some subgroup, rather than the whole group. This situation leads naturally to Euler–Poincaré equations on semidirect products. The general setup is as follows.

Let $\Phi : G \times V \to V$ be a representation of a Lie group G in a linear space V; we denote it by

$$\Phi(g) \cdot v \quad \text{for} \quad g \in G, \quad v \in V.$$

We also denote by ϕ the corresponding representation of the Lie algebra $\mathfrak{g}$ in V,

$$\phi(\xi) \cdot v = \frac{d}{d\epsilon}\Big(\Phi(e^{\epsilon\xi}) \cdot v\Big)\Big|_{\epsilon=0} \quad \text{for} \quad \xi \in \mathfrak{g}, \quad v \in V. \tag{83}$$

Map $\phi^* : \mathfrak{g} \times V^* \to V^*$ defined by

$$\langle \phi^*(\xi) \cdot y, v \rangle = \langle y, \phi(\xi) \cdot v \rangle \quad \forall v \in V, \ y \in V^*, \ \xi \in \mathfrak{g} \tag{84}$$

is an anti–representation of the Lie algebra $\mathfrak{g}$ in V^*. We shall also use the bilinear operation $\diamond : V^* \times V \to \mathfrak{g}^*$ defined as follows: let $v \in V$, $y \in V^*$, then

$$\langle y \diamond v, \xi \rangle = -\langle y, \phi(\xi) \cdot v \rangle \quad \forall \xi \in \mathfrak{g}. \tag{85}$$

Observe that the pairings on the left–hand side and on the right–hand side of the latter equation are defined on different spaces.

Fix an element $a \in V$, and consider the isotropy subgroup $G^{[a]}$ of a, i.e.,

$$G^{[a]} = \{h : \Phi(h) \cdot a = a\} \subset G. \tag{86}$$

Its Lie algebra is

$$\mathfrak{g}^{[a]} = \{\xi : \phi(\xi) \cdot a = 0\} \subset \mathfrak{g}. \tag{87}$$

Our subject in the present section will be the Lagrangian dynamics in the case of Lagrange functions invariant with respect to the action of $G^{[a]}$.

11.1 Continuous–Time Case

Suppose that the Lagrange function $\mathbf{L}(g, \dot{g})$ is invariant under the action of $G^{[a]}$ on TG induced by *left* translations on G:

$$\mathbf{L}(g_0 g, L_{g_0 *}\dot{g}) = \mathbf{L}(g, \dot{g}) , \quad g_0 \in G^{[a]} . \tag{88}$$

Denote the Lagrange function pushed forward by the left trivialization map (37) by $\mathbf{L}(g, \dot{g}) = \mathbf{L}^{(l)}(g, \Omega)$. The corresponding invariance property of $\mathbf{L}^{(l)}(g, \Omega)$ is expressed as

$$\mathbf{L}^{(l)}(g_0 g, \Omega) = \mathbf{L}^{(l)}(g, \Omega) , \quad g_0 \in G^{[a]} . \tag{89}$$

We want to reduce the Euler–Lagrange equations with respect to this left action. We realize the factor $TG/G^{[a]} \simeq (G \times \mathfrak{g})/G^{[a]}$ as the set $\mathfrak{g} \times O_a$, where O_a is the orbit of a under the action Φ:

$$O_a = \{\Phi(g) \cdot a , \ g \in G\} \subset V . \tag{90}$$

The reduction map is

$$(g, \Omega) \in G \times \mathfrak{g} \ \rightarrow \ (\Omega, P) \in \mathfrak{g} \times O_a , \quad \text{where} \quad P = \Phi(g^{-1}) \cdot a , \tag{91}$$

so that the reduced Lagrange function $\mathcal{L}^{(l)} \in \mathcal{F}(\mathfrak{g} \times O_a)$ is defined as

$$\mathcal{L}^{(l)}(\Omega, P) = \mathbf{L}^{(l)}(g, \Omega) , \quad \text{where} \quad P = \Phi(g^{-1}) \cdot a . \tag{92}$$

The reduced Lagrange function $\mathcal{L}^{(l)}(\Omega, P)$ is well defined because, from

$$P = \Phi(g_1^{-1}) \cdot a = \Phi(g_2^{-1}) \cdot a,$$

it follows that $\Phi(g_2 g_1^{-1}) \cdot a = a$, so that $g_2 g_1^{-1} \in G^{[a]}$, and $\mathbf{L}^{(l)}(g_1, \Omega) = \mathbf{L}^{(l)}(g_2, \Omega)$.

Theorem 11.1. a) *Consider the reduction* $(g, \Omega) \rightarrow (\Omega, P)$. *The reduced Euler–Lagrange equations are the following Euler–Poincaré equations:*

$$\begin{cases} \dot{M} = \mathrm{ad}^* \, \Omega \cdot M + \nabla_P \mathcal{L}^{(l)}(\Omega, P) \diamond P , \\ \dot{P} = -\phi(\Omega) \cdot P , \end{cases} \tag{93}$$

where

$$M = \nabla_\Omega \mathcal{L}^{(l)}(\Omega, P) \in \mathfrak{g}^* . \tag{94}$$

They describe the extremals of the constrained variational principle with functional

$$\mathbf{S}^{(l)} = \int_a^b \mathcal{L}^{(l)}(\Omega(t), P(t)) dt , \tag{95}$$

and admissible variations $(\Omega(t,\epsilon), P(t,\epsilon))$ *of* $(\Omega(t), P(t)) \in \mathfrak{g} \times O_a$ *of the form*

$$\Omega(t,\epsilon) = \Omega(t) + \epsilon\Big(\dot{\eta}(t) + [\Omega(t), \eta(t)]\Big), \quad P(t,\epsilon) = P(t) - \epsilon\phi(\eta(t)) \cdot P(t), \tag{96}$$

where $\eta : [a, b] \to \mathfrak{g}$ *is an arbitrary function taking values in the Lie algebra* $\mathfrak{g}$ *and satisfying* $\eta(a) = \eta(b) = 0$.

b) *If the Legendre transformation,*

$$(\Omega, P) \in \mathfrak{g} \times O_a \to (M, P) \in \mathfrak{g}^* \times O_a, \tag{97}$$

is invertible, then it transforms (93) *into a Hamiltonian system on* $\mathfrak{g}^* \times O_a$ *with respect to the Poisson bracket*

$$\{F_1, F_2\}^{(l)}(M, P) = \langle M, [\nabla_M F_1, \nabla_M F_2] \rangle$$
$$+ \langle \nabla_P F_1, \phi(\nabla_M F_2) \cdot P \rangle - \langle \nabla_P F_2, \phi(\nabla_M F_1) \cdot P \rangle, \tag{98}$$

and with the Hamilton function

$$H^{(l)}(M, P) = \langle M, \Omega \rangle - \mathcal{L}^{(l)}(\Omega, P), \tag{99}$$

where Ω *has to be expressed in terms of* (M, P).

c) *The full (non–reduced) Euler–Lagrange equations on* TG *have the integrals of motion,*

$$J_\xi^{(l)}(g, M) = \langle M, \operatorname{Ad} g^{-1} \cdot \xi \rangle, \quad \forall \xi \in \mathfrak{g}^{[a]}. \tag{100}$$

Remark 1. Formula (98) defines a Poisson bracket not only on $\mathfrak{g}^* \times O_a$, but on all of $\mathfrak{g}^* \times V$. Rewriting this formula as

$$\{F_1, F_2\}^{(l)} =$$
$$\langle M, [\nabla_M F_1, \nabla_M F_2] \rangle + \langle P, \phi^*(\nabla_M F_2) \cdot \nabla_P F_1 - \phi^*(\nabla_M F_1) \cdot \nabla_P F_2 \rangle$$

one immediately identifies this bracket with the Lie–Poisson bracket of the **semidirect product Lie algebra** $\mathfrak{g} \ltimes V^*$ corresponding to the representation $-\phi^*$ of $\mathfrak{g}$ in V^*. By definition, the Lie algebra $\mathfrak{g} \ltimes V^*$ corresponding to the representation ψ of $\mathfrak{g}$ in V^* coincides as a vector space with $\mathfrak{g} \times V^*$, and carries the Lie bracket,

$$[(\xi_1, y_1), (\xi_2, y_2)] = \Big([\xi_1, \xi_2], \psi(\xi_1) \cdot y_2 - \psi(\xi_2) \cdot y_1\Big).$$

Remark 2. An important particular case of the constructions of this section is the following: the vector space is chosen to be the Lie algebra of our basic Lie group, $V = \mathfrak{g}$, and the group representation is the adjoint, $\Phi(g) \cdot v = \operatorname{Ad} g \cdot v$, so that $\phi(\xi) \cdot v = \operatorname{ad} \xi \cdot v = [\xi, v]$. Then the bilinear operation, $\diamond$, is nothing but the coadjoint action of $\mathfrak{g}$ on $\mathfrak{g}^*$, $y \diamond v = \operatorname{ad}^* v \cdot y$.

Now assume that the function $\mathbf{L}(g, \dot{g})$ is invariant under the action of $G^{[a]}$ on TG induced by *right* translations on G,

$$\mathbf{L}(gg_0, R_{g_0*}\dot{g}) = \mathbf{L}(g, \dot{g}) , \quad g_0 \in G^{[a]} . \tag{101}$$

Under the right trivialization (41) of TG the Lagrange function becomes $\mathbf{L}(g, \dot{g}) = \mathbf{L}^{(r)}(g, \omega)$, and the invariance property reads:is

$$\mathbf{L}^{(r)}(gg_0, \omega) = \mathbf{L}^{(r)}(g, \omega) , \quad g_0 \in G^{[a]} . \tag{102}$$

The reduced Lagrange function, $\mathcal{L}^{(r)} \in \mathcal{F}(\mathfrak{g} \times O_a)$, is defined as

$$\mathcal{L}^{(r)}(\omega, p) = \mathbf{L}^{(r)}(g, \omega) , \quad \text{where} \quad p = \Phi(g) \cdot a . \tag{103}$$

Theorem 11.2. a) *Consider the reduction* $(g, \omega) \rightarrow (\omega, p)$. *The reduced Euler–Lagrange equations are the following Euler–Poincaré equations:*

$$\begin{cases} \dot{m} = -\mathrm{ad}^* \omega \cdot m - \nabla_p \mathcal{L}^{(r)}(\omega, p) \diamond p , \\ \dot{p} = \phi(\omega) \cdot p , \end{cases} \tag{104}$$

where

$$m = \nabla_\omega \mathcal{L}^{(r)}(\omega, p) \in \mathfrak{g}^* . \tag{105}$$

They describe the extremals of the constrained variational principle with functional

$$\mathbf{S}^{(r)} = \int_a^b \mathcal{L}^{(r)}(\omega(t), p(t)) dt , \tag{106}$$

and admissible variations $(\omega(t, \epsilon), p(t, \epsilon))$ *of* $(\omega(t), p(t))$ *of the form*

$$\omega(t, \epsilon) = \omega(t) + \epsilon\left(\dot{\eta}(t) + [\eta(t), \omega(t)]\right) , \quad p(t, \epsilon) = p(t) + \epsilon\phi(\eta(t)) \cdot p(t) , \tag{107}$$

where $\eta : [a, b] \rightarrow \mathfrak{g}$ *is an arbitrary function such that* $\eta(a) = \eta(b) = 0$.
   b) *If the Legendre transformation,*

$$(\omega, p) \in \mathfrak{g} \times O_a \rightarrow (m, p) \in \mathfrak{g}^* \times O_a , \tag{108}$$

is invertible, then it transforms (104) *into a Hamiltonian system on* $\mathfrak{g}^* \times O_a$ *with respect to the Poisson bracket*

$$\begin{aligned} \{F_1, F_2\}^{(r)}(m, p) = &-\langle m, [\nabla_m F_1, \nabla_m F_2] \rangle \\ &-\langle \nabla_p F_1, \phi(\nabla_m F_2) \cdot p \rangle + \langle \nabla_p F_2, \phi(\nabla_m F_1) \cdot p \rangle, \end{aligned} \tag{109}$$

with the Hamilton function

$$H^{(r)}(m, p) = \langle m, \omega \rangle - \mathcal{L}^{(r)}(\omega, p), \tag{110}$$

where ω has to be expressed in terms of (m, p).

c) *The full (non–reduced) Euler–Lagrange equations on TG have the integrals of motion,*

$$J_\xi^{(r)}(g, m) = \langle m, \mathrm{Ad}\, g \cdot \xi \rangle, \qquad \forall \xi \in \mathfrak{g}^{[a]}. \tag{111}$$

Remark. Formula (109) differs from (98) only in the overall minus sign. Therefore it also defines a Poisson bracket on all of $\mathfrak{g}^* \times V$, the opposite of the Lie–Poisson bracket of the semidirect product Lie algebra $\mathfrak{g} \ltimes V^*$ corresponding to the representation $-\phi^*$ of $\mathfrak{g}$ in V^*.

11.2 Discrete–Time Case

Turning to the discrete–time Lagrangian mechanics on G, suppose that the Lagrange function $\mathbb{L}(g_1, g_2)$ is invariant under the action of $G^{[a]}$ on $G \times G$ induced by left translations on G,

$$\mathbb{L}(g_0 g_1, g_0 g_2) = \mathbb{L}(g_1, g_2), \quad g_0 \in G^{[a]}. \tag{112}$$

Denote $\mathbb{L}(g_1, g_2) = \mathbb{L}^{(l)}(g_1, W)$, where $W = g_2^{-1} g_1$. The corresponding invariance property of $\mathbb{L}^{(l)}(g, W)$ is expressed as

$$\mathbb{L}^{(l)}(g_0 g, W) = \mathbb{L}^{(l)}(g, W), \quad g_0 \in G^{[a]}. \tag{113}$$

To reduce the discrete–time Euler–Lagrange equations with respect to this left action, we realize the factor $(G \times G)/G^{[a]}$ as the set $G \times O_a$, the reduction map being

$$(g, W) \in G \times G \;\to\; (W, P) \in G \times O_a, \qquad \text{where} \qquad P = \Phi(g^{-1}) \cdot a. \tag{114}$$

The reduced Lagrange function $\Lambda^{(l)} \in \mathcal{F}(G \times O_a)$ is defined as

$$\Lambda^{(l)}(W, P) = \mathbb{L}^{(l)}(g, W), \quad \text{where} \quad P = \Phi(g^{-1}) \cdot a. \tag{115}$$

Theorem 11.3. a) *Consider the reduction $(g, W) \to (W, P)$. The reduced Euler–Lagrange equations are the discrete–time Euler–Poincaré equations,*

$$\begin{cases} \mathrm{Ad}^*\, W_n^{-1} \cdot M_{n+1} = M_n + \nabla_P \Lambda^{(l)}(W_n, P_n) \diamond P_n, \\ P_{n+1} = \Phi(W_n^{-1}) \cdot P_n, \end{cases} \tag{116}$$

where

$$M_n = d'_W \Lambda^{(l)}(W_{n-1}, P_{n-1}) \in \mathfrak{g}^* . \tag{117}$$

They describe the extremals of the constrained variational principle with functional

$$\mathbb{S}^{(l)} = \sum_{n=a}^{b-1} \Lambda^{(l)}(W_n, P_n) , \tag{118}$$

and the admissible variations $\{(W_n(\epsilon), P_n(\epsilon))\}$ of $\{(W_n, P_n)\}$ of the form

$$W_n(\epsilon) = W_n e^{\epsilon \eta_{n+1} - \epsilon \mathrm{Ad}\, W_n^{-1} \cdot \eta_n} , \quad P_n(\epsilon) = P_n - \epsilon \phi(\eta_n) \cdot P_n , \tag{119}$$

where $\{\eta_n\}_{n=a}^{b}$ is an arbitrary sequence of elements of $\mathfrak{g}$ such that $\eta_a = \eta_b = 0$.

b) *If the Legendre transformation*

$$(W_{n-1}, P_{n-1}) \in G \times O_a \rightarrow (M_n, P_n) \in \mathfrak{g}^* \times O_a , \tag{120}$$

where $P_n = \Phi(W_{n-1}^{-1}) \cdot P_{n-1}$, is invertible, then equations (116) define a map $(M_n, P_n) \rightarrow (M_{n+1}, P_{n+1})$ on $\mathfrak{g}^* \times O_a$ which is Poisson with respect to bracket (98).

c) *The full (non–reduced) Euler–Lagrange equations on $G \times G$ have the integrals of motion,*

$$J_\xi^{(l)}(g_n, M_n) = \langle M_n, \mathrm{Ad}\, g_n^{-1} \cdot \xi \rangle , \qquad \forall \xi \in \mathfrak{g}^{[a]} . \tag{121}$$

Finally, assume that function $\mathbb{L}$ is invariant under the action of $G^{[a]}$ on $G \times G$ induced by right translations on G,

$$\mathbb{L}(g_1 g_0, g_2 g_0) = \mathbb{L}(g_1, g_2) , \quad g_0 \in G^{[a]} . \tag{122}$$

Let $\mathbb{L}(g_1, g_2) = \mathbb{L}^{(r)}(g_1, w)$, where $w = g_1 g_2^{-1}$. An equivalent formulation of the invariance property is then,

$$\mathbb{L}^{(r)}(g g_0, w) = \mathbb{L}^{(r)}(g, w) , \quad g_0 \in G^{[a]} . \tag{123}$$

Defining the reduced Lagrange function $\Lambda^{(r)} \in \mathcal{F}(G \times O_a)$ as

$$\Lambda^{(r)}(w, p) = \mathbb{L}^{(r)}(g, w) , \quad \text{where} \quad p = \Phi(g) \cdot a , \tag{124}$$

leads to the following statement.

Theorem 11.4. a) *Consider the reduction $(g, w) \to (w, p)$. The reduced Euler–Lagrange equations are the discrete–time Euler–Poincaré equations:*

$$\begin{cases} \mathrm{Ad}^* w_n \cdot m_{n+1} = m_n - \nabla_p \Lambda^{(r)}(w_n, p_n) \diamond p_n \,, \\ p_{n+1} = \Phi(w_n) \cdot p_n \,, \end{cases} \tag{125}$$

where

$$m_n = d_w \Lambda^{(r)}(w_{n-1}, p_{n-1}) \in \mathfrak{g}^* \,. \tag{126}$$

They describe the extremals of the constrained variational principle with functional

$$\mathbb{S}^{(r)} = \sum_{n=a}^{b-1} \Lambda^{(r)}(w_n, p_n) \,, \tag{127}$$

and admissible variations $\{(w_n(\epsilon), p_n(\epsilon))\}$ of $\{(w_n, p_n)\}$ of the form

$$w_n(\epsilon) = e^{\epsilon \eta_{n+1} - \epsilon \mathrm{Ad}\, w_n \cdot \eta_n} w_n \,, \quad p_n(\epsilon) = p_n + \epsilon \phi(\eta_n) \cdot p_n \,, \tag{128}$$

where $\{\eta_n\}_{n=a}^{b}$ is an arbitrary sequence of elements of $\mathfrak{g}$ such that $\eta_a = \eta_b = 0$.

b) *If the Legendre transformation,*

$$(w_{n-1}, p_{n-1}) \in G \times O_a \to (m_n, p_n) \in \mathfrak{g}^* \times O_a \,, \tag{129}$$

where $p_n = \Phi(w_{n-1}) \cdot p_{n-1}$, is invertible, then equations (125) define a map $(m_n, p_n) \to (m_{n+1}, p_{n+1})$ on $\mathfrak{g}^* \times O_a$ which is Poisson with respect to bracket (109).

c) *The full (non–reduced) Euler–Lagrange equations on $G \times G$ have the integrals of motion,*

$$J_\xi^{(r)}(g_n, m_n) = \langle m_n, \mathrm{Ad}\, g_n \cdot \xi \rangle \,, \qquad \forall \xi \in \mathfrak{g}^{[a]} \,. \tag{130}$$

The following table (opposite page) summarizes the information on Euler–Poincaré equations, both in the continuous and discrete time formulations.

Recall that the relation between the continuous time and the discrete time equations is established, if we set

$$g_n = g \,, \quad g_{n+1} = g + h\dot{g} + O(h^2) \,, \quad \mathbb{L}(g_n, g_{n+1}) = h\mathbf{L}(g, \dot{g}) + O(h^2) \,;$$
$$P_n = P \,, \quad W_n = 1 + h\Omega + O(h^2) \,, \quad \Lambda^{(l)}(W_n, P_n) = h\mathcal{L}^{(l)}(\Omega, P) + O(h^2) \,;$$
$$p_n = p \,, \quad w_n = 1 + h\omega + O(h^2) \,, \quad \Lambda^{(r)}(w_n, p_n) = h\mathcal{L}^{(r)}(\omega, p) + O(h^2) \,.$$

CONTINUOUS TIME	DISCRETE TIME
Left trivialization, left symmetry reduction	
$\mathbf{L}(g, \dot{g}) = \mathcal{L}^{(l)}(\Omega, P)$ $\Omega = L_{g^{-1} *}\dot{g}$ $P = \Phi(g^{-1}) \cdot a$ $M = L_g^* \Pi = \nabla_\Omega \mathcal{L}^{(l)}$ $\begin{cases} \dot{M} = \mathrm{ad}^* \, \Omega \cdot M + \nabla_P \mathcal{L}^{(l)} \diamond P \\ \dot{P} = -\phi(\Omega) \cdot P \end{cases}$	$\mathbb{L}(g_n, g_{n+1}) = \Lambda^{(l)}(W_n, P_n)$ $W_n = g_n^{-1} g_{n+1}$ $P_n = \Phi(g_n^{-1}) \cdot a$ $M_n = L_{g_n}^* \Pi_n = d'_W \Lambda^{(l)}(W_{n-1}, P_{n-1})$ $\begin{cases} \mathrm{Ad}^* \, W_n^{-1} \cdot M_{n+1} = M_n + \nabla_P \Lambda^{(l)}(W_n, P_n) \\ P_{n+1} = \Phi(W_n^{-1}) \cdot P_n \end{cases}$
Right trivialization, right symmetry reduction	
$\mathbf{L}(g, \dot{g}) = \mathcal{L}^{(r)}(\omega, p)$ $\omega = R_{g^{-1} *}\dot{g}$ $p = \Phi(g) \cdot a$ $m = R_g^* \Pi = \nabla_\omega \mathcal{L}^{(r)}$ $\begin{cases} \dot{m} = -\mathrm{ad}^* \, \omega \cdot m - \nabla_p \mathcal{L}^{(r)} \diamond p \\ \dot{p} = \phi(\omega) \cdot p \end{cases}$	$\mathbb{L}(g_n, g_{n+1}) = \Lambda^{(r)}(w_n, p_n)$ $w_n = g_{n+1} g_n^{-1}$ $p_n = \Phi(g_n) \cdot a$ $m_n = R_{g_n}^* \Pi_n = d_w \Lambda^{(r)}(w_{n-1}, p_{n-1})$ $\begin{cases} \mathrm{Ad}^* \, w_n \cdot m_{n+1} = m_n - \nabla_p \Lambda^{(r)}(w_n, p_n) \\ p_{n+1} = \Phi(w_n) \cdot p_n \end{cases}$

12 Neumann System

12.1 Continuous–Time Dynamics

Consider the motion of a point $x \in \mathbb{R}^N$ under the potential $\frac{1}{2}\langle \Omega x, x \rangle$, where $\Omega = \mathrm{diag}(\omega_1, \dots, \omega_N)$, constrained to the sphere

$$S = \left\{ x \in \mathbb{R}^N : \langle x, x \rangle = 1 \right\}. \tag{131}$$

This is the famous **Neumann system**.

The Lagrangian approach to this problem is as follows. The motions should correspond to local extrema for the action functional

$$S = \int_{t_0}^{t_1} \mathbf{L}\Big(x(t), \dot{x}(t)\Big) dt \ ,$$

where $L : TS \to \mathbb{R}$ is the Lagrange function, given by

$$L(x, \dot{x}) = \frac{1}{2}\langle \dot{x}, \dot{x} \rangle - \frac{1}{2}\langle \Omega x, x \rangle - \frac{1}{2}\alpha\left(\langle x, x \rangle - 1\right) . \tag{132}$$

Here the first two terms on the right–hand side represent the unconstrained Lagrange function, and the Lagrange multiplier α has to be chosen to assure that the solution of the variational problem lies on the constraint manifold, TS, which is described by the equations

$$\langle x, x \rangle = 1 , \qquad \langle \dot{x}, x \rangle = 0 . \tag{133}$$

The differential equations of the extremals of the above problem are

$$\ddot{x} = -\Omega x - \alpha x . \tag{134}$$

The value of α is determined by

$$0 = \langle \dot{x}, x \rangle^{\cdot} = \langle \dot{x}, \dot{x} \rangle + \langle \ddot{x}, x \rangle = \langle \dot{x}, \dot{x} \rangle - \langle \Omega x, x \rangle - \alpha .$$

Therefore

$$\alpha = \langle \dot{x}, \dot{x} \rangle - \langle \Omega x, x \rangle . \tag{135}$$

So, the complete description of the Neumann problem consists of the equations of motion (134), together with the expression (135).

The Legendre transformation, leading to the Hamiltonian interpretation of the above system, is given by

$$H(x, p) = \langle \dot{x}, p \rangle - L(x, \dot{x}) ,$$

where the canonically conjugate momenta p are defined by

$$p = \partial L / \partial \dot{x} = \dot{x} . \tag{136}$$

Hence we obtain

$$H(x, p) = \frac{1}{2}\langle p, p \rangle + \frac{1}{2}\langle \Omega x, x \rangle . \tag{137}$$

The corresponding symplectic structure is the restriction of the standard symplectic structure of the space $\mathbb{R}^{2N}(x, p)$,

$$\{p_k, x_j\} = \delta_{kj} , \tag{138}$$

to the submanifold T^*S which is singled out by the relations

$$\phi_1 = \langle x, x \rangle - 1 = 0 , \qquad \phi_2 = \langle p, x \rangle = 0 . \tag{139}$$

The Dirac Poisson bracket for this symplectic structure on T^*S is characterized by the following relations:

$$\{x_k, x_j\}_{\mathrm{D}} = 0 , \quad \{p_k, x_j\}_{\mathrm{D}} = \delta_{kj} - \frac{x_k x_j}{\langle x, x \rangle} , \quad \{p_k, p_j\}_{\mathrm{D}} = \frac{x_k p_j - p_k x_j}{\langle x, x \rangle} . \tag{140}$$

It is easy to show that the Hamiltonian vector field generated by the Hamilton function (137) with respect to the bracket $\{\cdot, \cdot\}_{\mathrm{D}}$ on T^*S, is given by

$$\dot{x}_k = \{\mathrm{H}, x_k\}_{\mathrm{D}} = p_k , \qquad \dot{p}_k = \{\mathrm{H}, p_k\}_{\mathrm{D}} = -\omega_k x_k - \alpha x_k , \tag{141}$$

where

$$\alpha = \langle p, p \rangle - \langle \Omega x, x \rangle . \tag{142}$$

This is nothing but the first–order form of (134) with multiplier (135).

It can be checked that the following functions are integrals of motion of the Neumann system, in the case when all ω_k's are distinct,

$$F_k = x_k^2 + \sum_{j \neq k} \frac{(p_k x_j - x_k p_j)^2}{\omega_k - \omega_j} , \qquad 1 \leq k \leq N . \tag{143}$$

Only $N - 1$ of these N integrals are functionally independent on T^*S, due to the relation

$$\sum_{k=1}^{N} F_k = \langle x, x \rangle , \tag{144}$$

which is equal to 1 on T^*S. The Hamilton function of the Neumann system can be represented as

$$\mathrm{H} = \frac{1}{2} \sum_{k=1}^{N} \omega_k F_k = \frac{1}{2} \left(\langle p, p \rangle \langle x, x \rangle - \langle p, x \rangle^2 \right) + \frac{1}{2} \langle \Omega x, x \rangle , \tag{145}$$

which coincides with (137) on T^*S.

Theorem 12.1. *Differential equations (141) are equivalent to the matrix equation,*

$$\dot{\mathcal{L}} = [\mathcal{M}, \mathcal{L}], \tag{146}$$

with $N \times N$ matrices depending on the spectral parameter λ:

$$\mathcal{L} = \mathcal{L}(x, p; \lambda) = \Omega + \lambda(px^{\mathrm{T}} - xp^{\mathrm{T}}) - \lambda^2 xx^{\mathrm{T}}, \tag{147}$$

$$\mathcal{M} = \mathcal{M}(x; \lambda) = \lambda xx^{\mathrm{T}}. \tag{148}$$

If all ω_k's are distinct, then differential equations (141) are also equivalent to the matrix equation,

$$\dot{\Lambda} = [\mathrm{M}, \Lambda] , \tag{149}$$

with 2×2 matrices depending on the spectral parameter z:

$$\Lambda = \Lambda(x, p; z) = \begin{pmatrix} \Phi_z(x, p) & -1 + \Phi_z(p, p) \\ -\Phi_z(x, x) & -\Phi_z(x, p) \end{pmatrix}, \tag{150}$$

$$M = M(x, p; z) = \begin{pmatrix} 0 & z + \alpha \\ -1 & 0 \end{pmatrix}. \tag{151}$$

Here

$$\Phi_z(x, p) = \left\langle (\Omega - zI)^{-1} x, p \right\rangle = \sum_{k=1}^{N} \frac{x_k p_k}{\omega_k - z}. \tag{152}$$

The relation between these two Lax representations is an example of duality. It can be best explained on the level of characteristic polynomials which are related, due to the Weinstein–Aronszajn formula,

$$\det \left(\mathcal{L}(\lambda) - zI_N \right) = \lambda^2 \prod_{j=1}^{N} (\omega_j - z) \det \left(\Lambda(z) + \lambda^{-1} I_2 \right). \tag{153}$$

Thus,

$$\det \left(\Lambda(z) + \lambda^{-1} I_2 \right) = \lambda^{-2} - \sum_{k=1}^{n} \frac{F_k}{\omega_k - z}. \tag{154}$$

12.2 Bäcklund Transformation for the Neumann System

Consider the following Lagrange function on $S \times S$,

$$h\mathbb{L}(\widetilde{x}, x) = -\left\langle (I + h^2 \Omega)^{1/2} \widetilde{x}, x \right\rangle, \quad x \in S, \ \widetilde{x} \in S, \tag{155}$$

and the corresponding discrete–time Newtonian system,

$$(I + h^2 \Omega)^{1/2} (\widetilde{x} + \underset{\sim}{x}) = \beta x. \tag{156}$$

Here β is the Lagrange multiplier, assuring that x remains on S during the discrete time evolution. Requiring $\langle \widetilde{x}, \widetilde{x} \rangle = 1$ provided that $\langle x, x \rangle = 1$ and $\langle \underset{\sim}{x}, \underset{\sim}{x} \rangle = 1$, we find that

$$\beta = \frac{2 \left\langle (I + h^2 \Omega)^{-1/2} x, \underset{\sim}{x} \right\rangle}{\left\langle (I + h^2 \Omega)^{-1} x, x \right\rangle} = \frac{2 \sum\limits_{k=1}^{N} (1 + h^2 \omega_k)^{-1/2} x_k \underset{\sim}{x}_k}{\sum\limits_{k=1}^{N} (1 + h^2 \omega_k)^{-1} x_k^2}. \tag{157}$$

Similarly, requiring that $\langle \underset{\sim}{x}, \underset{\sim}{x} \rangle = 1$ provided that $\langle x, x \rangle = 1$ and $\langle \widetilde{x}, \widetilde{x} \rangle = 1$, leads to the alternative expression,

$$\beta = \frac{2\left\langle (I + h^2\Omega)^{-1/2}\widetilde{x}, x \right\rangle}{\left\langle (I + h^2\Omega)^{-1}x, x \right\rangle} = \frac{2\sum_{k=1}^{N}(1 + h^2\omega_k)^{-1/2}\widetilde{x}_k x_k}{\sum_{k=1}^{N}(1 + h^2\omega_k)^{-1}x_k^2} . \tag{158}$$

Hence the following quantity is an integral of motion of our discrete–time Lagrangian system,

$$\sum_{k=1}^{N}(1 + h^2\omega_k)^{-1/2}\widetilde{x}_k x_k = \sum_{k=1}^{N}(1 + h^2\omega_k)^{-1/2}x_k \underset{\sim}{x}_k . \tag{159}$$

The momenta p canonically conjugate to x are given by

$$hp = (I + h^2\Omega)^{1/2}\widetilde{x} - \gamma x , \tag{160}$$
$$h\widetilde{p} = -(I + h^2\Omega)^{1/2}x + \gamma\widetilde{x} . \tag{161}$$

Here the scalar multiplier γ has to be chosen to assure the relations

$$p \in T_x^* S , \qquad \widetilde{p} \in T_{\widetilde{x}}^* S ,$$

or, in other words, the relations

$$\langle p, x \rangle = \langle \widetilde{p}, \widetilde{x} \rangle = 0 , \tag{162}$$

which are satisfied if

$$\gamma = \left\langle (I + h^2\Omega)^{1/2}\widetilde{x}, x \right\rangle . \tag{163}$$

The condition for $\widetilde{x}$ determined by (160) to lie on S is

$$\gamma^2 \sum_{k=1}^{N} \frac{x_k^2}{1 + h^2\omega_k} + 2h\gamma \sum_{k=1}^{N} \frac{x_k p_k}{1 + h^2\omega_k} + h^2 \sum_{k=1}^{N} \frac{p_k^2}{1 + h^2\omega_k} = 1 . \tag{164}$$

There is one root of this quadratic equation such that $\gamma = 1 + O(h^2)$ when $h \to 0$. It is

$$\gamma = \frac{-h\sum_{k=1}^{N} \dfrac{x_k p_k}{1 + h^2\omega_k} + \sqrt{\mathcal{F}}}{\sum_{k=1}^{N} \dfrac{x_k^2}{1 + h^2\omega_k}} , \tag{165}$$

where

$$\mathcal{F} = \sum_{k=1}^{N} \frac{\mathrm{F}_k}{1 + h^2\omega_k} , \qquad (166)$$

with the functions F_k given in (143).

Similarly, the condition that x determined by (161) lies on S is

$$\gamma^2 \sum_{k=1}^{N} \frac{\tilde{x}_k^2}{1 + h^2\omega_k} - 2h\gamma \sum_{k=1}^{N} \frac{\tilde{x}_k \tilde{p}_k}{1 + h^2\omega_k} + h^2 \sum_{k=1}^{N} \frac{\tilde{p}_k^2}{1 + h^2\omega_k} = 1 , \qquad (167)$$

and the root such that $\gamma = 1 + O(h^2)$ when $h \to 0$ is

$$\gamma = \frac{h \sum_{k=1}^{N} \dfrac{\tilde{x}_k \tilde{p}_k}{1 + h^2\omega_k} + \sqrt{\tilde{\mathcal{F}}}}{\sum_{k=1}^{N} \dfrac{\tilde{x}_k^2}{1 + h^2\omega_k}} , \qquad (168)$$

where $\tilde{\mathcal{F}}$ is obtained from $\mathcal{F}$ by replacing (x, p) by $(\tilde{x}, \tilde{p})$.

Lagrangian equations (160), (161) yield Newtonian ones (156) with

$$\beta = \gamma + \underset{\sim}{\gamma} . \qquad (169)$$

This, together with (164) and the downshifted version of (167) imply the following expression for the Lagrange multiplier β:

$$\beta = \frac{2\sqrt{\mathcal{F}}}{\sum_{k=1}^{N} (1 + h^2\omega_k)^{-1} x_k^2} . \qquad (170)$$

Comparing (170) with (157), (158) leads to the formula

$$\sqrt{\mathcal{F}} = \sum_{k=1}^{N} (1 + h^2\omega_k)^{-1/2} \tilde{x}_k x_k . \qquad (171)$$

Lax representations of the map (160), (161) are given in the following statement.

Theorem 12.2. *Equations of motion* (160), (161) *are equivalent to the following matrix factorizations,*

$$I + h^2 \mathcal{L} = I + h^2 \mathcal{L}(x, p; \lambda) = M^{\mathrm{T}}(\tilde{x}, x; -\lambda) M(\tilde{x}, x; \lambda) , \qquad (172)$$

$$I + h^2 \tilde{\mathcal{L}} = I + h^2 \mathcal{L}(\tilde{x}, \tilde{p}; \lambda) = M(\tilde{x}, x; \nu) M^{\mathrm{T}}(\tilde{x}, x; -\lambda) , \qquad (173)$$

where the matrix $\mathcal{L}$ is defined as in (147), and

$$M(\tilde{x}, x; \lambda) = (I + h^2\Omega)^{1/2} + h\lambda\tilde{x}x^{\mathrm{T}}. \tag{174}$$

Also, equations of motion (160), (161) are equivalent to the matrix equation,

$$\tilde{\Lambda}P = P\Lambda , \tag{175}$$

where the matrix $\Lambda = \Lambda(x, p; z)$ is given in (150), and

$$P = P(\tilde{x}, x; z) = \begin{pmatrix} \gamma & hz + h^{-1}(1 - \gamma^2) \\ -h & \gamma \end{pmatrix}. \tag{176}$$

Thus, discretization (160), (161) shares the Lax matrices, and therefore the integrals of motion, with the Neumann system itself.

12.3 Ragnisco's Discretization of the Neumann System

There exists an alternative integrable discretization of the Neumann system invented by Ragnisco. Its discrete–time Lagrange function on $S \times S$ is

$$h\mathbb{L}(\tilde{x}, x) = -\log\langle\tilde{x}, x\rangle - \frac{h^2}{2}\langle\Omega x, x\rangle , \quad x \in S, \ \tilde{x} \in S . \tag{177}$$

The momenta p canonically conjugate to x are given by

$$hp_k = \frac{\tilde{x}_k}{\langle\tilde{x}, x\rangle} + h^2\omega_k x_k - \gamma x_k , \qquad h\tilde{p}_k = -\frac{x_k}{\langle\tilde{x}, x\rangle} + \delta\tilde{x}_k . \tag{178}$$

Here the scalar multipliers γ, δ have to be chosen so as to assure that

$$p \in T_x^*S , \qquad \tilde{p} \in T_{\tilde{x}}^*S ,$$

or, in other words, the relations

$$\langle p, x\rangle = \langle\tilde{p}, \tilde{x}\rangle = 0 . \tag{179}$$

It is easy to see that this is achieved if

$$\gamma = 1 + h^2\langle\Omega x, x\rangle , \qquad \delta = 1 . \tag{180}$$

So, the following are the equations of motion of *Ragnisco's discrete–time Neumann system*:

$$hp_k = \frac{\tilde{x}_k}{\langle\tilde{x}, x\rangle} - x_k + h^2\omega_k x_k - h^2\langle\Omega x, x\rangle x_k , \tag{181}$$

$$h\tilde{p}_k = -\frac{x_k}{\langle\tilde{x}, x\rangle} + \tilde{x}_k . \tag{182}$$

The Newtonian form of the equations of motion is

$$\frac{\widetilde{x}_k}{\langle \widetilde{x}, x \rangle} - 2x_k + \frac{\underset{\approx}{x}_k}{\langle x, \underset{\approx}{x} \rangle} = -h^2 \omega_k x_k + h^2 \langle \Omega x, x \rangle x_k .$$
(183)

Equations (181), (182) define an explicit symplectic map $(x, p) \in T^*S \to (\widetilde{x}, \widetilde{p}) \in T^*S$. Indeed, the first equation (181) yields

$$\frac{\widetilde{x}_k}{\langle \widetilde{x}, x \rangle} = x_k + h p_k - h^2 \omega_k x_k + h^2 \langle \Omega x, x \rangle x_k ,$$

which allows us to determine

$$\frac{1}{\langle \widetilde{x}, x \rangle} = \| x + h p - h^2 \Omega x + h^2 \langle \Omega x, x \rangle x \| .$$
(184)

Now the previous equality defines

$$\widetilde{x}_k = \frac{x_k + h p_k - h^2 \omega_k x_k + h^2 \langle \Omega x, x \rangle x_k}{\| x + h p - h^2 \Omega x + h^2 \langle \Omega x, x \rangle x \|} ,$$

and, finally, the second equation of motion (182) defines $\widetilde{p}$.

Ragnisco's discrete–time Neumann system no longer shares the integrals of motion and the Lax matrices with its continuous–time counterpart.

Theorem 12.3. *Equations of motion (181), (182) are equivalent to the following matrix equation,*

$$\widetilde{\Lambda}_{\mathrm{R}} P_{\mathrm{R}} = P_{\mathrm{R}} \Lambda_{\mathrm{R}} ,$$
(185)

with the matrices

$$\Lambda_{\mathrm{R}} = \Lambda_{\mathrm{R}}(x, p; z) = \begin{pmatrix} -\dfrac{h}{2} + \Phi_z(x, p) & -1 + \Phi_z(p, p) \\ -\Phi_z(x, x) & \dfrac{h}{2} - \Phi_z(x, p) \end{pmatrix} .$$
(186)

$$P_{\mathrm{R}} = P_{\mathrm{R}}(x, p; z) = \begin{pmatrix} 1 & hz + h\alpha \\ -h & \gamma - h^2 z \end{pmatrix} ,$$
(187)

where

$$\gamma = 1 + h^2 \langle \Omega x, x \rangle ,$$
(188)

$$\alpha = \langle p - h\Omega x, p - h\Omega x \rangle - \langle \Omega x, x \rangle - h^2 \langle \Omega x, x \rangle^2 .$$
(189)

We do not know a "big" Lax representation of Ragnisco's discrete–time Neumann system. The characteristic polynomial of the matrix Λ_{R} yields the following integrals of motion of Ragnisco's discrete–time Neumann system:

$$\mathcal{F}_k = x_k^2 - h x_k p_k + \sum_{j \neq k} \frac{(p_k x_j - x_k p_j)^2}{\omega_k - \omega_j} .$$
(190)

12.4 Adler's Discretization of the Neumann System

It will be convenient to adopt the notation

$$\alpha_k = 1 - \frac{h^2}{4}\omega_k , \qquad A = \operatorname{diag}(\alpha_1, \dots, \alpha_N) = I - \frac{h^2}{4}\Omega . \tag{191}$$

Consider the discrete–time Lagrange function on $S \times S$,

$$h\mathbb{L}(\tilde{x}, x) = -2 \log(1 + \langle \tilde{x}, x \rangle) + 2 \log\langle Ax, x \rangle , \quad x \in S, \ \tilde{x} \in S. \tag{192}$$

The corresponding discrete–time Lagrangian equations of motion are

$$hp_k = \frac{2\tilde{x}_k}{1 + \langle \tilde{x}, x \rangle} - \frac{4\alpha_k x_k}{\langle Ax, x \rangle} - \gamma x_k , \qquad h\tilde{p}_k = -\frac{2x_k}{1 + \langle \tilde{x}, x \rangle} + \delta \tilde{x}_k . \tag{193}$$

Here the scalar multipliers γ, δ have to be chosen so as to assure that

$$p \in T_x^* S , \qquad \tilde{p} \in T_{\tilde{x}}^* S ,$$

or, in other words, to assure that the following relations hold,

$$\langle p, x \rangle = 0 , \qquad \langle \tilde{p}, \tilde{x} \rangle = 0 .$$

It is easy to see that this is achieved if

$$\gamma = \frac{2\langle \tilde{x}, x \rangle}{1 + \langle \tilde{x}, x \rangle} - 4 = -\frac{2}{1 + \langle \tilde{x}, x \rangle} - 2 , \quad \delta = \frac{2\langle \tilde{x}, x \rangle}{1 + \langle \tilde{x}, x \rangle} = -\frac{2}{1 + \langle \tilde{x}, x \rangle} + 2 . \tag{194}$$

So, we come to the following Lagrangian equations of motion of **Adler's discrete–time Neumann system,**

$$hp_k = \frac{2(\tilde{x}_k + x_k)}{1 + \langle \tilde{x}, x \rangle} - \frac{4\alpha_k x_k}{\langle Ax, x \rangle} + 2x_k , \tag{195}$$

$$h\tilde{p}_k = -\frac{2(\tilde{x}_k + x_k)}{1 + \langle \tilde{x}, x \rangle} + 2\tilde{x}_k . \tag{196}$$

These two equations yield also the Newtonian form of the equations of motion,

$$\frac{\tilde{x}_k + x_k}{1 + \langle \tilde{x}, x \rangle} + \frac{x_k + \underset{\sim}{x}_k}{1 + \langle x, \underset{\sim}{x} \rangle} = \frac{2\alpha_k x_k}{\langle Ax, x \rangle} . \tag{197}$$

Equations (195) and (196) define a symplectic map, $(x, p) \in T^*S \to (\tilde{x}, \tilde{p}) \in T^*S$. At a first glance, this definition looks implicit. However, this is actually not the case. Indeed, the first equation (195) yields

$$\frac{2(\tilde{x}_k + x_k)}{1 + \langle \tilde{x}, x \rangle} = \frac{4\alpha_k x_k}{\langle Ax, x \rangle} - 2x_k + hp_k ,$$

which implies

$$\frac{2}{1 + \langle \widetilde{x}, x \rangle} = \left\| \frac{2Ax}{\langle Ax, x \rangle} - x + \frac{hp}{2} \right\|^2 . \tag{198}$$

This, substituted back into (195), allows us to determine $\widetilde{x}$, and then, finally, the second equation of motion (196) defines $\widetilde{p}$.

A neat Lax representation for Adler's discrete–time Neumann system is not known. Nevertheless, it is completely integrable, as the following theorem shows.

Theorem 12.4. *If all α_k's are distinct, then the following functions are involutive integrals of motion of Adler's discrete–time Neumann system,*

$$\mathcal{F}_k = x_k^2 - hx_k p_k + \frac{h^2}{4} \sum_{j \neq k} \frac{(x_k p_j - x_j p_k)(\alpha_k x_k p_j - \alpha_j x_j p_k)}{\alpha_j - \alpha_k} . \tag{199}$$

13 Garnier System

13.1 Continuous–Time Dynamics

The equations of motion of the **Garnier system** are

$$\ddot{x}_k = -\omega_k x_k - 2x_k \langle x, x \rangle , \tag{200}$$

or, in the vector form,

$$\ddot{x} = -\Omega x - 2x \langle x, x \rangle . \tag{201}$$

Written as

$$\dot{x}_k = p_k , \qquad \dot{p}_k = -\omega_k x_k - 2x_k \langle x, x \rangle , \tag{202}$$

this is a Hamiltonian system on $\mathbb{R}^{2N}(x,p)$ equipped with the standard symplectic structure, with the Hamilton function

$$H(x,p) = \frac{1}{2} \sum_{k=1}^{N} (p_k^2 + \omega_k x_k^2) + \frac{1}{2} \left(\sum_{k=1}^{N} x_k^2 \right)^2 = \frac{1}{2} \langle p, p \rangle + \frac{1}{2} \langle \Omega x, x \rangle + \frac{1}{2} \langle x, x \rangle^2 . \tag{203}$$

If all ω_k's are distinct, then the full set of integrals of the Garnier system is given by the functions

$$F_k = p_k^2 + \omega_k x_k^2 + x_k^2 \langle x, x \rangle + \sum_{j \neq k} \frac{X_{kj}^2}{\omega_j - \omega_k} , \qquad 1 \leq k \leq N . \tag{204}$$

Then

$$H = \frac{1}{2} \sum_{k=1}^{N} F_k . \tag{205}$$

Theorem 13.1. *Differential equations (202) are equivalent to the matrix equation,*

$$\dot{\mathcal{L}} = [\mathcal{M}, \mathcal{L}] \tag{206}$$

with $(N+1) \times (N+1)$ matrices depending on the spectral parameter λ,

$$\mathcal{L} = \mathcal{L}(x, p; \lambda) = \begin{pmatrix} xx^{\mathrm{T}} + \Omega & p + \lambda x \\ p^{\mathrm{T}} - \lambda x^{\mathrm{T}} & -\langle x, x \rangle - \lambda^2 \end{pmatrix}, \tag{207}$$

$$\mathcal{M} = \mathcal{M}(x; \lambda) = \begin{pmatrix} 0 & x \\ -x^{\mathrm{T}} & -\lambda \end{pmatrix}. \tag{208}$$

If all ω_k's are distinct, then differential equations (202) are equivalent to the matrix equation,

$$\dot{\Lambda} = [\mathrm{M}, \Lambda], \tag{209}$$

with 2×2 matrices depending on the spectral parameter z,

$$\Lambda = \Lambda(x, p; z) = \begin{pmatrix} \Phi_z(x, p) & z + \langle x, x \rangle + \Phi_z(p, p) \\ -1 - \Phi_z(x, x) & -\Phi_z(x, p) \end{pmatrix}, \tag{210}$$

$$\mathrm{M} = \mathrm{M}(x; z) = \begin{pmatrix} 0 & z + 2\langle x, x \rangle \\ -1 & 0 \end{pmatrix}. \tag{211}$$

As for the Neumann system, also in the present case the characteristic polynomials of the Lax matrices yield the full set of integrals of motion.

13.2 Bäcklund Transformation for the Garnier System

Consider the discrete–time Lagrangian system,

$$hp = (E + h^2 \Omega)^{1/2} \tilde{x} - x(1 - h^2 \sigma)^{1/2}, \tag{212}$$

$$h\tilde{p} = \tilde{x}(1 - h^2 \sigma)^{1/2} - (E + h^2 \Omega)^{1/2} x, \tag{213}$$

where

$$\sigma = \langle \tilde{x}, \tilde{x} \rangle + \langle x, x \rangle. \tag{214}$$

The Lagrange function generating these equations of motion is equal to

$$h\mathrm{L}_0(\tilde{x}, x) = -\left\langle (E + h^2 \Omega)^{1/2} \tilde{x}, x \right\rangle - \frac{1}{3h^2}(1 - h^2 \sigma)^{3/2}. \tag{215}$$

The Newtonian form of the equations of motion is

$$(1 + h^2\omega_k)^{1/2}(\widetilde{x}_k + \underset{\sim}{x}_k) = x_k\left((1 - h^2\sigma)^{1/2} + (1 - h^2\underset{\sim}{\sigma})^{1/2}\right). \qquad (216)$$

Theorem 13.2. *Equations of motion* (212), (213) *are equivalent to the following matrix factorizations,*

$$I + h^2\mathcal{L} = I + h^2\mathcal{L}(x, p; \lambda) = M^{\mathrm{T}}(\widetilde{x}, x; -\lambda)M(\widetilde{x}, x; \lambda), \qquad (217)$$

$$I + h^2\widetilde{\mathcal{L}} = I + h^2\mathcal{L}(\widetilde{x}, \widetilde{p}; \lambda) = M(\widetilde{x}, x; \lambda)M^{\mathrm{T}}(\widetilde{x}, x; -\lambda), \qquad (218)$$

with the matrix $\mathcal{L}$ defined as in (207), *and*

$$M(\widetilde{x}, x; \lambda) = \begin{pmatrix} (E + h^2\Omega)^{1/2} & h\widetilde{x} \\ -hx^{\mathrm{T}} & (1 - h^2\sigma)^{1/2} - h\lambda \end{pmatrix}. \qquad (219)$$

As a consequence, the discrete–time evolution of the matrix $\mathcal{L}$ induced by (212), (213) *is isospectral. Also, equations of motion* (212), (213) *are equivalent to the 2×2 discrete–time Lax equation,*

$$\widetilde{\Lambda}\mathrm{P} = \mathrm{P}\Lambda, \qquad (220)$$

with the matrix $\Lambda(x, p; z)$ defined in (210), *and*

$$\mathrm{P} = \mathrm{P}(\widetilde{x}, x; z) = \begin{pmatrix} (1 - h^2\sigma)^{1/2} & hz + h\sigma \\ -h & (1 - h^2\sigma)^{1/2} \end{pmatrix}. \qquad (221)$$

13.3 Explicit Discretization of the Garnier System

There exists an alternative, explicit discretization of the Garnier system. Let $C = \mathrm{diag}(c_1, \ldots, c_N)$. Consider the system of second–order difference equations,

$$\widetilde{x} + \underset{\sim}{x} = \frac{2Cx}{1 + \langle Cx, x \rangle}. \qquad (222)$$

In the form of first–order difference equations,

$$\widetilde{y} = x, \quad \widetilde{x} = -y + \frac{2Cx}{1 + \langle Cx, x \rangle}, \qquad (223)$$

this system defines a symplectic map of $\mathbb{R}^{2N}(x, y)$ onto itself, provided the symplectic structure on this space is defined by means of the canonical Poisson bracket,

$$\{x_k, y_j\} = \delta_{kj}. \qquad (224)$$

To achieve the continuous limit, re-scale $x \to hx$, which leads to the equation

$$\widetilde{x} + \underset{\sim}{x} = \frac{2Cx}{1 + h^2 \langle Cx, x \rangle} \, ,$$

and then set $c_k = 1 - h^2 \omega_k / 2 + o(h^2)$, and finally let $h \to 0$.

It can be shown that, if all c_k's are distinct, the full set of involutive integrals of the discrete–time Garnier system (222) is given by N functions,

$$F_k = c_k^{-1}(x_k^2 + y_k^2) - 2x_k y_k + x_k^2 y_k^2$$
$$+ \sum_{j \neq k} \frac{c_k c_j}{c_k^2 - c_j^2} (x_k^2 y_j^2 + y_k^2 x_j^2) - \sum_{j \neq k} \frac{2c_j^2}{c_k^2 - c_j^2} x_k y_k x_j y_j \, , \quad 1 \leq k \leq N. \qquad (225)$$

Theorem 13.3. *Difference equations* (223) *are equivalent to the matrix equation,*

$$M\widetilde{L} = LM \qquad (226)$$

with $(N + 1) \times (N + 1)$ matrices depending on the spectral parameter λ,

$$L = L(x, y; \lambda) = \begin{pmatrix} \lambda^{-2} C^{-2} - E + xy^T & \lambda x - \lambda^{-1} C^{-1} y \\ \lambda^{-1} x^T C^{-1} - \lambda y^T & -\lambda^2 + 1 - \langle x, y \rangle \end{pmatrix}, \qquad (227)$$

$$M = M(x; \lambda) = \begin{pmatrix} \lambda^{-1} C^{-1} & -x \\ x^T & \lambda \end{pmatrix}. \qquad (228)$$

Also, difference equations (223) *are equivalent to the matrix equation,*

$$\widetilde{L}M = ML \, , \qquad (229)$$

with 2×2 matrices depending on the spectral parameter u,

$$L = L(x, y; u) = \begin{pmatrix} u^{-1} + u\Phi_u(x, y) & 1 + \Phi_u(x, C^{-1} x) \\ -1 - \Phi_u(y, C^{-1} y) & -u^{-1} - u\Phi_u(x, y) \end{pmatrix}, \qquad (230)$$

$$M = M(x; u) = \begin{pmatrix} \dfrac{2}{1 + \langle Cx, x \rangle} & u \\ -u & 0 \end{pmatrix}. \qquad (231)$$

14 Multi–dimensional Euler Top

14.1 Continuous–Time Dynamics

A natural configuration space for problems related to multidimensional analogues of the rigid body dynamics is the Lie group

$$G = \mathrm{SO}(n) \,,$$

with Lie algebra

$$\mathfrak{g} = \mathrm{so}(n) \,.$$

The "physical" rigid body corresponds to $n = 3$. Indeed, let $B(t) \subset \mathbb{R}^n$ be a rigid body with a fixed point $0 \in \mathbb{R}^n$. If $E(t) = (e_1(t), \dots, e_n(t))$ stands for the time evolution of a certain orthonormal frame attached to the moving body, called the body frame (all $e_k \in \mathbb{R}^n$), then

$$E(t) = g(t)E(0) \quad \Leftrightarrow \quad e_k(t) = g(t)e_k(0) \quad (1 \le k \le n) \,,$$

with some $g(t) \in G$. Let us derive the Lagrange function of a free rigid body. Consider some material point in the rigid body, and denote its trajectory by $x = x(t) \in \mathbb{R}^n$. We will write $x = (x_1, \dots, x_n)^{\mathrm{T}}$ for the position of the material point with respect to the rest frame. We assume that the coordinates $x_k = x_k(t)$ are calculated with respect to the frame $E(0)$, i.e.,

$$x(t) = \sum_{k=1}^{n} x_k(t)e_k(0) \,.$$

The rigidity of the body means that the coordinates of this vector with respect to the body frame $E(t)$ do not change with time,

$$x(t) = \sum_{k=1}^{n} X_k e_k(t) \,.$$

We will also denote by $X = (X_1, \dots, X_n)^{\mathrm{T}}$ the position of the corresponding material point with respect to the moving frame. Obviously,

$$x(t) = g(t)X \,.$$

Supposing that the density of the body at the point $x(t) = g(t)X$ is equal to $\rho(X)$, we can calculate the kinetic energy of the free rigid body, which will be simultaneously its Lagrange function,

$$\mathbf{L} = \frac{1}{2} \int_{B(t)} \langle \dot{x}, \dot{x} \rangle \rho(g^{-1}x) \mathrm{d}^n x \,.$$

Here $\langle \cdot, \cdot \rangle$ stands for the standard Euclidean scalar product in $\mathbb{R}^n$. Performing an orthogonal change of variables $x = gX$, we obtain

$$\mathbf{L} = \frac{1}{2} \int_{B(0)} \langle \dot{g}X, \dot{g}X \rangle \rho(X) \mathrm{d}^n X = \frac{1}{2} \int_{B(0)} \langle \Omega X, \Omega X \rangle \rho(X) \mathrm{d}^n X,$$

where

$$\Omega = g^{-1}\dot{g} \in \mathfrak{g} \tag{232}$$

is the matrix of the angular velocity of the rigid body in the body frame $E(t)$. Obviously, due to the skew–symmetry of Ω, the previous integral is equal to

$$\mathbf{L} = -\frac{1}{2}\operatorname{tr}(J\Omega^2),$$

where $J = (J_{jk})_{j,k=1}^n$ is a symmetric matrix with entries

$$J_{jk} = \int_{B(0)} X_j X_k \rho(X)\mathrm{d}^n X, \tag{233}$$

the so called inertia tensor of the rigid body. It is easy to understand that changing the initial frame leads to a conjugation of J by the corresponding orthogonal matrix. Hence, choosing the frame $E(0)$ properly, we can assure the matrix J to be diagonal,

$$J = \operatorname{diag}(J_1, \dots, J_n), \tag{234}$$

which will be supposed from now on. Let us introduce the scalar product on $\mathfrak{g}$ as

$$\langle \xi, \eta \rangle = -\frac{1}{2}\operatorname{tr}(\xi\eta), \quad \xi, \eta \in \mathfrak{g}.$$

This scalar product is used also to identify $\mathfrak{g}^*$ with $\mathfrak{g}$, so that the previous formula can also be considered as a pairing between the elements $\xi \in \mathfrak{g}$ and $\eta \in \mathfrak{g}^*$. Introduce, further, the symmetric operator $\mathcal{J} : so(n) \to so(n)$ acting as

$$\mathcal{J}(\Omega) = J\Omega + \Omega J. \tag{235}$$

In these notations we finally obtain the Lagrange function of a free rigid body rotating about a fixed point,

$$\mathbf{L}(g, \dot{g}) = \mathcal{L}^{(l)}(\Omega) = -\frac{1}{2}\operatorname{tr}(J\Omega^2) = \frac{1}{2}\langle \Omega, \mathcal{J}(\Omega) \rangle. \tag{236}$$

This function depends on $\Omega = g^{-1}\dot{g}$ alone, and is therefore obviously invariant under the left action of G. This allows us to apply Theorem 10.1, which delivers the following equations of motion,

$$\dot{M} = [M, \Omega], \tag{237}$$

where

$$M = \nabla_\Omega \mathcal{L}^{(l)} = \mathcal{J}(\Omega) \quad \Leftrightarrow \quad M_{jk} = (J_j + J_k)\Omega_{jk}. \tag{238}$$

According to the general theory, the system (237) is Hamiltonian on the dual $\mathfrak{g}^*$, with the Hamilton function

$$H(M) = \frac{1}{2}\langle M, \Omega \rangle = \frac{1}{2}\sum_{j<k}\frac{M_{jk}^2}{J_j + J_k}. \tag{239}$$

Recall that the Lie–Poisson bracket on $\mathfrak{g}^*$ is defined as

$$\{\varphi_1, \varphi_2\}(M) = \Big\langle M, [\nabla\varphi_1(M), \nabla\varphi_2(M)] \Big\rangle, \tag{240}$$

or, in coordinates,

$$\{M_{ij}, M_{k\ell}\} = M_{i\ell}\delta_{jk} - M_{kj}\delta_{\ell i} - M_{ik}\delta_{j\ell} + M_{\ell j}\delta_{ki}. \tag{241}$$

The generic orbits in this Poisson phase space have dimension $n(n-1)/2 - [n/2]$, since the dimension of $\mathfrak{g}^*$ is equal to $n(n-1)/2$, and the $[n/2]$ spectral invariants of $M \in \mathfrak{g}^*$ are Casimir functions of the Poisson bracket. In particular, for the "physical" case, $n = 3$, the dimension of the generic orbit is equal to 2, and in the case $n = 4$ this dimension is equal to 4. The number of independent involutive integrals of motion necessary for complete integrability is equal to one–half of the above dimension, so that any Lie–Poisson Hamiltonian system on so(3) is integrable, while for the integrability of a Lie–Poisson Hamiltonian system on so(4) one additional integral, independent of and involutive with the Hamilton function, is necessary.

Theorem 14.1. *Equations of motion* (237) *with* (238) *are equivalent to*

$$\dot{L}(\lambda) = [L(\lambda), B(\lambda)], \tag{242}$$

where

$$L(\lambda) = M + \lambda J^2, \tag{243}$$
$$B(\lambda) = \Omega + \lambda J. \tag{244}$$

The spectral invariants of the matrix $L(\lambda)$ provide us with the necessary number of independent integrals of motion, and their involutivity follows from the general r–matrix theory.

14.2 Discrete–Time Euler Top

To find a discrete analogue of the Lagrange function (236), we rewrite the latter once more as

$$\mathbf{L}(g, \dot{g}) = \frac{1}{2}\operatorname{tr}(\dot{g}J\dot{g}^\mathrm{T}). \tag{245}$$

Introduce the following discrete analogue,

$$\mathbb{L}(g_k, g_{k+1}) = \frac{1}{2h} \operatorname{tr}\left((g_{k+1} - g_k)J(g_{k+1} - g_k)^{\mathrm{T}}\right). \tag{246}$$

This function has the correct asymptotics in the continuous limit $h \to 0$, namely $\mathbb{L}(g_k, g_{k+1}) \approx h\mathbf{L}(g, \dot{g})$, when $g_k = g$ and $g_{k+1} \approx g + h\dot{g}$. Up to an additive constant, function (246) may be rewritten as

$$\mathbb{L}(g_k, g_{k+1}) = -\frac{1}{h} \operatorname{tr}(g_{k+1} J g_k^{\mathrm{T}}). \tag{247}$$

This is representable also in terms of $W_k = g_k^{\mathrm{T}} g_{k+1} \in G$ alone,

$$\mathbb{L}(g_k, g_{k+1}) = \Lambda^{(l)}(W_k) = -\frac{1}{h} \operatorname{tr}(W_k J). \tag{248}$$

(Recall that, in the continuous limit, $W_k \approx I + h\Omega$, so that $W_k^{\mathrm{T}} \approx I - h\Omega$.) Theorem 10.3 leads to the following equations of motion.

Theorem 14.2. *The discrete–time Euler–Lagrange equations for the Lagrange function* (248) *are equivalent to the following system,*

$$\begin{cases} hM_k = W_k J - J W_k^{\mathrm{T}}, \\ hM_{k+1} = J W_k - W_k^{\mathrm{T}} J. \end{cases} \tag{249}$$

The multi–valued map (correspondence) $M_k \to M_{k+1}$ described by (249) *is Poisson with respect to the Lie–Poisson bracket on $\mathfrak{g}^*$.*

The Lax representation of equations (249) is obtained in terms of the following matrices,

$$L_k(\lambda) = I - h^2 \lambda M_k - h^2 \lambda^2 J^2, \tag{250}$$
$$U_k(\lambda) = W_k + h\lambda J, \tag{251}$$

so that

$$U_k^{\mathrm{T}}(-\lambda) = W_k^{\mathrm{T}} - h\lambda J. \tag{252}$$

Theorem 14.3. *Equations of motion* (249) *are equivalent to the following matrix factorizations,*

$$\begin{cases} L_k(\lambda) = U_k(\lambda) U_k^{\mathrm{T}}(-\lambda), \\ L_{k+1}(\lambda) = U_k^{\mathrm{T}}(-\lambda) U_k(\lambda). \end{cases} \tag{253}$$

In particular, the matrix $L_k(\lambda)$ remains isospectral in the discrete–time evolution described by equations (249),

$$L_{k+1}(\lambda) = U_k^{-1}(\lambda) L_k(\lambda) U_k(\lambda).$$

Since $L_k(\lambda) = I - h^2 \lambda \mathcal{L}_k(\lambda)$, where

$$L_k(\lambda) = M_k + \lambda J^2,$$

an expression which formally coincides with (243), we see that the Lax matrix of the continuous–time Euler top undergoes an isospectral evolution also in the discrete–time dynamics. Theorem 14.3 provides us with a complete set of integrals of motion of our discrete–time Lagrangian map, the coefficients of the characteristic polynomial, $\det(L_k(\lambda) - \mu I)$. These integrals of motion clearly coincide with the integrals of motion of the continuous–time problem. In particular, they are in involution.

The definition of the above correspondence (249) depends crucially on the solvability of the first equation in (249) for $W_k \in G$. So, we have to discuss this point carefully. Dropping index k, we rewrite the first equation in (249) as

$$hM = WJ - JW^{\mathrm{T}}. \tag{254}$$

We are looking for a solution $W \in G$, i.e., such that

$$WW^{\mathrm{T}} = W^{\mathrm{T}}W = I. \tag{255}$$

Recall also that equation (254), together with constraint (255), is equivalent to a factorization problem for matrices depending on λ,

$$\boldsymbol{L}(\lambda) = \boldsymbol{U}(\lambda)\boldsymbol{U}^{\mathrm{T}}(-\lambda). \tag{256}$$

Define

$$p(\lambda) = \det(\boldsymbol{L}(\lambda)). \tag{257}$$

Obviously, $p(\lambda)$ is a polynomial of degree $2n$ with real coefficients, and, due to $\boldsymbol{L}(\lambda) = \boldsymbol{L}^{\mathrm{T}}(-\lambda)$, it satisfies

$$p(\lambda) = p(-\lambda). \tag{258}$$

Let Σ denote the set of all roots of $p(\lambda)$. Suppose that $p(\lambda)$ has no roots on the imaginary axis. Then we can find (many) splittings $\Sigma = \Sigma_+ \cup \Sigma_-$ into disjoint sets Σ_+, Σ_- satisfying

$$\Sigma_+ = -\Sigma_-, \qquad \overline{\Sigma}_+ = \Sigma_+, \qquad \overline{\Sigma}_- = \Sigma_-. \tag{258}$$

Such splittings are in a one–to–one correspondence with the factorizations

$$p(\lambda) = f(\lambda)f(-\lambda), \tag{259}$$

where $f(\lambda)$ is a polynomial of degree n with real coefficients such that Σ_+ is the set of the roots of $f(\lambda)$, so that $f(\lambda)$, $f(-\lambda)$ have no common roots.

Theorem 14.4. *To any polynomial factorization* (259) *there corresponds a matrix factorization* (256) *with* $f(\lambda) = \pm \det(\boldsymbol{U}(\lambda))$, *or, equivalently, a solution of matrix equation* (254) *under constraint* (255).

Remark. It is important to notice that, according to Theorem 14.3, the set Σ does not depend on k, and its splitting $\Sigma = \Sigma_+ \cup \Sigma_-$ can also be chosen independently of k.

15 Rigid Body in a Quadratic Potential

15.1 Continuous–Time Dynamics

Now we extend the considerations of Sect. 14.1 by including an arbitrary quadratic potential acting on the rigid body. To find the Lagrange function of a rigid body rotating about a fixed point $0 \in \mathbb{R}^n$ in a field with a quadratic potential,

$$\varphi(x) = -\frac{1}{2} \sum_{i,j=1}^{n} a_{ij} x_i x_j \,, \tag{260}$$

we have to calculate the body's kinetic and potential energy. Clearly, the kinetic energy coincides with the Lagrange function of a free rigid body. The potential energy is equal to

$$\int_{B(t)} \varphi(x) \rho(g^{-1}x) \mathrm{d}^n x = -\frac{1}{2} \sum_{i,j=1}^{n} a_{ij} \int_{B(t)} x_i x_j \rho(g^{-1}x) \mathrm{d}^n x \,.$$

Performing an orthogonal change of variables, $x = gX$, we find the following expression for the potential energy,

$$-\frac{1}{2} \sum_{i,j=1}^{n} a_{ij} \int_{B(0)} (gX)_i (gX)_j \rho(X) \mathrm{d}^n X =$$

$$-\frac{1}{2} \sum_{i,j=1}^{n} \sum_{\ell,m=1}^{n} a_{ij} g_{i\ell} g_{jm} \int_{B(0)} X_\ell X_m \rho(X) \mathrm{d}^n X.$$

Recalling definition (233), we write the potential energy as

$$-\frac{1}{2} \sum_{i,j=1}^{n} \sum_{\ell,m=1}^{n} a_{ij} g_{i\ell} g_{jm} J_{\ell m} = -\frac{1}{2} \operatorname{tr}(A g J g^{\mathrm{T}}) \,.$$

Here to the notations used above we add $A = (a_{ij})_{i,j=1}^{n}$, the symmetric matrix of coefficients of the quadratic form, $\varphi(x)$. So, finally, the Lagrange function of a rigid body rotating about a fixed point in a field with a quadratic potential is equal to

$$\mathbf{L}(g, \dot{g}) = -\frac{1}{2} \operatorname{tr}(\Omega J \Omega) + \frac{1}{2} \operatorname{tr}(A g J g^{\mathrm{T}}) \,. \tag{261}$$

To include this Lagrange function in the framework of Sect. 11, we make the following identifications:

- $V = \operatorname{Symm}(n)$, the linear space of all $n \times n$ symmetric matrices; we identify V^* with V via the following scalar product on V,

$$\langle v_1, v_2 \rangle = \frac{1}{2} \operatorname{tr}(v_1 v_2) \,, \quad v_1, v_2 \in V.$$

- The representation Φ of G on V is defined as

$$\Phi(g) \cdot v = gvg^{-1} = gvg^{\mathrm{T}} \quad \text{for} \quad g \in G, \ v \in V.$$

- Therefore the representation ϕ of $\mathfrak{g}$ on V is given by

$$\phi(\xi) \cdot v = [\xi, v] \quad \text{for} \quad \xi \in \mathfrak{g}, \ v \in V,$$

while the anti–representation ϕ^* of $\mathfrak{g}$ in V^* is given by

$$\phi^*(\xi) \cdot y = -[\xi, y] \quad \text{for} \quad \xi \in \mathfrak{g}, \ y \in V^*,$$

- Finally, the bilinear operation $\diamond : V^* \times V \to \mathfrak{g}^*$ is given by

$$y \diamond v = -[y, v] \quad \text{for} \quad y \in V^*, \ v \in V.$$

Now defining $P = g^{\mathrm{T}} A g = \Phi(g^{-1}) \cdot A$, we represent (261) in the form

$$\mathbf{L}(g, \dot{g}) = \mathcal{L}^{(l)}(\Omega, P) = -\frac{1}{2} \operatorname{tr}(\Omega J \Omega) + \frac{1}{2} \operatorname{tr}(JP) = \frac{1}{2} \langle \Omega, \mathcal{J}(\Omega) \rangle + \langle J, P \rangle, \tag{262}$$

which is manifestly invariant under the left action of the isotropy subgroup $G^{[A]}$. Now Theorem 11.1 is applicable, which yields the following equations of motion,

$$\begin{cases} \dot{M} = [M, \Omega] + [P, J], \\ \dot{P} = [P, \Omega], \end{cases} \tag{263}$$

where M is given by the formula

$$M = \nabla_\Omega \mathcal{L}^{(l)} = \mathcal{J}(\Omega) \quad \Leftrightarrow \quad M_{jk} = (J_j + J_k)\Omega_{jk}, \tag{264}$$

identical with (238) for the Euler top. According to the general theory, the system (263) is Hamiltonian on the dual of the semidirect product Lie algebra $\mathfrak{g} \ltimes V^*$, with the Hamilton function

$$H(M, P) = \frac{1}{2} \langle M, \Omega \rangle - \langle J, P \rangle = \frac{1}{2} \sum_{j<k} \frac{M_{jk}^2}{J_j + J_k} - \frac{1}{2} \sum_{k=1}^n J_k P_{kk}. \tag{265}$$

The Lie–Poisson bracket on $\mathfrak{g}^* \times V$ is given by:

$$\{\varphi_1, \varphi_2\} = \Big\langle M, [\nabla_M \varphi_1, \nabla_M \varphi_2] \Big\rangle + \Big\langle P, [\nabla_P \varphi_1, \nabla_M \varphi_2] - [\nabla_P \varphi_2, \nabla_M \varphi_1] \Big\rangle, \tag{266}$$

or, in coordinates,

$$\{M_{ij}, M_{k\ell}\} = M_{i\ell}\delta_{jk} - M_{kj}\delta_{\ell i} - M_{ik}\delta_{j\ell} + M_{\ell j}\delta_{ki}, \tag{267}$$

$$\{M_{ij}, P_{k\ell}\} = P_{i\ell}\delta_{jk} - P_{kj}\delta_{\ell i} + P_{ik}\delta_{j\ell} - P_{\ell j}\delta_{ki}, \tag{268}$$

Generic orbits in this Poisson phase space have dimension $n^2 - n$ (the dimension of $\mathfrak{g}^* \times V$ is equal to $n(n-1)/2 + n(n+1)/2 = n^2$; the n spectral invariants of $P \in V$ are Casimir functions of the Poisson bracket). In particular, for the "physical" case, $n = 3$, the dimension of the generic orbit is equal to 6. Therefore the number of independent involutive integrals of motion necessary for complete integrability is equal to $n(n-1)/2$ (equal to 3 for $n = 3$).

Theorem 15.1. *Equations of motion* (263) *with* (264) *are equivalent to*

$$\dot{L}(\lambda) = [L(\lambda), B(\lambda)], \tag{269}$$

where

$$L(\lambda) = \lambda^{-1}P + M + \lambda J^2, \tag{270}$$
$$B(\lambda) = \Omega + \lambda J. \tag{271}$$

The spectral invariants of the matrix $L(\lambda)$ provide us with the necessary number of independent integrals of motion. Their involutivity follows from the r–matrix interpretation of the above Lax equation.

15.2 Discrete–Time Top in a Quadratic Potential

To find a discrete analogue of the Lagrange function (261), we rewrite the latter once more as

$$\mathbf{L}(g, \dot{g}) = \frac{1}{2}\operatorname{tr}(\dot{g}J\dot{g}^{\mathrm{T}}) + \frac{1}{2}\operatorname{tr}(gJg^{\mathrm{T}}A). \tag{272}$$

Introduce the following discrete analogue,

$$\mathbb{L}(g_k, g_{k+1}) = \frac{1}{2h}\operatorname{tr}\left((g_{k+1} - g_k)J(g_{k+1} - g_k)^{\mathrm{T}}\right) + \frac{h}{2}\operatorname{tr}(g_{k+1}Jg_k^{\mathrm{T}}A). \tag{273}$$

The powers of h are introduced in a way that assures the correct limit when $h \to 0$, namely $\mathbb{L}(g_k, g_{k+1}) \approx h\mathbf{L}(g, \dot{g})$, if $g_k = g$ and $g_{k+1} \approx g + h\dot{g}$. Up to an additive constant, function (273) may be rewritten as

$$\mathbb{L}(g_k, g_{k+1}) = -\frac{1}{h}\operatorname{tr}(g_{k+1}Jg_k^{\mathrm{T}}) + \frac{h}{2}\operatorname{tr}(g_{k+1}Jg_k^{\mathrm{T}}A). \tag{274}$$

This is representable also in terms of $W_k = g_k^{\mathrm{T}}g_{k+1} \in G$ and $P_k = g_k^{\mathrm{T}}Ag_k \in O_A \subset V$,

$$\mathbb{L}(g_k, g_{k+1}) = \Lambda^{(l)}(W_k, P_k) = -\frac{1}{h}\operatorname{tr}(W_k J) + \frac{h}{2}\operatorname{tr}(W_k JP_k). \tag{275}$$

Recall that, in the continuous limit, $W_k \approx I + h\Omega$, and $W_k^{\mathrm{T}} \approx I - h\Omega$. Theorem 11.1 allows us to come to the following conclusions.

Theorem 15.2. *Discrete–time Euler–Lagrange equations for the Lagrange function* (275) *are equivalent to the following system,*

$$
\begin{cases}
M_k = \dfrac{1}{h}(W_k J - J W_k^{\mathrm{T}}) - \dfrac{h}{2}(P_k W_k J - J W_k^{\mathrm{T}} P_k)\,, \\[2mm]
M_{k+1} = \dfrac{1}{h}(J W_k - W_k^{\mathrm{T}} J) - \dfrac{h}{2}(J P_k W_k - W_k^{\mathrm{T}} P_k J)\,, \\[2mm]
P_{k+1} = W_k^{\mathrm{T}} P_k W_k\,.
\end{cases}
\tag{276}
$$

The multivalued map (correspondence) $(M_k, P_k) \to (M_{k+1}, P_{k+1})$ *described by* (276) *is Poisson with respect to bracket* (266).

Map (276) admits a Lax representation with matrices depending on a spectral parameter λ,

$$
\begin{aligned}
\boldsymbol{L}_k(\lambda) &= \left(I - \frac{h^2}{2} P_k\right)^2 - h^2 \lambda M_k - h^2 \lambda^2 J^2 \\
&= I - h^2\left(P_k + \lambda M_k + \lambda^2 J^2\right) + \frac{h^4}{4} P_k^2\,,
\end{aligned}
\tag{277}
$$

$$
\boldsymbol{U}_k(\lambda) = \left(I - \frac{h^2}{2} P_k\right) W_k + h \lambda J,
\tag{278}
$$

$$
\boldsymbol{U}_k^{\mathrm{T}}(-\lambda) = W_k^{\mathrm{T}}\left(I - \frac{h^2}{2} P_k\right) - h \lambda J.
\tag{279}
$$

Theorem 15.3. *Equations of motion* (276) *are equivalent to the following matrix factorizations,*

$$
\begin{cases}
\boldsymbol{L}_k(\lambda) = \boldsymbol{U}_k(\lambda)\,\boldsymbol{U}_k^{\mathrm{T}}(-\lambda)\,, \\[1mm]
\boldsymbol{L}_{k+1}(\lambda) = \boldsymbol{U}_k^{\mathrm{T}}(-\lambda)\,\boldsymbol{U}_k(\lambda)\,.
\end{cases}
\tag{280}
$$

In particular, matrix $\boldsymbol{L}_k(\lambda)$ remains isospectral in the discrete–time evolution described by equations (276):

$$
\boldsymbol{L}_{k+1}(\lambda) = \boldsymbol{U}_k^{-1}(\lambda)\boldsymbol{L}_k(\lambda)\boldsymbol{U}_k(\lambda)\,.
$$

Hence, the same holds for the matrix

$$
L_k(\lambda) = \lambda^{-1}\left(P_k - \frac{h^2}{4} P_k^2\right) + M_k + \lambda J^2.
\tag{281}
$$

Theorem 15.3 provides us with a complete set of integrals of motion of our discrete–time Lagrangian system. These are the coefficients of the characteristic polynomial $\det(L_k(\lambda) - \mu I)$. Notice that these integrals of motion do not coincide with the integrals of motion of the continuous–time problem (with the only exception of the free rigid body motion, i.e., $P_k = 0$ considered above). To be more concrete, the integrals of motion of our map are obtained from the integrals of the continuous time problem by replacing P

by $\widehat{P} = P - \frac{1}{4}h^2 P^2$. A simple calculation shows that (268) also holds for the quantities $\widehat{P}_{k\ell}$. Therefore, the integrals of motion of the discrete–time Lagrangian system are in involution with respect to the bracket (267), (268).

The definition of the above correspondence (276) depends crucially on the solvability of the first equation in (276) for $W_k \in G$. This problem is very similar to the one treated in Sect. 14.2. Dropping index k, and introducing the notation

$$U = \left(I - \frac{h^2}{2}P\right)W,$$

we rewrite the first equation in (276) as

$$hM = UJ - JU^{\mathrm{T}}. \tag{282}$$

We are looking for $W \in G$, which means $WW^{\mathrm{T}} = W^{\mathrm{T}}W = I$, hence the solution of (282) for which we are looking has to satisfy the constraint

$$UU^{\mathrm{T}} = B, \tag{283}$$

where, for the sake of brevity, we also use the abbreviation

$$B = \left(I - \frac{h^2}{2}P\right)^2. \tag{284}$$

Recall that equation (282), together with constraint (283), is equivalent to the factorization

$$\boldsymbol{L}(\lambda) = \boldsymbol{U}(\lambda)\boldsymbol{U}^{\mathrm{T}}(-\lambda) \tag{285}$$

with the matrices

$$\boldsymbol{L}(\lambda) = B - h^2\lambda M - h^2\lambda^2 J^2, \tag{286}$$

and

$$\boldsymbol{U}(\lambda) = U + h\lambda J, \qquad \boldsymbol{U}^{\mathrm{T}}(-\lambda) = U^{\mathrm{T}} - h\lambda J. \tag{287}$$

Introducing the polynomial (257), and arguing as at the end of Sect. 14.2, we come to a statement analogous to the one given there.

Theorem 15.4. *To any polynomial factorization* (259) *there corresponds a matrix factorization* (285) *with* $f(\lambda) = \pm \det(\boldsymbol{U}(\lambda))$, *or, equivalently, a solution of matrix equation* (282) *under constraint* (283).

Again, according to Theorem 15.3, the set Σ does not depend on k, and its splitting $\Sigma = \Sigma_+ \cup \Sigma_-$ can also be chosen independently of k.

16 Multi–dimensional Lagrange Top

16.1 Body Frame Formulation

Our next object will be a heavy top, i.e., a rigid body with a fixed point, subject to a gravitational force with a linear potential,

$$\varphi(x) = \langle p, x \rangle \,,$$

where p is some constant vector. Calculating the potential energy of such a body, we find

$$\int_{B(t)} \varphi(x)\rho(g^{-1}x)\mathrm{d}^n x = \int_{B(0)} \varphi(gX)\rho(X)\mathrm{d}^n X = \langle p, gA \rangle \,,$$

where

$$A = \int_{B(0)} X\rho(X)\mathrm{d}^n X \qquad (288)$$

is the vector in $\mathbb{R}^n$ pointing from the fixed point to the center of mass of the rigid body, calculated with respect to the frame moving with the body. Therefore,

$$\mathbf{L}(g, \dot{g}) = \frac{1}{2} \langle \Omega, \mathcal{J}(\Omega) \rangle - \langle p, a \rangle \,. \qquad (289)$$

Here, on the right–hand side,

$$a = gA \qquad (290)$$

is the vector in $\mathbb{R}^n$ pointing from the fixed point to the center of mass of the rigid body, calculated with respect to the rest frame. Introducing

$$P = g^{-1}p \,, \qquad (291)$$

the gravity vector calculated in the moving frame, we can rewrite the above Lagrange function as

$$\mathbf{L}(g, \dot{g}) = \frac{1}{2} \langle \Omega, \mathcal{J}(\Omega) \rangle - \langle P, A \rangle \,. \qquad (292)$$

This function is in the framework of Sect. 11, if the following identifications are made:

- $V = V^* = \mathbb{R}^n$ with the standard Euclidean scalar product;
- The representation Φ of G in V is defined as

$$\Phi(g) \cdot v = gv \quad \text{for} \quad g \in G, \, v \in V.$$

- Therefore the representation ϕ of $\mathfrak{g}$ on V is given by

$$\phi(\xi) \cdot v = \xi v \text{ for } \xi \in \mathfrak{g}, \, v \in V,$$

while the anti–representation ϕ^* of $\mathfrak{g}$ on V^* is given by

$$\phi^*(\xi) \cdot y = -\xi y \text{ for } \xi \in \mathfrak{g}, \, y \in V^* = V.$$

- Finally, the bilinear operation $\diamond : V^* \times V \to \mathfrak{g}^*$ is given by

$$y \diamond v = vy^{\mathrm{T}} - yv^{\mathrm{T}} = v \wedge y \text{ for } y \in V^*, \, v \in V.$$

The Lagrange function (292) is manifestly invariant under the left action of the isotropy subgroup $G^{[p]}$ (rotations about the axis of the gravity field). Theorem 11.1 is applicable, and delivers the following equations of motion of a heavy top in the moving frame,

$$\begin{cases} \dot{M} = [M, \Omega] + A \wedge P, \\ \dot{P} = -\Omega P, \end{cases} \tag{293}$$

where

$$M = \nabla_\Omega \mathcal{L}^{(l)} = \mathcal{J}(\Omega) \quad \Leftrightarrow \quad M_{jk} = (J_j + J_k)\Omega_{jk}. \tag{294}$$

According to the general theory, system (293) is Hamiltonian on the dual of the semidirect product Lie algebra, $e(n) = so(n) \ltimes \mathbb{R}^n$, with the Hamilton function

$$H(M, P) = \frac{1}{2}\langle M, \Omega \rangle + \langle P, A \rangle = \frac{1}{2}\sum_{j<k} \frac{M_{jk}^2}{J_j + J_k} + \sum_{k=1}^{n} P_k A_k. \tag{295}$$

The corresponding invariant Poisson bracket is

$$\{M_{ij}, M_{k\ell}\} = M_{i\ell}\delta_{jk} - M_{kj}\delta_{\ell i} - M_{ik}\delta_{j\ell} + M_{\ell j}\delta_{ki}, \tag{296}$$

$$\{M_{ij}, P_k\} = P_i\delta_{jk} - P_j\delta_{ik}. \tag{297}$$

The **multi–dimensional Lagrange top** is characterized by the following data, $J_1 = J_2 = \ldots = J_{n-1}$, which means that the body is rotationally symmetric with respect to the nth coordinate axis, and $A_1 = A_2 = \ldots = A_{n-1} = 0$, which means that the fixed point lies on the symmetry axis. Choosing units properly, we may assume that

$$J_1 = J_2 = \ldots = J_{n-1} = \frac{\alpha}{2}, \quad J_n = 1 - \frac{\alpha}{2}, \quad A = (0, 0, \ldots, 0, 1)^{\mathrm{T}}. \tag{298}$$

The action of the operator $\mathcal{J}$ is given by

$$M_{ij} = \mathcal{J}(\Omega)_{ij} = \begin{cases} \alpha\Omega_{ij}, \, 1 \leq i, j \leq n-1, \\ \Omega_{ij}, \, i = n \text{ or } j = n, \end{cases} \tag{299}$$

or, in a more invariant fashion,

$$M = \mathcal{J}(\Omega) = \alpha\Omega + (1 - \alpha)(\Omega AA^{\mathrm{T}} + AA^{\mathrm{T}}\Omega) \tag{300}$$

$$= \alpha\Omega - (1 - \alpha)A \wedge (\Omega A). \tag{301}$$

Therefore, in the particular situation of the Lagrange top, (292) may be written as

$$\mathbf{L}(g, \dot{g}) = \frac{\alpha}{2}\langle\Omega, \Omega\rangle + \frac{1 - \alpha}{2}\langle\Omega A, \Omega A\rangle - \langle P, A\rangle. \tag{302}$$

The Legendre transformation (300), (301) is easily invertible. Notice that from (300) there follows immediately

$$MA = \Omega A, \tag{303}$$

and introducing this into the second term of the right–hand side of (301), we find

$$\Omega = \frac{1}{\alpha}M + \frac{1 - \alpha}{\alpha}A \wedge (MA). \tag{304}$$

The integrability of the multi–dimensional Lagrange top follows from its Lax representation in the loop algebra $\mathrm{sl}(n+1)[\lambda, \lambda^{-1}]$ twisted by the Cartan automorphism.

Theorem 16.1. *For the Lagrange top, the moving frame equations (293) are equivalent to the matrix equation,*

$$\dot{L}(\lambda) = [L(\lambda), U(\lambda)], \tag{305}$$

where

$$L(\lambda) = L(M, P; \lambda) = \begin{pmatrix} M & \lambda A - \lambda^{-1}P \\ \lambda A^{\mathrm{T}} - \lambda^{-1}P^{\mathrm{T}} & 0 \end{pmatrix}, \tag{306}$$

$$U(\lambda) = U(M; \lambda) = \begin{pmatrix} \Omega & \lambda A \\ \lambda A^{\mathrm{T}} & 0 \end{pmatrix}. \tag{307}$$

The involutivity of integrals of motion follows, as usual, from the r–matrix interpretation of the previous result.

16.2 Rest Frame Formulation

Integrability is not the only distinctive feature of the Lagrange top. Another one is the existence of a nice Euler–Poincaré description in the rest frame. Rewriting (302) as

$$\mathbf{L}(g, \dot{g}) = \frac{\alpha}{2}\langle\omega, \omega\rangle + \frac{1 - \alpha}{2}\langle\omega a, \omega a\rangle - \langle p, a\rangle, \tag{308}$$

where $\omega = \dot{g}g^{-1}$, we observe that the Lagrange function of the Lagrange top is not only left–invariant with respect to $G^{[p]}$, i.e., rotations about the axis of the gravity field, but also right–invariant with respect to $G^{[A]}$, i.e, rotations about the symmetry axis of the body. Therefore, we may apply (104), which, in the present setup, yields

$$\begin{cases} \dot{m} = [\omega, m] + \nabla_a \mathcal{L}^{(r)} \wedge a\,, \\ \dot{a} = \omega a\,. \end{cases} \tag{309}$$

Straightforward calculations based on (308) yield

$$\nabla_a \mathcal{L}^{(r)} = -(1-\alpha)\omega^2 a - p,$$

$$m = \nabla_\omega \mathcal{L}^{(r)} = \alpha\omega + (1-\alpha)(\omega a a^{\mathrm{T}} + a a^{\mathrm{T}}\omega).$$

The last formula implies, first, that

$$ma = \alpha\omega a + (1-\alpha)\omega a\langle a, a\rangle = \omega a,$$

and, second, that

$$[\omega, m] - (1-\alpha)(\omega^2 a) \wedge a = \Big[\omega, m - (1-\alpha)(\omega a a^{\mathrm{T}} + a a^{\mathrm{T}}\omega)\Big] = [\omega, \alpha\omega] = 0.$$

Introducing these results into (309), we finally arrive at the following nice system,

$$\begin{cases} \dot{m} = a \wedge p\,, \\ \dot{a} = ma\,, \end{cases} \tag{310}$$

where $m = gMg^{-1}$ is the kinetic moment in the rest frame. This is a Hamiltonian system with respect to the opposite of the Lie–Poisson bracket of $e(n)$,

$$\{m_{ij}, m_{k\ell}\} = -m_{i\ell}\delta_{jk} + m_{kj}\delta_{\ell i} + m_{ik}\delta_{j\ell} - m_{\ell j}\delta_{ki}\,, \tag{311}$$

$$\{m_{ij}, a_k\} = -a_i\delta_{jk} + a_j\delta_{ik}\,, \tag{312}$$

with the Hamilton function

$$H(m, a) = \frac{1}{2}\langle m, m\rangle + \langle a, p\rangle. \tag{313}$$

Remarkable in the system (310) is its independence of the anisotropy parameter α.

Theorem 16.2. *The rest frame equations (310) of the Lagrange top are equivalent to the matrix equation*

$$\dot{\ell}(\lambda) = [\ell(\lambda), u(\lambda)], \tag{314}$$

where

$$\ell(\lambda) = \ell(m, a; \lambda) = \begin{pmatrix} m & \lambda a - \lambda^{-1} p \\ \lambda a^{\mathrm{T}} - \lambda^{-1} p^{\mathrm{T}} & 0 \end{pmatrix}, \tag{315}$$

$$u(\lambda) = u(a; \lambda) = \begin{pmatrix} 0 & \lambda a \\ \lambda a^{\mathrm{T}} & 0 \end{pmatrix}. \tag{316}$$

16.3 Discrete–Time Analogue of the Lagrange Top: Rest Frame Formulation

Consider the following discrete analogue of the Lagrange function (308),

$$\mathbb{L}(g_k, g_{k+1}) =$$
$$-\frac{\alpha}{h} \operatorname{tr} \log \left(2I + w_k + w_k^{-1}\right) - \frac{2(1-\alpha)}{h} \log\left(1 + \langle a_k, a_{k+1}\rangle\right) - h\langle p, a_k\rangle,$$
$$(317)$$

where w_k, a_k are defined by

$$w_k = g_{k+1} g_k^{-1}, \quad a_k = g_k A, \quad \text{so that} \quad a_{k+1} = w_k a_k.$$

The powers of h are introduced in a way assuring the correct limit when $h \to 0$, namely $\mathbb{L}(g_k, g_{k+1}) \approx h\mathbf{L}(g, \dot{g})$, if $g_k = g$ and $g_{k+1} \approx g + h\dot{g}$. This is seen with the help of the following simple lemma.

Lemma 16.1. *Let* $w(h) = I + h\omega + O(h^2) \in \mathrm{SO}(n)$ *be a smooth curve,* $\omega \in \mathrm{so}(n)$. *Then*

$$\operatorname{tr} \log \left(2I + w(h) + w^{-1}(h)\right) = \text{const} - \frac{h^2}{2} \langle \omega, \omega \rangle + O(h^3). \qquad (318)$$

For an arbitrary $a \in \mathbb{R}^n$,

$$\langle a, w(h)a \rangle = \langle a, a \rangle - \frac{h^2}{2} \langle \omega a, \omega a \rangle + O(h^3). \qquad (319)$$

Substituting $w_k a_k$ for a_{k+1} in (317), we find the Lagrange function $\Lambda^{(r)}(a_k, w_k) = \mathbb{L}(g_k, g_{k+1})$ depending only on a_k, w_k, hence invariant with respect to the right action of $G^{[A]}$. It may be reduced following the procedure of Sect. 11.

Theorem 16.3. *The Euler–Lagrange equations of motion for the Lagrange function (317) are equivalent to the following system:*

$$\begin{cases} m_{k+1} = m_k + ha_k \wedge p, \\ a_{k+1} = \dfrac{I + (h/2)m_{k+1}}{I - (h/2)m_{k+1}} a_k. \end{cases} \qquad (320)$$

The map $(m_k, a_k) \to (m_{k+1}, a_{k+1})$ *is Poisson with respect to bracket (311), (312), and the following function is an integral of motion of this map,*

$$H_h(m, a) = \frac{1}{2}\langle m, m \rangle + \langle a, p \rangle - \frac{h}{2}\langle m, p \wedge a \rangle. \qquad (321)$$

An amazing feature of this result is that (320) is a genuine map, and not a multi–valued correspondence, like the discretizations of the Euler top and the top in a quadratic potential. It would be highly desirable to be able to distinguish between these two situations by just looking at the discrete Lagrange function, but at present we do not have methods for solving this problem.

Actually, map (320) possesses not only integral (321) but the full set of involutive integrals necessary for the complete integrability. The most direct way to the proof of this statement is, as usual, through the Lax representation which lives, just as in the continuous time situation, in the loop algebra $sl(n+1)[\lambda, \lambda^{-1}]$ twisted by the Cartan automorphism.

Theorem 16.4. *Map* (320) *admits the following Lax representation,*

$$\ell_{k+1}(\lambda) = v_k^{-1}(\lambda)\ell_k(\lambda)v_k(\lambda), \tag{322}$$

with the matrices

$$\ell_k(\lambda) = \begin{pmatrix} m_k & \lambda b_k - \lambda^{-1}p \\ \lambda b_k^{\mathrm{T}} - \lambda^{-1}p^{\mathrm{T}} & 0 \end{pmatrix}, \tag{323}$$

$$v_k(\lambda) = \frac{I + (h/2)u_k(\lambda)}{I - (h/2)u_k(\lambda)}, \quad u_k(\lambda) = \begin{pmatrix} 0 & \lambda a_k \\ \lambda a_k^{\mathrm{T}} & 0 \end{pmatrix}, \tag{324}$$

where the following abbreviation is used,

$$b_k = \left(I - \frac{h}{2}m_k\right)a_k + \frac{h^2}{4}p. \tag{325}$$

Thus, we get a complete set of involutive integrals of motion for our discrete–time Lagrangian map, as the coefficients of the characteristic polynomial $\det(\ell_k(\lambda) - \mu I)$. Observe that the integrals of map (320) are obtained from the integrals of the continuous–time problem by the replacement of a by $b = a + O(h)$ given in (325). Therefore, the former are $O(h)$–perturbations of the latter.

16.4 Discrete–Time Analogue of the Lagrange Top: Moving Frame Formulation

It turns out that the equations of the discrete–time Lagrange top in the moving frame formulation become a bit nicer under a modest change of the Lagrange function (317), namely the replacement of $\langle p, a_k \rangle$ in the last term on the right–hand side by $\langle p, a_{k+1} \rangle$. This modification does not influence the discrete action functional, $\mathbb{S} = \sum \mathbb{L}(g_k, g_{k+1})$, apart from the boundary terms. On the other hand, it is not difficult to see that this modification is equivalent to exchanging $w_k \leftrightarrow w_k^{-1}$, $a_k \leftrightarrow a_{k+1}$, which in turn is equivalent to considering the evolution backwards in time with the simultaneous change

$h \to -h$. Now express the discrete Lagrange function (317) with the above modification in terms of $P_k = g_k^{-1} p$ and $W_k = g_k^{-1} g_{k+1}$,

$$\mathbb{L}(g_k, g_{k+1}) = \Lambda^{(l)}(P_k, W_k) =$$
$$= -\frac{\alpha}{h} \operatorname{tr} \log \left(2I + W_k + W_k^{-1}\right) - \frac{2(1-\alpha)}{h} \log \left(1 + \langle A, W_k A \rangle\right)$$
$$-h \langle P_k, W_k A \rangle. \tag{326}$$

Since $W_k = I + h\Omega + O(h^2)$, we can apply Lemma 16.1 to show that

$$\Lambda^{(l)}(P_k, W_k) = h\mathcal{L}^{(l)}(P, \Omega) + O(h^2),$$

where $\mathcal{L}^{(l)}(P, \Omega)$ is the Lagrange function (302) of the continuous–time Lagrange top. Now, one can derive all the results concerning the discrete–time Lagrange top in the body frame from those for the rest frame by performing the change of frames so that

$$M_k = g_k^{-1} m_k g_k, \quad P_k = g_k^{-1} p, \quad A = g_k^{-1} a_k,$$

and taking into account the modification mentioned above. Alternatively, one can derive them independently from, and similarly to, the rest frame results, applying Theorem 11.3. Anyway, the corresponding results are

Theorem 16.5. *The Euler–Lagrange equations for Lagrange function (326) are equivalent to the following system,*

$$\begin{cases} M_{k+1} = W_k^{-1} M_k W_k + hA \wedge P_{k+1}, \\ P_{k+1} = W_k^{-1} P_k, \end{cases} \tag{327}$$

where the "angular velocity" $W_k \in \mathrm{SO}(n)$ is related to the "angular momentum" $M_k \in \mathrm{so}(n)$ by the Legendre transformation,

$$M_k = \frac{2\alpha}{h} \cdot \frac{W_k - I}{W_k + I} - \frac{2(1-\alpha)}{h} \cdot \frac{A \wedge (W_k A)}{1 + \langle A, W_k A \rangle}. \tag{328}$$

Map (327), (328) is Poisson with respect to bracket (296), (297), and has a complete set of involutive integrals assuring its complete integrability. One of these integrals is

$$\bar{H}_h(M, P) = \frac{1}{2} \langle M, M \rangle + \langle P, A \rangle + \frac{h}{2} \langle M, P \wedge A \rangle. \tag{329}$$

We close this section with a Lax representation of map (327), (328).

Theorem 16.6. *Map (327), (328) has the following Lax representation,*

$$L_{k+1}(\lambda) = V_k^{-1}(\lambda) L_k(\lambda) V_k(\lambda), \tag{330}$$

with the matrices

$$L_k(\lambda) = \begin{pmatrix} M_k & \lambda B_k - \lambda^{-1} P_k \\ \lambda B_k^{\mathrm{T}} - \lambda^{-1} P_k^{\mathrm{T}} & 0 \end{pmatrix}, \tag{331}$$

$$V_k(\lambda) = \begin{pmatrix} W_k & 0 \\ 0 & 1 \end{pmatrix} \frac{I + (h/2)U(\lambda)}{I - (h/2)U(\lambda)}, \quad U(\lambda) = \begin{pmatrix} 0 & \lambda A \\ \lambda A^{\mathrm{T}} & 0 \end{pmatrix}, \tag{332}$$

where the following abbreviation is used,

$$B_k = \left(I + \frac{h}{2} M_k\right) A + \frac{h^2}{4} P_k. \tag{333}$$

17 Rigid Body Motion in an Ideal Fluid: The Clebsch Case

17.1 Continuous–Time Dynamics

We next turn to the problem of the motion of an n–dimensional rigid body in an ideal fluid. This problem is traditionally described by a Hamiltonian system on e$^*(n)$ with the Hamilton function

$$H(M, Y) = \frac{1}{2} \sum_{j<k} c_{jk}^{-1} M_{jk}^2 - \frac{1}{2} \sum_{k=1}^{n} b_k Y_k^2. \tag{334}$$

Here $(M, Y) \in$ e$^*(n)$, so that $M \in$ so(n), $Y \in \mathbb{R}^n$, and $C = \{c_{jk}\}_{j,k=1}^n$, $B = \mathrm{diag}(b_k)$ are symmetric matrices. The physical meaning of M and Y is the total angular momentum and the total linear momentum, respectively, of the system "rigid body plus fluid". The equations of motion (*Kirchhoff equations*) are

$$\begin{cases} \dot{M} = [M, \Omega] + Y \wedge (BY), \\ \dot{Y} = -\Omega Y, \end{cases} \tag{335}$$

where the matrix $\Omega \in$ so(n) and the linear operator $\mathcal{J}$: so$(n) \to$ so(n) are defined by the formula

$$M_{jk} = \mathcal{J}(\Omega)_{jk} = c_{jk} \Omega_{jk}. \tag{336}$$

The "physical" Lagrangian formulation of this problem involves a Lagrange function on the group E(n), that is left–invariant under the action of a whole group. However, a Lagrangian formulation may be also obtained in the framework of Sect. 11, from a Lagrange function on SO(n), that is left–invariant under the action of the isotropy subgroup of some element $y \in \mathbb{R}^n$. These two different settings lead to formally identical results.

So, one considers the Lagrange function

$$\mathbf{L}(g,\dot{g}) = \mathcal{L}^{(l)}(\Omega, Y) = \frac{1}{2}\langle\Omega, \mathcal{J}(\Omega)\rangle + \frac{1}{2}\langle Y, BY\rangle. \tag{337}$$

Here $(g,\dot{g}) \in TSO(n)$, and $\Omega = g^T\dot{g} \in so(n)$, $Y = g^Ty \in \mathbb{R}^n$. The reduced equations of motion for this Lagrange function yielded by Theorem 11.1 coincide with (335).

The integrable *Clebsch case* of the Kirchhoff equations is characterized by the relations

$$c_{ij}(b_i - b_j) + c_{jk}(b_j - b_k) + c_{ki}(b_k - b_i) = 0, \tag{338}$$

which implies that

$$c_{jk} = \frac{a_j - a_k}{b_j - b_k} \tag{339}$$

for some matrix $A = \mathrm{diag}(a_k)$. An invariant way of expressing this is $[B, M] = [A, \Omega]$. This relation is critical in proving the following theorem.

Theorem 17.1. *Under condition* (339), *equations* (335) *are equivalent to the matrix equation*

$$\dot{L}(\lambda) = [L(\lambda), U(\lambda)], \tag{340}$$

where

$$L(\lambda) = \lambda A + M + \lambda^{-1}YY^T, \qquad U(\lambda) = \lambda B + \Omega. \tag{341}$$

There are two important particular subcases of the Clebsch case, for which we will present integrable discretizations.

Case $A = B^2$ is characterized by the following choice of coefficients,

$$b_j = J_j, \quad a_j = J_j^2, \quad \text{so that} \quad c_{jk} = J_j + J_k.$$

The Hamilton function (334) takes the form

$$H(M, Y) = \frac{1}{2}\sum_{j<k}\frac{M_{jk}^2}{J_j + J_k} - \frac{1}{2}\sum_{j=1}^{n}J_jY_j^2. \tag{342}$$

Observe that the kinetic energy term in (337) is, in this case, that of the heavy top (236), since in both cases $\mathcal{J}(\Omega) = J\Omega + \Omega J$. The Lagrange function (337) takes the form

$$\mathbf{L}(g,\dot{g}) = \mathcal{L}^{(l)}(\Omega, Y) = -\frac{1}{2}\mathrm{tr}(\Omega J\Omega) + \frac{1}{2}\langle Y, JY\rangle. \tag{343}$$

Clearly, this is a particular case of the Lagrange function (261) of the rigid body in a quadratic potential, appearing when the matrix A of coefficients of the quadratic potential is of rank 1, $A = yy^T$ for some $y \in \mathbb{R}^n$. (For example, this is the case when the quadratic potential (260) represents the quadratic terms in the expansion of the potential of a single point mass; it is supposed that the distance from the rigid body to this point mass is much larger than the size of the body itself, and the ratio of these two length scales is the small parameter of the above mentioned expansion.) Defining $Y = g^T y$, we obtain $P = YY^T$, i.e., the orbit O_A in $\mathrm{Symm}(n)$ consists of matrices of rank 1. Differential equations (263) are, in this particular case, formally identical with the Kirchhoff equations (335).

Case $A = B$ is characterized by the following choice of coefficients,

$$a_k = b_k , \quad \text{so that} \quad c_{jk} = 1 .$$

The Hamilton function (334) takes the form

$$\mathrm{H}(M,Y) = \frac{1}{2}\langle M, M\rangle - \frac{1}{2}\langle Y, BY\rangle . \tag{344}$$

The Lagrange function (337) takes the form

$$\mathbf{L}(g,\dot{g}) = \mathcal{L}^{(l)}(\Omega,Y) = \frac{1}{2}\langle\Omega,\Omega\rangle + \frac{1}{2}\langle Y, BY\rangle , \tag{345}$$

so that

$$M = \Omega , \tag{346}$$

and the Kirchhoff equations of motion simplify to

$$\begin{cases} \dot{M} = Y \wedge (BY) , \\ \dot{Y} = -MY. \end{cases} \tag{347}$$

The Lax representation of Theorem 17.1 can be written for this flow in two equivalent forms,

$$\dot{L}(\lambda) = [L(\lambda), U_+(\lambda)] = [U_-(\lambda), L(\lambda)] , \tag{348}$$

where

$$L(\lambda) = \lambda B + M + \lambda^{-1} YY^T, \tag{349}$$

$$U_+(\lambda) = \lambda B + M , \qquad U_-(\lambda) = \lambda^{-1} YY^T. \tag{350}$$

17.2 Discretization of the Clebsch Problem, Case $A = B^2$

We obtain the corresponding results from Sect. 15.2, simply by replacing P_k with $Y_k Y_k^T$. The integrable discretization of the system under consideration is given by the discrete–time Lagrange function,

$$\mathbb{L}(g_k, g_{k+1}) = \Lambda^{(l)}(W_k, Y_k) = -\frac{1}{h}\operatorname{tr}(W_k J) + \frac{h}{2}\langle Y_{k+1}, JY_k\rangle$$

$$= -\frac{1}{h}\operatorname{tr}(W_k J) + \frac{h}{2}\langle Y_k, W_k JY_k\rangle, \qquad (351)$$

where, as usual, $W_k = g_k^{\mathrm{T}} g_{k+1} \in G$ and $Y_k = g_k^{\mathrm{T}} y \in O_y$. The equations of motion of this discretization are

$$\begin{cases} M_k = \frac{1}{h}(W_k J - JW_k^{\mathrm{T}}) - \frac{h}{2}Y_k \wedge (JY_{k+1}), \\ M_{k+1} = \frac{1}{h}(JW_k - W_k^{\mathrm{T}} J) + \frac{h}{2}Y_{k+1} \wedge (JY_k), \\ Y_{k+1} = W_k^{\mathrm{T}} Y_k. \end{cases} \qquad (352)$$

The Lax matrix and the Lax representation of this map are obtained from (277) and Theorem 15.3 by replacing P_k with $Y_k Y_k^{\mathrm{T}}$. In particular, (281) asserts that the following is the Lax matrix of map (352),

$$L(M, Y; \lambda) = \lambda J^2 + M + \lambda^{-1} YY^{\mathrm{T}}\left(1 - \frac{h^2}{4}\langle Y, Y\rangle\right). \qquad (353)$$

Observe that $\langle Y, Y\rangle$ is a Casimir function, in particular, that it is an integral of motion.

17.3 Discretization of the Clebsch Problem, Case $A = B$

The following discrete–time Lagrange function approximates (345),

$$\mathbb{L}(g_k, g_{k+1}) = \Lambda^{(l)}(W_k, Y_k)$$

$$= -\frac{1}{h}\operatorname{tr}\log\left(2I + W_k + W_k^{-1}\right) + \frac{4}{h}\log\left(1 + \frac{h^2}{4}\langle Y_k, BY_k\rangle\right). \qquad (354)$$

Here, as usual, $W_k = g_k^{\mathrm{T}} g_{k+1}$ and $Y_k = g_k^{\mathrm{T}} y$.

Theorem 17.2. *The reduced Euler–Lagrange equations of motion for Lagrange function (354) are*

$$\begin{cases} M_{k+1} = M_k + h\dfrac{Y_k \wedge (BY_k)}{1 + (h^2/4)\langle Y_k, BY_k\rangle}, \\ Y_{k+1} = \dfrac{I - (h/2)M_{k+1}}{I + (h/2)M_{k+1}} Y_k. \end{cases} \qquad (355)$$

Map $(M_k, Y_k) \to (M_{k+1}, Y_{k+1})$ is Poisson with respect to bracket (296), (297). This map is completely integrable and admits the following Lax representation,

$$L_{k+1}(\lambda) = V_k(\lambda)L_k(\lambda)V_k^{-1}(\lambda), \qquad (356)$$

with the matrices

$$L_k(\lambda) = \left(I + \frac{h^2}{4}B\right)^{-1}\left(\lambda B + M_k + \lambda^{-1}\mathcal{P}_k\right), \tag{357}$$

$$V_k(\lambda) = \frac{I + (h\lambda^{-1}/2)\mathcal{Q}_k}{I - (h\lambda^{-1}/2)\mathcal{Q}_k}, \tag{358}$$

where

$$\mathcal{P}_k = \left(I + \frac{h}{2}M_k\right)Y_kY_k^{\mathrm{T}}\left(I - \frac{h}{2}M_k\right) \tag{359}$$

and

$$\mathcal{Q}_k = \frac{1}{1 + (h^2/4)\langle Y_k, BY_k\rangle}\, Y_kY_k^{\mathrm{T}}\left(I + \frac{h^2}{4}B\right). \tag{360}$$

18 Systems of the Toda Type

There exist large classes of integrable lattice systems of the so-called Toda type,

$$\ddot{x}_k = f(\dot{x}_k)\Big(g(x_{k+1} - x_k) - g(x_k - x_{k-1})\Big),$$

and of the relativistic Toda type,

$$\ddot{x}_k = r(\dot{x}_k)\Big(\dot{x}_{k+1}f(x_{k+1} - x_k) - \dot{x}_{k-1}f(x_k - x_{k-1})$$
$$+g(x_{k+1} - x_k) - g(x_k - x_{k-1})\Big).$$

Both are characterized by the fact that each lattice site only interacts with its nearest neighbors. (The classical Toda lattice is a particular case of the first system when $f(v) = 1$, $g(\xi) = \exp(\xi)$.) We restrict ourselves here to one representative of each of these two classes.

18.1 Toda Type System

Theorem 18.1. *Consider the following Newtonian equations of motion,*

$$\ddot{x}_k = -\dot{x}_k^2\left(\frac{1}{x_{k+1} - x_k} - \frac{1}{x_k - x_{k-1}}\right). \tag{361}$$

They are Lagrangian with the Lagrange function

$$\mathcal{L}(x, \dot{x}) = \sum_{k=1}^{N}\log(\dot{x}_k) - \sum_{k=1}^{N}\log(x_{k+1} - x_k). \tag{362}$$

The corresponding Hamilton function is

$$H(x,p) = \sum_{k=1}^{N} \log(p_k) + \sum_{k=1}^{N} \log(x_{k+1} - x_k) . \tag{363}$$

The Hamiltonian equations of motion are

$$\begin{cases} \dot{x}_k = 1/p_k , \\ \dot{p}_k = \dfrac{1}{x_{k+1} - x_k} - \dfrac{1}{x_k - x_{k-1}} . \end{cases} \tag{364}$$

They admit a 2×2 Lax representation,

$$\dot{L}_k = M_k L_k - L_k M_{k-1} , \tag{365}$$

with the matrices

$$L_k = I + \lambda \begin{pmatrix} p_k x_k & -p_k x_k^2 \\ p_k & -p_k x_k \end{pmatrix} , \tag{366}$$

$$M_k = \frac{\lambda}{x_{k+1} - x_k} \begin{pmatrix} x_k & -x_k x_{k+1} \\ 1 & -x_{k+1} \end{pmatrix} . \tag{367}$$

An integrable discretization of this system is given in the following statement.

Theorem 18.2. *Consider the discrete–time Lagrange function,*

$$\Lambda(\widetilde{x}, x) = h \sum_{k=1}^{N} \log(\widetilde{x}_k - x_k) - h \sum_{k=1}^{N} \log(x_{k+1} - \widetilde{x}_k) , \tag{368}$$

with the corresponding Newtonian equations of motion,

$$\frac{1}{\widetilde{x}_k - x_k} - \frac{1}{x_k - \underset{\sim}{x}_k} = \frac{1}{x_{k+1} - x_k} - \frac{1}{x_k - \widetilde{x}_{k-1}} . \tag{369}$$

Its Lagrangian form is

$$p_k = h \left(\frac{1}{\widetilde{x}_k - x_k} + \frac{1}{x_k - \widetilde{x}_{k-1}} \right) , \tag{370}$$

$$\widetilde{p}_k = h \left(\frac{1}{\widetilde{x}_k - x_k} + \frac{1}{x_{k+1} - \widetilde{x}_k} \right) . \tag{371}$$

This map admits a 2×2 Lax representation,

$$V_k \widetilde{L}_k = L_k V_{k-1} \tag{372}$$

with the same Lax matrix (366) and

$$V_k = I + \frac{h\lambda}{x_{k+1} - \widetilde{x}_k} \begin{pmatrix} \widetilde{x}_k & -\widetilde{x}_k x_{k+1} \\ 1 & -x_{k+1} \end{pmatrix}. \tag{373}$$

A simple change of variables, $x_k(t) \to x_k(t + kh)$, turns the implicit discretization (369) into an explicit one. However, the Hamiltonian and the Lax aspects of this change are less trivial.

Theorem 18.3. *Consider the discrete–time Lagrange function,*

$$\Lambda(\widetilde{x}, x) = h \sum_{k=1}^{N} \log(\widetilde{x}_k - x_k) - h \sum_{k=1}^{N} \log(x_{k+1} - x_k), \tag{374}$$

with the corresponding Newtonian equations of motion,

$$\frac{1}{\widetilde{x}_k - x_k} - \frac{1}{x_k - \underset{\sim}{x}_k} = \frac{1}{x_{k+1} - x_k} - \frac{1}{x_k - x_{k-1}}. \tag{375}$$

Their Lagrangian form is

$$p_k = \frac{h}{\widetilde{x}_k - x_k} - \frac{h}{x_{k+1} - x_k} + \frac{h}{x_k - x_{k-1}}, \tag{376}$$

$$\widetilde{p}_k = \frac{h}{\widetilde{x}_k - x_k}. \tag{377}$$

They admit a 2×2 Lax representation (372) with the matrices

$$L_k = I + \lambda \begin{pmatrix} x_k p_k - h & -x_k(x_k p_k - h) \\ p_k & -x_k p_k \end{pmatrix}, \tag{378}$$

$$V_k = I + \frac{h\lambda}{x_{k+1} - x_k} \begin{pmatrix} x_k & -x_{k+1} x_k \\ 1 & -x_{k+1} \end{pmatrix}. \tag{379}$$

18.2 Relativistic Toda Type System

The "relativistic" generalization of these systems is formulated as follows.

Theorem 18.4. *Consider the following Newtonian equations of motion,*

$$\ddot{x}_k = -\dot{x}_k^2 \left(\frac{1}{x_{k+1} - x_k} - \frac{1}{x_k - x_{k-1}} - \frac{\alpha \dot{x}_{k+1}}{(x_{k+1} - x_k)^2} + \frac{\alpha \dot{x}_{k-1}}{(x_k - x_{k-1})^2} \right). \tag{380}$$

They are Lagrangian with the Lagrange function

$$\mathcal{L}(x, \dot{x}) = \sum_{k=1}^{N} \log(\dot{x}_k) - \sum_{k=1}^{N} \log(x_{k+1} - x_k) - \alpha \sum_{k=1}^{N} \frac{\dot{x}_k}{x_{k+1} - x_k} . \tag{381}$$

The corresponding Hamilton function is

$$H(x, p) = \sum_{k=1}^{N} \log\left(p_k(x_{k+1} - x_k) + \alpha\right). \tag{382}$$

The Lagrangian form of the equations of motion is

$$p_k = \frac{1}{\dot{x}_k} - \frac{\alpha}{x_{k+1} - x_k} , \tag{383}$$

$$\dot{p}_k = \frac{x_{k+1} - x_k - \alpha\dot{x}_k}{(x_{k+1} - x_k)^2} - \frac{x_k - x_{k-1} - \alpha\dot{x}_{k-1}}{(x_k - x_{k-1})^2} \tag{384}$$

They admit a 2×2 Lax representation (365) with the matrices

$$L_k = I + \lambda \begin{pmatrix} x_k p_k - \alpha & -x_k(x_k p_k - \alpha) \\ \\ p_k & -x_k p_k \end{pmatrix}, \tag{385}$$

$$M_k = \frac{\lambda}{x_{k+1} - x_k + \alpha/p_k} \begin{pmatrix} x_k - \alpha/p_k & -x_{k+1}(x_k - \alpha/p_k) \\ \\ 1 & -x_{k+1} \end{pmatrix}. \tag{386}$$

The discretization of this system is given as follows.

Theorem 18.5. *Consider the discrete–time Lagrange function,*

$$\Lambda(\widetilde{x}, x) =$$

$$h \sum_{k=1}^{N} \log(\widetilde{x}_k - x_k) - \alpha \sum_{k=1}^{N} \log(x_{k+1} - x_k) + (\alpha - h) \sum_{k=1}^{N} \log(x_{k+1} - \widetilde{x}_k),$$

$$\tag{387}$$

with the corresponding Newtonian equations of motion,

$$\frac{h}{\widetilde{x}_k - x_k} - \frac{h}{x_k - \underset{\sim}{x}_k} = \frac{\alpha}{x_{k+1} - x_k} - \frac{\alpha}{x_k - x_{k-1}} - \frac{\alpha - h}{\underset{\sim}{x}_{k+1} - x_k} + \frac{\alpha - h}{x_k - \widetilde{x}_{k-1}} . \tag{388}$$

Their Lagrangian form is

$$p_k = \frac{h}{\widetilde{x}_k - x_k} - \frac{\alpha - h}{x_k - \widetilde{x}_{k-1}} - \frac{\alpha}{x_{k+1} - x_k} + \frac{\alpha}{x_k - x_{k-1}} , \tag{389}$$

$$\widetilde{p}_k = \frac{h}{\widetilde{x}_k - x_k} - \frac{\alpha - h}{x_{k+1} - \widetilde{x}_k} . \tag{390}$$

They admit a 2×2 Lax representation (372) with the matrices (385) and

$$V_k = I + \frac{h\lambda}{x_{k+1} - \widetilde{x}_k + \alpha/\widetilde{p}_k} \begin{pmatrix} \widetilde{x}_k - \alpha/\widetilde{p}_k & -x_{k+1}(\widetilde{x}_k - \alpha/\widetilde{p}_k) \\ 1 & -x_{k+1} \end{pmatrix}. \quad (391)$$

It is easy to see that the explicit discretization (375) of the non–relativistic system (361) is in fact the particular case $\alpha = h$ of the discrete–time relativistic system (388), and, therefore, shares the Lax matrix, the integrals of motion etc. (or belongs to the hierarchy) of the relativistic system with the relativistic parameter chosen to be the time step.

19 Bibliographical Remarks

Sects. 2–4. There exist several comprehensive textbooks covering all the material reviewed in these sections. Our presentation is based mainly on [1,9,29]. These excellent books treat also the historical aspects of the development of mathematical methods in classical mechanics, and provide a rich bibliography.

Sect. 5. The modern version of the integrability concept, known under the name of the Liouville–Arnold integrability, appeared in the first edition of [9].

Sect. 6. A survey of the r–matrix approach to the Lax formalism is given in [43, 26].

Sects. 7–8. Continuous–time Lagrangian mechanics is a classical subject, see, e.g., [1, 9, 29]. Discrete–time Lagrangian mechanics on manifolds was developped in [54]. A generalization to partial differential equations and their variational discretization is given in [28].

Sect. 9. Specific features of Lagrangian mechanics on Lie groups in the continuous–time case are elaborated in the above–mentioned textbooks [1, 9, 29], and see also [19], where (47), (48) for the Poisson bracket on the trivialized cotangent bundle of a Lie group are called "perhaps the basic formula of Hamiltonian mechanics" (p. 405).

Sects. 10–11. For historical remarks on the Lie–Poisson bracket and the Euler–Poincaré equations, which were known already to Lie and Poincaré, but rediscovered independently in the 60's by Berezin, Kirillov, Souriau, and Kostant, see, e.g., [29]. The Hamiltonian version of the semi–direct product reduction was developed in the 80's, see, e.g., [30,31]. The Lagrangian version was, strangely enough, formulated only recently, see [23, 16]. The discrete–time counterparts of these constructions were elaborated in [12, 13].

Sect. 12.1. The system studied here was discovered in [36] (of course, in the "physical" dimension $N = 3$ only). In this classical paper the system was integrated in hyperelliptic functions with the help of separation of variables. The modern period in studying the Neumann system starts with the work by Moser [32,33]. He formulated the Hamiltonian structure in terms of the constrained Dirac bracket, stated all integrals of motion explicitly, and discovered the $N \times N$ Lax representation. The 2×2 Lax representation of the Neumann system was given in [35]. The existence of such different Lax representations is a particular case of the general "duality" explored in [2].

Sect. 12.2. The discretization of the Neumann system was achieved for the first time in [54], and its relation to the matrix factorization was established in [34]. A different approach to these discretizations, based on the interpretation of the Neumann system as a restricted flow of the KdV hierarchy, and therefore on 2×2 Lax representations, was proposed in [24]. This approach led also to the interpretation of these discretizations as Bäcklund transformations.

Sect. 12.3. This discretization was proposed in [38], and its r–matrix interpretation was given in [39].

Sect. 12.4. This discretization was proposed in [4], as a result of the discretization of the Landau–Lifshits equation. Also two integrals of motion were found there in the case $N = 3$. Integrability in the general case was proved in [48].

Sect. 13.1. The Garnier system was discovered in [21]. It should be mentioned that this paper contains one of the earliest examples of a Lax representation of a mechanical system. The $N \times N$ Lax matrix (207) is written there explicitly! This remarkable discovery remained forgotten for a long time, and was rediscovered in [17], where it was derived from the Lax representation of the vector nonlinear Schrödinger equation, for which the Garnier system is a stationary flow. The 2×2 Lax representation was derived, as a by–product of the interpretation of the Garnier system as a restricted flow of the KdV hierarchy, in [7].

Sect. 13.2. The Bäcklund transformation for the Garnier system was found in [24] on the base of the 2×2 Lax representation. We add here to their results the $N \times N$ Lax representation.

Sect. 13.3. Explicit discretization of the Garnier system was found in [45]. This discrete–time system can be interpreted as a stationary "flow" of the vector analogue of the Ablowitz–Ladik discrete nonlinear Schrödinger equation. Interestingly enough, the integrability of this latter system was established only recently, in [53]. The corresponding Lax representation is not the non–stationary version of ours, in particular, the Lax matrix is not $N \times N$, but rather lives in the Clifford algebra $\mathcal{C}\ell_N$.

Sect. 14.1. According to bibliographical remarks in [15], the multidimensional generalization of the Euler top is due to Frahm [20] and Schottky [44], and was rediscovered independently several times later. The modern interest in this system started with the general theory of geodesic flows on Lie groups, due to Arnold [8], as well as with the paper by Manakov [27], who found the n–dimensional Euler top as a reduction of the so called n–wave equations, along with a spectral–parameter–dependent Lax representation which we reproduce in Theorem 14.1.

Sect. 14.2. A discretization of the Euler top is due to Veselov [54], an interpretation in terms of matrix factorization was given in [34].

Sect. 15.1. Integrability of a rigid body in an arbitrary quadratic potential was discovered by Reyman [42], who gave also the r–matrix interpretation, and independently by Bogoyavlensky [14], who integrated this system in terms of Prym theta functions.

Sect. 15.2. Integrable discretization of the top in a quadratic potential was found in [47].

Sect. 16.1. The Lagrange integrable case of the heavy top motion was discovered by Lagrange in 1788. For a modern approach to its integration based on the Lax representation and an algebro–geometric technique see [41], [22]. As for the multi–dimensional case, there are two possible ways to generalize the notion of a heavy top, and, correspondingly, two different generalizations of the Lagrange top. The first one, living on $so(n) \ltimes \mathbb{R}^n$, was proposed in [10]; the Lax representation of Theorem 16.1 is due to [11]. The second one, living on $so(n) \ltimes so(n)$, was introduced in [40].

Sect. 16.3. Integrable discretization of the three–dimensional Lagrange top in the SU(2) framework was performed in [12]. This paper also contains the motivation, which lies in the theory of elastic curves and their discrete analogues. Discretization of the multi–dimensional Lagrange top was achieved in [49].

Sect. 17.1. The integrable case of the motion of a 3–dimensional rigid body in an ideal fluid was found by Clebsch [18]. His work was a development of results of Kirchhoff [25], who gave a Lagrangian derivation of equations of motion in the general case, and pointed out an integrable case connected with a rotational symmetry. The multi–dimensional generalization, along with the Lax representation, was discovered by Perelomov [37].

Sect. 17.2. The integrable discretization of the $A = B^2$ case of the multi–dimensional Clebsch problem was found in [47], that of the $A = B$ case in [49].

Sect. 18. The exponential Toda lattice was discovered by Toda [51], see also [52]. The system (361) was found by Yamilov [55], who also gave a complete classification of integrable systems of the Toda type. Its discretizations (369), (375) were found in [46]. The relativistic system (380) was first given in [5,6]. In the subsequent publication [3], a classification of integrable discrete–time

Lagrangian systems of the relativistic Toda type was accomplished. Among others, the discrete–time lattice (388) was found there. It was pointed out that such systems are best interpreted as Lagrangian systems on the regular triangular lattice.

References

1. Abraham, R. and Marsden, J.E. (1978) *Foundations of mechanics*, Addison–Wesley.
2. Adams, M.R., Harnad, J. and Hurtubise, J. (1990) Dual moment maps into loop algebras. *Lett. Math. Phys.*, **20**, 299–308.
3. Adler, V.E. (1999) Legendre transformations on the triangular lattice. *Funct. Anal. Appl.*, **34** (2000), 1–9.
4. Adler, V.E. (2000) On discretizations of the Landau–Lifshits equation. *Theor. Math. Phys.*, **124**, 897–908.
5. Adler, V.E. and Shabat, A.B. (1997) On a class of Toda chains. *Theor. Math. Phys.*, **111**, 647–657.
6. Adler, V.E. and Shabat, A.B. (1997) Generalized Legendre transformations. *Theor. Math. Phys.*, **112**, 935–948.
7. Antonowicz, M. (1992) Gelfand–Dikii hierarchies with sources and Lax representations for restricted flows. *Phys. Lett. A*, **165**, 47–52.
8. Arnold, V.I. (1966) Sur la géometrie différentielle des groupes de Lie de dimension infinie et ses applications à l'hydrodynamique des fluides parfaits. *Ann. Inst. Fourier*, **16**, 319–361.
9. Arnold, V.I. (1989) *Mathematical methods of classical mechanics.* Berlin etc: Springer–Verlag.
10. Belyaev, A.V. (1981) On the motion of a multidimensional body with a fixed point in a gravitational field. *Math. USSR Sbornik*, **42**, 413–418.
11. Bobenko, A.I., Reyman, A.G., and Semenov-Tian-Shansky, M.A. (1989) The Kowalewski top 99 years later: a Lax pair, generalizations and explicit solutions. *Commun. Math. Phys.*, **122**, 321–354.
12. Bobenko, A.I. and Suris, Yu.B. (1999) Discrete time Lagrangian mechanics on Lie groups, with an application to the Lagrange top. *Commun. Math. Phys.*, **204**, 147–188.
13. Bobenko, A.I. and Suris, Yu.B. (1999) Discrete Lagrangian reduction, discrete Euler–Poincaré equations, and semi–direct products. *Lett. Math. Phys.*, **49**, 79–93.
14. Bogoyavlensky, O.I. (1984) Integrable Euler equations on Lie algebras arising in problems of mathematical physics. *Math. USSR Izv.*, **25**, 207–257.
15. Bogoyavlensky, O.I. (1992) Euler equations on finite-dimensional Lie coalgebras, arising in problems of mathematical physics. *Russ. Math. Surv.*, **47**, 117–189.
16. Cendra, H., Holm, D.D., Marsden, J.E. and Ratiu, T.S. (1998) Lagrangian reduction, the Euler-Poincaré equations and semidirect products. – In: Geometry of differential equations. *Amer. Math. Soc. Transl.*, **186**, 1-25.
17. Choodnovsky, D.V. and Choodnovsky, G.V. (1979) Completely integrable class of mechanical systems connected with Korteweg–de Vries and multicomponent Schrödinger equations. *Lett. Nuovo Cim.*, **22**, Nr. 2, 47–51.

18. Clebsch, A. (1870) Über die Bewegung eines Körpers in einer Flüssigkeit. *Math. Annalen*, **3**, 238–262.

19. Cushman, R.H., Bates, L.M. (1997) *Global aspects of classical integrable systems.* Boston etc.: Birkhäuser.

20. Frahm, W. (1875) Über gewisse Differentialgleichungen. *Math. Ann.*, **8**, 35–44.

21. Garnier, R. (1919) Sur une classe de systèmes différentiels abéliens déduits de la théorie des équations linéaires. *Rend. Circ. Matem. Palermo*, **43**, Nr. 4, 155–191.

22. Gavrilov, L. and Zhivkov, A. (1998) The complex geometry of the Lagrange top. *L'Enseign. Math.*, **44**, 133–170.

23. Holm, D.D., Marsden, J.E. and Ratiu, T.S. (1998) The Euler-Poincaré equations and semidirect products, with applications to continuum theories. *Adv. in Math.*, **137**, 1–81.

24. Hone, A.N.W., Kuznetsov, V.B., and Ragnisco, O. (1999) Bäcklund transformations for many-body systems related to KdV. *J. Phys. A: Math. Gen.*, **32**, L299–L306.

25. Kirchhoff, G. (1869) Über die Bewegung eines Rotationskörpers in einer Flüssigkeit. *J. Reine Angew. Math.*, **71**, 237–262.

26. Kosmann-Schwarzbach, Y. (2004) Lie Bialgebras, Poisson Lie Groups, and Dressing Transformations, Lect. Notes Phys. **638**, pp. 107–173.

27. Manakov, S.V. (1976) Note on the integration of Euler's equations of the dynamics of an n–dimensional rigid body. *Funct. Anal. Appl.*, **10**, 328–329.

28. Marsden, J.E., Patrick, G.W. and Shkoller, S. (1998) Multisymplectic geometry, variational integrators and nonlinear PDEs. *Comm. Math. Phys.*, **199**, 351–395.

29. Marsden, J.E. and Ratiu, T.S. (1999) *Introduction to mechanics and symmetry.* Berlin etc: Springer (2nd edition).

30. Marsden, J.E., Ratiu, T.S., and Weinstein, A. (1984a) Semi–direct products and reduction in mechanics. *Trans. Amer. Math. Soc.*, **281**, 147–177.

31. Marsden, J.E., Ratiu, T.S., and Weinstein, A. (1984b) Reduction and Hamiltonian structures on duals of semidirect product Lie algebras. *Contemp. Math.*, **28**, 55–100.

32. Moser, J. (1980a) Various aspects of integrable Hamiltonian systems. – In: *Dynamical systems, C.I.M.E. Lectures, Progress in Math.* **8**. Boston etc.: Birkhäuser, 233–290.

33. Moser, J. (1980b) Geometry of quadrics and spectral theory. – In: *The Chern symposium, Berkeley, June 1979.* Berlin etc.: Springer, 147–188.

34. Moser, J. and Veselov, A.P. (1991) Discrete versions of some classical integrable systems and factorization of matrix polynomials. *Commun. Math. Phys.*, **139**, 217–243.

35. Mumford, D. (1984) *Tata lectures on theta.* Boston etc.: Birkhäuser.

36. Neumann, C. (1859) De problemate quodam mechanica, quod ad primam integralium ultraellipticorum classem revocatur. *J. Reine Angew. Math.*, **56**, 46–69.

37. Perelomov, A.M. (1980) A few remarks about integrability of the equations of motion of a rigid body in ideal fluid. *Phys. Lett. A*, **80**, 156–158.

38. Ragnisco, O. (1992) A discrete Neumann system. *Phys. Lett. A*, **167**, 165–171.

39. Ragnisco, O. (1995) Dynamical r-matrices for integrable maps. *Phys. Lett. A*, **198**, 295–305.

40. Ratiu, T. (1982) Euler–Poisson equations on Lie algebras and the n–dimensional heavy rigid body. *Amer. J. Math.*, **104**, 409–448.

41. Ratiu, T. and van Moerbeke, P. (1982) The Lagrange rigid body motion. *Ann. Inst. Fourier*, **32**, 211–234.

42. Reyman, A.G. (1980) Integrable Hamiltonian systems connected with graded Lie algebras. *J. Sov. Math.*, **19**, 1507–1545.

43. Reyman, A.G. and Semenov-Tian-Shansky, M.A. (1994) Group theoretical methods in the theory of finite dimensional integrable systems. – In: *Encyclopaedia of mathematical science, v.16: Dynamical Systems VII*, Springer, 116–225.

44. Schottky, F. (1891) Über das analytische Problem der Rotation eines starren Körpers in Raume von vier Dimensionen. *Sitzungsber. Königl. Preuss. Akad. Wiss. Berlin*, **13**, 227–232.

45. Suris, Yu.B. (1994) A discrete–time Garnier system. *Phys. Lett. A*, **189**, 281–289.

46. Suris, Yu.B. (1997) On some integrable systems related to the Toda lattice. *J. Phys. A*, **30**, 2235–2249.

47. Suris, Yu.B. (2000) The motion of a rigid body in a quadratic potential: an integrable discretization. *Intern. Math. Research Notices*, **12**, 643–663.

48. Suris, Yu.B. (2001) Integrability of V. Adler's discretization of the Neumann system. *Phys. Lett. A*, **279**, 327–332.

49. Suris, Yu.B. (2001) Integrable discretizations of some cases of the rigid body dynamics. *J. Nonlin. Math. Phys.*, **8**, 534–560.

50. Suris, Yu.B. (2003) *The problem of integrable discretization: Hamiltonian approach.* Progress in Mathematics, Vol. 219, Basel etc.: Birkhäuser.

51. Toda, M. (1967) Vibration of a chain with nonlinear interaction. *J. Phys. Soc. Japan*, **22**, 431–436.

52. Toda, M. (1989) *Theory of nonlinear lattices*, Berlin etc: Springer–Verlag.

53. Tsuchida, T., Ujino, H., and Wadati, M. (1999) Integrable semi–discretization of the coupled nonlinear Schrödinger equations. *J. Phys. A: Math. Gen.*, **32**, 2239–2262.

54. Veselov, A.P. (1988) Integrable discrete time systems and difference operators. *Funct. Anal. Appl.*, **22**, 83–93.

55. Yamilov, R.I. (1989) Generalizations of the Toda chain, and conservation laws. *Preprint Inst. of Math., Ufa* (in Russian). English version: Classification of Toda–type scalar lattices. – In: *Nonlinear evolution equations and dynamical systems, NEEDS'92*, Eds. V. Makhankov, I. Puzynin, O. Pashaev (1993). Singapore: World Scientific, 423–431.

Symmetries of Discrete Systems

Pavel Winternitz

Centre de recherches mathématiques and Département de mathématiques et de statistique, Université de Montréal, C.P. 6128, Succursale Centre-Ville, Montréal, Qc, H3C 3J7, Canada, `wintern@crm.umontreal.ca`

Abstract. In this series of lectures, we review the application of Lie point symmetries, and their generalizations, to the study of difference equations. The overall theme could be called "continuous symmetries of discrete equations".

1 Introduction

1.1 Symmetries of Differential Equations

Before studying the symmetries of difference equations, let us very briefly review the theory of the symmetries of differential equations. For all details, proofs and further information we refer to the many excellent books on the subject, *e.g.*, $[52, 7, 53, 32, 23, 3, 57, 63]$.

Let us consider a general system of differential equations

$$E_a(x, u, u_x, u_{xx}, \ldots u_{nx}) = 0, \quad x \in \mathbf{R}^p, u \in \mathbf{R}^q, a = 1, \ldots N, \qquad (1)$$

where u_{nx} denotes all (partial) derivatives of u of order n. The numbers p, q, n and N are all nonnegative integers.

We are interested in the symmetry group G of system (1), *i.e.*, in the local Lie group of local point transformations taking solutions of (1) into solutions of the same equation. Point transformations in the space $X \times U$ of independent and dependent variables have the form

$$\tilde{x} = \Lambda_\lambda(x, u), \quad \tilde{u} = \Omega_\lambda(x, u), \qquad (2)$$

where λ denotes the group parameters. Thus

$$\Lambda_0(x, u) = x, \quad \Omega_0(x, u) = u,$$

and the inverse transformation $(\tilde{x}, \tilde{u}) \mapsto (x, u)$ exists, at least locally.

The transformations (2) of local coordinates in $X \times U$ also determine the transformations of functions $u = f(x)$ and of derivatives of functions. A group G of local point transformations of $X \times U$ will be a symmetry group of system (1) if the fact that $u(x)$ is a solution implies that $\tilde{u}(\tilde{x})$ is also a solution.

P. Winternitz, Symmetries of Discrete Systems, Lect. Notes Phys. **644**, 185–243 (2004)
`http://www.springerlink.com/`

The two fundamental questions are:

1. How to find the maximal symmetry group G for a given system of equations (1)?
2. Once the group G is found, what do we do with it?

Let us first discuss the question of motivation. The symmetry group G allows us to do the following:

1. Generate new solutions from known ones. Sometimes trivial solutions can be boosted into interesting ones.
2. Identify equations with isomorphic symmetry groups. Such equations may be transformable into each other. Sometimes nonlinear equations can be transformed into linear ones.
3. Perform symmetry reduction. For partial differential equations, we can reduce the number of variables and obtain particular solutions, satisfying particular boundary conditions, called group-invariant solutions. For ODEs of order n, we can reduce the order of the equation. In this reduction, there is no loss of information. If we can reduce the order to zero, we obtain a general solution depending on n constants, or a general integral (an algebraic function depending on n constants).

How does one find the symmetry group G? Instead of looking for "global" transformations as in (2) one looks for infinitesimal ones, *i.e.*, one looks for the Lie algebra L that corresponds to G. A one-parameter group of infinitesimal point transformations will have the form

$$\tilde{x}_i = x_i + \lambda \xi_i(x, u), \quad |\lambda| \ll 1 \tag{3}$$
$$\tilde{u}_\alpha = u_\alpha + \lambda \phi_\alpha(x, u), \quad 1 \le i \le p, \quad 1 \le \alpha \le q.$$

The functions ξ_i and ϕ_α must be found from the condition that $\tilde{u}(\tilde{x})$ is a solution whenever $u(x)$ is one. The derivatives $\tilde{u}_{\alpha,\tilde{x}_i}$ must be calculated using (3) and will involve derivatives of ξ_i and ϕ_α. A K-th derivative of $\tilde{u}_\alpha$ with respect to the variable $\tilde{x}_i$ will involve derivatives of ξ_i and ϕ_α up to order K. We then substitute the transformed quantities into (1) and require that the equation be satisfied for $\tilde{u}(\tilde{x})$, whenever it is satisfied for $u(x)$. Thus, terms of order λ^0 will drop out. Terms of order λ will provide a system of determining equations for ξ_i and ϕ_α. Terms of order λ^k, $k = 2, 3, \ldots$ are to be ignored, since we are looking for infinitesimal symmetries.

The functions ξ_i and ϕ_α depend only on x and u, not on first, or higher derivatives, u_{α,x_i}, $u_{\alpha,x_i x_k}$, etc. This is actually the definition of "point" symmetries. The determining equations will explicitly involve derivatives of u_α, up to order n (the order of the studied equation). The coefficients of all linearly independent expressions in the derivatives must vanish separately. This provides a system of determining equations for the functions $\xi_i(x, u)$ and $\phi_\alpha(x, u)$. This is a system of linear partial differential equations of order n. The determining equations are linear, even if the original system (1) is nonlinear. This "linearization" is due to the fact that all terms of order λ^j, $j \geq 2$, are ignored.

The system of determining equations is usually overdetermined, $i.e.$, there are usually more determining equations than unknown functions ξ_i and ϕ_α ($p+q$ functions). The independent variables in the determining equations are $x \in \mathbf{R}^p$, $u \in \mathbf{R}^q$.

For an overdetermined system, there are three possibilities:

1. The only solution is the trivial one $\xi_i = 0$, $\phi_\alpha = 0$, $i = 1, \ldots p, \alpha = 1, \ldots, q$. In this case the symmetry algebra is $L = \{0\}$, the symmetry group is $G = I$, and the symmetry method is to no avail.
2. The general solution of the determining equations depends on a finite number K of constants. In this case, the studied system (1) has a finite-dimensional Lie point symmetry group and $\dim G = K$.
3. The general solution depends on a finite number of arbitrary functions of some of the variables $\{x_i, u_\alpha\}$. In this case the symmetry group is infinite dimensional. This last case is of particular interest.

The search for the symmetry algebra L of a system of differential equations is best formulated in terms of vector fields acting on the space $X \times U$ of independent and dependent variables. Indeed, consider the vector field

$$X = \sum_{i=1}^{p} \xi_i(x,u)\partial_{x_i} + \sum_{\alpha=1}^{q} \phi_\alpha(x,u)\partial_{u_\alpha}, \qquad (4)$$

where the coefficients ξ_i and ϕ_α are the same as in (3). If these functions are known, the vector field (4) can be integrated to obtain the finite transformations (2). Indeed, all we have to do is to integrate the equations

$$\frac{d\tilde{x}_i}{d\lambda} = \xi_i(\tilde{x}, \tilde{u}), \quad \frac{d\tilde{u}_\alpha}{d\lambda} = \phi_\alpha(\tilde{x}, \tilde{u}), \qquad (5)$$

subject to the initial conditions

$$\tilde{x}_i \mid_{\lambda=0} = x_i \quad \tilde{u}_\alpha \mid_{\lambda=0} = u_\alpha. \qquad (6)$$

This provides us with a one-parameter group of local Lie point transformations of the form (2) where λ is the group parameter.

The vector field (4) tells us how the variables x and u transform. We also need to know how derivatives like u_x, u_{xx}, $\ldots$ transform. This is given by the prolongation of the vector field X.

We have

$$\mathrm{pr}\, X = X + \sum_\alpha \left\{ \sum_i \phi_\alpha^{x_i} \partial_{u_{x_i}} + \sum_{i,k} \phi_\alpha^{x_i x_k} \partial_{u_{x_i x_k}} \right. \qquad (7)$$

$$\left. + \sum_{i,k,l} \phi_\alpha^{x_i x_k x_l} \partial_{u_{x_i x_k x_l}} + \cdots \right\},$$

where the coefficients in the prolongation can be calculated recursively, using the total derivative operator,

$$D_{x_i} = \partial_{x_i} + u_{\alpha,x_i}\partial_{u_\alpha} + u_{\alpha,x_a x_i}\partial_{u_{\alpha,x_a}} + u_{\alpha,x_a x_b x_i}\partial_{u_{\alpha,x_a x_b}} + \cdots \qquad (8)$$

(a summation over repeated indices is to be understood). The recursive formulas are

$$\phi_\alpha^{x_i} = D_{x_i}\phi_\alpha - (D_{x_i}\xi_a)u_{\alpha,x_a}$$
$$\phi_\alpha^{x_i x_k} = D_{x_k}\phi_\alpha^{x_i} - (D_{x_k}\xi_a)u_{\alpha,x_i x_a}$$
$$\phi_\alpha^{x_i x_k x_l} = D_{x_l}\phi_\alpha^{x_i x_k} - (D_{x_l}\xi_a)u_{\alpha,x_i x_k x_a}, \qquad (9)$$

etc.

The n-th prolongation of X acts on functions of x, u and all derivatives of u up to order n. It also tells us how derivatives transform. Thus, to obtain the transformed quantities $\tilde{u}_{\tilde{x}_i}$ we must integrate (5) with conditions (6), together with

$$\frac{d\tilde{u}_{\tilde{x}_i}}{d\lambda} = \phi^{x_i}(\tilde{x}, \tilde{u}, \tilde{u}_{\tilde{x}}), \quad \tilde{u}_{\tilde{x}}\,|_{\lambda=0} = u_x. \qquad (10)$$

We see that the coefficients of the prolonged vector field are expressed in terms of derivatives of ξ_i and ϕ_α, the coefficients of the original vector field. They carry no new information. The transformation of derivatives is completely determined, once the transformations of functions are known.

The invariance condition for system (1) is expressed in terms of the operator (7) as

$$\mathrm{pr}^{(n)} X E_a\,|_{E_1 = \cdots = E_N = 0} = 0, \quad a = 1, \ldots N, \qquad (11)$$

where $\mathrm{pr}^{(n)} X$ is the prolongation (7) calculated up to order n where n is the order of system (1).

In practice the symmetry algorithm consists of several steps, most of which can be carried out on a computer. For early computer programs calculating symmetry algebras, see [61,9]. For a more recent review, see [26].

The individual steps are:

1. Calculate all the coefficients in the n-th prolongation of X. This depends only on the order of system (1), $i.e.$, n, and on the number of independent and dependent variables, $i.e.$, p and q.
2. Consider system (1) as a system of algebraic equations for x, u, u_x, u_{xx}, etc. Choose N variables v_1, v_2, $\ldots v_N$ and solve system (1) for these variables. The v_i must satisfy the following conditions:
   i) Each v_i is a derivative of one u_α of at least order 1.
   ii) The variables v_i are all independent; none of them is a derivative of any other one.
   iii) No derivatives of any of the v_i figure in the system (1).

3. Apply $\mathrm{pr}^{(n)} X$ to all the equations in (1) and eliminate all expressions v_i from the result. This yields system (11).
4. Determine all linearly independent expressions in the derivatives remaining in (11), once the quantities v_i are eliminated. Set the coefficients of these expressions equal to zero. This yields the determining equations, a system of linear partial differential equations of order n for $\phi_\alpha(x, u)$ and $\xi_i(x, u)$.
5. Solve the determining equations to obtain the symmetry algebra.
6. Integrate the obtained vector fields to obtain the one-parameter subgroups of the symmetry group. Compose them appropriately to obtain the connected component G_o of the symmetry group G.
7. Extend the connected component G_o to the full group G by adding all discrete transformations leaving system (1) invariant. These discrete transformations will form a finite, or discrete group G_D. Then

$$G = G_D \rtimes G_o , \tag{12}$$

i.e., G_o is an invariant subgroup of G.

Let us consider the case when, at Step 5, we obtain a finite dimensional Lie algebra L, *i.e.*, a vector field X depending on K parameters, $K \in \mathbf{Z}^>$, $K < \infty$. We can then choose a basis

$$\{X_1, X_2, \ldots, X_K\} \tag{13}$$

of the Lie algebra L. The basis that is naturally obtained in this manner depends on our integration procedure, though the algebra L itself does not. It is useful to transform the basis (13) into a canonical form in which all the properties of L which are independent of the choice of basis are evident. Thus, if L can be decomposed into a direct sum of indecomposable components,

$$L = L_1 \oplus L_2 \oplus \cdots \oplus L_M, \tag{14}$$

a basis should be chosen that respects this decomposition. The components L_i that are simple should be identified according to the Cartan classification (over $\mathbf{C}$) or the Gantmakher classification (over $\mathbf{R}$) [50, 25]. The components that are solvable should be so organized that their nilradical [56, 33] is evident. For those components that are neither simple, nor solvable, the basis should be chosen so as to respect the Levi decomposition [56, 33].

So far we have considered only point transformations, as in (2), in which the new variables $\tilde{x}$ and $\tilde{u}$ depend only on the old ones, x and u. More general transformations are "contact transformations", where $\tilde{x}$ and $\tilde{u}$ also depend on first derivatives of u. A still more general class of transformations are generalized transformations, also called "Lie-Bäcklund" transformations [52, 4]. For these,

$$\tilde{x} = \Lambda_\lambda(x, u, u_x, u_{xx}, \ldots) \tag{15}$$
$$\tilde{u} = \Omega_\lambda(x, u, u_x, u_{xx}, \ldots)$$

involving derivatives up to an arbitrary, but finite order. The coefficients ξ_i and ϕ_α of the vector fields (4) will then also depend on derivatives of u_α.

When studying generalized symmetries, and sometimes also point symmetries, it is convenient to use a different formalism, namely that of evolutionary vector fields.

Let us first consider the case of Lie point symmetries, i.e., vector fields of the form (4) and their prolongations (7). To each vector field (4) we can associate its evolutionary counterpart X_e, defined as

$$X_e = Q_\alpha(x, u, u_x)\partial_{u_\alpha}, \tag{16}$$

$$Q_\alpha = \phi_\alpha - \xi_i \frac{\partial u_\alpha}{\partial x_i}. \tag{17}$$

The prolongation of the evolutionary vector field (16) is defined as

$$\mathrm{pr}\, X_e = Q_\alpha \partial_{u_\alpha} + Q_\alpha^{x_i} \partial_{u_{\alpha,x_i}} + Q_\alpha^{x_i x_k} \partial_{u_{\alpha,x_i x_k}} + \dots \tag{18}$$
$$Q_\alpha^{x_i} = D_{x_i} Q_\alpha, \quad Q_\alpha^{x_i x_k} = D_{x_i} D_{x_k} Q_\alpha, \dots.$$

The functions Q_α are called the characteristics of the vector field. Observe that X_e and $\mathrm{pr}\, X_e$ do not act on the independent variables x_i.

For Lie point symmetries evolutionary and ordinary vector fields are entirely equivalent and it is easy to pass from one to the other. Indeed, (17) gives the connection between the two.

The symmetry algorithm for calculating the symmetry algebra L in terms of evolutionary vector fields is also equivalent. Equation (11) is simply replaced by

$$\mathrm{pr}^{(n)}\, X_e E_a \,|_{E_1 = \dots = E_N = 0}, = 0, \quad a = 1, \dots N. \tag{19}$$

The reason that (11) and (19) are equivalent is the following. It is easy to show that

$$\mathrm{pr}^{(n)}\, X_e = \mathrm{pr}^{(n)}\, X - \xi_i D_i. \tag{20}$$

The total derivative D_i is itself a generalized symmetry of (1), i.e.,

$$D_i E_a \,|_{E_1 = E_2 = \dots = E_N = 0}, = 0 \quad i = 1, \dots p, \quad a = 1, \dots N. \tag{21}$$

Equations (20) and (21) prove that systems (11) and (19) are equivalent. Equation (21) itself follows from the fact that $D_i E_a = 0$ is a differential consequence of (1), hence every solution of (1) is also a solution of (21).

To find generalized symmetries of order k, we use (16) but allow the characteristics Q_α to depend on all derivatives of u_α up to order k. The prolongation is calculated using (18). The symmetry algorithm is again (19).

A very useful property of evolutionary symmetries is that they provide compatible flows. This means that the system of equations

$$\frac{\partial u_\alpha}{\partial \lambda} = Q_\alpha \tag{22}$$

is compatible with system (1). In particular, group invariant solutions, *i.e.*, solutions invariant under a subgroup of G, are obtained as fixed points

$$Q_\alpha = 0. \tag{23}$$

If Q_α is the characteristic of a point transformation, then (23) is a system of quasilinear first order partial differential equations. They can be solved and their solutions can be substituted into (1), yielding the invariant solutions explicitly.

1.2 Comments on Symmetries of Difference Equations

The study of symmetries of difference equations is much more recent than that of differential equations. Early work in this direction is due to Maeda [47, 48] who mainly studied transformations acting on the dependent variables only. A more recent series of papers was devoted to Lie point symmetries of differential-difference equations on fixed regular lattices [43–45, 24, 35, 34, 49, 54, 55, 8]. A different approach was developed mainly for linear or linearizable difference equations and involved transformations acting on more than one point of the lattice [21, 22, 42, 29, 38]. The symmetries considered in this approach are really generalized ones, however they reduce to point symmetries in the continuous limit.

A more general class of generalized symmetries has also been studied for difference equations and differential-difference equations on fixed regular lattices [27, 28, 30, 36].

A different approach to symmetries of discrete equations was originally suggested by V. Dorodnitsyn and collaborators [12, 14, 13, 5, 16, 15, 20, 17, 19, 18]. The main aim of this series of papers is to discretize differential equations while preserving their Lie point symmetries.

Symmetries of ordinary and partial difference schemes on lattices that are a priori given, but are allowed to transform under point transformations, were studied in Ref. [39, 40, 46].

While the study of symmetries of difference equations is a relatively recent subject, it has already acquired a life of its own. Indeed, a series of biannual conferences, "Symmetries and Integrability of Difference Equations", is dedicated to this topic. The first was held in Esterel, near Montreal in 1994 [41]. The subsequent SIDE meetings were held in Canterbury (UK) [10], Sabaudia (Italy) [37], Tokyo (Japan) [60] and Giens (France) [51]. The proceedings contain much information on this interesting subject.

2 Ordinary Difference Schemes and Their Point Symmetries

2.1 Ordinary Difference Schemes

An ordinary differential equation (ODE) of order n is a relation involving one independent variable, x, one dependent variable, $u = u(x)$, and n derivatives of the function $u, u', u'', \ldots u^{(n)}$,

$$E(x, u, u', u'', \ldots, u^{(n)}) = 0, \quad \frac{\partial E}{\partial u^{(n)}} \neq 0. \tag{24}$$

An ordinary difference scheme (OΔS) involves two objects, a difference equation and a lattice. We shall specify an OΔS by a system of two equations, both involving two continuous variables x and $u(x)$, evaluated at a discrete set of points $\{x_n\}$.

Thus, a difference scheme of order K will have the form

$$\begin{aligned}
E_a(\{x_k\}_{k=n+M}^{n+N}, \{u_k\}_{k=n+M}^{n+N}) = 0, a = 1, 2 \\
K = N - M + 1, \quad n, M, N \in \mathbf{Z}, \quad u_k \equiv u(x_k).
\end{aligned} \tag{25}$$

At this stage we are not imposing any boundary conditions, so the reference point x_n can be arbitrarily shifted to the left, or to the right. The order K of the system is the number of points involved in the scheme (25) and it is assumed to be finite. We also assume that if the values of x_k and u_k are specified at $(N - M)$ neighbouring point, we can calculate their values at the point to the right, or to the left of the given set, using equations (25).

A continuous limit, when the spacings between all neighbouring points go to zero, if it exists, will take one of the equations (25) into a differential equation of order $K' \leq K$, the other into an identity (like $0 = 0$).

When taking the continuous limit it is convenient to introduce different quantities, namely differences between neighbouring points and discrete derivatives like

$$\begin{aligned}
h_+(x_n) &= x_{n+1} - x_n, \quad h_-(x_n) = x_n - x_{n-1}, \\
u_{,x} &= \frac{u_{n+1} - u_n}{x_{n+1} - x_n}, \quad u_{,\underline{x}} = \frac{u_n - u_{n-1}}{x_n - x_{n-1}}, \\
u_{,x\underline{x}} &= 2\frac{u_{,x} - u_{,\underline{x}}}{x_{n+1} - x_{n-1}}, \ldots
\end{aligned} \tag{26}$$

In the continuous limit,

$$h_+ \to 0, \quad h_- \to 0, \quad u_{,x} \to u', \quad u_{,\underline{x}} \to u', \quad u_{,x\underline{x}} \to u''.$$

As a clarifying example of the meaning of the difference scheme (25), let us consider a three-point scheme that will approximate a second-order linear difference equation:

$$E_1 = \frac{u_{n+1} - 2u_n + u_{n-1}}{(x_{n+1} - x_n)^2} - u_n = 0, \tag{27}$$

$$E_2 = x_{n+1} - 2x_n + x_{n-1} = 0. \tag{28}$$

The solution of $E_2 = 0$, determines a uniform lattice

$$x_n = hn + x_0. \tag{29}$$

The scale h and the origin x_0 in (29) are not fixed by (28), instead they appear as integration constants, *i.e.*, they are *a priori* arbitrary. Once they are chosen, (27) reduces to a linear difference equation with constant coefficients, since we have $x_{n+1} - x_n = h$. Thus, a solution of (27) will have the form

$$u_n = \lambda^{x_n}. \tag{30}$$

Substituting (30) into (27) we obtain the general solution of the difference scheme (27), (28),

$$u(x_n) = c_1 \lambda_1^{x_n} + c_2 \lambda_2^{x_n}, \quad x_n = hn + x_0, \tag{31}$$

$$\lambda_{1,2} = \left(\frac{2 + h^2 \pm h\sqrt{4 + h^2}}{2} \right)^{1/2}.$$

The solution (31) of system (27)–(28) depends on 4 arbitrary constants c_1, c_2, h and x_0.

Now let us consider a general three-point scheme of the form

$$E_a(x_{n-1}, x_n, x_{n+1}, u_{n-1}, u_n, u_{n+1}) = 0, \quad a = 1, 2, \tag{32}$$

satisfying

$$\det \left(\frac{\partial(E_1, E_2)}{\partial(x_{n+1}, u_{n+1})} \right) \neq 0, \quad \det \left(\frac{\partial(E_1, E_2)}{\partial(x_{n-1}, u_{n-1})} \right) \neq 0, \tag{33}$$

(possibly after an up or down shifting). The two conditions on the Jacobians (33) are sufficient to allow us to calculate (x_{n+1}, u_{n+1}) if $(x_{n-1}, u_{n-1}, x_n, u_n)$ are known. Similarly, (x_{n-1}, u_{n-1}) can be calculated if $(x_n, u_n, x_{n+1}, u_{n+1})$ are known. The general solution of the scheme (32) will hence depend on 4 arbitrary constants and will have the form

$$u_n = f(x_n, c_1, c_2, c_3, c_4) \tag{34}$$

$$x_n = \phi(n, c_1, c_2, c_3, c_4). \tag{35}$$

A more standard approach to difference equations would be to consider a fixed equally spaced lattice, *e.g.*, with spacing $h = 1$. We can then identify the continuous variable x, sampled at discrete points x_n, with the discrete variable n,

$$x_n = n. \tag{36}$$

Instead of a difference scheme we then have a difference equation,

$$E(\{u_k\}_{k=n+M}^{n+N}) = 0, \tag{37}$$

involving $K = N - M + 1$ points. Its general solution has the form

$$u_n = f(n, c_1, c_2, \ldots c_{N-M}) , \tag{38}$$

i.e., it depends on $N - M$ constants.

Below, when studying point symmetries of discrete equations we will see the advantage of considering difference systems like system (25).

2.2 Point Symmetries of Ordinary Difference Schemes

In this section we shall follow rather closely the article [39]. We shall define the symmetry group of an ordinary difference scheme in the same manner as for ODEs: it is, a group of continuous local point transformations of the form (2) taking solutions of the OΔS (25) into solutions of the same scheme. The transformations considered being continuous, we will adopt an infinitesimal approach, as in (3). We drop the labels i and α, since we are considering the case of one independent and one dependent variable only.

As in the case of differential equations, our basic tool will be vector fields of the form (4). In the case of OΔS they will have the form

$$X = \xi(x, u)\partial_x + \phi(x, u)\partial_u, \tag{39}$$

with

$$x \equiv x_n, \quad u \equiv u_n = u(x_n).$$

Because we are considering point transformations, ξ and ϕ in (39) depend on x and u at one point only.

The prolongation of the vector field X is different from that of the case of ODEs. Instead of prolonging to derivatives, we prolong to all points of the lattice figuring in scheme (25). Thus we set

$$\mathrm{pr}\, X = \sum_{k=n+M}^{n+N} \xi(x_k, u_k)\partial_{x_k} + \sum_{k=n+M}^{n+N} \phi(x_k, u_k)\partial_{u_k}. \tag{40}$$

In these terms the requirement that the transformed function $\tilde{u}(\tilde{x})$ should satisfy the same OΔS as the original $u(x)$ is expressed by the requirement

$$\mathrm{pr}\, X E_a \,|_{E_1 = E_2 = 0} = 0, \quad a = 1, 2. \tag{41}$$

Since we must respect both the difference equation and the lattice, we have two conditions (41) from which to determine $\xi(x, u)$ and $\phi(x, u)$. Since each of these functions depends on a single point (x, u) and the prolongation (40) introduces $N - M + 1$ points in space $X \times U$, equation (41) will imply a system of determining equations for ξ and ϕ. Moreover, in general this will be an overdetermined system of linear functional equations that we transform into an overdetermined system of linear differential equations [1, 2].

To illustrate the method and the role of the choice of the lattice, let us start from a simple example. The example will be that of difference equations that approximate the ODE

$$u'' = 0 \tag{42}$$

on several different lattices.

Let us find the Lie point symmetry group of ODE (42), $i.e.$, the equation of a free particle on a line. Following the algorithm of Sect. 1, we set

$$\mathrm{pr}^{(2)} X = \xi \partial_x + \phi \partial_u + \phi^x \partial_{u'} + \phi^{xx} \partial_{u''} \tag{43}$$
$$\phi^x = D_x \phi - (D_x \xi) u' = \phi_x + (\phi_u - \xi_x) u' - \xi_u u'^2$$
$$\phi^{xx} = D_x \phi^x - (D_x \xi) u'' = \phi_{xx} + (2\phi_{xu} - \xi_{xx}) u'$$
$$+ (\phi_{uu} - 2\xi_{xu}) u'^2 - \xi_{uu} u'^3 + (\phi_u - 2\xi_x) u''$$
$$- 3\xi_u u' u''.$$

The symmetry formula (11) in this case reduces to

$$\phi^{xx} \mid_{u''=0} = 0. \tag{44}$$

Setting the coefficients of u'^3, u'^2, u' and u'^0 equal to zero, we obtain an 8-dimensional Lie algebra, isomorphic to $\mathrm{sl}(3, \mathbf{R})$ with basis

$$X_1 = \partial_x, \quad X_2 = x \partial_x, \quad X_3 = u \partial_x \tag{45}$$
$$X_4 = \partial_u, \quad X_5 = x \partial_u, \quad X_6 = u \partial_u,$$
$$X_7 = x(x \partial_x + u \partial_u), \quad X_8 = u(x \partial_x + u \partial_u).$$

This result was of course already known to Sophus Lie. Moreover, any second-order ODE that is linear, or linearizable by a point transformation has a symmetry algebra isomorphic to $\mathrm{sl}(3, \mathbf{R})$. The group $\mathrm{SL}(3, \mathbf{R})$ acts as the group of projective transformations of the Euclidean space E_2 (with coordinates x, u).

Now let us consider some difference schemes that have (42) as their continuous limit. We shall take the equation to be

$$\frac{u_{n+1} - 2u_n + u_{n-1}}{(x_{n+1} - x_n)^2} = 0. \tag{46}$$

However before looking for the symmetry algebra, we multiply out the denominator and study the equivalent equation,

$$E_1 = u_{n+1} - 2u_n + u_{n-1} = 0. \tag{47}$$

To this equation we must add a second equation, specifying the lattice. We consider three different examples at first glance quite similar, but leading to different symmetry algebras.

Example 2.1. Free particle (47) on a fixed uniform lattice. We consider

$$E_2 = x_n - hn - x_0 = 0, \tag{48}$$

where h and x_0 are fixed constants, that are not transformed by the group (*e.g.*, $h = 1$, $x_0 = 0$).

Applying the prolonged vector field (40) to (48) we obtain

$$\xi(x_n, u_n) = 0 \tag{49}$$

for all x_n and u_n. Next, let us apply (40) to (47) and replace x_n, using (48) and u_{n+1}, using (47). We obtain

$$\phi\big(h(n+1) + x_0, 2u_n - u_{n-1}\big) - 2\phi(hn + x_0, u_n)$$
$$+\phi\big(h(n-1) + x_0, u_{n-1}\big) = 0. \tag{50}$$

Differentiating (50) twice, once with respect to u_{n-1}, and then with respect to u_n, we obtain

$$\frac{\partial^2}{\partial u_{n+1}^2}\phi(x_{n+1}, u_{n+1}) = 0 \tag{51}$$

and hence

$$\phi(x_n, u_n) = A(x_n)u_n + B(x_n). \tag{52}$$

We substitute (52) back into (50) and equate coefficients of u_n, u_{n-1} and 1. The result is

$$A(n+1) = A(n), \quad B(n+1) - 2B(n) + B(n-1) = 0. \tag{53}$$

Hence

$$A = A_0, \quad B = B_1 n + B_0 = b_1 x + b_0, \tag{54}$$

where A_0, B_1, B_0, b_1 and b_0 are constants. We obtain the symmetry algebra of the OΔS (47), (48) and it is only three-dimensional, spanned by

$$X_1 = \partial_u, \quad X_2 = x\partial_u, \quad X_3 = u\partial_u. \tag{55}$$

The corresponding one parameter transformation groups are obtained by integrating these vector fields (see (5) and (6)),

$$G_1 : \tilde{x} = x$$
$$\tilde{u}(\tilde{x}) = u(x) + \lambda$$
$$G_2 : \tilde{x} = x \tag{56}$$
$$\tilde{u}(\tilde{x}) = u(x) + \lambda x$$
$$G_3 : \tilde{x} = x$$
$$\tilde{u}(\tilde{x}) = e^{\lambda} u(x).$$

G_1 and G_2 just tell us that adding an arbitrary solution of the scheme to any given solution yields a new solution, G_3 corresponds to the scale invariance of (47).

Example 2.2. Free particle (47) on a uniform two point lattice. Instead of (48) we define a lattice by setting

$$E_2 = x_{n+1} - x_n = h, \tag{57}$$

where h is a fixed (non-transforming) constant. Note that (57) tells us the distance between any two neighbouring points but does not fix an origin (as opposed to (48)).

Applying the prolonged vector field (40) to (57) and using (57), we obtain

$$\xi(x_n + h, u_{n+1}) - \xi(x_n, u_n) = 0. \tag{58}$$

Since u_{n+1} and u_n are independent, (58) implies $\xi = \xi(x)$. Moreover $\xi(x_n + h) = \xi(x)$, so

$$\xi = \xi_0 = \text{const}. \tag{59}$$

Further, we apply pr X to (47), and set $u_{n+1} = 2u_n - u_{n-1}$, $x_{n+1} = x_n + h$, $x_{n-1} = x_n - h$ in the obtained expressions. As in Example 2.1 we find that $\phi(x, u)$ is linear in u as in (52) and ultimately satisfies

$$\phi(x, u) = au + bx + c. \tag{60}$$

The symmetry algebra in this case is four-dimensional. To the basis elements (55) we add translational invariance,

$$X_4 = \partial_x. \tag{61}$$

Example 2.3. Free particle (47) on a uniform three-point lattice. Let us choose the lattice equation to be

$$E_2 = x_{n+1} - 2x_n + x_{n-1} = 0. \tag{62}$$

Applying pr X to E_2 and substituting for x_{n+1} and u_{n+1}, we find

$$\xi(2x_n - x_{n-1}, 2u_n - u_{n-1}) - 2\xi(x_n, u_n) + \xi(x_{n-1}, u_{n-1}) = 0. \tag{63}$$

Differentiating twice with respect to u_n and u_{n-1}, we find that ξ is linear in u. Substituting $\xi = A(x)u + B(x)$ into (63) we obtain

$$\xi(x_n, u_n) = Au_n + Bx_n + C, \tag{64}$$

and similarly, applying $\mathrm{pr}\, X$ to (47), we obtain

$$\phi(x_n, u_n) = Du_n + Ex_n + F. \tag{65}$$

where $A, \ldots, F$ are constants. Finally, we obtain a six-dimensional symmetry algebra for the O$\varDelta$S (47), (62) with basis $X_1, \ldots, X_6$ as in (45). It has been shown [17] that the entire $\mathrm{sl}(3, \mathbf{R})$ algebra cannot be obtained as the symmetry algebra of any 3-point O$\varDelta$S.

From the above examples we can draw the following conclusions:

1. The Lie point symmetry group of an O$\varDelta$S depends crucially on both equations in system (25). In particular, if we choose a fixed lattice, as in (48) (a "one-point lattice") we are left with point transformations that act on the dependent variable only.
   If we wish to preserve anything like the power of symmetry analysis for differential equations, we must either go beyond point symmetries to generalized ones, or use lattices that are also transformed and that are adapted to the symmetries we consider.
2. The method for calculating symmetries of O$\varDelta$S is reasonably straightforward. It will however involve solving functional equations.

The method can be summarized as follows:

1. Solve equations (25) for two of the variables it contains, to make the equations explicit. For instance, take system (32)–(33). We can solve, e.g., for x_{n+1} and u_{n+1}, and obtain

$$x_{n+1} = f_1(x_{n-1}, x_n, u_{n-1}, u_n) \tag{66}$$
$$u_{n+1} = f_2(x_{n-1}, y_n, u_{n-1}, u_n)$$

2. Apply the prolonged vector field (40) to (25) and substitute (66) for x_{n+1}, u_{n+1}. We obtain two functional equations for ξ and ϕ of the form

$$\left\{ \xi(f_1, f_2) \frac{\partial E_a}{\partial x_{n+1}} + \xi(x_n, u_n) \frac{\partial E_a}{\partial x_n} + \xi(x_{n-1}, u_{n-1}) \frac{\partial E_a}{\partial x_{n-1}} \right.$$
$$+ \phi(f_1, f_2) \frac{\partial E_a}{\partial u_{n+1}} + \phi(x_u, u_n) \frac{\partial E_a}{\partial u_n} \tag{67}$$
$$\left. + \phi(x_{n-1}, u_{n-1}) \frac{\partial E_a}{\partial u_{n-1}} \right\} \bigg|_{\substack{x_{n+1}=f_1 \\ u_{n+1}=f_2}} = 0, \quad a = 1, 2.$$

3. Assume that the functions ξ, ϕ, E_1 and E_2 are sufficiently smooth, and differentiate (67) with respect to the variables x_k and u_k so as to obtain

differential equations for ξ and ϕ. If the original equations are polynomial in all quantities we can thus obtain single-term differential equations from (67). These we must solve, and then substitute back into (67) and solve this equation.

We will illustrate the procedure on several examples in Sect. 2.3.

2.3 Examples of Symmetry Algebras of OΔS

Example 2.4. Monomial nonlinearity on a uniform lattice. Let us first consider the nonlinear ODE,

$$u'' - u^N = 0, \quad N \neq 0, 1. \tag{68}$$

For $N \neq -3$, (68) is invariant under a two-dimensional Lie group whose Lie algebra is given by

$$X_1 = \partial_x, \quad X_2 = (N-1)x\partial_x - 2u\partial_u \tag{69}$$

(translations and dilations). For $N = -3$ the symmetry algebra is three-dimensional, isomorphic to $\mathrm{sl}(2, \mathbf{R})$, *i.e.*, it contains a third element in addition to (69). A convenient basis for the symmetry algebra of the equation

$$u'' - u^{-3} = 0 \tag{70}$$

is

$$X_1 = \partial_x, \quad X_2 = 2x\partial_x + u\partial_u, \quad X_3 = x(x\partial_x + u\partial_u). \tag{71}$$

A very natural OΔS that has (68) as its continuous limit is

$$E_1 = \frac{u_{n+1} - 2u_n + u_{n-1}}{(x_{n+1} - x_n)^2} - u_n^N = 0 \quad N \neq 0, 1 \tag{72}$$

$$E_2 = x_{n+1} - 2x_n + x_{n-1} = 0. \tag{73}$$

Let us now apply the symmetry algorithm described in Sect. 2.2 to system (72)–(73). To illustrate the method, we shall present all calculations in detail.

First, we choose two variables that will be substituted in (41), once the prolonged vector field (40) is applied to system (72)–(73), namely

$$x_{n+1} = 2x_n - x_{n-1} \tag{74}$$

$$u_{n+1} = (x_n - x_{n-1})^2 u_n^N + 2u_n - u_{n-1}.$$

We apply $\mathrm{pr}\, X$ of (40) to (73) and obtain

$$\xi(x_{n+1}, u_{n+1}) - 2\xi(x_n, u_n) + \xi(x_{n-1}, u_{n-1}) = 0, \tag{75}$$

where, x_n, u_n x_{n-1}, u_{n-1} are independent, but x_{n+1}, u_{n+1} are expressed in terms of these quantities, as in (74). Taking this into acccount, we differentiate (75) first with respect to u_{n-1}, then with respect to u_n. We obtain successively

$$-\xi_{,u_{n+1}}(x_{n+1}, u_{n+1}) + \xi_{,u_{n-1}}(x_{n-1}, u_{n-1}) = 0 \tag{76}$$

$$(N(x_n - x_{n-1})^2 u_n^{N-1} + 2)\xi_{,u_{n+1}u_{n+1}}(x_{n+1}, u_{n+1}) = 0. \tag{77}$$

Equation (77) is the desired one-term equation. It implies that

$$\xi(x, u) = a(x)u + b(x). \tag{78}$$

Substituting (78) into (76) we obtain

$$-a(2x_n - x_{n-1}) + a(x_{n-1}) = 0. \tag{79}$$

Differentiating with respect to x_n, we obtain $a = a_0 = $ const. Finally, we substitute (78) with $a = a_0$ into (75) and obtain

$$a = 0, \quad b(2x_n - x_{n-1}) - 2b(x_n) + b(x_{n-1}) = 0, \tag{80}$$

and hence

$$\xi = b = b_1 x + b_0, \tag{81}$$

where b_0 and b_1 are constants. To obtain the function $\phi(x_n, u_n)$, we apply pr X to (72) and obtain

$$\phi(x_{n+1}, u_{n+1}) - 2\phi(x_n, u_n) + \phi(x_{n-1}, u_{n-1})$$
$$-(x_n - x_{n-1})^2 (N\phi(x_n, u_n)u_n^{N-1} + 2b_1 u_n^N) = 0. \tag{82}$$

Differentiating successively with respect to u_{n-1} and u_n (taking (74) into account), we obtain

$$-\phi_{,u_{n+1}}(x_{n+1}, u_{n+1}) + \phi_{,u_{n-1}}(x_{n-1}, u_{n-1}) = 0 \tag{83}$$

$$(N(x_n - x_{n-1})^2 u_n^N + 2)\phi_{,u_{n+1}u_{n+1}} = 0, \tag{84}$$

and hence

$$\phi = \phi_1 u + \phi_0(x), \quad \phi_1 = \text{const}. \tag{85}$$

Equation (82) now reduces to

$$\phi_0(2x_n - x_{n-1}) - 2\phi_0(x_n) + \phi_0(x_{n-1})$$
$$-(x_n - x_{n-1})^2((N-1)\phi_1 + 2b_1)u_n^N$$
$$-N(x_n - x_{n-1})^2 \phi_0 u_n^{N-1} = 0. \tag{86}$$

We have $N \neq 0, 1$ and hence (86) implies that

$$\phi_0 = 0, \quad (N - 1)\phi_1 + 2b_1 = 0. \tag{87}$$

We have thus proven that the symmetry algebra of the OΔS (72)–(73) is the same as that of the ODE (68), namely the algebra (69).

We observe that the value $N = -3$ is not distinguished here and that system (72)–(73) is not invariant under SL(2, $\mathbf{R}$) for $N = -3$. Actually, a difference scheme invariant under SL(2, $\mathbf{R}$) does exist and it will have (70) as its continuous limit. It will not however have the form (72)–(73), and the lattice will not be uniform [17, 19].

Had we taken a two-point lattice, $x_{n+1} - x_n = h$ with h fixed, instead of $E_2 = 0$ as in (73), we would only have obtained translational invariance for the equation (72) and lost the dilational invariance represented by X_2 of (69).

Example 2.5. A nonlinear OΔS on a uniform lattice. We consider

$$E_1 = \frac{u_{n+1} - 2u_n + u_{n-1}}{(x_{n+1} - x_n)^2} - f\left(\frac{u_n - u_{n-1}}{x_n - x_{n-1}}\right) = 0, \tag{88}$$

$$E_2 = x_{n+1} - 2x_n + x_{n-1} = 0, \tag{89}$$

where $f(z)$ is some sufficiently smooth function satisfying

$$f''(z) \neq 0. \tag{90}$$

The continuous limit of (88) and (89) is

$$u'' - f(u') = 0, \tag{91}$$

and it is invariant under a two-dimensional group with Lie algebra,

$$X_1 = \partial_x, \quad X_2 = \partial_u, \tag{92}$$

for any function $f(u')$. For certain functions f the symmetry group is three-dimensional, where the additional basis element of the Lie algebra is

$$X_3 = (ax + bu)\partial_x + (cx + du)\partial_u. \tag{93}$$

The matrix

$$M = \begin{pmatrix} a & b \\ c & d \end{pmatrix} \tag{94}$$

can be transformed into Jordan canonical form, and a different function $f(z)$ is obtained for each canonical form.

Now let us consider the discrete system (88)–(89). Before applying pr X to this system we choose two variables to substitute in (41), namely

$$x_{n+1} = 2x_n - x_{n-1} \tag{95}$$

$$u_{n+1} = 2u_n - u_{n-1} + (x_n - x_{n-1})^2 f\left(\frac{u_n - u_{n-1}}{x_n - x_{n-1}}\right).$$

Applying pr X to (89) we obtain (75) with x_{n+1} and u_{n+1} as in (95). Differentiating twice, with respect to u_{n-1} and u_n successively, we obtain

$$\xi_{,u_{n+1}u_{n+1}}(1 + (x_n - x_{n-1})f')(2 + (x_n - x_{n-1})f') + \xi_{,u_{n+1}}f'' = 0. \quad (96)$$

For $f'' \neq 0$, the only solution is $\xi_{,u_{n+1}} = 0$, i.e., $\xi = \xi(x)$. Substituting back into (75), we obtain

$$\xi = \alpha x + \beta, \quad (97)$$

with $\alpha = $ const, $\beta = $ const.

Now let us apply pr X to E_1 of (88) and (89) and replace x_{n+1}, u_{n+1} as in (95). We obtain the equation

$$\phi(x_{n+1}, u_{n+1}) - 2\phi(x_n, u_n) + \phi(x_{n-1}, u_{n-1}) = 2\alpha(x_n - x_{n-1})^2 f(z)$$
$$+(x_n - x_{n-1})^2 f'(z) \left(\frac{\phi(x_n, u_n) - \phi(x_{n-1}, u_{n-1})}{x_n - x_{n-1}} - \alpha z \right) (98)$$

with α as in (97). Thus, we only need to distinguish between $\alpha = 0$ and $\alpha = 1$. Equation (98) is a functional equation, involving two unknown functions ϕ and f. There are only four independent variables involved, x_n, x_{n-1}, u_n and u_{n-1}. We simplify (98) by introducing new variables $\{x, u, h, z\}$, setting

$$x_n = x, \quad x_{n+1} = x + h, \quad x_{n-1} = x - h \quad (99)$$
$$u_n = u, \quad u_{n-1} = u - hz, \quad u_{n+1} = u + hz + h^2 f(z),$$

where we have used (95) and defined

$$z = \frac{u_n - u_{n-1}}{x_n - x_{n-1}}, \quad h = x_{n+1} - x_n. \quad (100)$$

Equation (98) in these variables is

$$\phi\big(x + h, u + hz + h^2 f(z)\big) - 2\phi(x, u) + \phi(x - h, u - hz)$$
$$= 2\alpha h^2 f(z) + h^2 f'(z) \left(\frac{\phi(x, u) - \phi(x - h, u - hz)}{h} - \alpha z \right). \quad (101)$$

First of all, we observe that for any function $f(z)$ we have two obvious symmetry elements, namely X_1 and X_2 of (92), corresponding to $\alpha = 0$, $\beta = 1$ in (101) (and (97)) and $\phi = 0$ and $\phi = 1$, respectively. Equation (101) is quite difficult to solve directly. However, any three-dimensional Lie algebra of vector fields in 2 variables, containing $\{X_1, X_2\}$ of (92) as a subalgebra, must have X_3 of (93) as its third element. Moreover, (97) shows that we have $b = 0$ in (93) and (94). We set $\alpha = a$ and

$$\phi(x, u) = cx + du. \quad (102)$$

Substituting into (101) we obtain

$$(d - 2a)f(z) = (c + (d - a)z)f'(z). \tag{103}$$

From (103) we obtain two types of solutions:
For $d \neq a$,

$$f = f_0((d - a)z + c)^{(d-2a)/(d-a)}, \quad c \neq 0. \tag{104}$$

For $d = a$,

$$f = f_0 e^{-(a/c)x}. \tag{105}$$

With no loss of generality we could have transformed the matrix (94) with $b = 0$ into Jordan canonical form and we would have obtained two different cases, simplifying (104) and (105), respectively. They are

$$f = f_0 z^N, \quad X_3 = x\partial_x + \frac{N - 2}{N - 1}u\partial_u, \quad N \neq 1, \tag{106}$$

$$f = f_0 e^{-z}, \quad X_3 = x\partial_x + (x + u)\partial_u. \tag{107}$$

The result can be stated as follows. The OΔS (88)–(89) is always invariant under the group generated by $\{X_1, X_2\}$ as in (92). It is invariant under a three-dimensional group with algebra including X_3 as in (93) if f satisfies (103), i.e., has the form (106) or (107). These two cases also exist in the continuous limit. However, one more case exists in the continuous limit, namely

$$u'' = \left(1 + (u')^2\right)^{3/2} e^{k \arctan u'}, \tag{108}$$

with

$$X_3 = (kx + u)\partial_x + (ku - x)\partial_u. \tag{109}$$

This equation can also be discretized in a symmetry preserving way [17], not however on the uniform lattice (89).

3 Lie Point Symmetries of Partial Difference Schemes

3.1 Partial Difference Schemes

In this section we generalize the results of Sect. 2 to the case of two discretely varying independent variables. We follow the ideas and notation of [40]. The generalization to n variables is immediate, though cumbersome. Thus, we will consider a partial difference scheme (PΔS), involving one continuous function of two continuous variables $u(x, t)$. The variables (x, t) are sampled on a two-dimensional lattice, itself defined by a system of compatible relations

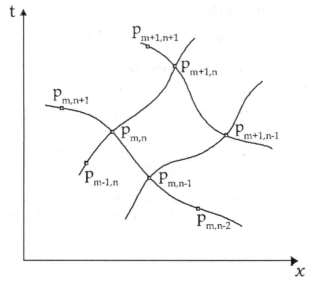

Fig. 1.

between points. Thus, a lattice will be an infinite set of points P_i lying in the real plane $\mathbf{R}^2$. The points will be labelled by two discrete subscripts $P_{m,n}$ with $-\infty < m < \infty$, $-\infty < n < \infty$. The cartesian coordinates of the point $P_{m,n}$ will be denoted $(x_{m,n}, t_{m,n})$, or similarly any other coordinates $(\alpha_{m,n}, \beta_{m,n})$.

A two-variable P$\varDelta$S will be a set of five relations between the quantities $\{x, t, u\}$ at a finite number of points. We choose a reference point $P_{m,n} \equiv P$ and two families of curves intersecting at the points of the lattice. The labels $m = m_0$ and $n = n_0$ will parametrize these curves (see Fig. 1). To define an orientation of the curves, we specify

$$x_{m+1,n} - x_{m,n} \equiv h_m > 0, \quad t_{m,n+1} - t_{m,n} \equiv h_n > 0 \qquad (110)$$

at the original reference point.

The actual curves and the entire P$\varDelta$S are specified by the 5 relations,

$$E_a(\{x_{m+i,n+j}, t_{m+i,n+j}, u_{m+i,n+j}\}) = 0$$
$$1 \le a \le 5 \quad i_1 \le i \le i_2 \quad j_i \le j \le j_2. \qquad (111)$$

In the continuous limit, if one exists, all five equations (111) are supposed to reduce to a single PDE, e.g., $E_1 = 0$ can reduce to the PDE and $E_a = 0$, $a \ge 2$ to $0 = 0$. The orthogonal uniform lattice of Fig. 2 is clearly a special case of that on Fig. 1.

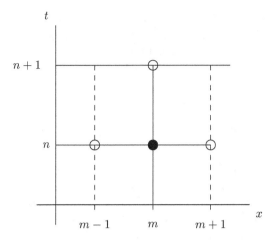

Fig. 2.

Some independence conditions must be imposed on system (111), *e.g.*,

$$|J| = \left| \frac{\partial(E_1, \ldots, E_5)}{\partial(x_{m+i_2,n}, t_{m+i_2n}, x_{m,n+j_2}, t_{m,n+j_2}, u_{m+i_2,n+j_2})} \right| \neq 0. \qquad (112)$$

This condition allows us to move upward and to the right along the curves passing through $P_{m,n} = (x_{m,n}, t_{m,n})$. Moreover, compatibility of the five equations (111) must be assured.

As an example of a PΔS, let us consider the linear heat equation on a uniform and orthogonal lattice. The heat equation in the continuous case is

$$u_t = u_{xx}. \qquad (113)$$

An approximation on a uniform orthogonal lattice is given by the five equations,

$$E_1 = \frac{u_{m,n+1} - u_{m,n}}{h_2} - \frac{u_{m+1,n} - 2u_{m,n} + u_{m-1,n}}{(h_1)^2} = 0 \qquad (114)$$

$$E_2 = x_{m+1,n} - x_{m,n} - h_1 = 0 \qquad E_3 = t_{m+1,n} - t_{m,n} = 0 \qquad (115)$$

$$E_4 = x_{m,n+1} - x_{m,n} = 0 \qquad E_5 = t_{m,n+1} - t_{m,n} - h_2 = 0. \qquad (116)$$

Equations (115) and (116) can of course be integrated to give the standard expressions

$$x_{m,n} = h_1 m + x_0 \qquad t_{m,n} = h_2 n + t_0. \qquad (117)$$

Observe that h_1 and h_2 are constants that cannot be scaled, they are fixed in eqs. (115), (116). On the other hand x_0 and t_0 are integration constants and are thus not fixed by system (115)–(116). As written, these equations are invariant under translations, but not under dilations.

Finally, we observe that the usual fixed lattice condition is obtained from (117) by setting $x_0 = t_0 = 0$, $h_1 = h_2 = 1$ and identifying

$$x = m, \quad t = n. \tag{118}$$

Though the above example is essentially trivial, it brings out several points.

1. Four equations are indeed needed to specify a two-dimensional lattice and to allow us to move along the coordinate lines.
2. In order to solve the P$\varDelta$S (114), (116) for h_1 and h_2 given, we must specify for instance $\{x_{m,n}, t_{m,n}, u_{m,n}, u_{m+1,n}, u_{m-1,n}\}$. Then we can directly calculate $\{x_{m+1,n}, t_{m+1,n}\}$, $\{x_{m,n+1}, t_{m,n+1}\}$. In order to calculate the coordinates of the fourth point figuring in (114), namely $\{x_{m,n-1}, t_{m,n-1}\}$ we must shift (115) down by one unit in m.
3. The Jacobian condition (112), allowing us to perform these calculations, is obviously satisfied, since we have

$$\left| \frac{\partial(E_1, E_2, E_3, E_4, E_5)}{\partial(x_{m+1,n}, t_{m+1,n}, x_{m,n+1}, t_{m,n+1}, u_{m,n+1})} \right| = 1. \tag{119}$$

A partial difference scheme with one dependent and n independent variables will involve $n^2 + 1$ relations between the variables $(x_1, x_2, \ldots x_n, u)$, evaluated at a finite number of points.

3.2 Symmetries of Partial Difference Schemes

As in the case of O$\varDelta$S treated in Sect. 2, we shall restrict ourselves to point transformations

$$\tilde{x} = F_\lambda(x, t, u), \quad \tilde{t} = G_\lambda(x, t, u), \quad \tilde{u} = H_\lambda(x, t, u). \tag{120}$$

The requirement is that $\tilde{u}_\lambda(\tilde{x}, \tilde{t})$ should be a solution, whenever it is defined and whenever $u(x, t)$ is a solution. The group action (120) should be defined and invertible, at least locally, in some neighbourhood of the reference point $P_{m,n}$, including all points $P_{m+i,n+j}$ involved in system (111).

As in the case of a single independent variable, we shall consider infinitesimal transformations that allow us to use Lie algebraic techniques. Instead of transformations (120) we consider

$$\begin{aligned} \tilde{x} &= x + \lambda \xi(x, t, u), \\ \tilde{t} &= t + \lambda \tau(x, t, u), \\ \tilde{u} &= u + \lambda \phi(x, t, u), \quad |\lambda| \ll 1. \end{aligned} \tag{121}$$

Once the functions ξ, τ and ϕ are determined from the invariance requirement, then the actual transformations (120) are determined by integration, as in (5), (6).

The transformations act on the entire space (x, t, u), at least locally. This means that the same functions F, G and H in (120), or ξ, τ and ϕ in (121) determine the transformations of all points.

We formulate the problem of determining the symmetries (121), and ultimately (120), in terms of a Lie algebra of vector fields of the form

$$X = \xi(x, t, u)\partial_x + \tau(x, t, u)\partial_t + \phi(x, t, u)\partial_u, \qquad (122)$$

where ξ, τ and ϕ are the same as in (121). The operator (122) acts at one point only, namely $(x, t, u) \equiv (x_{m,n}, t_{m,n}, u_{m,n})$. Its prolongation will act at all points figuring in system(111) and we set

$$\mathrm{pr}\, X = \sum_{j,k}(\xi(x_{jk}, t_{jk}, u_{jk})\partial_{x_{jk}} + \tau(x_{jk}, t_{jk}, u_{jk})\partial_{t_{jk}} \qquad (123)$$

$$+ \phi(x_{jk}, t_{jk}, u_{jk})\partial_{u_{jk}}),$$

where the sum is over all points figuring in (111). To simplify notation we set

$$\xi_{jk} \equiv \xi(x_{jk}, t_{jk}, u_{jk}), \quad \tau_{jk} \equiv \tau(x_{jk}, t_{jk}, u_{jk}) \qquad (124)$$
$$\phi_{jk} \equiv \phi(x_{jk}, t_{jk}, u_{jk}).$$

The functions ξ, τ, and ϕ figuring in (122) and (123) are determined from the invariance condition

$$\mathrm{pr}\, X E_a \mid_{E_1 = \cdots = E_5 = 0} = 0, \quad a = 1, \ldots 5. \qquad (125)$$

It is (125) that provides an algorithm for determining the symmetry algebra, $i.e.$, the coefficients ξ, τ and ϕ.

The procedure is the same as in the case of ordinary difference schemes, described in Sect. 2. In the case of system (111), we proceed as follows:

1. Choose 5 variables v_a to eliminate from the condition (125) and express them in terms of the other variables, using system (111) and the Jacobian condition (112). For instance, we can choose

$$v_1 = x_{m+i_2,n}, \quad v_2 = t_{m+i_2,n}, \qquad (126)$$

$$v_3 = x_{m,n+j_2}, \quad v_4 = t_{m,n+j_2}, \quad v_5 = u_{m+i_2,j+i_2}$$

and use (111) to express

$$v_a = v_a(x_{m+i,n+j}, t_{m+i,n+j}, u_{m+i,n+j})$$
$$i_1 \le i \le i_2 - 1, \quad j_1 \le j \le j_2 - 1.$$

The quanties v_a must be chosen consistently. None of them can be a shifted value of another one (in the same direction). No relations between the quantities v_a should follow from system (111). Once eliminated from (124), they should not reappear due to shifts. For instance, the choice (126) is consistent if $m + i_2$ and $n + j_2$ are the highest values of these labels that figure in (111).

2. Once the quantities v_a are eliminated from system (125), using (3.2), each remaining value of $x_{i,k}$, $t_{i,k}$ and $u_{i,k}$ is independent. Each of them can figure in the corresponding functions $\xi_{i,k}$, $\tau_{i,k}$, $\phi_{i,k}$ (see (124)), in the functions E_a directly, or by means of the expressions v_a, in the functions ξ, τ and ϕ with different labels. This provides a system of five functional equations for ξ, τ and ϕ.

3. Assume that the dependence of ξ, τ and ϕ on x, t and u is analytic. Convert the obtained functional equations into differential equations by differentiating with respect to $x_{i,k}$, $t_{i,k}$, or $u_{i,k}$. This provides an overdetermined system of differential equations that we must solve. If possible, use multiple differentiations to obtain single-term differential equations that are easy to solve.

4. Substitute the solution of the differential equations back into the original functional equations and solve these. The differential equations are consequences of the functional ones and hence have more solutions. The functional equations impose further restrictions on the constants and arbitrary functions appearing in the integration of the differential consequences.

Let us now consider examples on different lattices.

3.3 The Discrete Heat Equation

The Continuous Heat Equation. The symmetry group of the continuous heat equation (113) is well known [52]. Its symmetry algebra has the structure of a semidirect sum

$$L = L_0 \uplus L_1, \tag{127}$$

where L_0 is six-dimensional and L_1 is an infinite-dimensional ideal corresponding to the linear superposition principle (present for any linear PDE). A convenient basis for this algebra is given by the vector fields

$$
\begin{aligned}
&P_0 = \partial_t, \quad D = 4t\partial_t + 2x\partial_x + u\partial_u, \\
&K = 4t(t\partial_t + x\partial_x) + (x^2 + 2t)u\partial_u, \\
&P_1 = \partial_x, \quad B = 2t\partial_x + xu\partial_u, \quad W = u\partial_u, \\
&S = S(x,t)\partial_u, \quad S_t - S_{xx} = 0.
\end{aligned}
\tag{128}
$$

$$\text{(129)}$$

The $sl(2, \mathbf{R})$ subalgebra $\{P_0, D, K\}$ represents time translations, dilations and "expansions". The Heisenberg subalgebra $\{P_1, B, W\}$ represents space translations, Galilei boosts and the possibility of multiplying a solution u by a constant. The presence of S simply tells us that we can add a solution to any given solution. Thus, S and W correspond to the linearity, P_0 and P_1 the fact that the equation is autonomous, $i.e.$, has constant coefficients.

Discrete Heat Equation on Fixed Rectangular Lattice. Let us consider the discrete heat equation (114) on the four-point uniform orthogonal lattice (115), (116). We apply the prolonged operator (113) to the equations for the lattice and obtain

$$\xi(x_{m+1,n}, t_{m+1,n}, u_{m+1,n}) - \xi(x_{m,n}, t_{m,n}, u_{m,n}) = 0$$
$$\xi(x_{m,n+1}, t_{m,n+1}, u_{m,n+1}) - \xi(x_{m,n}, t_{m,n}, u_{m,n}) = 0 \tag{130}$$

and similarly for $\tau(x, t, u)$. The quantities v_i of (126) can be chosen to be

$$v_1 = x_{m+1,n} \quad v_2 = t_{m+1,n} \quad v_3 = x_{m,n+1},$$
$$v_4 = t_{m,n+1}, \quad v_5 = u_{m,n+1}. \tag{131}$$

However, in (130) $u_{m+1,n}$ and $u_{m,n+1}$ cannot be expressed in terms of $u_{m,n}$, since (114) also involves $u_{m-1,n}$. Differentiating (130) with respect to, e.g., $u_{m,n}$. we find that ξ cannot depend on u:

$$\frac{\partial \xi(x_{m,n}, t_{m,n}, u_{m,n})}{\partial u_{m,n}} = 0. \tag{132}$$

Since we have $t_{n+1,n} = t_{m,n}$ and $x_{m,n+1} = x_{m,n}$ the two equations (130) yield

$$\frac{\partial \xi_{m,n}}{\partial x_{m,n}} = 0, \quad \frac{\partial \xi_{m,n}}{\partial t_{m,n}} = 0, \tag{133}$$

respectively. The same is obtained for the coefficient τ, so finally we have

$$\xi = \xi_0, \quad \tau = \tau_0, \tag{134}$$

where ξ_0 and τ_0 are constants.

Now let us apply pr X to (114). We obtain

$$\phi_{m,n+1} - \phi_{m,n} - \frac{h_2}{(h_1)^2}(\phi_{m+1,n} - 2\phi_{m,n} + \phi_{m-1,n}) = 0. \tag{135}$$

In more detail, eliminating the quantities v_a in eq (131) we have

$$\phi(x_{m,n}, t_{m,n} + h_2, u_{m,n} + \frac{h_2}{h_1^2}(u_{m+1,n} - 2u_{m,n} + u_{m-1,n}))$$

$$-\phi(x_{m,n}, t_{m,n}, u_{m,n}) - \frac{h_2}{h_1^2}(\phi(x_{m,n} + h_1, t_{m,n}, u_{m+1,n}) - 2\phi(x_{m,n}, t_{m,n}, u_{m,n})$$

$$+\phi(x_{m,n} - h_1, t_{m,n}, u_{m-1,n})) = 0. \tag{136}$$

We differentiate (136) twice, with respect to $u_{m+1,n}$ and $u_{m-1,n}$ respectively. We obtain

$$\frac{\partial^2 \phi_{m,n+1}}{\partial u_{m,n+1}^2} = 0, \tag{137}$$

that is

$$\phi_{m,n} = A(x_{m,n}, t_{m,n})u_{m,n} + B(x_{m,n}, t_{m,n}). \tag{138}$$

We substitute $\phi_{m,n}$ of (138) back into (136) and set the coefficients of $u_{m+1,n}$, $u_{m,n}$, $u_{m-1,n}$ and 1 equal to zero separately. From the resulting determining equations we find that $A(x_{m,n}, t_{m,n}) = A_0$ must be constant and that $B(x,t)$ must satisfy the discrete heat equation (114). The result is that the symmetry algebra of system (114)–(116) is very restricted. It is generated by

$$P_0 = \partial_t, \quad P_1 = \partial_x, \quad W = u\partial_u, \quad S = S(x,t)\partial_u \tag{139}$$

and reflects only the linearity of the system and the fact that it is autonomous.

The dilations, expansions and Galilei boosts, generated by D, K and B of (128) in the continuous case are absent on the lattice (115) and (116). Other lattices will allow other symmetries.

Discrete Heat Equation Invariant Under Dilations. Let us now consider a five-point lattice that is also uniform and orthogonal. We set

$$\frac{u_{m,n+1} - u_{m,n}}{t_{m,n+1} - t_{m,n}} = \frac{u_{m+1,n} - 2u_{m,n} + u_{m-1,n}}{(x_{m+1,n} - x_{m,n})^2} \tag{140}$$

$$x_{m+1,n} - 2x_{m,n} + x_{m-1,n} = 0 \qquad x_{m,n+1} - x_{m,n} = 0 \tag{141}$$

$$t_{m+1,n} - t_{m,n} = 0 \qquad t_{m,n+1} - 2t_{m,n} + t_{m,n-1} = 0. \tag{142}$$

The variables v_a that we shall substitute from (140), (141) and (142) are $x_{m+1,n}, t_{m+1,n}, x_{m,n+1}, t_{m,n+1}$ and $u_{m,n+1}$. Applying pr X to (141) we obtain

$$\xi(2x_{m,n} - x_{m-1,n}, t_{m,n}, u_{m+1,n}) - 2\xi(x_{m,n}, t_{m,n}, u_{m,n})$$
$$+\xi(x_{m-1,n}, t_{m,n}, u_{m-1,n}) = 0 \tag{143}$$

$$\xi(x_{m,n}, 2t_{m,n} - t_{m,n-1}, u_{m,n+1}) - \xi(x_{m,n}, t_{m,n}, u_{m,n}) = 0. \tag{144}$$

In (144) $u_{m,n}$ and $u_{m,n+1}$ are independent. Differentiating with respect to $u_{m,n}$ we find $\partial\xi_{m,n}/\partial u_{m,n} = 0$ and hence ξ does not depend on u. Differentiating (144) with respect to $t_{m,n-1}$ we obtain $\partial\xi_{m,n+1}/\partial t_{m,n+1} = 0$. Thus, ξ depends on x alone. Equation (143) can then be solved and we find that ξ is linear in x. Applying pr X to (142) we obtain similar results for $\tau(x, t, u)$. Finally, invariance of the lattice equations (141) and (142) implies:

$$\xi = ax + b, \quad \tau = ct + d. \tag{145}$$

Let us now apply pr X to (140). We obtain, after using the P$\varDelta$S (140) - (142)

$$\frac{\phi_{m,n+1} - \phi_{m,n}}{t_{m,n+1} - t_{m,n}} - \frac{\phi_{m+1,n} - 2\phi_{m,n} + \phi_{m-1,n}}{(x_{m+1,n} - x_{m,n})^2}$$
$$+(2a - c)\frac{u_{m+1,n} - 2u_{m,n} + u_{m-1,n}}{(x_{m+1,n} - x_{m,n})^2} = 0. \tag{146}$$

Observe that $u_{m,n+1}$ (and hence $\phi_{m,n+1}$) depends on $u_{m+1,n}$ and $u_{m-1,n}$, whereas all terms in (146) depend on at most one of these quantities. Taking the second derivative $\partial_{u_{m+1,n}}\partial_{u_{m-1,n}}$ of (146), we find

$$\frac{\partial^2 \phi_{m,n+1}}{\partial u^2_{m,n+1}} = 0, \ i.e., \ \phi = A(x,t)u + B(x,t). \tag{147}$$

We substitute this expression back into (146) and find

$$A(x,t) = A_0 = \text{const} \tag{148}$$

and see that $B(x,t)$ must satisfy system (140)–(142). Moreover, we find $c = 2a$ in (145). Finally, the symmetry algebra has the basis

$$P_0 = \partial_t, \quad P_1 = \partial_x, \quad W = u\partial_u, \quad D = x\partial_x + 2t\partial_t, \tag{149}$$
$$S = S(x,t)\partial_u. \tag{150}$$

Thus, the dilational invariance is recovered, but not the Galilei invariance. Other symmetries can be recovered on other lattices.

3.4 Lorentz Invariant Difference Schemes

The Continuous Case. Let us consider the PDE

$$u_{xx} - u_{tt} = 4f(u). \tag{151}$$

Equation (151) is invariant under the Poincaré group of $1 + 1$ dimensional Minkowski space for any function $f(u)$. Its Lie algebra is represented by

$$P_0 = \partial_t, \quad P_1 = \partial_x, \quad L = t\partial_x + x\partial_t. \tag{152}$$

For specific interactions $f(u)$ the symmetry algebra may be larger, in particular for $f = e^u$, $f = u^N$, or $f = \alpha u + \beta$.

Before presenting a discrete version of (151), we find it convenient to change to light-cone coordinates

$$y = x + t, \quad z = x - t \tag{153}$$

in which (151) is rewritten as

$$u_{yz} = f(u) \tag{154}$$

and the Poincaré symmetry algebra (152) is

$$P_1 = \partial_y, \quad P_2 = \partial_z, \quad L = y\partial_y - z\partial_z. \tag{155}$$

A Discrete Lorentz Invariant Scheme. A particular PΔS that has (154) as its continuous limit is

$$\frac{u_{m+1,n+1} - u_{m,n+1} - u_{m+1,n} + u_{m,n}}{(y_{m+1,n} - y_{m,n})(z_{m,n+1} - z_{m,n})} = f(u_{m,n}) \tag{156}$$

$$y_{m+1,n} - 2y_{m,n} + y_{m-1,n} = 0, \quad y_{m,n+1} - y_{m,n} = 0 \tag{157}$$

$$z_{m+1,n} - z_{m,n} = 0, \quad z_{m,n+1} - 2z_{m,n} + z_{m,n-1} = 0. \tag{158}$$

To find the Lie point symmetries of this difference scheme, we set

$$X = \eta(y, z, u)\partial_y + \xi(y, z, u)\partial_z + \phi(y, z, u)\partial_u. \tag{159}$$

We apply the prolonged vector field pr X first to (157) and (158), eliminate $y_{m+1,n}$, $y_{m,n+1}$, $z_{m+1,n}$ and $z_{m,n+1}$, using system (157)–(158), and observe that all values of u_{ik} that figure in the equations obtained for η_{ik} and ξ_{ik} are independent.

The result that we obtain is that η and ξ must be independent of u and linear in y and z, respectively. Finally we obtain

$$\xi = \alpha y + \gamma, \quad \eta = \beta z + \delta, \tag{160}$$

where $\alpha, \ldots, \delta$ are constants. Invariance of (156) implies that the coefficient ϕ in the vector field (159) must be linear in u and moreover have the form

$$\phi = Au + B(y, z), \tag{161}$$

where A is a constant. Taking (160) and (161) into account and applying pr X to (156), we obtain

$$(A - \alpha - \beta)f(u_{m,n}) + \frac{B_{m+1,n+1} - B_{m,n+1} - B_{m+1,n} + B_{m,n}}{(y_{m+1,n} - y_{m,n})(z_{m,n+1} - z_{m,n})}$$
$$= (Au_{m,n} + B_{m,n})f'(u_{m,n}). \tag{162}$$

Differentiating (162) with respect to $u_{m,n}$ we finally obtain the following determining equation:

$$(\alpha + \beta)\frac{df}{du_{m,n}} + (Au_{m,n} + B(y_{m,n}, z_{m,n}))\frac{d^2 f}{du_{m,n}^2} = 0. \tag{163}$$

For $f(u_{m,n})$ arbitrary, we find $\beta = -\alpha$, $A = B = 0$. Thus for arbitrary $f(u)$ the scheme (156)–(157) has the same symmetries as its continuous limit. The point symmetry algebra is given by (155), *i.e.*, it generates translations and Lorentz transformations.

Now let us seek special cases of $f(u)$ when further symmetries exist. That means that (163) must be solved in a nontrivial manner. Let us restrict ourselves to the case when the interaction is nonlinear, *i.e.*,

$$\frac{d^2}{du_{m,n}^2} f(u_{m,n}) \neq 0. \tag{164}$$

Then we must have

$$B(y_{m,n}, z_{m,n}) = B = \text{const}. \tag{165}$$

The equation to be solved for $f(u)$ is actually (162) which simplifies to

$$(Au + B)f'(u) = (A - \alpha - \beta)f(u). \tag{166}$$

For $A \neq 0$ the solution of (166) is

$$f = f_0 u^p, \tag{167}$$

and the symmetry is

$$D_1 = y\partial_y + z\partial_z - \frac{2}{p-1} u\partial_u. \tag{168}$$

For $A = 0$, $B \neq 0$ we obtain

$$f = f_0 e^{pu} \tag{169}$$

and the additional symmetry is

$$D_2 = y\partial_y + z\partial_z - 2\partial_u. \tag{170}$$

Thus, for nonlinear interactions $f(u)$, $f'' \neq 0$, the PΔS (156)–(158) has exactly the same point symmetries as its continuous limit (154).

The linear case

$$f(u) = Ru + T \tag{171}$$

is different. The PDE (154) in this case is conformally invariant. This infinite-dimensional symmetry algebra is not present for the discrete case considered in this section.

4 Symmetries of Discrete Dynamical Systems

4.1 General Formalism

In this section we shall discuss differential-difference equations on a fixed one-dimensional lattice. Thus, time t will be a continuous variable, $n \in \mathbf{Z}$ a discrete one. We will be modeling discrete monoatomic or diatomic molecular chains with equally-spaced rest positions. The individual atoms will be vibrating around their rest positions. For monoatomic chains the actual position of the n-th atom is described by one continuous variable $u_n(t)$. For diatomic atoms there will be two such functions, $u_n(t)$ and $v_n(t)$.

Only nearest neighbour interaction will be considered. The interactions are described by a priori unspecified functions. Our aim is to classify these functions according to their symmetries.

Three different models have been studied [45, 24, 35]. They correspond to Figs. 3, 4 and 5, respectively.

The model illustrated in Fig. 3 corresponds to the equation [45]

$$\ddot{u}_n(t) - F_n\big(t, u_{n-1}(t), u_n(t), u_{n+1}(t)\big) = 0. \tag{172}$$

Figure 4 could correspond to a very primitive model of the DNA molecule. The equations are [24]

$$\ddot{u}_n = F_n\big(t, u_{n-1}(t), u_n(t), u_{n+1}(t), v_{n-1}(t), v_n(t), v_{n+1}(t)\big) = 0$$
$$\ddot{v}_n = G_n\big(t, u_{n-1}(t), u_n(t), u_{n+1}(t), v_{n-1}(t), v_n(t), v_{n+1}(t)\big) = 0. \tag{173}$$

The model corresponding to Fig. 5 already took translational and Galilei invariance into account, so the equations are

$$\ddot{u}_n = F_n(\xi_n, t) + G_n(\eta_{n-1}, t)$$
$$\ddot{v}_n = K_n(\xi_n, t) + P_n(\eta_n, t) \tag{174}$$
$$\xi_n = y_n - x_n, \quad \eta_n = x_{n+1} - y_n.$$

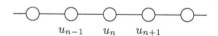

Fig. 3. A monoatomic chain

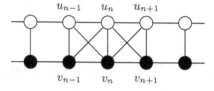

Fig. 4. A diatomic molecule with two types of atoms on parallel chains

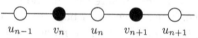

Fig. 5. A diatomic molecule with two types of atoms alternating along one chain

Dissipation was ignored in all three cases, so no first derivatives are present.

In these lectures we shall only treat the case (172). The lattice is fixed, i.e., it is given by the relation

$$x_n = hn + x_0, \tag{175}$$

with h and x_0 given constants. With no loss of generality we can choose $h = 1$, $x_0 = 0$, so that $x_n = n$.

Our aim is to find all functions F_n for which (172) admits a nontrivial group of local Lie point transformations. We shall also assume that the interaction is nonlinear and that it does indeed couple neighbouring states.

Let us summarize the conditions imposed on the model (172) in order to study its symmetries.

1. The lattice is fixed and regular ($x_n = n$).
2. The interaction involves nearest neighbours only, is nonlinear and coupled, i.e.,

$$\frac{\partial^2 F_n}{\partial u_i \partial u_k} \neq 0, \quad \frac{\partial F_n}{\partial u_{n-1}} \neq 0, \quad \frac{\partial F_n}{\partial u_{n+1}} \neq 0. \tag{176}$$

3. We consider point symmetries only. Since the lattice is fixed, the transformations are generated by vector fields of the form [40]

$$X = \tau(t)\partial_t + \phi_n(t, u_n)\partial_{u_n}. \tag{177}$$

We also assume that $\tau(t)$ is an analytic function of t and $\phi_n(t, u_n)$ is also analytic as a function of t and u_n.

The symmetry algorithm is the usual one, namely

$$\text{pr } XE_n \,|_{E_n=0} = 0. \tag{178}$$

The prolongation in (178) involves a prolongation to t-derivatives $\dot{u}_n$ and $\ddot{u}_n$, and to all values of n figuring in (172), i.e., $n \pm 1$.

The terms that we actually need in the prolongation are

$$\text{pr}^{(2)} X = \tau\partial_t + \sum_{k=n-1}^{n+1} \phi_k(t, u_k)\partial_{u_k} + \phi_n^{tt}\partial_{\ddot{u}_n}. \tag{179}$$

The coefficient ϕ_n^{tt} is calculated using the formulas of Sect. 1 (or e.g., [52]). We have

$$\phi_n^{tt} = D_t^2\phi_n - (D_t^2\tau)u_n - 2(D_t\tau)\ddot{u}_n. \tag{180}$$

Applying $\text{pr}^{(2)} X$ to (172) and replacing $\ddot{u}$ from that equation, we obtain an expression involving $(\dot{u}_n)^3$, $(\dot{u}_n)^2$, $(\dot{u}_n)^1$ and $(\dot{u}_n)^0$. The coefficients of all these terms must vanish separately. The first three of these equations do not depend on F_n and can be solved easily. They imply

$$\phi_n(t, u_n) = \left(\frac{1}{2}\dot{\tau}(t) + a_n\right) u_n + \beta_n(t), \quad \tau = \tau(t), \quad \dot{a}_n = 0. \quad (181)$$

The remaining determining equation is

$$\frac{1}{2}\tau_{ttt}u_n + \beta_{n,tt} + \left(a_n - \frac{3}{2}\tau_t\right) F_n$$

$$-\tau F_{n,t} - \sum_\alpha \left(\left(\frac{1}{2}\tau_t + a_\alpha\right) u_\alpha + \beta_\alpha\right) F_{n,u_\alpha} = 0, \quad (182)$$

and the vector fields realizing the symmetry algebra are

$$X = \tau(t)\partial_t + \left(\left(\frac{1}{2}\dot{\tau}(t) + a_n\right) u_n + \beta_n(t)\right) \partial_{u_n}. \quad (183)$$

Since we are classifying the interactions F_n, we must decide which functions F_n will be considered to be equivalent. To do this we introduce a group of "allowed transformations", or a "classifying group". We define this to be a group of fiber-preserving point transformations

$$u_n(t) = \Omega_n(\tilde{u}_n(\tilde{t}), \tilde{t}, g), \quad \tilde{t} = \tilde{t}(t, g), \quad \tilde{n} = n, \quad (184)$$

taking (172) into an equation of the same form

$$\ddot{\tilde{u}}_n(\tilde{t}) = \tilde{F}_n\left(\tilde{t}, \tilde{u}_{n-1}(\tilde{t}), \tilde{u}_n(\tilde{t}), \tilde{u}_{n+1}(\tilde{t})\right) = 0. \quad (185)$$

That is, the allowed transformations can change the function F_n (as opposed to symmetry transformations), but cannot introduce first derivatives, or terms other than nearest-neighbour terms. These conditions narrow down the transformations (184) to linear ones of the form

$$u_n(t) = \frac{A_n}{\sqrt{\tilde{t}_t}}\tilde{u}_n(\tilde{t}) + B_n(t), \quad \tilde{t} = \tilde{t}(t),$$

$$A_{n,t} = 0, \quad \tilde{t}_t \neq 0, \quad A_n \neq 0, \quad \tilde{n} = n. \quad (186)$$

Equation (172) is transformed into

$$\tilde{u}_{n,\tilde{t}\tilde{t}} = A_n^{-1}(\tilde{t}_t)^{-3/2}\Bigg\{ F_n(t, u_{n-1}, u_n, u_{n+1})$$

$$+ \left(-\frac{3}{4}A_n(\tilde{t}_t)^{-5/2}(\tilde{t}_{tt})^2\right.$$

$$\left. + \frac{A_n}{2}(\tilde{t}_t)^{-3/2}\tilde{t}_{ttt}\right)\tilde{u}_n(\tilde{t}) - B_{n,tt}\Bigg\}. \quad (187)$$

The transformed vector field (183) is

$$\widehat{X} = \tau(t)\tilde{t}_t(t)\partial_{\tilde{t}} + \left\{ \left(\frac{1}{2}(\tau(t)\tilde{t}_t)_{\tilde{t}} + a_n \right) \tilde{u}_n \right.$$

$$\left. + (\tilde{t}_t)^{1/2} A_n^{-1} \left(\left(\frac{1}{2}\tau_t + a_n \right) B_n + \beta_n - \tau B_{n,t} \right) \right\} \partial_{\tilde{u}_n}. \quad (188)$$

In (187) and (188), $\tau(t)$, a_n, β_n and F_n are given, whereas $\tilde{t}(t)$, A_n and $B_n(t)$ are ours to choose. We use these quantities to simplify the expression of the vector field $\widehat{X}$.

Our classification strategy is the following. We first classify one-dimensional subalgebras. Thus, we have one vector field of the form (183). If $\tau(t)$ satisfies $\tau(t) \neq 0$ in some open interval, we use $\tilde{t}(t)$ to normalize $\tau(t) = 1$ and $B_n(t)$ to transform $\beta_n(t)$ into $\beta_n(t) = 0$. If we have $\tau(t) = 0$, $a_n \neq 0$, we use $B_n(t)$ to annul $\beta_n(t)$. The last possibility is $\tau(t) = 0$, $a_n = 0$, $\beta_n(t) \neq 0$. Then we cannot simplify further. The same transformations will also simplify the determining equation (182) and we can, in each case, solve it for the interaction $F_n(t, u_{n-1}, u_n, u_{n+1})$.

Once all interactions allowing one-dimensional symmetry algebras are determined, we proceed using results on the structure of Lie algebras, to be described in Sect. 4.4. We first find all Abelian symmetry groups and the corresponding interactions allowing them. We run through our list of one-dimensional algebras and take them in an already established "canonical" form. Let us call this element X_1 (in each case). We then find all elements X of the form (183) that satisfy $[X_1, X] = 0$. We classify the obtained operators X under the action of a subgroup of the group of allowed transformations, namely the isotropy group of X_1 (the group that leaves the one-dimensional subalgebra generated by X_1 invariant). For each Abelian group we find the invariant interaction.

From Abelian symmetry algebras we proceed to nilpotent ones, then to solvable ones and finally to nonsolvable ones. These can be semisimple, or they may have a nontrivial Levi decomposition.

All details can be found in the original article [45], here we shall present the main results.

4.2 One-Dimensional Symmetry Algebras

Three classes of one-dimensional symmetry algebras exist. Together with their invariant interactions, they can be represented by

$$A_{1,1} \qquad X = \partial_t + a_n u_n \partial_{u_n}$$
$$F_n(t, u_k) = f_n(\xi_{n-1}, \xi_n, \xi_{n+1})e^{a_n t} \qquad (189)$$
$$\xi_k = u_k e^{-a_k t}, \quad k = n - 1, n, n + 1.$$

$$A_{1,2} \qquad X = a_n u_n \partial_{u_n}$$

$$F_n(t, u_k) = u_n f_n(t, \xi_{n-1}, \xi_{n+1}) \qquad (190)$$

$$\xi_{n\pm1} = u_{n\pm1}^{a_n} u_n^{-a_{n\pm1}}.$$

$$A_{1,3} \qquad X = \beta_n(t)\partial_{u_n}$$

$$F_n(t, u_k) = \frac{\ddot{\beta}_n}{\beta_n} u_n + f_n(t, \xi_{n-1}, \xi_{n+1}) \qquad (191)$$

$$\xi_{n\pm1} = \beta_n(t)u_{n\pm1} - \beta_{n\pm1}(t)u_n.$$

We see that the existence of a one-dimensional Lie algebra implies that the interaction F is an arbitrary function of three variables, rather than the original four. The actual form of the interaction in (189), (190) and (191) was obtained by solving (182), once the canonical form of vector field X in (189), (190), or (191) was taken into account.

4.3 Abelian Lie Algebras of Dimension $N \geq 2$

Without proof we state several theorems.

Theorem 4.1. *An Abelian symmetry algebra of* (172) *can have dimension N satisfying $1 \leq N \leq 4$.*

Comment: For $N = 1$ these are the algebras $A_{1,1}$, $A_{1,2}$ and $A_{1,3}$ of (189), (190) and (191).

Theorem 4.2. *Five distinct classes of interactions F_n exist having symmetry algebras of dimension $N = 2$. For four of them the interaction will involve an arbitrary function of two variables, for the fifth a function of three variables.*

The five classes can be represented by the following algebras and interactions.

$$A_{2,1}: \quad X_1 = \partial_t + a_{1,n}u_n\partial_{u_n}, \quad X_2 = a_{2n}u_n\partial_{u_n}$$

$$F_n = u_n f_n(\xi_{n-1}, \xi_{n+1}), \quad a_{2n} \neq 0 \qquad (192)$$

$$\xi_k = u_k^{a_{2n}} u_n^{-a_{2k}} e^{(a_{1,n}a_{2k} - a_{1k}a_{2n})t}, \quad k = n \pm 1$$

$$A_{2,2}: \quad X_1 = \partial_t + a_n u_n\partial_{u_n} \quad X_2 = e^{a_n t}\partial_{u_n}$$

$$F_n = a_n^2 u_n + e^{a_n t} f_n(\xi_{n-1}, \xi_{n+1}) \qquad (193)$$

$$\xi_k = u_k e^{-a_k t} - u_n e^{-a_n t}, \quad k = n \pm 1$$

$A_{2,3}:$ $\quad X_1 = a_{1,n}u_n\partial_{u_n}\quad X_2 = a_{2n}u_n\partial_{u_n}$

$$F_n = u_n f_n(t,\xi) \tag{194}$$

$$\xi = u_{n-1}^{\alpha_{n+1,n}} u_n^{\alpha_{n-1,n+1}} u_{n+1}^{\alpha_{n,n-1}}$$

$$\alpha_{kl} = a_{1k}a_{2l} - a_{1l}a_{2k} \neq 0$$

$A_{2,4}:$ $\quad X_1 = \beta_{1,n}(t)\partial_{u_n},\quad X_2 = \beta_{2n}(t)\partial_{u_n}$

$$\beta_{1,n}\beta_{2n+1} - \beta_{1,n+1}\beta_{2n} \neq 0$$

$$F_n = \frac{(\beta_{1,n}\ddot\beta_{2n} - \ddot\beta_{1,n}\beta_{2n})u_{n+1} - (\beta_{1,n+1}\ddot\beta_{2n} - \ddot\beta_{1,n}\beta_{2n+1})}{\beta_{1,n}\beta_{2n+1} - \beta_{1,n+1}\beta_{2n}} \tag{195}$$

$$+ f_n(t,\xi)$$

$$\xi = (\beta_{1,n}\beta_{2n+1} - \beta_{1,n+1}\beta_{2n})u_{n-1} + (\beta_{1,n+1}\beta_{2n-1} - \beta_{1,n-1}\beta_{2n+1})u_n$$

$$+ (\beta_{1,n-1}\beta_{2n} - \beta_{1,n}\beta_{2n-1})u_{n+1}$$

$A_{2,5}:$ $\quad X_1 = \partial_{u_n},\quad X_2 = t\partial_{u_n}$

$$F_n = f_n(t,\xi_{n-1},\xi_{n+1}),\quad \xi_k = u_k - u_n,\quad k = n \pm 1 \tag{196}$$

The algebra $A_{2,5}$ is of particular physical significance since X_1 and X_2 in (196) correspond to translation and Galilei invariance for the considered chain. Unless we are considering a molecular chain in an external field, or unless an external geometry is imposed, the symmetry algebra $A_{2,5}$ should always be present, possibly as a subalgebra of a larger symmetry algebra.

Theorem 4.3. *Four classes of three-dimensional Abelian symmetry algebras exist. Only one of them contains the $A_{2,5}$ subalgebra and can be presented as*

$$A_{3,4}\quad X_1 = \partial_{u_n},\quad X_2 = t\partial_{u_n},$$
$$X_3 = \beta_n(t)\partial_{u_n},\quad \beta_{n+1} \neq \beta_n,\; \ddot\beta_n \neq 0. \tag{197}$$

The invariant interaction is

$$F_n = \frac{\ddot\beta_n}{\beta_{n+1} - \beta_n}(u_{n+1} - u_n) + f_n(t,\xi), \tag{198}$$

$$\xi = (\beta_n - \beta_{n+1})u_{n-1} + (\beta_{n+1} - \beta_{n-1})u_n + (\beta_{n-1} - \beta_n)u_{n+1}. \tag{199}$$

For $A_{3,1}$, $A_{3,2}$ and $A_{3,4}$, see the original article [45].

Theorem 4.4. *There exist two classes of interactions F_n in (172) satisfying conditions (176), allowing four-dimensional Abelian symmetry algebras. Only one of them contains the subalgebra $A_{2,5}$. It is represented by the following.*

$$A_1\quad F_n = \frac{B_n(t)\gamma_n}{\gamma_n - \gamma_{n+1}}(u_n - u_{n+1}) + f_n(t,\xi),\quad f_{n,\xi\xi} \neq 0 \tag{200}$$

$$X_1 = \partial_{u_n},\quad X_2 = t\partial_{u_n},\quad X_3 = \psi_1(t)\gamma_n\partial_{u_n},$$
$$X_4 = \psi_2(t)\gamma_n\partial_{u_n}$$

$$\gamma_{n+1} \neq \gamma_n,\quad \dot\gamma_n = 0,\quad \psi_1\dot\psi_2 - \dot\psi_1\psi_2 = \text{const} \neq 0,$$

with

$$\xi = (\gamma_n - \gamma_{n+1})u_{n-1} + (\gamma_{n+1} - \gamma_{n-1})u_n + (\gamma_{n-1} - \gamma_n)u_{n+1},$$

ψ_1, ψ_2 *satisfying*

$$\ddot{\psi}_i - B(t)\psi_i = 0, \quad i = 1, 2.$$

4.4 Some Results on the Structure of Lie Algebras

Let us recall some basic properties of finite-dimensional Lie algebras. Consider a Lie algebra L with basis $X_1, X_2, \ldots, X_n$. To each algebra L one associates two series of subalgebras.

The *derived series* consists of the algebras

$$L^0 \equiv L, \quad L^1 \equiv DL = [L, L], \quad L^2 \equiv D^2 L = [DL, DL], \ldots,$$
$$L^N \equiv D^N L = [D^{N-1}L, D^{N-1}L], \ldots \quad (201)$$

The algebra of commutators DL is called the *derived algebra*. If $DL = L$, the algebra L is called *perfect*. If an integer N exists for which $D^N L = \{0\}$, the algebra L is called *solvable*.

The *central series* consists of the algebras

$$L_0 \equiv L, \quad L_1 = L^1 = [L, L], \quad L_2 = [L, L_1], \ldots, L_N = [L, L_{N-1}], \ldots \quad (202)$$

If there exists an integer N for which $L_N = \{0\}$, the algebra L is called *nilpotent*. Clearly, every nilpotent algebra is solvable, but the converse is not true.

Let us consider two examples.

1. The Lie algebra of the Euclidean group of a plane: $e(2) \sim \{L_3, P_1, P_2\}$. The commmutation relations are

$$[L_3, P_1] = P_2, \quad [L_3, P_2] = -P_1, \quad [P_1, P_2] = 0. \quad (203)$$

   The derived series is

$$L = \{L_3, P_1, P_2\} \supset DL = \{P_1, P_2\}, \quad D^2 L = \{0\}$$

   and the central series is

$$L \supset L_1 = \{P_1, P_2\} = L_2 = L_3 = \ldots$$

   Hence $e(2)$ is solvable but not nilpotent.
2. The Heisenberg algebra $H_1 \sim \{X_1, X_2, X_3\}$ where the basis can be realized by the derivation operator, the coordinate x and the identity 1:

$$X_1 = \partial_x, \quad X_2 = x, \quad X_3 = 1.$$

We have

$$[X_1, X_2] = X_3, \quad [X_1, X_3] = [X_2, X_3] = 0, \tag{204}$$

hence

$$DL = \{X_3\}, \quad D^2L = 0.$$
$$L_1 = \{X_3\}, \quad L_2 = 0.$$

We see that the Heisenberg algebra is nilpotent (and therefore solvable).

An Abelian Lie algebra is of course also nilpotent.

We shall need some results concerning nilpotent Lie algebras (by nilpotent we mean nilpotent non-Abelian).

1. Nilpotent Lie algebras always contain Abelian ideals.
2. All nilpotent Lie algebras contain the three-dimensional Heisenberg algebra as a subalgebra.

We shall also use some basic properties of solvable Lie algebras, where by solvable we mean solvable, non-nilpotent.

1. Every solvable Lie algebra L contains a unique maximal nilpotent ideal called the nilradical $NR(L)$. The dimension of the nilradical satisfies

$$\frac{1}{2} \dim(L) \leq \dim NR(L) \leq \dim(L) - 1. \tag{205}$$

2. If the nilradical $NR(L)$ is Abelian, then we can choose a basis for L in the form $\{X_1, \ldots, X_n, Y_1, \ldots, Y_m\}, \quad m \leq n$, with commutation relations

$$[X_i, X_k] = 0, \quad [X_i, Y_k] = (A_k)_{ij} X_j, \quad [Y_i, Y_k] = C_{ik}^l X_l. \tag{206}$$

The matrices A_k commute and are linearly nilindependent, *i.e.*, no linear combination of them is a nilpotent matrix.

If a Lie algebra L is not solvable, it can be simple, semisimple, or it may have a nontrivial Levi decomposition [33]. A *simple* Lie algebra L has no nontrivial ideals, *i.e.*,

$$I \subseteq L, \quad [L, I] \subseteq I \tag{207}$$

implies $I = \{0\}$, or $I = L$.

A *semisimple* Lie algebra L is a direct sum of simple Lie algebras L_i

$$L \sim L_1 \oplus L_2 \oplus \cdots \oplus L_p, \quad [L_i, L_k] = 0. \tag{208}$$

If L is not simple, semisimple, or solvable, then it has a unique *Levi decomposition* into a semidirect sum

$$L \sim S \rtimes R, \quad [S, S] = S, \quad [R, R] \subset R, \quad [S, R] \subseteq R \tag{209}$$

where S is semisimple and R is solvable; R is called the radical of L. It is the maximal solvable ideal.

Let us now return to the symmetry classification of discrete dynamical systems.

4.5 Nilpotent Non-Abelian Symmetry Algebras

Since every nilpotent Lie algebra contains the three-dimensional Heisenberg algebra, we start by constructing this algebra, $H_1 \sim \{X_1, X_2, X_3\}$. The central element X_3 of (204) is uniquely defined. We start from this element, take it in one of the standard forms (189), (190), or (191), then construct the two complementary elements X_1 and X_2. The result is that two inequivalent realizations of H_1 exist:

$$N_{3,1}: \quad X_1 = \partial_{u_n}, \quad X_2 = \partial_t, \quad X_3 = t\partial_{u_n}$$
$$F_n = f_n(\xi_{n+1}, \xi_{n-1}), \quad \xi_k = u_k - u_n, \quad k = n \pm 1 \qquad (210)$$

$$N_{3,2}: \quad X_1 = e^{a_n t}\partial_{u_n}, \quad X_2 = \partial_t + a_n u_n \partial_{u_n}$$
$$X_3 = (t + \gamma_n)e^{a_n t}\partial_{u_n}, \quad \dot{a}_n = 0, \quad \dot{\gamma}_n = 0, \gamma_{n+1} \neq \gamma_n$$
$$F_n = \frac{a_n^2(\gamma_{n+1} - \gamma_n) - 2a_n}{\gamma_{n+1} - \gamma_n}u_n \qquad (211)$$
$$+ \frac{2a_n}{\gamma_{n+1} - \gamma_n}u_{n+1}e^{(a_n - a_{n+1})t} + e^{a_n t}f_n(\xi)$$
$$\xi = (\gamma_n - \gamma_{n+1})u_{n-1}e^{-a_{n-1}t} + (\gamma_{n+1} - \gamma_{n-1}), u_n e^{-a_n t}$$
$$+ (\gamma_{n-1} - \gamma_n)u_{n+1}e^{-a_{n+1}t}.$$

Observe that $N_{3,1}$ contains the physically important subalgebra $A_{2,5}$, whereas $N_{3,2}$ does not.

Extending the algebras $N_{3,1}$ and $N_{3,2}$ by further elements, we find that $N_{3,1}$ gives rise to two five-dimensional nilpotent symmetry algebras $N_{5,k}$ and $N_{3,2}$ to a four-dimensional one $N_{4,1}$.

Here we shall only list $N_{5,1}$ and $N_{5,2}$ which contain $N_{3,1}$ and hence $A_{2,5}$:

$$N_{5,k}: \quad X_1 = \partial_{u_n}, \quad X_2 = t\partial_{u_n}, \quad X_3 = \left(\frac{(k-1)t^2}{2} + \gamma_n\right)\partial_{u_n},$$
$$X_4 = \left(\frac{(k-1)t^3}{6} + \gamma_n t\right)\partial_{u_n}, \quad X_5 = \partial_t, \quad k = 1, 2 \qquad (212)$$
$$F_n = \frac{2(k-1)}{\gamma_{n+1} - \gamma_n}(u_{n+1} - u_n) + f_n(\xi)$$

with ξ as in (200).

4.6 Solvable Symmetry Algebras with Non-Abelian Nilradicals

We already know all nilpotent symmetry algebras, so we can start from the nilradical and extend it by further non-nilpotent elements. The result can be stated as a theorem.

Theorem 4.5. *Seven classes of solvable symmetry algebras with non-Abelian nilradicals exist for (172). Four of them have $N_{3,1}$ as nilradical, three have $N_{5,1}$.*

For $N_{3,1}$ we can add just one more element Y, one of the following

$$SN_{4,1}: \quad Y = t\partial_t + \left(\frac{1}{2} + a\right)u_n\partial_{u_n}, \quad a \neq -\frac{1}{2}$$
$$F_n = (u_{n+1} - u_n)e^{(a-3/2)/(a+1/2)}f_n(\xi) \tag{213}$$

$$SN_{4,2}: \quad Y = t\partial_t + (2u_n + t^2)\partial_{u_n}$$
$$F_n = \ln(u_{n+1} - u_n) + f_n(\xi) \tag{214}$$

$$SN_{4,3}: \quad Y = u_n\partial_{u_n}$$
$$F_n = (u_{n+1} - u_n)f_n(\xi). \tag{215}$$

In all the above cases we have

$$\xi = \frac{u_{n-1} - u_n}{u_{n+1} - u_n}. \tag{216}$$

$$SN_{4,4}: \quad Y = t\partial_t + \gamma_n\partial_{u_n}, \quad \gamma_{n+1} \neq \gamma_n, \quad \dot{\gamma}_n = 0$$
$$F_n = \exp\left(-2\frac{u_{n+1} - u_n}{\gamma_{n+1} - \gamma_n}\right)f_n(\xi) \tag{217}$$

with ξ as in (200). For $N_{5,1}$ we can also add at most one non-nilpotent element and we obtain

$$SN_{6,1}: \quad Y = t\partial_t + \left(\frac{1}{2} + a\right)u_n\partial_{u_n}$$
$$F_n = c_n\xi^{(a-3/2)/(a+1/2)}, \quad a \neq -\frac{1}{2}, \quad a \neq \frac{3}{2} \tag{218}$$

$$SN_{6,2}: \quad Y = t\partial_t + (2u_n + (a + b\gamma_n)t^2)\partial_{u_n}, \quad a^2 + b^2 \neq 0$$
$$F_n = c_n + (a + b\gamma_n)\ln\xi \tag{219}$$

$$SN_{6,3}: \quad Y = t\partial_t + \rho_n\partial_{u_n}, \quad \rho_n \neq A + B\gamma_n, \quad \dot{\rho}_n = 0$$
$$F_n = c_ne^\zeta \tag{220}$$

$$\zeta = \frac{-2\xi}{(\gamma_n - \gamma_{n+1})\rho_{n-1} + (\delta_{n+1} - \gamma_{n-1})\rho_n + (\gamma_{n-1} - \gamma_n)\rho_{n+1}}.$$

In all cases ξ is as in (200).

4.7 Solvable Symmetry Algebras with Abelian Nilradicals

The results in this case are very rich. There exist 31 such symmetry algebras and their dimensions satisfy $2 \leq d \leq 5$.

For all details and a complete list of results we refer to the original article [45]. Here we give just one example of a five-dimensional Lie algebra with $NR(L) = A_{4,1}$.

$$SA_{5,1}: \quad X_1 = \partial_{u_n}, \quad X_2 = t\partial_{u_n}, \quad X_3 = e^t \gamma_n \partial_{u_n},$$
$$X_4 = e^{-t} \gamma_n \partial_{u_n}$$
$$Y = \partial_t + a u_n \partial_{u_n} \quad a \neq 0, \quad \gamma_n \neq \gamma_{n+1}, \quad \ddot{\gamma}_n = 0$$
$$F_n = \frac{\gamma_n(u_{n+1} - u_n)}{\gamma_{n+1} - \gamma_n} + e^{at} f_n(\xi) \tag{221}$$
$$\xi = ((\gamma_n - \gamma_{n+1})u_{n-1} + (\gamma_{n+1} - \gamma_{n-1})u_n$$
$$+ (\gamma_{n-1} - \gamma_n)u_{n+1})e^{-at}.$$

We mention that there exist 7 five-dimensional algebras with $A_{4,1}$ as their nilradical [45].

4.8 Nonsolvable Symmetry Algebras

A Lie algebra that is not solvable must have a simple subalgebra. The only simple algebra that can be constructed out of vector fields of the form (183) is $sl(2, \mathbf{R})$. The algebra and the corresponding invariant interaction can be represented as:

$$NS_{3,1}: \quad X_1 = \partial_t, \quad X_2 = t\partial_t + \frac{1}{2} u_n \partial_{u_n}$$
$$X_3 = t^2 \partial_t + t u_n \partial_{u_n} \tag{222}$$
$$F_n = \frac{1}{u_n^3} f_n(\xi_{n-1}, \xi_{n+1}), \quad \xi_k = \frac{u_k}{u_n}, \quad k = n \pm 1.$$

This algebra can be further extended to a four, five or seven-dimensional symmetry algebra. In two cases the algebra will have an $A_{2,5}$ subalgebra, namely
$NS_{5,1}$: In addition to X_1, X_2, X_3 of (222) we have

$$X_4 = \partial_{u_n}, \quad X_5 = t\partial_{u_n}$$
$$F_n = (u_{n+1} - u_n)^{-3} f_n(\xi), \quad \xi = \frac{u_{n+1} - u_n}{u_{n-1} - u_n}. \tag{223}$$

$NS_{7,1}$: The additional elements are

$$X_n = \partial_{u_n}, \quad X_5 = t\partial_{u_n}, \quad X_6 = \gamma_n \partial_{u_n}, \quad X_7 = t\gamma_n \partial_{u_n}$$
$$\gamma_{n+1} \neq \gamma_n, \quad \ddot{\gamma}_n = 0. \tag{224}$$

The invariant interaction is

$$F_n = s_n((\gamma_n - \gamma_{n+1})u_{n-1} + (\gamma_{n+1} - \gamma_{n-1})u_n$$
$$+ (\gamma_{n-1} - \gamma_n)u_{n+1})^{-3}, \quad \dot{s}_n = 0, \quad s_n \neq 0. \quad (225)$$

4.9 Final Comments on the Classification

Let us first of all sum up the discrete dynamical systems of the type (172) with the largest symmetry algebras.

We set

$$\xi = (\gamma_n - \gamma_{n+1})u_{n-1} + (\gamma_{n+1} - \gamma_{n-1})u_n + (\gamma_{n-1} - \gamma_n)u_{n+1} \quad (226)$$

and find that this variable is involved in all cases with 7-, or 6-dimensional symmetry algebras.

The algebras and interactions are given in (224), (218), (219) and (220), respectively.

A natural question is where is the Toda lattice in this classification. The Toda lattice is described by the equation

$$u_{n,tt} = e^{u_{n-1} - u_n} - e^{u_n - u_{n+1}}. \quad (227)$$

This equation is of the form (172). It is integrable [62] and has many interesting properties. In our classification it appears as a special case of the algebra $SN_{4,4}$. Indeed, (227) is obtained from

$$\ddot{u}_n = \exp\left(-2\frac{u_{n+1} - u_n}{\gamma_{n+1} - \gamma_n}\right)f_n(\xi), \quad (228)$$

taking

$$f_n(\xi) = -1 + e^{\xi/2}, \quad \gamma_n = 2n. \quad (229)$$

Thus, its symmetry group is four-dimensional. We see that the Toda lattice is not particularly distinguished by its point symmetries: other interactions have larger symmetry groups. Even in the $SN_{4,4}$ class two functions have to be specialized (see (229)) to reduce (228) to (227).

5 Generalized Point Symmetries of Linear and Linearizable Systems

5.1 Umbral Calculus

In this section we take a different point of view than in the previous ones. Instead of purely point symmetries, we shall consider a specific class of generalized symmetries of difference equations that we shall call "generalised point symmetries". They act simultaneously at several or even infinitely many points of a lattice, but they reduce to point symmetries of a differential equation in the continuous limit.

The approach that we shall discuss here is at this stage applicable either to linear difference equations, or to nonlinear equations that can be linearized by a transformation of variables (not necessarily only point transformations).

The mathematical basis for this type of study is the so-called "umbral calculus" reviewed in recent books and articles by G.G. Rota and his collaborators [59, 58, 11]. Umbral calculus provides a unified basis for studying symmetries of linear differential and difference equations.

Let us introduce several fundamental concepts.

Definition 5.1. *A shift operator T_δ is a linear operator acting on polynomials or formal power series in the following manner*

$$T_\delta f(x) = f(x + \delta), \quad x \in \mathbf{R}, \quad \delta \in \mathbf{R}. \tag{230}$$

For functions of several variables we introduce shift operators in the same manner

$$T_{\delta_i} f(x_1, \ldots x_{i-1}, x_i, x_{i+1} \ldots x_n)$$
$$= f(x_1, \ldots, x_{i-1}, x_i + \delta_i, x_{i+1}, \ldots, x_n). \tag{231}$$

In this section we restrict the exposition to the case of one real variable $x \in \mathbf{R}$. The extension to n variables and other fields is obvious. We will sometimes drop the subscript on the shift operator T when that does not give rise to misinterpretations.

Definition 5.2. *An operator U is called a* delta operator *if it satisfies the following properties,*

1) *It is shift invariant;*

$$T_\delta U = U T_\delta, \quad \forall \delta \in \mathbf{R}, \tag{232}$$

2)

$$Ux = c \neq 0, \quad c = \text{const}, \tag{233}$$

3)

$$Ua = 0, \quad a = \text{const}, , \tag{234}$$

and the kernel of U consists precisely of all constants.

Important properties of delta operators are:

1. For every delta operator U there exists a unique series of basic polynomials $\{P_n(x)\}$ satisfying

$$P_0(x) = 1 \quad P_n(0) = 0, \quad n \geq 1, \quad U P_n(x) = n P_{n-1}(x). \tag{235}$$

2. For every delta operator U there exists a conjugate operator β, such that

$$[U, x\beta] = 1. \tag{236}$$

The operator β satisfies

$$\beta = (U')^{-1}, \quad U' = [U, x]. \tag{237}$$

The expression

$$U' \equiv U * x \equiv [U, x] \tag{238}$$

is called the "Pincherle derivative" of U [59, 58, 11].

For us the fundamental fact is that the pair of operators, U and $x\beta$, satisfies the Heisenberg relation (236).

Before going further, let us give the two simplest possible examples.

Example 5.1. The (continuous) derivative

$$\begin{aligned} U = \partial_x, \quad \beta = 1 \\ P_0 = 1, \quad P_1 = x, \quad \dots, \quad P_n = x^n, \dots \end{aligned} \tag{239}$$

Example 5.2. The right discrete derivative

$$\begin{aligned} U = \Delta^+ = \tfrac{T-1}{\delta}, \quad \beta = T^{-1} \\ P_0 = 1, \quad P_1 = x, \quad P_2 = x(x - \delta) \\ P_n = x(x - \delta) \dots \big(x - (n-1)\delta\big). \end{aligned} \tag{240}$$

For any operator U one can construct β and the basic series will be

$$P_n = (x\beta)^n \cdot 1, \quad n \in \mathbf{N}. \tag{241}$$

5.2 Umbral Calculus and Linear Difference Equations

Let us consider a Lie algebra L realized by vector fields

$$X_a = f_a(x_1, \dots, x_n)\partial_{x_a} \tag{242}$$
$$[X_a, X_b] = C_{ab}^c X_c. \tag{243}$$

The Heisenberg relation (236) allows us to realize the same abstract Lie algebra by difference operators

$$X_a^D = f_a(x_1\beta_1, x_2\beta_2, \dots, x_n\beta_n)\Delta_{x_a}, \quad [\Delta_{x_a}, x_a\beta_a] = 1, \quad a = 1, \dots n. \tag{244}$$

As long as the functions f_a are polynomials, or formal power series in the variables x_a, the substitution

$$x_a \to x_a\beta_a, \quad \partial_{x_a} \to \Delta_{x_a} \tag{245}$$

preserves the commutation relations (243).

We shall call the substitution (245) and more generally any substitution

$$\{U, \beta\} \leftrightarrow \{\tilde{U}, \tilde{\beta}\} \tag{246}$$

an "umbral correspondence". This correspondence will also take the set of basic polynomials related to $\{U, \beta\}$ into the set related to the pair $\{\tilde{U}, \tilde{\beta}\}$.

We shall use two types of delta operators. The first is simply the derivative $U = \partial_x$, for which we have $\beta = 1$. The second is a general difference operator $U = \Delta$ that has ∂_x as its continuous limit. We set

$$\Delta = \frac{1}{\delta} \sum_{k=l}^{m} a_k T_\delta^k, \quad l, m \in \mathbf{Z}, \quad l < m, \tag{247}$$

where a_k and δ are real constants and T_δ is a shift operator as in (230). Condition (232) is satisfied. Condition (234) implies

$$\sum_{k=l}^{m} a_k = 0. \tag{248}$$

We also require that when δ goes to 0, Δ goes to ∂_x. This requires a further restriction on the coefficients a_k, namely

$$\sum_{k=l}^{m} a_k k = 1. \tag{249}$$

Then relation (233) is also satisfied, with $c = 1$.

More generally, for Δ as in (247),

$$\Delta f(x) = \frac{1}{\delta} \sum_{k=l}^{m} a_k f(x + k\delta)$$

$$= \frac{1}{\delta} \sum_{q=0}^{\infty} \frac{f^{(q)}(x)}{q!} \delta^q \sum_{k=l}^{m} a_k k^q.$$

We define

$$\gamma_q = \sum_{k=l}^{m} a_k k^q, \quad q \in \mathbf{Z}, \tag{250}$$

and thus

$$\Delta f(x) = \frac{df}{dx} + \sum_{q=2}^{\infty} \gamma_q \frac{f^{(q)}(x)}{q!} \delta^{q-1} f. \tag{251}$$

Thus the limit of Δ is the derivative at least to order δ. We can also impose

$$\gamma_q = 0, \quad q = 2, 3, \ldots m - l. \tag{252}$$

Then we have

$$\Delta = \frac{d}{dx} + O(\delta^{m-l}).$$

Definition 5.3. *A* difference operator *of degree* $m - l$ *is a* delta operator *of the form*

$$U \equiv \Delta = \frac{1}{\delta} \sum_{k=l}^{m} a_k T_\delta^k, \tag{253}$$

where a_k *and* δ *are constants,* T_δ *is a shift operator and*

$$\sum_{k=l}^{m} a_k = 0, \quad \sum_{k=l}^{m} a_k k = 1. \tag{254}$$

Comment: $\tilde{\Delta} = T^j \Delta$ is a difference operator of the same degree as Δ.

Theorem 5.1. *The operator* β *conjugate to* $\Delta = (1/\delta) \sum_{k=l}^{m} a_k T_\delta^k$ *is*

$$\beta = \left(\sum_{k=l}^{m} a_k k T^k \right)^{-1}. \tag{255}$$

Proof. By definition, $\beta = (\Delta')^{-1} = [\Delta, x]^{-1}$ and

$$[\Delta, x] = \frac{1}{\delta} \left(\sum_{k=l}^{m} a_k (x + k\delta) T^k - x \sum_{k=l}^{m} a_k T^k \right)$$

$$= \sum_{k=l}^{m} a_k k T^k.$$

Examples:

$$\Delta^s = \frac{T - T^{-1}}{2\delta}, \quad \beta = \left(\frac{T + T^{-1}}{2} \right)^{-1} \tag{256}$$

$$\Delta^3 = -\frac{1}{6\delta}(T^2 - 6T + 3 + 2T^{-1}), \quad \beta = \left(-\frac{T^2 - 3T - T^{-1}}{3} \right)^{-1} \tag{257}$$

Comment:

$$\Delta^s = \frac{\partial}{\partial x} + O(\delta^2) \quad \Delta^3 = \frac{\partial}{\partial x} + O(\delta^3).$$

Now let us apply the above considerations to the study of symmetries of linear difference equations.

Definition 5.4. *An* umbral equation *of order n is an operator equation of the form*

$$\sum_{k=0}^{n} \hat{a}_k(x\beta) \Delta^k \hat{f} = \hat{g}, \tag{258}$$

where $\hat{a}_k(x\beta)$ and $\hat{g}(x\beta)$ are given formal power series in $x\beta$ and $\hat{f}(x\beta)$ is the unknown operator function.

For $\Delta = \partial_x$, $\beta = 1$ this is a differential equation. For Δ as in (247), (258) is an operator equation. Applying both sides of (258) to 1 we get a difference equation. Its solution is

$$f(x) = \hat{f}(x\beta) \cdot 1. \tag{259}$$

Thus $f(x)$ is a function, obtained by applying the operator $\hat{f}(x\beta)$ to the constant 1.

More generally, an umbral equation of order n in p variables is

$$\sum_{k_1=0,\ldots,k_p=0}^{n_1,\ldots,n_p} \hat{a}_{k_1\ldots k_p}(x_1\beta_1, x_2\beta_2, \ldots, x_p\beta_p) \Delta_{\delta_1}^{k_1}, \ldots \Delta_{\delta_p}^{k_p} \hat{f}(x_1\beta_1, \ldots, x_p\beta_p)$$

$$= \hat{g}(x_1\beta_1 \ldots x_p\beta_p), \quad \sum_{i=1}^{p} n_i = n. \tag{260}$$

As an example, we study the umbral equation,

$$\Delta\hat{f} = a\hat{f}, \quad a \neq 0. \tag{261}$$

(i)   If $\Delta = \partial_x$, then $f(x) = e^{ax}$.
(ii) If $\Delta = \Delta^+ = \frac{T-1}{\delta}$, $\beta = T^{-1}$, and

$$f(x + \delta) - f(x) = a\delta f(x). \tag{262}$$

If $f(x) = \lambda^x$, then

$$\lambda^{x+\delta} - \lambda^x = a\delta\lambda^x$$

so that

$$\lambda = (1 + a\delta)^{1/\delta}.$$

We get a single "umbral" solution

$$f_1(x) = (1 + a\delta)^{x/\delta}. \tag{263}$$

The umbral correspondence yields:

$$f_2(x) = e^{axT^{-1}} \cdot 1. \tag{264}$$

If we expand into power series, we obtain $f_1(x) = f_2(x)$, and

$$\lim_{\delta \to 0} f_{1,2}(x) = e^{ax}.$$

From now on we shall call *umbral solutions* those that we can get from solutions of the corresponding differential equation by the umbral correspondance. Other solutions, if they exist, will be called *non-umbral solutions*.

(iii) For comparison, take $\Delta = \Delta^s = (T - T^{-1})/2\delta$, $\beta = ((T + T^{-1})/2)^{-1}$. Then

$$f(x + \delta) - f(x - \delta) = 2\delta a f(x). \tag{265}$$

Setting $f(x) = \lambda^x$, we get two values of λ and

$$\begin{aligned} f &= A_1(a\delta + \sqrt{a^2\delta^2 + 1})^{x/\delta} + A_2(a\delta - \sqrt{a^2\delta^2 + 1})^{x/\delta} \\ &\equiv A_1 f_1 + A_2 f_2. \end{aligned} \tag{266}$$

We have

$$\lim_{\delta \to 0} f_1(x) = e^{ax}, \tag{267}$$

but the limit of $f_2(x)$ does not exist. The umbral correspondence yields

$$f_u(x) = \exp\left(ax\left(\frac{T + T^{-1}}{2}\right)^{-1}\right) \cdot 1.$$

Expanding into power series, we find $f_u = f_1$. The solution f_2 is a non–umbral solution.

Theorem 5.2. *Let Δ be a difference operator of order p. Then the linear umbral equation of order n (258) has np linearly independent solutions, n of them umbral ones.*

There may be convergence problems for the formal series. Consider the exponential

$$\hat{f}(x) = e^{ax\beta}, \quad \beta = \left(\sum_{k=l}^{m} a_k k T^k\right)^{-1}. \tag{268}$$

For $m - l \geq 3$, β will involve infinitely many shifts, *i.e.*, each term in the expansion (268) could involve infinitely many shifts. However

$$P_n(x) = (x\beta)^n \cdot 1 \tag{269}$$

is a well defined polynomial. For a proof see [38].

Let us assume that we know the solution of an umbral equation for $\Delta = \partial_x$ and that it has the form

$$f(x) = \sum_{n=0}^{\infty} \frac{f^{(n)}(0)}{n!} x^n. \tag{270}$$

Then for any difference operator Δ there will exist a corresponding umbral solution

$$\hat{f}(x)1 = \sum_{n=0}^{\infty} \frac{f^{(n)}(0)}{n!} P_n(x), \tag{271}$$

where $P_n(x) = (x\beta)^n \cdot 1$ are the basic polynomials corresponding to Δ.

5.3 Symmetries of Linear Umbral Equations

Let us consider a linear differential equation

$$Lu = 0, \quad L = \sum_{k_1,\ldots,k_p} a_{k_1,\ldots,k_p}(x_1,\ldots,x_p) \frac{\partial^{k_1}}{\partial x_1^{k_1}} \cdots \frac{\partial^{k_p}}{\partial x_p^{k_p}}. \tag{272}$$

The Lie point symmetries of (272) can be realized by evolutionary vector fields of the form

$$X = Q(x_i, u, u_{x_i})\partial_u,$$

$$Q = \phi - \sum_{i=1}^{p} \xi_i u_{x_i}. \tag{273}$$

The following theorem holds for these symmetries.

Theorem 5.3. *All Lie point symmetries for an ODE of order $n \geq 3$, or a PDE of order $n \geq 2$ are generated by evolutionary vector fields of the form (273) with the characteristic Q satisfying*

$$Q = Xu + \chi(x_1,\ldots,x_p), \tag{274}$$

where χ is a solution of (272) and X is a linear operator

$$X = \sum_{i=1}^{p} \xi_i(x_1,\ldots,x_p)\partial_{x_i} \tag{275}$$

satisfying

$$[L, X] = \lambda(x_1,\ldots,x_p)L, \tag{276}$$

i.e., commuting with L on the solution set of L. In (276), λ is an arbitrary function.

For a proof we refer to the literature [6].

In other words, if the conditions of Theorem 5.3 apply, then all symmetries of (272) beyond those representing the linear superposition principle, are generated by linear operators of the form (275), commuting with L on the solution set of (272).

Now let us turn to the umbral equation (260) with $\hat{g} = 0$, *i.e.*,

$$\sum_{k_1,\ldots,k_p} \hat{a}_{k_1\ldots k_p}(x_1\beta_1,\ldots,x_p\beta_p)\Delta_{\delta_1}^{k_1}\ldots\Delta_{\delta_p}^{k_p}\hat{u}(x_1\beta_1,\ldots,x_p\beta_p) = 0. \quad (277)$$

We shall realize the symmetries of (277) by evolutionary vector fields of the form

$$v_E = Q_D\partial_u, \quad Q_D = \phi_D - \sum_{i=1}^{p}\xi_{D,i}\Delta_i u \quad (278)$$

where ϕ_D and $\xi_{D,i}$ are functions of $x_i\beta_i$ and u. The prolongation of v_E will also act on the discrete derivatives $\Delta_{\delta_i}^{k_i}u$. We are now considering transformations on a fixed (nontransforming) lattice. In the evolutionary formalism the transformed variables satisfy

$$\tilde{x}_k\tilde{\beta}_k = x_k\beta_k, \quad \tilde{\beta}_k = \beta_k$$
$$\tilde{u}(\tilde{x}_k\tilde{\beta}_k) = u(x_k\beta_k) + \lambda Q_D, \quad |\lambda| << 1, \quad (279)$$

and we request that $\tilde{u}$ be a solution whenever u is one. The transformation of the discrete derivatives is given by

$$\Delta_{\tilde{x}_k}\tilde{u} = \Delta_{x_k}u + \lambda\Delta_{x_k}Q$$
$$\Delta_{\tilde{x}_k\tilde{x}_k}\tilde{u} = \Delta_{x_kx_k}u + \lambda\Delta_{x_kx_k}Q \quad (280)$$

etc., where Δ_{x_k} are discrete total derivatives.

In terms of the vector field (278),

$$\text{pr}\, v_E = Q_D\partial_u + Q_D^{x_i}\partial_{\Delta_i u} + Q_D^{x_ix_k}\partial_{\Delta_i\Delta_k u} + \cdots$$
$$Q_D^{x_i} = \Delta_i Q_D, \quad Q_D^{x_ix_k} = \Delta_i\Delta_k Q_D \quad (281)$$

(we have set $\Delta_{x_i} \equiv \Delta_i$).

As in the continuous case, we obtain determining equations by requiring

$$\text{pr}\, v_E(L_D\hat{u})\,|_{L_D\hat{u}} = 0 \quad (282)$$

where $L_D\hat{u}$ is the left hand side of (277).

The determining equations will be an umbral version of the determining equations in the continuous case, *i.e.*, are obtained by the umbral correspondence, $\partial_{x_i} \mapsto \Delta_i$, $x_i \mapsto x_i\beta_i$.

The symmetries of the umbral equation (277) will hence have the form (278) with

$$Q_D = X_D u + \chi(x_1\beta_1, \ldots x_p\beta_p),$$
(283)

where X_D is a difference operator commuting with L_D on the solutions of (277). Moreover, X_D is obtained from X by the umbral correspondence.

We shall call such symmetries "generalized point symmetries". Because of the presence of the operators β_i they are not really point symmetries. In the continuous limit they become point symmetries.

We consider examples in the next section.

5.4 The Discrete Heat Equation

The (continuous) linear heat equation in $(1+1)$ dimensions is

$$u_t - u_{xx} = 0.$$
(284)

Its symmetry group is of course well-known. Factoring out the infinite-dimensional pseudo-group corresponding to the linear superposition principle we have a 6-dimensional symmetry group. We write its Lie algebra in evolutionary form as

$$P_0 = u_t \partial_u, \quad P_1 = u_x \partial_u, \quad W = u\partial_u,$$
$$B = (2tu_x + xu)\partial_u, \quad D = \left(2tu_t + xu_x + \frac{1}{2}u\right)\partial_u,$$
(285)
$$K = \left(t^2 u_t + txu_x + \frac{1}{4}(x^2 + 2t)u\right)\partial_u,$$

where P_0, P_1, B, D, K and W generate time and space translations, Galilei boosts, dilations, "expansions" and the multiplication of u by a constant, respectively.

A very natural discretization of (284) is the discrete heat equation

$$\Delta_t u - (\Delta_x)^2 u = 0,$$
(286)

where Δ_t and Δ_x are the difference operators considered in Sect. 5.2. We use the corresponding conjugate operators β_t and β_x. The umbral correspondence determines the symmetry algebra of (286), starting from the algebra (285). Namely,

$$P_0^D = (\Delta_t u)\partial_u, \quad P_1^D = (\Delta_x u)\partial_u, \quad W^D = u\partial_u,$$
$$B^D = (2(t\beta_t)\Delta_x u + (x\beta_x)u)\partial_u$$
$$D^D = \left(2t\beta_t\Delta_t u + x\beta_x\Delta_x u + \frac{1}{2}u\right)\partial_u,$$
(287)
$$K^D = \left((t\beta_t)^2 \Delta_t u + (t\beta_t)(x\beta_x)\Delta_x u + \frac{1}{4}((x\beta_x)^2 + 2t\beta_t)u\right)\partial_u.$$

In particular, we can choose both Δ_t and Δ_x to be right derivatives

$$\Delta_t = \frac{T_t - 1}{\delta_t}, \quad \beta_t = T_t^{-1}, \quad \Delta_x = \frac{T_x - 1}{\delta_x}, \quad \beta_x = T_x^{-1}, \tag{288}$$

The characteristic Q_K of the element K^D is X_K such that

$$Q_K = X_K u, \quad X_K = (t^2 - \delta_t t) T_t^{-2} \Delta_t + t x T_x^{-1} T_t^{-1} \Delta_x$$
$$+ \frac{1}{4}((x^2 - \delta_x) T_x^{-2} + 2t T_t^{-1}), \tag{289}$$

so it is not a point transformation: it involves u evaluated at several points. Each of the basis elements (287) (or any linear combination of them) provides a flow commuting with (286):

$$u_\lambda = X u. \tag{290}$$

Equations (286) and (290) can be solved simultaneously and this fact constitutes an anlogue for equations of the separation of variables in PDEs, and a tool for studying new types of special functions.

5.5 The Discrete Burgers Equation and Its Symmetries

The Continuous Case. The Burgers equation

$$u_t = u_{xx} + 2u u_x \tag{291}$$

is the simplest equation that combines nonlinearity and dissipative effects. It is also the prototype of an equation linearizable by a coordinate transformation, that is C-linearizable in Calogero's terminology [64].

We set $u = v_x$ and obtain the potential Burgers equation for v:

$$v_t = v_{xx} + v_x^2. \tag{292}$$

Setting $w = e^v$, we find

$$w_t = w_{xx}. \tag{293}$$

In other words, the usual Burgers equation (291) is linearized into the heat equation (293) by the Cole-Hopf transformation

$$u = \frac{w_x}{w} \tag{294}$$

(which is not a point transformation).

A possible way of viewing the Cole-Hopf transformation is that it provides a Lax pair for the Burgers equation:

$$w_t = w_{xx}, \quad w_x = uw. \tag{295}$$

Setting

$$w_t = Aw, \quad w_x = Bw, \quad A = u_x + u^2, \quad B = u$$

we obtain the Burgers equation as a compatibility condition

$$A_x - B_t + [A, B] = 0. \tag{296}$$

Our aim is to discretize the Burgers equation in such a way as to preserve its linearizability and also its five-dimensional Lie point symmetry algebra. We already know the symmetries of the discrete heat equation and we will use them to obtain the symmetry algebra of the discrete Burgers equation. This will be an indirect application of umbral calculus to a nonlinear equation.

The Discrete Burgers Equation as a Compatibility Condition. Let us write a discrete version of the pair (295) in the form:

$$\Delta_t \phi = \Delta_{xx} \phi, \quad \Delta_x \phi = u\phi, \tag{297}$$

where

$$\Delta_t = \frac{T_t - 1}{\delta_t}, \quad \Delta_x = \frac{T_x - 1}{\delta_x}. \tag{298}$$

The pair (297) can be rewritten as

$$\Delta_t \phi = (\Delta_x u + uT_x u)\phi, \quad \Delta_x \phi = u\phi. \tag{299}$$

We have used the Leibnitz rule for the discrete derivative Δ_x of (298), namely

$$\Delta_x fg = f(x)\Delta_x g + (T_x g)\Delta_x f. \tag{300}$$

Compatibility of (299), i.e., $\Delta_x \Delta_t \phi = \Delta_t \Delta_x \phi$, yields the discrete Burgers equation

$$\Delta_t u = \frac{1 + \delta_x u}{1 + \delta_t(\Delta_x \Delta_x u + uT_x u)} \Delta_x(\Delta_x u + uT_x u). \tag{301}$$

In the continuous limit $\Delta_t \to \partial/\partial t$, $\Delta_x \to \partial/\partial x$, $T_x \to 1, \delta_x = 0$, $\delta_t = 0$ we obtain the Burgers equation (291) [31,29]. This is not a "naive" discretization like

$$\Delta_t u = (\Delta_x)^2 u + 2u\Delta_x u \tag{302}$$

which would loose all integrability properties.

Symmetries of the Discrete Burgers Equation. We are looking for "generalized point symmetries" on a fixed lattice. We write them in evolutionary form

$$X_e = Q(x, t, T_x^a T_t^b u, T_x^c T_t^d \Delta_x u, T_x^e T_t^f \Delta_t u, \dots)\partial_u \tag{303}$$

and each symmetry will provide a commuting flow

$$u_\lambda = Q.$$

We shall use the Cole-Hopf transformation to transform the symmetry algebra of the discrete heat equation into that of the discrete Burgers equation.

All the symmetries of the discrete heat equation given in (287) can be written as

$$\phi_\lambda = S\phi, \quad S = S(x, t, \phi, T_x, T_x \dots) \tag{304}$$

where S is a linear operator. The same is true for any linear difference equation.

For the discrete heat equation

$$\Delta_t \phi - (\Delta_x)^2 \phi = 0 \tag{305}$$

with Δ_t and Δ_x as in (298) we rewrite the flows corresponding to (287) as

$$\phi_{\lambda_1} = \Delta_t \phi, \quad \phi_{\lambda_2} = \Delta_x \phi, \quad \phi_{\lambda_3} = \left(2tT_t^{-1}\Delta_x + xT_x^{-1} + \frac{1}{2}\delta_x T_x^{-1}\right)\phi$$

$$\phi_{\lambda_4} = \left(2tT_t^{-1}\Delta_t + xT_x^{-1}\Delta_x + \frac{1}{2}\right)\phi$$

$$\phi_{\lambda_5} = \left(t^2 T_t^{-2}\Delta_t + txT_t^{-1}T_x^{-1}\Delta_x + \frac{1}{4}x^2 T_x^{-2}\right.$$

$$\left. +t\left(T_t^{-2} - \frac{1}{2}T_t^{-1}T_x^{-1}\right) - \frac{1}{16}\delta_x^2 T_x^{-2}\right)\phi \tag{306}$$

$$\phi_{\lambda_6} = \phi.$$

Let us first prove a general result.

Theorem 5.4. *Let* (304) *represent a symmetry of the discrete heat equation* (305). *Then the same linear operator* S *provides a symmetry of the discrete Burgers equation* (301), *the flow of which is given by*

$$u_\lambda = (1 + \delta_x u)\Delta_x\left(\frac{S\phi}{\phi}\right), \tag{307}$$

where $(S\phi)/\phi$ *can be expressed in terms of* $u(x,t)$.

Proof. We require that (304) and the Cole-Hopf transformation in (297) be compatible

$$\frac{\partial}{\partial\lambda}(\Delta_x\phi) = \Delta_x\phi_\lambda. \tag{308}$$

This implies

$$u_\lambda = \frac{\Delta_x(S\phi) - uS\phi}{\phi}. \tag{309}$$

A direct calculation yields

$$\Delta_x\left(\frac{S\phi}{\phi}\right) = \frac{1}{T_x\phi}(\Delta_x(S\phi) - u(S\phi)), \tag{310}$$

and (307) follows. It is still necessary to show that $S\phi/\phi$ depends only on $u(x,t)$. The expressions for $S\phi$ can be found from (306). We thus see that all expressions involved can be expressed in terms of $u(x,t)$ and its shifted values, using the Cole-Hopf transformation again. Indeed,

$$\begin{aligned}
\Delta_x\phi &= u\phi, \quad \Delta_t\phi = v\phi, \\
T_x\phi &= (1 + \delta_x u)\phi, \quad T_t\phi = (1 + \delta_t v)\phi, \\
T_x^{-1}\phi &= \left(T_x^{-1}\frac{1}{1 + \delta_x u}\right)\phi, \quad T_t^{-1}\phi = \left(T_t^{-1}\frac{1}{1 + \delta_t v}\right)\phi,
\end{aligned} \tag{311}$$

where we define

$$v = \Delta_x u + uT_x u. \tag{312}$$

Explicitly, (307) maps the 6-dimensional symmetry algebra of the discrete heat equation into the 5-dimensional Lie algebra of the discrete Burgers equation. The corresponding flows are

$$u_{\lambda_1} = (1 + \delta_t v)\Delta_t u$$

$$u_{\lambda_2} = (1 + \delta_x u)\Delta_x u$$

$$u_{\lambda_3} = (1 + \delta_x u)\Delta_x \left(2tT_t^{-1}\frac{u}{1 + \delta_t v} + \left(x + \frac{1}{2} - \delta_x\right)T_x^{-1}\frac{1}{1 + \delta_x u} \right)$$

$$u_{\lambda_4} = (1 + \delta_x u)\Delta_x \left(2tT_t^{-1}\frac{v}{1 + \delta_t v} + xT_x^{-1}\frac{u}{1 + \delta_x u} - \frac{1}{2}T_x^{-1}\frac{1}{1 + \delta_x u} \right)$$

$$u_{\lambda_5} = (1 + \delta_x u)\Delta_x \left(t^2 T_t^{-1}\left(\frac{1}{1 + \delta_t v}T_t^{-1}\frac{v}{1 + \delta_t v}\right) \right. \tag{313}$$

$$+ txT^{-1}\left(\frac{1}{1 + \delta_x u}T_t^{-1}\frac{u}{1 + \delta_t v}\right)$$

$$+ \frac{1}{4}\left(x^2 - \frac{\delta_x^2}{4}\right)T_x^{-1}\left(\frac{1}{1 + \delta_x u}T_x^{-1}\frac{1}{1 + \delta_x u}\right)$$

$$+ tT_t^{-1}\left(\frac{1}{1 + \delta_t v}T_t^{-1}\frac{1}{1 + \delta_t v}\right)$$

$$\left. - \frac{1}{2}tT_x^{-1}\left(\frac{1}{1 + \delta_x u}T_t^{-1}\frac{1}{1 + \delta_t v}\right) \right)$$

$$u_{\lambda_6} = 0.$$

Symmetry Reduction for the Discrete Burgers Equation. We first treat the symmetry reduction for the continuous Burgers equation. We add a compatible equation to the Burgers equation

$$u_t = u_{xx} + 2uu_x$$
$$u_\lambda = Q(x, t, u, u_{x,t}) = 0 \tag{314}$$

and solve the two equations simultaneously. An example is furnished by the time translations,

$$u_\lambda = u_t = 0. \tag{315}$$

Then $u = u(x)$ and

$$u_{xx} + 2uu_x = 0 \tag{316}$$

implies

$$u_x + u^2 = K.$$

We therefore obtain three types of solutions,

$$u = \frac{1}{x}, \quad u = k\arctanh kx, \quad u = k\arctan kx. \tag{317}$$

We now consider the discrete case. All flows have the form (307). Condition $u_\lambda = 0$ hence implies

$$S\phi = K(t)\phi, \tag{318}$$

where $K(t)$ is an arbitrary function. This equation must be solved together with the discrete Burgers equation in order to obtain group-invariant solutions.

Let us consider just one example, that of time translations, the first equation in (313). Equation (318) reduces to

$$\Delta_t\phi = K(t)\phi, \tag{319}$$

i.e.,

$$v = \Delta_x u + uT_x u = K(t). \tag{320}$$

We rewrite the Burgers equation as

$$\Delta_t u = \frac{1 + \delta_x u}{1 + \delta_t v}\Delta_x v, \quad v \equiv \Delta_x u + uT_x u. \tag{321}$$

However, from (320), we obtain $v = K(t)$ and hence $\Delta_t u = 0$, $K(t) = K_0 =$ const. Since ϕ satisfies the heat equation, we can rewrite (319) as

$$\Delta_{xx}\phi = K_0\phi. \tag{322}$$

The general solution of (322) is obtained for $K_0 \neq 0$ by setting $\phi = a^x$ and solving (322) for a. We find

$$\phi = c_1(1 + \sqrt{K_0}\delta_x)^{x/\delta_x} + c_2(1 - \sqrt{K_0}\delta_x)^{x/\delta_x}. \tag{323}$$

For $K_0 = 0$,

$$\phi = c_1 + c_2 x. \tag{324}$$

Solutions of the discrete Burgers equation are obtained via the Cole-Hopf transformation

$$u = \frac{\Delta_x\phi}{\phi}. \tag{325}$$

The same procedure can be followed for all other symmetries. We obtain linear second order difference equations for ϕ involving one variable only. However, the equations have variable coefficients and are hard to solve. They can be reexpressed as equations for $u(x,t)$, again involving only one independent variable. Thus, a reduction takes place, but it is not easy to solve the reduced equations explicitly.

For instance, Galilei-invariant solutions of the discrete Burgers equation must satisfy the ordinary difference equation

$$2tT_x u + x - K(t) + 2t\delta_x uT_x u + \delta_t\left(\frac{7}{2}T_x u + \frac{7}{2}\delta_x uT_x u\right.$$
$$\left. + xuT_x u + x\Delta_x u - \frac{3}{2}u\right) + \frac{3}{2}\delta_x - K(t)(\delta_x u + \delta_t(T_x\Delta_x u \tag{326}$$
$$+ uT_x^2 u - uT_x u) + T_x uT_x^2 u + \delta_x uT_x uT_x^2 u) = 0$$

where t figures as a parameter.

Acknowledgements

I thank the organizers of the CIMPA School for giving me the opportunity to present these lectures and one of the editors of this volume for her useful proof-reading of my manuscript. Special thanks go to the Tamizhmani family for making my visit to Pondicherry both pleasant and memorable.

The research reported upon in these lectures was partly supported by research grants from NSERC of Canada, FQRNT du Québec and the NATO collaborative grant n. PST.CLG.978431.

References

1. J. Aczel: *Lectures on Functional Equations and their Applications* (Academic Press, New York 1966)
2. J. Aczel (Editor): *Functional Equations: History, Applications and Theory* (Reidel, 1984)
3. W.F. Ames: *Nonlinear Partial Differential Equations in Engineering* (Academic Press, New York 1972)
4. R.L. Anderson and N.H. Ibragimov: *Lie-Bäcklund Transformations in Applications*, SIAM Studies in Appl. Math., No 1 (SIAM, Philadelphia 1979)
5. M. Bakirova, V. Dorodnitsyn, and R. Kozlov: *Invariant difference schemes for heat transfer equations with a source*, J. Phys. A **30** (1997) pp 8139–8155
6. G. Bluman: *Simplifying the form of Lie groups admitted by a given differential equation*, J. Math. Anal. Appl. **145** (1990) pp 52–62
7. G. Bluman and S. Kumei: *Symmetries of Differential Equations* (Springer-Verlag, New York 1989)
8. G.B. Byrnes, R. Sahadevan, and G.R.W. Quispel: *Factorizable Lie symmetries and the linearization of difference equations*, Nonlinearity **8** (1995) pp 443–459
9. B. Champagne, W. Hereman, and P. Winternitz: *The computer calculation of Lie point symmetries of large systems of differential equations*, Comput. Phys. Commun. **66** (1991) pp 319–340
10. P. Clarkson and F. Nijhoff (Editors): *Symmetries and integrability of difference equations*, LMS Lect. Note Series, vol. 255 (Cambridge Univ. Press, 1999)
11. A. DiBucchianico and D. Loeb: *Umbral Calculus*, Electr. J. Combin. DS3 (2000) http://www.combinatorics.org
12. V.A. Dorodnitsyn: *Transformation groups in a space of difference variables*, J. Sov. Math. **55** (1991) pp 1490–1517
13. V.A. Dorodnitsyn: *Finite difference analog of the Noether theorem*, Dokl. Ak. Nauk **328** (1993) pp 678–682
14. V.A. Dorodnitsyn: *Finite difference models entirely inheriting continuous symmetries of differential equations*, Int. J. Mod. Phys. C **5** (1994) pp 723–734
15. V. Dorodnitsyn: *Group Properties of Difference Equations* (Fizmatlit, Moscow 2001) (in Russian)
16. V. Dorodnitsyn and R. Kozlov: *A heat transfer with a source: the complete set of invariant difference schemes*, J. Nonlinear Math. Phys. **10** (2003) pp 16–50
17. V. Dorodnitsyn, R. Kozlov, and P. Winternitz: *Lie group classification of second order difference equations*, J. Math. Phys. **41** (2000) pp 480–509

18. V. Dorodnitsyn, R. Kozlov, and P. Winternitz: *Continuous symmetries of Lagrangians and exact solutions of discrete equations*, J. Math. Phys. **45** (2004) pp 336–359

19. V. Dorodnitsyn, R. Kozlov, and P. Winternitz: *Symmetries, Lagrangian formalism and integration of second order ordinary difference equations*, J. Nonlinear Math. Phys. **10**, Suppl. 2 (2003) pp 41–56

20. V. Dorodnitsyn and P. Winternitz: *Lie point symmetry preserving discretizations for variable coefficient Korteweg-de Vries equations*, Nonlinear Dynamics **22** (2000) pp 49–59

21. R. Floreanini, J. Negro, L.M. Nieto, and L. Vinet: *Symmetries of the heat equation on a lattice*, Lett. Math. Phys. **36** (1996) pp 351–355

22. R. Floreanini and L. Vinet: *Lie symmetries of finite difference equations*, J. Math. Phys. **36** (1995) pp 7024–7042

23. G. Gaeta: *Nonlinear Symmetries and Nonlinear Equations* (Kluwer, Dordrecht 1994)

24. D. Gomez-Ullate, S. Lafortune, and P. Winternitz: *Symmetries of discrete dynamical systems involving two species*, J. Math. Phys. **40** (1999) pp 2782–2804

25. S. Helgason: *Differential Geometry and Symmetric Spaces* (Academic Press, New York 1962)

26. W. Hereman: Symbolic software for Lie symmetry analysis. In: *Lie Group Analysis of Differential Equations*, vol. 3, ed. by N.H. Ibragimov (CRC Press, Boca Raton 1996) pp 367–413

27. R. Hernandez Heredero, D. Levi, M.A. Rodriguez, and P. Winternitz: *Lie algebra contractions and symmetries of the Toda hierarchy*, J. Phys. **A33** (2000) pp 5025–5040

28. R. Hernandez Heredero, D. Levi, M.A. Rodriguez, and P. Winternitz: *Relation between Bäcklund transformations and higher continuous symmetries of the Toda equation*, J. Phys. **A34** (2001) pp 2459–2465

29. R. Hernandez Heredero, D. Levi, and P. Winternitz: *Symmetries of the discrete Burgers equation*, J. Phys. **A32** (1999) pp 2685–2695

30. R. Hernandez Heredero, D. Levi, and P. Winternitz: *Symmetries of the discrete nonlinear Schrödinger equation*, Theor. Math. Phys. **127** (2001) pp 729–737

31. R. Hirota: *Nonlinear partial difference equations V. Nonlinear equations reducible to linear equations*, J. Phys. Soc. Japan **46** (1979) pp 312–319

32. N.H. Ibragimov: *Transformation Groups Applied to Mathematical Physics* (Reidel, Boston 1985)

33. N. Jacobson: *Lie Algebras* (Interscience, New York 1962)

34. S. Lafortune, L. Martina, and P. Winternitz: *Point symmetries of generalized Toda field theories*, J. Phys. **A33** (2000) pp 2419–2435

35. S. Lafortune, S. Tremblay, and P. Winternitz: *Symmetry classification of diatomic molecular chains*, J. Math. Phys. **42** (2001) pp 5341–5357

36. D. Levi and L. Martina: *Integrable hierarchies of nonlinear difference-difference equations and symmetries*, J. Phys. **A34** (2001) pp 10357–10368

37. D. Levi and O. Ragnisco (Editors): *SIDE III–Symmetries and integrability of difference equations*, CRM Proc. and Lect. Notes, vol. 25 (AMS Publ., 2000)

38. D. Levi, P. Tempesta, and P. Winternitz: *Umbral calculus, difference equations and the discrete Schrödinger equation*, Preprint nlin S1/0305047

39. D. Levi, S. Tremblay, and P. Winternitz: *Lie point symmetries of difference equations and lattices*, J. Phys. A **33** (2000) pp 8507–8524

40. D. Levi, S. Tremblay, and P. Winternitz: *Lie Symmetries of multidimensional difference equations*, J. Phys. A. **34** (2001) pp 9507–9524
41. D. Levi, L. Vinet, and P. Winternitz (Editors): *Symmetries and integrability of difference equations*, CRM Proc. and Lect. Notes, vol. 9 (AMS Publ., 1996)
42. D. Levi, L. Vinet and P. Winternitz: *Lie group formalism for difference equations*, J. Phys. **A30** (1997) pp 663–649
43. D. Levi and P. Winternitz: *Continuous symmetries of discrete equations*, Phys. Lett. **A152** (1991) pp 335–338
44. D. Levi and P. Winternitz: *Symmetries and conditional symmetries of differential-difference equations*, J. Math. Phys. **34** (1993) pp 3713–3730
45. D. Levi and P. Winternitz: *Symmetries of discrete dynamical systems*, J. Math. Phys. **37** (1996) pp 5551–5576
46. D. Levi and P. Winternitz: *Lie point symmetries and commuting flows for equations on lattices*, J. Phys. A **35** (2002) pp 2249–2262
47. S. Maeda: *Canonical structure and symmetries for discrete systems*, Math. Japan **25** (1980) pp 405–420
48. S. Maeda: *The similarity method for difference equations*, IMA J. Appl. Math. **38**, 129 (1987) pp 129–134
49. L. Martina, S. Lafortune, and P. Winternitz: *Point symmetries of generalized field theories II. Symmetry reduction*, J. Phys. **A33** (2000) pp 6431–6446
50. M.A. Naimark and A.I. Stern: *Theory of Group Representations* (Springer-Verlag, New York 1982)
51. F.W. Nijhoff, Yu. Suris, C. Viallet (Editors): J. Nonlinear Math. Phys. **10** (2003), Supplement 2, pp 1–245
52. P.J. Olver: *Applications of Lie Groups to Differential Equations* (Springer-Verlag, New York 1993)
53. L.V. Ovsiannikov: *Group Analysis of Differential Equations* (Academic Press, New York 1982)
54. G.R.W. Quispel, H.W. Capel and R. Sahadevan: *Continuous symmetries of difference equations; the Kac-van Moerbeke equation and Painlevé reduction*, Phys. Lett. **A170** (1992) pp 379-383
55. G.R.W. Quispel and R. Sahadevan: *Lie symmetries and integration of difference equations*, Phys. Lett. **A184** (1993) pp 64–70
56. D.W. Rand, P. Winternitz, and H. Zassenhaus: *On the identification of a Lie algebra given by its structure constants. I. Direct decompositions, Levi decomposition and nilradicals*, Lin. Algebra Appl. **109**, pp 197–246 (1988)
57. C. Rogers and W.F. Ames: *Nonlinear Boundary Value Problems in Science and Engineering* (Academic Press, Boston 1989)
58. S. Roman: *The Umbral Calculus* (Academic Press, San Diego 1984)
59. G.C. Rota: *Finite Operator Calculus* (Academic Press, San Diego 1975)
60. J. Satsuma and T. Tokihiro (Editors): J. Phys. A: Math. Gen. **34** (7 Dec 2001) pp 10337–10774
61. F. Schwarz: *A REDUCE package for determining Lie symmetries of ordinary and partial differential equations*, Comput. Phys. Commun. **27**, 179 (1982)
62. M. Toda: *Theory of Nonlinear Lattices* (Springer, Berlin 1991)
63. P. Winternitz: Lie Groups and solutions of nonlinear partial differential equations. In: *Integrable Systems, Quantum Groups and Quantum Field Theories*, ed. by A. Ibort and M.A. Rodriguez (Kluwer, Dordrecht 1993) pp 429–495
64. V.E. Zakharov (Editor): *What is Integrability?* (Springer, New York 1991)

Discrete Painlevé Equations: A Review

B. Grammaticos[1] and A. Ramani[2]

[1] GMPIB, Université Paris VII, Tour 24-14, 5^{e}étage, case 7021,
75251 Paris, France, `grammati@paris7.jussieu.fr`
[2] CPT, Ecole Polytechnique, CNRS, UMR 7644, 91128 Palaiseau, France,
`ramani@cpht.polytechnique.fr`

Abstract. We present a review of what is current knowledge about discrete Painlevé equations. We start with a historical introduction which explains how the discrete Painlevé equations made their appearance. Since the most widely used derivation method was the one based on integrability detectors, we start by presenting a brief review of the existing discrete integrability detectors, and proceed to show how they were indeed used in deriving the discrete Painlevé equations. Given the profusion of different forms of these mappings, we examine the various approaches to their classification, from the first, crude ones, to the complete classification based on affine Weyl groups. The properties of the discrete Painlevé equations, which make them so special, are also reviewed with ample details. Finally, we present a series of results which are peripherally related to the discrete Painlevé equations but help cast more light on these (very) special systems.

$\Pi \rho o \lambda \epsilon \gamma \acute{o} \mu \epsilon \nu o \nu$

Prologue

Why Painlevé? Why discrete? What makes Painlevé equations so special? The best way to formulate the answer is Kruskal's [1]: Painlevé equations are the most complicated systems that one can solve (in a nontrivial way). Anything simpler becomes trivially integrable, anything more complicated becomes hopelessly nonintegrable. Thus Painlevé equations are at the borderline between trivial integrability (usually linearisability) and nonintegrability, a fact that bestows upon them a plethora of interesting properties [2]. The keyword here is integrability: Painlevé equations are the paradigmatic integrable systems and they are omnipresent in the study of integrability [3].

This review will be devoted to the study of discrete Painlevé equations. The important notion here is discreteness. What we will be talking about are equations in which the *independent* variable assumes discrete values. The simplest case is that of difference equations, but it is not the only one. In many physical situations the independent variable is time, and discreteness in this context would correspond to a time which is discrete. At this point the reader would start wondering about the pertinence of these ideas: space-time as perceived by our senses and interpreted by our intuition looks smooth, homogeneous, isotropic, leaving no place for discreteness. But what is the

B. Grammaticos and A. Ramani, Discrete Painlevé Equations: A Review, Lect. Notes Phys. **644**, 245–321 (2004)
`http://www.springerlink.com/`

value of intuition as far as fundamental physical notions are concerned? In the light of modern physics (where, as explained by Mills [4], counterintuitive notions abound) the fact that a notion may appear nonphysical at first sight is not really reason enough to discard it. As a matter of fact, recent progress in field theory [5] leads toward a fundamentally discrete description of nature and thus, to counter a famous saying by Einstein [6], it indeed looks as though we are learning how to breathe in empty space.

The program for describing physical phenomena through discrete equations is not without difficulties. Even setting aside the fact that our arsenal of methods for dealing with continuous systems overpowers the meager genuinely discrete tools at our disposal, the problems appear at the very first step of modelling. It is in fact quite uncertain what the proper discrete equation of a model may be. The continuous limit, being in fact a reductive procedure, discards information and thus quite often leads to a unique choice for the equation which would model the system. This is not the case for discrete equations, and the profusion of discrete Painlevé equations with the same continuous limit (and the possible nomenclature confusion that this may entail) is a convincing illustration of this problem. So we must seek a way to deal with these difficulties. What is needed is a way to do away with the ambiguity in mathematical modelling while leading to models for which a great deal of theory can be developed. Fortunately, the property of integrability is a very good and effective criterium. Integrable systems are those that, albeit nonlinear and highly nontrivial, can be dealt with by systematic and rigorous approaches. And while one can argue that they are too special to represent the entire wealth of physical phenomena, this is a quality rather than a drawback. Integrable systems are so very special that we lose all ambiguity in the search for the governing equations for our models.

Of course, all these considerations are typically those of a physicist. From a purely mathematical point of view, one can very well study discrete equations for their own sake, without having to care about their applicability in real-world modelling. (However the authors, being die-hard physicists, do care about these questions, unwilling to get entrapped into mathematics for mathematics' sake).

What do we really mean by discrete Painlevé equations? To put it simply, a discrete Painlevé equation (d-$\mathbb{P}$) is an integrable (nonautonomous) discrete equation the continuous limit of which is a (continuous) Painlevé equation. As in the continuous case we reserve the name of Painlevé equations ($\mathbb{P}$'s) for second-order systems. Higher-order systems (which, by the way, are almost *terra incognita*) will be distinguished by the appropriate qualifier, "higher $\mathbb{P}$'s".

Why are d-$\mathbb{P}$'s (and their study) interesting? First of all, as we said before, the continuous equations are the result of some reduction procedure. Discrete systems are the genuine, unadulterated entities. They are indeed the most fundamental objects. Having established their properties we can obtain those of the continuous ones through a simple limiting procedure. This is of course an idealised, utopic view of things: a whole century of studies devoted to continuous integrable systems (and in particular $\mathbb{P}$'s) has made it possible to establish most of the properties of continuous systems. Still the study of the properties of their discrete analogues has its usefulness: it makes it possible to establish a perfect parallel between the properties of the continuous and those of the discrete integrable systems. On a more practical level, the d-$\mathbb{P}$'s present another interest: they constitute integrable discretisations of the continuous $\mathbb{P}$'s [6]. Thus we can use the d-$\mathbb{P}$'s for numerical studies of their celebrated continuous analogues. Given the paucity of numerical algorithms for the computation of the solutions of $\mathbb{P}$'s, and the inexistence of tabulations, the existence of these ready-made simulators is invaluable. (Still, curiously, very few studies have used this approach to date).

Having set the framework, we are ready now to embark upon the detailed study of these fascinating systems.

Ιστορικά

1 The (Incomplete) History of Discrete Painlevé Equations

Pretending to achieve completeness (and/or objectivity) when one embarks upon writing the history of past (even recent) events is at best a delusion. There is just no way to give the full history of discrete Painlevé equations. There is no way to capture the intentions and intuitions of past authors when the only thing at our disposal (most of the times) are dry scientific accounts of their findings. With this *caveat* out of the way, we may proceed now to the history of the d-$\mathbb{P}$'s as it is known to the authors of the present review with the additional warning that our views of the matter are time-dependent.

The first author to have worked on integrable, discrete nonautonomous systems (and to have produced examples thereof) is Laguerre [7]. He obtained integrable difference equations from the recursion relations established when working with orthogonal polynomials. A casual analysis of the systems of Laguerre shows that they are of order higher than two and thus will define higher-order d-$\mathbb{P}$'s (although no continuous limits have been computed, neither has there been any effort to reduce these systems as much as possible so as to obtain their lowest-order recursion). The most interesting point is that these integrable discrete systems predate the discovery of the continuous Painlevé equations (c-$\mathbb{P}$'s) [8]. This historical coincidence reestablishes the essence of justice since the discrete systems are more fundamental than the continuous ones.

Much (much) later (1939) a discrete Painlevé equation, easily recognizable today, was derived, in essentially the same framework as Laguerre's. Indeed Shohat [9], working on orthogonal polynomials obtained the integrable recursion relation

$$x_{n+1} + x_{n-1} + x_n = \frac{z_n}{x_n} + 1, \qquad (1.1)$$

where $z_n = \alpha n + \beta$. The fate of this second order difference equation was to resurface many years later (and on this last occasion be properly recognized as a d-$\mathbb{P}$ [10]).

The next apparition of a d-$\mathbb{P}$ was in 1981 in the work of Jimbo and Miwa on continuous $\mathbb{P}$'s [11]. In the appendix of this fundamental and thorough study of Painlevé equations, the authors addressed the question of the contiguity relations of c-$\mathbb{P}$'s, i.e., of relations between solutions of a given Painlevé equation for different values of some parameter. From the solutions of P_{II}: $w'' = 2w^3 + tw + \alpha$ they obtained the contiguity relation

$$\frac{\alpha_n + 1/2}{x_{n+1} + x_n} + \frac{\alpha_n - 1/2}{x_n + x_{n-1}} = -(2x_n^2 + t), \qquad (1.2)$$

where $x_n = w(t, \alpha_n)$ and $\alpha_n = n + \alpha_0$. Curiously, no continuous limit was derived, and thus the naming of d-$\mathbb{P}$'s had to wait for ten more years. During that period, integrable difference (and differential-difference) recursion relations had made their appearence in field-theoretical models. It would probably be possible, with hindsight, to establish today their relations to (possibly higher-order) d-$\mathbb{P}$'s (but this would make this historical review less incomplete), so let us move to the climax.

While investigating a field-theoretical model of 2-dimensional gravity, and having to compute a partition function, Brézin and Kazakov resorted to the method of moments. They obtained for their partition function a recursion relation which was precisely the one obtained forty years before by Shohat, namely (1.1). The main difference this time was that Brézin and Kazakov did compute the continuous limit of (1.1) and obtained $w'' = 6w^2 + t$, i.e., Painlevé I. The discrete $\mathbb{P}$'s were born!

This breakthrough kindled an unprecedented interest in the domain. Shortly afterwards, and in a similar framework, Periwal and Shevitz discovered an integrable discrete system which had P_{II} as its continuous limit. This mapping had the form

$$x_{n+1} + x_{n-1} = \frac{z_n x_n}{1 - x_n^2}, \qquad (1.3)$$

and, at the continuous limit, becomes $w'' = 2w^3 + tw$, i.e., the P_{II} equation for the zero value of its parameter α.

Almost simultaneously, the same equation was derived by a completely independent approach by Nijhoff and Papageorgiou [13]. These authors, in collaboration with Capel, had already derived discrete forms for such well-known 1+1 dimensional evolution equations as KdV and mKdV [14]. (Incidentally, these discrete forms were no other than the ones that had been obtained by Hirota [15] in his groundbreaking work in the 70's and early 80's.) The originality in the approach of Nijhoff and Papageorgiou lies in the fact that, starting from the well-established fact that the similarity reduction of mKdV leads to P_{II}, they obtained the similarity constraint of discrete mKdV. Contrary to what happens in the continuous case, where this constraint is linear, for discrete evolution equations the similarity constraint is a nonlinear, nonautonomous evolution equation, integrable in its own right [16]. The similarity reduction results from the application of the compatibility between the autonomous equation and the nonautonomous one. The result of this similarity reduction was, as expected, discrete P_{II}, precisely the one derived by Periwal and Shevitz, eq. (1.3).

It is really noteworthy that, at that point in time, not only did examples of d-$\mathbb{P}$'s exist but also they had been obtained using three of the four main methods for their derivation: orthogonal polynomials (essentially spectral methods), reductions, and contiguity relations. The time was ripe for the introduction of a fourth method.

The main idea was to apply in a direct way an integrability detector to some postulated functional form and to select the integrable cases. The starting point was a family of second-order integrable autonomous mappings proposed by Quispel, Roberts and Thompson (QRT) [17].

Let us now depart briefly from the purely historical narrative and present a brief account of the QRT mapping. There exist two families of QRT mappings [17] which are dubbed respectively symmetric and asymmetric for reasons which will become obvious below. One starts by introducing two 3×3 matrices, A_0 and A_1, of the form

$$A_i = \begin{pmatrix} \alpha_i & \beta_i & \gamma_i \\ \delta_i & \epsilon_i & \zeta_i \\ \kappa_i & \lambda_i & \mu_i. \end{pmatrix} \tag{1.4}$$

If both these matrices are symmetric the mapping is called symmetric. Otherwise it is called asymmetric. Next one introduces the vector $\boldsymbol{X} = \begin{pmatrix} x^2 \\ x \\ 1 \end{pmatrix}$ and constructs the two vectors $\boldsymbol{F} \equiv \begin{pmatrix} f_1 \\ f_2 \\ f_3 \end{pmatrix}$ and $\boldsymbol{G} \equiv \begin{pmatrix} g_1 \\ g_2 \\ g_3 \end{pmatrix}$ through

$$\boldsymbol{F} = (A_0 \boldsymbol{X}) \times (A_1 \boldsymbol{X}) \tag{1.5a}$$

$$\boldsymbol{G} = (\tilde{A}_0 \boldsymbol{X}) \times (\tilde{A}_1 \boldsymbol{X}) \tag{1.5b}$$

where the tilde denotes the transpose of the matrix. The components f_i, g_i of the vectors $\boldsymbol{F}, \boldsymbol{G}$ are, in general, quartic polynomials in x. Given f_i, g_i, the mapping assumes the form

$$x_{n+1} = \frac{f_1(y_n) - x_n f_2(y_n)}{f_2(y_n) - x_n f_3(y_n)} \tag{1.6a}$$

$$y_{n+1} = \frac{g_1(x_{n+1}) - y_n g_2(x_{n+1})}{g_2(x_{n+1}) - y_n g_3(x_{n+1})}. \tag{1.6b}$$

In the symmetric case $g_i = f_i$, and (1.6) reduces to a single equation,

$$x_{m+1} = \frac{f_1(x_m) - x_{m-1}f_2(x_m)}{f_2(x_m) - x_{m-1}f_3(x_m)}, \tag{1.7}$$

with the identification $x_n \to x_{2m}$, $y_n \to x_{2m+1}$.

Since $\boldsymbol{F}$ and $\boldsymbol{G}$ are obtained as vector products, it is clear that the result will be the same if one replaces the matrices A_0 and A_1 by the linear combinations $\rho_0 A_0 + \sigma_0 A_1$ and $\rho_1 A_0 + \sigma_1 A_1$, where $\rho_0, \sigma_0, \rho_1, \sigma_1$ are four free parameters (the only constraint is that $\rho_0 \sigma_1 \neq \rho_1 \sigma_0$). This transformation can be used in order to reduce the effective number of the parameters of the system to 14 in the asymmetric case, and to 8 in the symmetric one. However this is still not the number of the effective parameters since we have the full freedom of a homographic transformation, which amounts to three parameters, separately for x and y in the asymmetric case and for x alone in the symmetric one. Thus the final number of genuine parameters in this system is 8 for the asymmetric mapping and 5 for the symmetric one.

The QRT mapping possesses an invariant which is biquadratic in x and y

$$(\alpha_0 + K\alpha_1)x_n^2 y_n^2 + (\beta_0 + K\beta_1)x_n^2 y_n + (\gamma_0 + K\gamma_1)x_n^2 + (\delta_0 + K\delta_1)x_n y_n^2$$
$$+(\epsilon_0 + K\epsilon_1)x_n y_n + (\zeta_0 + K\zeta_1)x_n + (\kappa_0 + K\kappa_1)y_n^2 + (\lambda_0 + K\lambda_1)y_n + (\mu_0 + K\mu_1) = 0, \tag{1.8}$$

where K plays the role of the integration constant. In the symmetric case the invariant becomes just

$$(\alpha_0 + K\alpha_1)x_{n+1}^2 x_n^2 + (\beta_0 + K\beta_1)x_{n+1}x_n(x_{n+1} + x_n) + (\gamma_0 + K\gamma_1)(x_{n+1}^2 + x_n^2)$$
$$+(\epsilon_0 + K\epsilon_1)x_{n+1}x_n + (\zeta_0 + K\zeta_1)(x_{n+1} + x_n) + (\mu_0 + K\mu_1) = 0, \tag{1.9}$$

Viewed as a relation between x_n and y_n, (1.8) is a 2-2 correspondence (and similarly for (1.9)). While the generic biquadratic correspondence is not in general integrable [18], leading to an exponential growth of the number of images and preimages of a given point, this is not the case for (1.8), which has linear growth [19]. The integration of this biquadratic correspondence in the symmetric case, in terms of elliptic functions, goes in fact as far back as Euler. A clear pedagogical presentation can be found in the book by Baxter [20]. Curiously, the integration of the asymmetric mapping had to wait for the 21st century [21, 22]. Thus, the solution of the QRT mapping is given by the values of an elliptic function at equidistant points on a line in the complex plane.

The pertinence of QRT to the derivation of d-$\mathbb{P}$'s is clear. The continuous $\mathbb{P}$'s can be viewed as nonautonomous extensions of second-order, ordinary differential equations the solutions of which are given by elliptic functions. (Practically, this means that the $\mathbb{P}$'s have the same functional forms as the above mentioned autonomous equations, with some of the coefficients having a specific dependence on the independent variable). Thus it seems natural to start from a mapping with constant coefficients, and with elliptic function solutions, to allow the coefficients to depend on the independent variable, and then to select the integrable cases through the application of some integrability detector.

In the very first study [23] introducing the singularity confinement discrete integrability criterion, we analysed the possible integrability of two mappings with forms close to (1.1) and (1.3) *i.e.*, d-P_I and d-P_{II}. First we considered the most classical example of an integrable mapping, the McMillan mapping

$$x_{n+1} + x_{n-1} = \frac{2\mu x_n}{1 - x_n^2}. \tag{1.10}$$

A singularity may appear in the recursion (1.10) whenever x passes through the value 1 (or -1). So we assumed that x_0 was finite and that $x_1 = 1 + \epsilon$. (This could be obtained from a perfectly regular x_{-1}). We found then the following values: $x_2 = -\mu/\epsilon - (x_0 + \mu/2) + \mathcal{O}(\epsilon)$, $x_3 = -1 + \epsilon + \mathcal{O}(\epsilon^2)$ and $x_4 = x_0 + \mathcal{O}(\epsilon)$. Thus, not only was the singularity confined at this step, but also the mapping had recovered the memory of the initial conditions through x_0.

Next we generalised the McMillan mapping (1.10) to the nonautonomous case

$$x_{n+1} + x_{n-1} = \frac{a + bx_n}{1 - x_n^2} \tag{1.11}$$

where we considered a and b to be functions of n, and determined its integrable nonautonomous form, based on the singularity confinement property. Assuming that, for some n, we had a regular x_n, and that $x_{n+1} = \sigma + \epsilon$, where $\sigma = \pm 1$ (in order to cover the two possibilities of x going through a root of the denominator of the right-hand side), and iterating further we found that

$$x_{n+2} = -\frac{b_{n+1} + \sigma a_{n+1}}{2\epsilon} + \frac{a_{n+1} - \sigma b_{n+1}}{4} - x_n + \mathcal{O}(\epsilon),$$

and

$$x_{n+3} = -\sigma + \frac{2b_{n+2} - b_{n+1} - \sigma a_{n+1}}{b_{n+1} + \sigma a_{n+1}}\epsilon + \mathcal{O}(\epsilon^2). \tag{1.12}$$

The condition for x_{n+4} to be finite was:

$$b_{n+1} - 2b_{n+2} + b_{n+3} + \sigma(a_{n+1} - a_{n+3}) = 0, \tag{1.13}$$

which led to $a_{n+1} = a_{n+3}$ and $b_{n+1} - 2b_{n+2} + b_{n+3} = 0$. Thus $b_n(\equiv z_n) = \alpha n + \beta$ and $a_n = \delta + \gamma(-1)^n$. Again we ignored, at that time, the even-odd dependence and took a as a strict constant. We then obtained

$$x_{n+1} + x_{n-1} = \frac{a + z_n x_n}{1 - x_n^2} \tag{1.14},$$

a form of discrete P_{II} in agreement with previous results derived by different approaches.

Similarly one can start from a nonautonomous form of (1.1)

$$x_{n+1} + x_n + x_{n-1} = a + \frac{b}{x_n} \tag{1.15}$$

where we again consider a and b to be functions of n. Our assumption is that at some iteration step n, x_n is regular while x_{n+1} vanishes. Following the ideas we sketched in the introduction of this lecture, we set $x_{n+1} = \epsilon$. We obtain thus the following sequence of values

$$x_{n+2} = \frac{b_{n+1}}{\epsilon} + a_{n+1} - x_n + \mathcal{O}(\epsilon) \tag{1.16}$$

and

$$x_{n+3} = -\frac{b_{n+1}}{\epsilon} + a_{n+2} - a_{n+1} + x_n + \mathcal{O}(\epsilon). \tag{1.17}$$

(Note that there is no way one can make the divergence of x_{n+3} disappear.) Computing x_{n+4} we find that it diverges unless $a_{n+3} - a_{n+2} = 0$. Thus, for confinement, a must be constant. Implementing this constraint we find that

$$x_{n+4} = \frac{b_{n+1} - b_{n+2} - b_{n+3}}{b_{n+1}} \epsilon + \mathcal{O}(\epsilon^2). \tag{1.18}$$

We now ask for x_{n+5} to be finite, and we obtain the second condition

$$b_{n+1} - b_{n+2} - b_{n+3} + b_{n+4} = 0. \tag{1.19}$$

The solution of (1.19) is $b_n = \alpha n + \beta + \gamma(-1)^n$, i.e., b_n is linear in n, up to a parity-dependent constant. For the time being we can ignore this even-odd dependence (we shall come back to it later). Setting $b_n \equiv z_n = \alpha n + \beta$ we find that

$$x_{n+1} + x_n + x_{n-1} = a + \frac{z_n}{x_n}. \tag{1.20}$$

This is a discrete form of P$_\mathrm{I}$.

An even more interesting result was obtained in [24], which constituted the very first attempt at a systematic derivation of the d-P's. While examining the de-autonomisation of the $f_2 = 0$ family of the QRT mapping, we obtained the following integrability candidate (its integrability having been confirmed later through the explicit constuction of its Lax pair)

$$x_{n+1} x_{n-1} = \frac{ab(x_n - cq_n)(x_n - dq_n)}{(x_n - a)(x_n - b)}, \tag{1.21}$$

where $a, b, c,$ and d are constants and $q_n = q_0 \lambda^n$. Thus this mapping, the continuous limit of which is P$_\mathrm{III}$, is not a difference equation, as all the previous examples of d-P's, but a q- (multiplicative) mapping. We now know that q-discrete forms exist for all d-P's and in some sense they are more fundamental than the difference-P's.

After this incomplete (and certainly biased) account of the remote and recent past of the d-P's, we are now ready to embark upon more technical matters.

Ανιχνευτές

2 Detectors, Predictors, and Prognosticators (of Integrability)

There exist several reasons why discrete $\mathbb{P}$'s have remained in limbo for almost a century and then mushroomed all of a sudden. First, discrete systems started attracting the interest of physicists (and, invariably, with a short time-lag, that of mathematicians) at the beginning of the 90's. Thus a critical mass of results on integrable discrete systems was attained. Second, specific techniques for the study of discrete systems started being proposed. Foremost among them were criteria for the detection of discrete integrability. The problem was already nontrivial for continuous systems. Still the use of complex analysis has made possible the development of specific and efficient tools for the prediction of integrability, and actual integration of systems expressed as ordinary or partial differential equations. According to Poincaré, to integrate a differential equation is to find, for the general solution, a finite expression, possibly multivalued, in terms of a finite number of functions. The word "finite" indicates that integrability is related to a *global* rather than *local* knowledge of the solution. However, this definition is not very useful unless one defines more precisely what is meant by "function". By extending the solution of a given ordinary differential equation (ODE) in the complex domain, one has the possibility, instead of asking for a global solution for an ODE, of looking for solutions locally and of obtaining a more global result by analytic continuation. If we wish to define a function, we must find a way to treat branch points, *i.e.*, points around which two (or more) determinations are exchanged. This can be done by various uniformisation procedures, provided the branch points are fixed. Linear ODE's are such that all the singularities of their solutions are fixed, and these equations are thus considered integrable. In the case of nonlinear ODE's, the situation is not so simple due to the fact that the position of the singular points in this case may depend on the initial conditions: they are movable. The approach of Painlevé [8] and his school, which, to be fair, was based on ideas of Fuchs and Kovalevskaya, was simple: they decided to look for those of the nonlinear ODE's the solutions of which were free from movable branch points. Painlevé managed to take up the challenge of Picard and to determine the functions defined by the solutions of second-order nonlinear equations. The success of this approach is well-known. The Painlevé transcendents have been discovered in that way, and their importance in mathematical physics is ever growing. The Painlevé property, *i.e.*, absence of movable branch points, has since been used with great success in the detection of integrability [25].

We must stress one important point here. The Painlevé property, as introduced by Painlevé, is not just a *predictor* of integrability but practically a definition of integrability. As such, it becomes a tautology rather than a criterion. It is thus crucial to make the distinction between the Painlevé property and the algorithm for its investigation. This algorithm can only search for movable branch points *subject to certain assumptions*. The search can thus lead to a conclusion the validity of which is questionable: if we find that the system passes what is usually referred to as the Painlevé test (in one of its several variants), this does not necessarily mean that the system possesses the Painlevé property. Thus at least as far as its usual practical application is concerned, the Painlevé test may not be sufficient for integrability. The situation becomes still more complicated if we consider systems that are integrable by quadratures and/or cascade linearisation. If we extend the notion of integrability in order to include such systems, it turns out [26] that the Painlevé property is no longer related to it. Thus the criterion based on the singularity structure is not a necessary one in this case.

Despite these considerations, the Painlevé test has been of great heuristic value for the study of the integrability of continuous systems [27], leading to the discovery of a host of new integrable systems. The question thus naturally arose, whether these techniques could be transposed *mutatis mutandis*, to the study of discrete systems. The discrete systems to which we are referring here (and which play an important role in physical applications) are systems that are cast into a rational form, perhaps after some transformation of the dependent variable. Since these systems have singularities, it is natural to assume that singularities would play an important role in connection with integrability. While this is quite plausible, the approach based on singularities would be unable to deal with polynomial mappings which do not possess any. Still, one would not expect all polynomial mappings to be integrable, in particular in view of the fact that many of them exhibit chaotic behaviour. Moreover, any argument based on singularities in the discrete domain can only bear a superficial resemblance to the situation in the continuous case. One cannot hope to relate directly the singularities of mappings to those of ODE's for the simple reason that there exist discrete systems which do not have any nontrivial continuous limit. Having set the frame, we can now present a review of the various discrete integrability detectors.

Singularity Confinement

Singularity confinement is the name of a discrete integrability criterion we introduced in the early 90's [23, 28] and which has been instrumental in the derivation of the d-$\mathbb{P}$'s. The role played by singularities is best illustrated through some concrete example.

Consider the mapping

$$x_{n+1} + x_{n-1} = \frac{a}{x_n} + \frac{1}{x_n^2}. \tag{2.1}$$

Obviously, a singularity appears whenever the value of x_n becomes 0. Iterating this value, one obtains the sequence $\{0, \infty, 0\}$ and then the indeterminate form $\infty - \infty$. As Kruskal points out the real problem lies in the latter, while the occurrence of a simple infinity is something with which one can easily deal by going to projective space. The way to treat this difficulty is to use an argument of continuity with respect to the initial conditions and to introduce a small parameter ϵ. In this case, if we assume that $x_n = \epsilon$, we obtain for the first values of x, $x_{n+1} \approx 1/\epsilon^2$, $x_{n+2} \approx -\epsilon$, and when we compute carefully the next one we find not only that it is finite but also that it contains a memory of the initial condition x_{n-1}. The singularity has disappeared.

This is the property that we have dubbed *singularity confinement* and, after having analysed a host of discrete systems, we concluded that it was characteristic of systems integrable through spectral methods. By a bold move, singularity confinement has been elevated to the rank of an integrability criterion. In what follows, we shall comment on its necessary and sufficient character.

Several questions had to be answered for singularity confinement to be really operative. The first, that we encountered above, was related to the fact that the iteration of a mapping may not be defined uniquely in both directions. Thus we proposed the criterion of preimage non-proliferation [29], which had the advantage of eliminating *en masse* all polynomial nonlinear mappings. One remark, unavoidable at this point, is related to the existence of integrable mappings involving two variables. The typical example is what we call *asymmetric discrete Painlevé equations*. It can be argued that, in a such a mapping, one of the variables can always be eliminated, leading to a single mapping or rather a correspondence for the other one. Indeed, for a generic second-order system, the resulting relation will be one where the variables x_{n+1} and x_{n-1} appear at powers higher than unity. Its evolution leads in general to an exponential number of images, and preimages, of the initial point. This non-singlevalued system cannot be integrable [30]. This is not in contradiction to the fact that we can obtain one solution for the mapping, namely the one furnished by the evolution of the two-variable system. This is *the only solution that we know how to describe*, while the full system, with an exponentially increasing number of branches, eludes a full description.

The second point is that the notion of 'singularity' had to be refined. Clearly the simple appearance of an infinity in the iteration of a mapping is not really a problem. What is crucial is that a mapping may at some point 'lose a degree of freedom'. In a mapping of the form $x_{n+1} = f(x_n, x_{n-1})$, this simply means that $\partial x_{n+1}/\partial x_{n-1} = 0$ and the memory of the initial condition x_{n-1} disappears from the iteration. What does 'confinement' mean in this case? Clearly, the mapping must recover the lost degree of freedom and the only way to do this is by the appearence of an indeterminate form $0/0$, $\infty - \infty$, etc., in the subsequent iterations.

Over the years singularity confinement has turned out to be a very convenient discrete integrability detector. We know of a large domain of mappings for which the confinement property is satisfied: they are those that are integrable through inverse scattering tranform (IST) methods. No counterexample to this is known. On the other hand, there exists a whole class of systems, the ones integrable by linearisation, the integrability of which is not associated to the singularity confinement property (just as in the continuous case, the linearisability is not related to the Painlevé property).

Perturbative Painlevé

The main idea of the perturbative Painlevé approach is the following. Suppose that we start from a discrete system which contains a small parameter. Typically, one considers the lattice spacing δ as small, whenever one is interested in the continuous limit of the mapping. Next, we expand in a power series in this small parameter. If the initial mapping is integrable then the equations obtained at each order of the series must equally be integrable. (This is something that we learned from J. Satsuma [31], but no practical application of this idea was suggested at the time.)

The colleagues who rediscovered this idea (which in fact goes back to Poincaré) and use it in practice, were Conte and Musette [32]. The way they did this was to work with expansions in the lattice spacing, obtain a sequence of coupled differential equations, and investigate the integrability of the latter using the Painlevé algorithm. The advantage of this approach is that one can treat polynomial mappings on the same footing as rational ones. Let us illustrate this approach through an example. We choose the well-known logistic map

$$x_{n+1} = \lambda x_n (1 - x_n). \tag{2.2}$$

We introduce the lattice parameter δ and expand everything in power series in it. Let $\lambda = \lambda_0 + \delta \lambda_1 + \delta^2 \lambda_2 + \ldots$ and $x_n = w_0 + \delta w_1 + \delta^2 w_2 + \ldots$. Similarly, $x_{n+1} = (w_0 + \delta w_0' + \delta^2 w_0''/2 + \ldots) + \delta(w_1 + \delta w_1' + \delta^2 w_1''/2 + \ldots) + \delta^2(w_2 + \ldots) + \ldots$, where the continuous variable is $t = n\delta$. In this particular case, we take $w_0 \equiv 0$, $\lambda_0 = 1$, and the first equation (at order δ^2) is nonlinear in terms of the quantity w_1,

$$w_1' = -w_1^2 + \lambda_1 w_1 \tag{2.3}$$

which is a Riccati equation and has movable poles as its only singularities. Then at order δ^3 we have an equation for w_2

$$w_2' = -w_2(\lambda_1 - 2w_1) - w_1^3 + \frac{\lambda_1}{2} w_1^2 + (\lambda_2 - \frac{\lambda_1^2}{2}) w_1, \tag{2.4}$$

and similarly at higher orders. Notice that (2.4) is linear in w_2. The same applies to all the subsequent equations. Indeed at order δ^{n+1} we find a differential equation for the new quantity w_n in terms of the w's that have already been obtained. Since this equation is linear, it cannot have movable singularities when considered as an equation for w_n, everything else being supposed to be known. However, when we consider the cascade of equations, the subsequent objects will in general have singularities whenever the earlier ones are singular, and these singularities are *movable* in terms of the whole cascade. Moreover they are not poles. Already (2.4) shows that, in the neighbourhood of a pole of w_1, where $w_1 \approx 1/s$ with $s = t - t_0$ (t_0 being the location of the movable singularity of w_1), w_2 has logarithmic singularities $w_2 \approx -\log(s)/s^2$. This singularity is a critical one which must be considered to be movable in terms of the *cascade*, and therefore the perturbative Painlevé property is not satisfied. This is consistent with the fact that the logistic map is known to be nonintegrable.

Although the method of the perturbative Painlevé approach is powerful enough, it is not without drawbacks. The main disadventage is due to the fact that not all discrete systems possess nontrivial continuous limits. In this case, if one does not have a valid starting point, the whole approach collapses. Moreover, the way this method was applied in the literature was to discretise a given continuous equation by introducing some freedom and using the perturbative Painlevé approach in order to pinpoint the integrable subcases. However, this method is only as good as one's imagination, and if the proposed discretisation is not rich enough, one may miss very interesting cases.

Algebraic Entropy

A most powerful discrete integrability detector was the one based on the ideas of Arnold [33] and Veselov [34]. Arnold introduced the notion of complexity which (for mappings in the plane) is the number of intersection points of a fixed curve with the image of a second curve obtained under the iterations of the mapping at hand. While the complexity grows exponentially with the iteration for generic mappings, it can be shown to grow only polynomially for a large class of integrable mappings.

Veselov has elaborated further on the notion of slow growth, applying it to the investigation of the integrability of mappings and correspondences. His main idea is summarised in the catch phrase "integrability has an essential correlation with the weak growth of certain characteristics". He studied the integrability of polynomial mappings and showed that the mapping $x_{n+1} = P(x_n, y_n)$, $y_{n+1} = Q(x_n, y_n)$, is integrable only if there exists a polynomial change of variables transforming the mapping to triangular form $x_{n+1} = \alpha x_n + R(y_n)$, $y_{n+1} = \beta y_n + \gamma$, with polynomial R.

The notion of complexity was further extended in the works of Viallet and his collaborators who focused on rational mappings [35]. They introduced what they called *algebraic entropy*, which is a global index of the complexity of the mapping. The main idea is that there exists a link between the dynamical complexity of a mapping and the degree of its iterates. If we consider a mapping of degree d, then the n-th iterate will have a degree d^n, unless common factors lead to simplifications. It turns out that when the mapping is integrable, such simplifications do occur in a massive way, leading to a degree growth which is polynomial in n, instead of exponential. Thus, while the generic, nonintegrable, mapping has exponential degree growth, polynomial growth is an indication of integrability.

Let us illustrate this approach with a practical application on a mapping that we already encountered

$$x_{n+1} + x_{n-1} = \frac{a}{x_n} + \frac{1}{x_n^2}. \tag{2.5}$$

In order to compute the degree of the iterates, we introduce the homogeneous coordinates by taking $x_0 = p$, $x_1 = q/r$, assigning the degree zero to p, and computing the degree of homogeneity in q and r at every iteration. We could of course have introduced a different choice for x_0 but it turns out that the choice of a zero-degree x_0 considerably simplifies the calculations. We thus obtain the degrees 0, 1, 2, 5, 8, 13, 18, 25, 32, 41, ... , . Clearly, the degree growth is polynomial, $d_{2m} = 2m^2$ and $d_{2m+1} = 2m^2 + 2m + 1$. This is in perfect agreement with the fact that mapping (2.5) is integrable (in terms of elliptic functions), being a member of the QRT family of integrable mappings. (A remark is necessary at this point. In order to obtain a closed-form expression for the degrees of the iterates, we start by computing a sufficient number of them. Once the expression of the degree has been heuristically established we compute the next few ones and verify that they agree with the predicted analytical expression). As a matter of fact, the precise values of the degrees are not important: they are not invariant under coordinate changes. However, the *type of growth* is invariant and can be used as an indication of whether or not the mapping is integrable.

Let us show what happens in the case of a nonintegrable mapping. As such we choose the one proposed by Hietarinta and Viallet [36] which has the form

$$x_{n+1} + x_{n-1} = x_n + \frac{1}{x_n^2}. \tag{2.6}$$

The particularity of this mapping lies in the fact that it satisfies the singularity confinement criterion. Its unique singularity pattern is $\{0, \infty, \infty, 0\}$. Nevertheless, as shown by Hietarinta and Viallet, this mapping behaves chaotically. With the same initial conditions as for the mapping above, we obtain the following succession of degrees: 0, 1, 3, 8, 23, 61, 162, 425, ... , . The degree growth is here exponential. Hietarinta and Viallet have found that the degree obeys the recursion relation $d_{n+4} = 3(d_{n+3} - d_{n+1}) + d_n$, leading to a asymptotic ratio $d_{n+1}/d_n \to (3 + \sqrt{5})/2$. The same result was obtained in a rigorous approach by Takenawa [37].

As we have shown in [38], the detailed study of the degree growth of the iterates of a mapping is an indication, not just of integrability, but also of the precise integration method for the mapping.

Nevanlinna Theory

As we have explained above, we expect the integrability of a mapping to be conditioned by the behavsingularity confinement criterion of its solutions when the independent variable goes to infinity. The tools for the study of the growth of a given function are furnished by the theory of meromorphic functions [39]. The reason why such an approach would apply to discrete systems has to do with the formal identity which exists between discrete systems and delay equations [40]. Thus one starts from a difference equation and considers it to be a delay equation in the complex plane of the independent variable [41]. The natural framework for the study of the behaviour near infinity of the solutions of a given mapping is Nevanlinna theory [42] which provides tools for the study of the value distribution of meromorphic functions. In particular, it introduces the notion of order. The latter is infinite for very-fast-growing functions, while a finite order indicates moderate growth. It would be reasonable to surmise that an infinite order is an indication of nonintegrability *for discrete systems*. (We do not make any statement here concerning continuous systems). The Nevanlinna theory provides an estimation of the growth of the solutions of a given discrete system. However, since the order may depend on the precise coefficients of the equation and their dependence on the independent variable, the starting point of our application of the Nevanlinna theory is to consider first only autonomous mappings, *i.e.*, mappings the coefficients of which are constants.

The main tool for the study of the value distribution of entire and meromorphic functions is the Nevanlinna characteristic (and various quantities related to it). The Nevanlinna characteristic of a function f, denoted by $T(r; f)$, measures the 'affinity' of f for the value ∞. It is usually represented as the sum of two terms, the frequency of poles, and the contribution from the arcs $|z| = r$ where $|f(z)|$ is large. From the characteristic one can define the order of a meromorphic function, $\sigma = \limsup_{r \to \infty} \log T(r; f)/\log r$. When f is rational, $T(r; f) \propto \log r$, and $\sigma = 0$. When f is of the type $e^{P_n(z)}$, where P_n is a polynomial of degree n, one finds $T \propto r^n$ and $\sigma = n$. A fast growing function like e^{e^z} leads to $T \propto e^r$ and thus to $\sigma = \infty$.

In what follows, we shall introduce the symbols $\asymp$, $\preceq$ and $\prec$ which will denote equality, inequality and strict inequality, respectively, *up to a function of r which remains bounded when $r \to \infty$*. The two basic relations, which express the fact that the affinities of f for ∞, 0 or a are the same, are

$$T(r; 1/f) \asymp T(r; f) \tag{2.7}$$

$$T(r; f - a) \asymp T(r; f). \tag{2.8}$$

Using those two identities, we can easily prove that the characteristic of a homographic transformation of f (with constant coefficients) is equal to $T(r; f)$ up to a bounded quantity. From a theorem due to Valiron [43] we have

$$T\left(r; \frac{P(f)}{Q(f)}\right) \asymp \sup(p, q)T(r; f), \tag{2.9}$$

where P and Q are polynomials with constant coefficients, of degrees p and q respectively, provided the rational expression P/Q is irreducible.

Let us also give some useful classical inequalities:

$$T(r; fg) \preceq T(r; f) + T(r; g) \tag{2.10}$$

$$T(r; f + g) \preceq T(r; f) + T(r; g) \tag{2.11}$$

Another inequality, which was proven in [44], is

$$T(r; fg + gh + hf) \preceq T(r; f) + T(r; g) + T(r; h). \tag{2.12}$$

One last property of the Nevanlinna characteristic was obtained by Ablowitz et al [41] (AHH). In our notation it reads:

$$T(r; f(z \pm 1)) \preceq (1 + \epsilon)T(r + 1; f(z)). \tag{2.13}$$

This relation (which is valid for any given ϵ if r is large enough) makes possible to have access to the characteristic, and thus the order, of the solution of some difference equations.

The discrete equations satisfied by the d-$\mathbb{P}$'s have the general form

$$A(x_n, x_{n-1}, x_{n+1}) = B(x_n), \tag{2.14}$$

where, usually, A is polynomial and B is rational with coefficients which do not depend on the independent variable n (something to which we shall come back). Moreover, in the cases we shall consider, A is linear separately in x_{n-1} and x_{n+1}. Following the approach of AHH we consider (2.14) to be a delay equation in the complex domain, and evaluate the Nevanlinna characteristic of both members of the equality, using (2.8) and (2.9). We find that

$$(1 + \epsilon)uT(r + 1; x) + vT(r; x) \succeq wT(r; x), \tag{2.15}$$

(with $u = 2$ if A is linear separately in x_{n-1} and x_{n+1}), for appropriate values of v and w. From (2.15) we see that

$$T(r + 1; x) \succeq \frac{w - v}{(1 + \epsilon)u} T(r; x) \tag{2.16}$$

Now if $w > u + v$, one can always choose ϵ small enough so that $\lambda \equiv \frac{w-v}{(1+\epsilon)u} > 1$. The precise meaning of (2.16) is that, for r large enough, we have

$$T(r + 1; x) \geq \lambda T(r; x) - C \tag{2.17}$$

for some C independent of r. The case where C is negative is trivial, $T(r + k; x) \geq \lambda^k T(r; x)$. For positive C,

$$T(r + 1; x) - \frac{C}{\lambda - 1} \geq \lambda\left(T(r; x) - \frac{C}{\lambda - 1}\right). \tag{2.18}$$

Thus, whenever $T(r; x)$ is an unbounded increasing function of r, i.e., $T \succ 0$, then, for some r large enough, the right hand side of this inequality becomes strictly positive. Iterating (2.18) we see that $T(r + k; x)$ diverges at least as fast as λ^k, thus $\log T(r; x) > r \log \lambda$, and the order σ of x is infinite. Thus, according to the AHH hypothesis, the mapping cannot be integrable. The only way out is if $T(r; x)$ is a constant, which means that x is itself a constant, since the slowest possible growth of the Nevanlinna characteristic for a non-constant meromorphic function is $T(r; f) \asymp \log r$, for f a homographic function of z. Giving that the mapping is rational, there can only be a finite number of constant solutions. We could, in principle, have had an infinite number of constant solutions if the identity $A(x_n, x_n, x_n) \equiv B(x_n)$ held true. However this would imply $w \leq u + v$. Thus, when $w > u + v$ in (2.15), the only possible finite order solutions are (a finite number of) constant solutions, *all* the remaining ones having $\sigma = \infty$.

The advantage of working with autonomous mappings lies in the fact that we can control precisely the corrective terms in the inequalities for T. Had we worked with nonautonomous systems, we would have had unbounded corrective terms. For instance if the coefficients depend rationally on z, there would be corrective terms of order $\mathcal{O}(\log r)$ and we would have been unable to exclude (finite-order) rational solutions. Though one may suspect that the *generic* solution is not rational, one could not easily disprove this possibility in the nonautonomous case. However in our approach, we consider nonautonomous equations as obtained from autonomous ones by a de-autonomisation procedure. This procedure will never transform a $\sigma = \infty$ solution into a finite σ one. So the generic solution will have $\sigma = \infty$ whenever $w > u + v$ in (2.15) even in the nonautonomous case. The rational solutions that we cannot exclude can only come, through the de-autonomisation procedure, from the finite- (in effect, zero-)order constant solutions, of which there is a finite number.

The way we have implemented the Nevanlinna theory was through a three-tiered approach. The first step, given a mapping, is to use the Nevanlinna characteristic techniques in order to estimate the rate of growth of the solutions. Since for nonautonomous equations this rate depends on the rate of growth of the coefficients of the equation, we opt for a simple approach. At this first step we consider *only autonomous* mappings. This first step puts severe constraints on the discrete equations at hand. However, usually, these constraints are not restrictive enough so as to determine completely the form of the mapping, hence the necessity of the second step. Once the constraints of the first step have been implemented, we pursue, using singularity confinement, in order to constrain further our discrete equation. Thus all autonomous equations that do not satisfy confinement are rejected at this second step.

The third step consists in the de-autonomisation of the system, once again using the confinement criterion. We thus obtain a mapping which (hopefully) satisfies the Nevanlinna criterion for slow-growth of the solutions and the singularity confinement as well. The major difficulty lies in the fact that the practical evaluation of the Nevanlinna characteristic gives a clear-cut answer as to mappings the solutions of which must be (generically) of infinite-order, but this does not mean that all the remaining ones have their generic solution of finite-order. Particular care is needed in the application of this criterion, lest one proclaim of finite-order systems which have in fact infinite-order solutions.

Παραγωγή

3 Discrete ℙ's Galore

In what follows, we shall present the derivation of specific examples of d-ℙ's, using the methods of singularity confinement and algebraic entropy.

Derivation of d-ℙ's

In order to obtain the discrete Painlevé equations through the application of the singularity confinement method, we start from the general QRT mapping,

$$x_{n+1} = \frac{f_1(x_n) - x_{n-1}f_2(x_n)}{f_2(x_n) - x_{n-1}f_3(x_n)}. \tag{3.1}$$

As mentioned above, one would expect to find the discrete forms by de-autonomising the QRT map. In order to gain some insight into the choice of the f_i's, we rewrite the QRT map as

$$f_3(x_n)\Pi - f_2(x_n)\Sigma + f_1(x_n) = 0, \tag{3.2}$$

where $\Sigma = x_{n+1} + x_{n-1}$, $\Pi = x_{n+1}x_{n-1}$, and the f_i's are quartic polynomials. We require that this equation become the continuous Painlevé under consideration at the continuous limit. For this purpose, we introduce a lattice parameter δ and obtain

$$\Sigma = 2x + \delta^2 x'' + \mathcal{O}(\delta^4)$$

$$\Pi = x^2 + \delta^2(xx'' - x'^2) + \mathcal{O}(\delta^4), \tag{3.3}$$

and when we extract from (3.3) the part involving derivatives, we obtain a continuous limit (as δ goes to zero) of the form

$$x'' = \frac{f_3(x)}{xf_3(x) - f_2(x)}x'^2 + g(x). \tag{3.4}$$

If we are looking for a specific Painlevé equation, we must first choose f_2, f_3 so as to force $\frac{f_3(x)}{xf_3(x) - f_2(x)}$ to coincide with the factor multiplying x'^2 in that equation.

For P_I and P_{II} we have, clearly, $f_3 = 0$. The derivation for those two d-ℙ's was presented in a previous section. We obtained the mappings

$$x_{n+1} + x_{n-1} + x_n = a + \frac{an + \beta + \gamma(-1)^n}{x_n} \qquad (3.5)$$

and

$$x_{n+1} + x_{n-1} = \frac{x_n(an + \beta) + \delta + \gamma(-1)^n}{1 - x_n^2} \qquad (3.6)$$

To derive the continuous limit of (3.5) we start by setting to zero the parity-dependent term γ, and further set $x = 1 + \epsilon^2 w$, $z = -3 - \epsilon^4 t$, $a = 6$. At the limit $\epsilon \to 0$, (3.5) becomes $w'' + 3w^2 + t = 0$. Similarly for (3.6), we take $\gamma = 0$, set $x = \epsilon w$, $z = 2 + \epsilon^3 t$, $\delta = \epsilon^3 \mu$, and obtain the continuous P_{II}: $w'' = 2w^3 + tw + \mu = 0$.

In the case of P_{III}, $x'' = \frac{x'^2}{x} + g(x)$. Instead of deriving the discrete form of the "usual" P_{III}, we will work with the more convenient form

$$w'' = \frac{w'^2}{w} + e^z(aw^2 + b) + e^{2z}(cw^3 + \frac{d}{w}), \qquad (3.7)$$

obtained from the usual one by the transformation $z \to e^z$ that absorbs the $\frac{w'}{z}$ term. This form agrees with (3.4) if we simply take $f_2 = 0$. In that case, the mapping takes the form

$$x_{n+1}x_{n-1} = \frac{\kappa(n)x_n^2 + \zeta(n)x_n + \mu(n)}{x_n^2 + \beta(n)x_n + \gamma(n)}. \qquad (3.8)$$

In order to fix the n-dependent coefficients, we will study the singularity behaviour as described before. When one solves for x_{n+1}, there are two possible sources of singularity for this mapping. Either x_n is a zero of the denominator or x_{n-1} becomes zero. In the first case, the singularity sequence is the following: x_{n+1} diverges, x_{n+2} has a finite value $\frac{\kappa(n+1)}{x_n}$ and x_{n+3} would in principle be proportional to $\frac{1}{x_{n+1}}$ and thus zero. This would lead to a new divergence. The only way out is to require that x_{n+2} also be a zero of the appropriate denominator, so that x_{n+3} does not vanish. Expressing x_{n+2} in terms of x_n and taking into account that this must be true for both zeros of x_n, we obtain $\beta(n) = \frac{\beta(n+2)\kappa(n+1)}{\gamma(n+2)}$ and $\gamma(n) = \frac{\kappa^2(n+1)}{\gamma(n+2)}$. Multiplying x_n by an arbitrary function of n does not change the form of (3.8), but only affects the coefficients. This scaling freedom allows us to take a constant value β for $\beta(n)$, resulting in $\kappa(n+1) = \gamma(n+2)$, $\gamma(n) = \gamma(n+2)$. Thus the γ's and κ's must be constants within a given parity, $\gamma(\text{even})=\kappa(\text{odd})=\gamma_+$, $\gamma(\text{odd})=\kappa(\text{even})=\gamma_-$. In order to study the second kind of singularity, we start with x_n such that x_{n+1} vanishes, i.e., $\kappa(n)x_n^2 + \zeta(n)x_n + \mu(n) = 0$. We then find that x_{n+2} has a finite value $\frac{\mu(n+1)}{\gamma(n+1)x_n}$ and this would lead to a divergent x_{n+3} unless the numerator also vanish. Substituting the expression for x_{n+2} and using the fact that once again this must be true for both zeros of $\kappa(n)x_n^2 + \zeta(n)x_n + \mu(n)$, we obtain $\mu(n) = \frac{\zeta(n)\mu(n+1)}{\zeta(n+2)} = \frac{\mu^2(n+1)}{\mu(n+2)}$. The solution to these equations is straightforward, $\mu(n) = \mu_0 \lambda^{2n}$ and $\zeta(n) = \zeta_{0,\pm} \lambda^n$, where μ_0, $\zeta_{0,\pm}$, are constants, the $\pm$ sign being related to the parity of n. Note that, in that case, there is no second kind of singularity at all! Indeed x_{n+3} is not allowed to diverge even though $x_{n+1} = 0$. (This is reminiscent of the case of continuous equations where, if a denominator appears, one must consider the values of the dependent variable that makes this denominator vanish to ascertain that this does not generate a singularity.) Since at the continuous limit we decided to neglect the distinction between even and odd, we can rewrite d-P_{III}, after a change of the variable, as

$$x_{n+1}x_{n-1} = \frac{cd(x_n - aq_n)(x_n - bq_n)}{(x_n - c)(x_n - d)} \qquad (3.9)$$

where $q_n = \lambda^n$. As expected, the continuous variable is $w(= w_0) = x$. We find moreover that $\lambda = 1 + \epsilon$, $c = 1/\epsilon - \alpha/2$, $d = -1/\epsilon - \alpha/2$, $a = \beta\epsilon - \gamma\epsilon^2/2$, $b = -\beta\epsilon - \gamma\epsilon^2/2$, leading to the equation:

$$w'' = \frac{w'^2}{w} + w^3 + \alpha w^2 + \gamma t - \frac{\beta^2 t^2}{w}, \qquad (3.10)$$

which is P_{III}, albeit in a slightly noncanonical form.

We now turn to the algebraic entropy, slow growth, approach. Let us start with a simple case. We consider the mapping

$$x_{n+1} + x_{n-1} = \frac{ax_n + b}{x_n^2}, \qquad (3.11)$$

where a and b are constants. In the previous section we have computed the degrees of the iterates and found, starting with $x_0 = p$ and $x_1 = q/r$, that the common degree of homogeneity in q and r of the numerator and denominator of the iterates was 0, 1, 2, 5, 8, 13, 18, 25, 32, 41, ... , a clearly polynomial growth. We now turn to the de-autonomisation of the mapping. The singularity confinement result is that a and b must satisfy the conditions $a_{n+1} - 2a_n + a_{n-1} = 0$, $b_{n+1} = b_{n-1}$, i.e., a is linear in n while b is a constant with an even/odd dependence. Assuming now that a and b are arbitrary functions of n, we compute the degrees of the iterates of (3.11). We obtain successively 0, 1, 2, 5, 10, 21, 42, 85, ... The growth is now exponential, the degrees behaving like $d_{2m-1} = (2^{2m} - 1)/3$ and $d_{2m} = 2d_{2m-1}$, a clear indication that the mapping is not integrable in general. Already at the fourth iteration the degrees differ in the autonomous and nonautonomous cases. Our approach consists in requiring that the degree in the nonautonomous case be *identical* to the one obtained in the autonomous one. If we implement the requirement that d_4 be 8 instead of 10, we find two conditions $a_{n+1} - 2a_n + a_{n-1} = 0$, $b_{n+1} = b_{n-1}$, i.e., precisely the ones obtained through singularity confinement. Moreover, once these two conditions are satisfied, the subsequent degrees of the nonautonomous case coincide with that of the autonomous one. Thus this mapping, leading to polynomial growth, should be integrable, and, in fact, it is. As we have shown in [45], where we presented its Lax pair, (3.11) with $a(n) = \alpha n + \beta$ and b constant - the even-odd dependence can be gauged out by a parity-dependent rescaling of the variable x - is a discrete form of the Painlevé I equation. In the examples that follow, we shall show that in all cases the nonautonomous form of an integrable mapping obtained through singularity confinement leads to exactly the same degrees of the iterates as the autonomous one.

Our second example is a multiplicative mapping,

$$x_{n+1}x_{n-1} = \frac{a_n x_n + b}{x_n^2}, \tag{3.12}$$

where one can set $b = 1$ by an appropriate gauge. In the autonomous case, starting with $x_0 = p$ and $x_1 = q/r$, we obtain successively the degrees: 0, 1, 2, 3, 4, 7, 10, 13, 16, 21, 26, ... , i.e., again quadratic growth. In fact, if n is of the form $4m + k$, ($k = 0,1,2,3$), the degree is given by $d_n = 4m^2 + (2m+1)k$. The de-autonomisation of (3.12) is straightforward. We compute the successive degrees in the generic case and find 0, 1, 2, 3, 4, 7, 11, ... At this stage we require that a factorisation occur in order to bring the degree d_6 from 11 to 10. The condition for this is that $a_{n+2}a_{n-2} = a_n^2$, i.e., a of the form $a_{e,o}\lambda_{e,o}^n$ with an even-odd dependence which can be easily gauged away. This condition is sufficient to bring the degrees of the successive iterates down to the values obtained in the autonomous case. Quite expectedly, the condition on a is precisely the one obtained by singularity confinement.

For the discrete Painlevé equations, for which the Lax pairs are not yet known, it is important to have one more check of their integrability provided by the algebraic entropy approach. We start with d-P_{IV} in the form

$$(x_{n+1} + x_n)(x_{n-1} + x_n) = \frac{(x_n^2 - a^2)(x_n^2 - b^2)}{(x_n + z_n)^2 - c^2}, \tag{3.13}$$

where a, b and c are constants. If z_n is constant, we obtain $d_n = 0$, 1, 3, 6, 11, 17, 24, ... , for the degrees of the successive iterates. The general expression of the growth is $d_n = 6m^2$ if $n = 3m$, $d_n = 6m^2 + 4m + 1$ if $n = 3m + 1$ and $d_n = 6m^2 + 8m + 3$ if $n = 3m + 2$. This polynomial (quadratic) growth is expected since in the autonomous case this equation is integrable, its solution being given in terms of elliptic functions. For a generic z_n we obtain the sequence $d_n = 0$, 1, 3, 6, 13, ... ,. The condition for the extra factorisations to occur in the last case, bringing down the degree d_4 to 11, is for z_n to be linear in n. We can check that the subsequent degrees coincide with those of the autonomous case.

For the q-P_V we start from

$$(x_{n+1}x_n - 1)(x_{n-1}x_n - 1) = \frac{(x_n^2 + ax_n + 1)(x_n^2 + bx_n + 1)}{(1 - z_n c x_n)(1 - z_n d x_n)}, \tag{3.14}$$

where a, b, c and d are constants. If, moreover, z_n is also a constant, we obtain exactly the same sequence of degrees $d_n = 0$, 1, 3, 6, 11, 17, 24, ... , as in the d-P$_{IV}$ case. Again, this polynomial (quadratic) growth is expected since this mapping is also integrable in terms of elliptic functions. For the generic nonautonomous case, we again find the sequence d_n=0, 1, 3, 6, 13, ... , . Once more we require a factorisation to bring down d_4 to 11. It turns out that this entails a z_n which is exponential in n, and generates the same sequence of degrees as the autonomous case. In both the d-P$_{IV}$ and q-P$_V$ cases, we find the n-dependence already obtained through singularity confinement. Since this results in a vanishing algebraic entropy, we expect both equations to be integrable.

An important remark is in order here. As we have seen in the derivations above, two different kinds of d-$\mathbb{P}$'s exist. The first corresponds to difference equations, where the independent variable enters linearly, through $z_n = \alpha n + \beta$. Thus, if x_n denotes the variable x at point z_n, then x_{n+1} is the variable at point $z_{n+1} = z_n + \alpha$. The second corresponds to multiplicative, q-equations. Here the independent variable enters exponentially, through $q_n = q_0\lambda^n$. So if the symbol x_n is used for the variable at point q_n, x_{n+1} denotes the variable at point $q_{n+1} = \lambda q_n$. It can be argued that there is no fundamental difference between the two types of equations, and that a change of variable suffices to go from one to the other. However this argument is fallacious. The difference between the two types of d-$\mathbb{P}$'s, additive and multiplicative, is far deeper. It is in fact related to their precise integration methods. A look at the Lax pairs of the two kinds of d-$\mathbb{P}$'s shows immediately their fundamental difference.

Lax Pairs

For additive equations there is a linear isospectral deformation problem of the form

$$\zeta\Phi_{n,\zeta} = L_n(\zeta)\Phi_n \tag{3.15a}$$

$$\Phi_{n+1} = M_n(\zeta)\Phi_n \tag{3.15b}$$

leading to the compatibility condition:

$$\zeta M_{n,\zeta} = L_{n+1}M_n - M_nL_n. \tag{3.16}$$

Examples of Lax pairs for additive discrete Painlevé equations were presented in [46]. In the case of d-P$_I$ the Lax pair is

$$L_n = \begin{pmatrix} 0 & x_n & 1 \\ \zeta & z_n & x_{n+1} + z_n/x_n \\ \zeta x_{n-1} & \zeta & z_{n+1} \end{pmatrix} \tag{3.17a}$$

and

$$M_n = \begin{pmatrix} -z_n/x_n & 1 & 0 \\ 0 & 0 & 1 \\ \zeta & 0 & 0 \end{pmatrix}, \tag{3.17b}$$

where $z_n = n/2 + \beta$. From the consistency conditions we obtain the mapping $x_{n+2} + z_{n+1}/x_{n+1} = x_{n-1} + z_n/x_n$. Adding x_n to both sides of this equality, we find

$$x_{n+1} + x_{n-1} + x_n - \frac{z_n}{x_n} = a \tag{3.18}$$

where a is a constant, *i.e.*, precisely d-P$_{\mathrm{I}}$.

For d-P$_{\mathrm{II}}$ the Lax pair is

$$L_n = \begin{pmatrix} a\ x_{n+1} & 1 & 0 & \\ 0 & z_n & 2b - x_n & 1 \\ \zeta & 0 & 0 & 2b - x_{n+1} \\ \zeta x_n & \zeta & 0 & z_{n+1} \end{pmatrix} \tag{3.19a}$$

and

$$M_n = \begin{pmatrix} (a - z_n)/x_{n+1} & 1 & 0 & 0 \\ 0 & 0 & 1 & 0 \\ 0 & 0 & z_{n+1}/(2b - x_{n+1}) & 1 \\ \zeta & 0 & 0 & 0 \end{pmatrix}, \tag{3.19b}$$

where $z_n = n/2 + \beta$ and a, b are constant. (The latter could have had an even-odd dependence but we choose to neglect it here.) From the consistency condition we find an equation for x_n which is best written for the translated variable $X = b - x$

$$X_{n+1} + X_{n-1} = \frac{(z_n + z_{n-1} - a)X_n + b(z_{n-1} - z_n - a)}{b^2 - X_n^2} \tag{3.20}$$

More examples of Lax pairs for additive d-$\mathbb{P}$'s can be found in [46].

For multiplicative equations, the isospectral problem is a q-difference one rather than a differential one,

$$\Phi_n(q\zeta) = L_n(\zeta)\Phi_n(\zeta) \tag{3.21a}$$

$$\Phi_{n+1} = M_n(\zeta)\Phi_n(\zeta), \tag{3.21b}$$

leading to

$$M_n(q\zeta)L_n(\zeta) = L_{n+1}(\zeta)M_n(\zeta). \tag{3.22}$$

The Lax pair of q-P$_{\mathrm{I}}$ can be easily obtained from our results in [46]. Let us introduce the matrices

$$L_n = \begin{pmatrix} 0 & 0 & \frac{k}{x_n} & 0 \\ 0 & 0 & x_{n-1} & qx_{n-1} \\ hx_n & 0 & 1 & q \\ 0 & \frac{hk_{n-1}}{x_{n-1}} & 0 & 0 \end{pmatrix} \tag{3.23a}$$

and

$$M_n = \begin{pmatrix} 0 & \frac{x_n}{k(x_n+1)} & 0 & 0 \\ 0 & 0 & 1 & 0 \\ 0 & 0 & \frac{1}{x_n} & \frac{q}{x_n} \\ h & 0 & 0 & 0 \end{pmatrix}. \tag{3.23b}$$

From the compatibility condition (3.22) we obtain the equation $x_{n+1}x_{n-1} = k_n k_{n+1}(x_n + 1)/x_n^2$, where $k_{n+1} = q k_{n-1}$, which is q-P_I, up to a gauge transformation.

For q-P_{III} we take $q = \alpha^2$. We introduce the matrices

$$L_n = \begin{pmatrix} \lambda_1 & \lambda_1 + \frac{\kappa}{x_n} & \frac{\kappa}{x_n} & 0 \\ 0 & \lambda_2 & \lambda_2 + x_{n-1} & x_{n-1} \\ \zeta x_n & 0 & \lambda_3 & \lambda_3 + x_n \\ \zeta(\lambda_4 + \alpha \frac{\kappa}{x_n}) & h\alpha \frac{\kappa}{x_n} & 0 & \lambda_4 \end{pmatrix} \tag{3.24a}$$

$$M_n = \begin{pmatrix} \frac{(\alpha\lambda_1 - \lambda_4)x_n}{\lambda_4 x_n + \alpha\kappa} & \frac{\lambda_1 x_n + \kappa}{\lambda_2 x_n + \kappa} & 0 & 0 \\ 0 & 0 & 1 & 0 \\ 0 & 0 & \frac{\lambda_3 - q\lambda_2}{x_n + q\lambda_2} & \frac{x_n + \lambda_3}{x_n + \lambda_4} \\ \zeta & 0 & 0 & 0 \end{pmatrix} \tag{3.24b}$$

where $\lambda_1 = const.$, $\lambda_3 = const.$, $\lambda_2 = \lambda\alpha^{n-1}$, $\lambda_4 = \lambda\alpha^n$, $\kappa = C\alpha^n$. From the compatibility condition (3.22) we obtain q-P_{III} in the form

$$x_{n+1}x_{n-1} = \frac{\alpha\kappa(x_n + \lambda_3)(\kappa + \lambda_2 x_n)}{(\kappa + \lambda_1 x_n)(x_n + \lambda_4)}. \tag{3.25}$$

As we can see from these examples, there exists a fundamental difference between additive and multiplicative d-$\mathbb{P}$'s. In what follows, we shall see (based on the relation between continuous and additive $\mathbb{P}$'s), that this difference goes even deeper.

Οργάνωσις

4 Introducing Some Order into the d-$\mathbb{P}$ Chaos

By now it must have been clear to the careful reader that the domain of d-$\mathbb{P}$'s is vast. When we started our investigations, we immediately realised that some order was needed lest the discrete $\mathbb{P}$'s degenerate into an impenetrable jungle. The first attempt to introduce some order was the definition of the standard family [47]. We use the term "order" instead of "classification" because it was clear that this first lumping together of d-$\mathbb{P}$'s was not based on any (coherent,) explicit criterion. Rather it was related to a historical derivation and the names of the equations was based on their continuous limits. (A much more accurate classification was to follow, albeit years later.) This was, admittedly, a most unfortunate choice. It led to a profusion of equations with the same name (in a recent publication [48], we have presented over a dozen discrete P_I's). Moreover, it does not take into account the fact that a given d-$\mathbb{P}$ may have more than one continuous limit (or none at all!). Finally, for equations with more parameters than P_{VI}, – and there exist many such equations – the continuous limit collapses everything to P_{VI}. Still this method of naming the d-$\mathbb{P}$'s has the advantage of familiarity and since it has been used traditionally, it remains in use, though with some adequate qualifiers.

When discrete Painlevé equations were first systematically derived we established what we called the 'standard' d-$\mathbb{P}$'s which fall into a degeneration cascade, $i.e.$, an equation with a given number of parameters can be obtained from one with more parameters by an appropriate coalescence procedure. This list comprised three-point mappings for one dependent variable and was initially incomplete since the discrete 'symmetric' (in the QRT terminology) form of P_{VI} was missing. This gap has been recently filled in [49] and we can now give the full list of standard d-$\mathbb{P}$'s:

δ-P_I
$$x_{n+1} + x_{n-1} = -x_n + \frac{z_n}{x_n} + 1$$

δ-P_{II}
$$x_{n+1} + x_{n-1} = \frac{z_n x_n + a}{1 - x_n^2}$$

q-P_{III}
$$x_{n+1} x_{n-1} = \frac{(x_n - aq_n)(x_n - bq_n)}{(1 - cx_n)(1 - x_n/c)}$$

δ-P_{IV}
$$(x_{n+1} + x_n)(x_n + x_{n-1}) = \frac{(x_n^2 - a^2)(x_n^2 - b^2)}{(x_n - z_n)^2 - c^2}$$

q-P_V
$$(x_{n+1}x_n - 1)(x_n x_{n-1} - 1) =$$
$$\frac{(x_n - a)(x_n - 1/a)(x_n - b)(x_n - 1/b)}{(1 - cx_n q_n)(1 - x_n q_n/c)}$$

δ-P_V
$$\frac{(x_n + x_{n+1} - z_n - z_{n+1})(x_n + x_{n-1} - z_n - z_{n-1})}{(x_n + x_{n+1})(x_n + x_{n-1})} =$$
$$\frac{(x_n - z_n - a)(x_n - z_n + a)(x_n - z_n - b)(x_n - z_n + b)}{(x_n - c)(x_n + c)(x_n - d)(x_n + d)}$$

q-P_{VI}
$$\frac{(x_n x_{n+1} - q_n q_{n+1})(x_n x_{n-1} - q_n q_{n-1})}{(x_n x_{n+1} - 1)(x_n x_{n-1} - 1)} =$$
$$\frac{(x_n - aq_n)(x_n - q_n/a)(x_n - bq_n)(x_n - q_n/b)}{(x_n - c)(x_n - 1/c)(x_n - d)(x_n - 1/d)}$$

where $z_n = \alpha n + \beta$, $q_n = q_0 \lambda^n$ and a, b, c, d are constants. We distinguish difference and multiplicative equations through the use of the prefixes δ and q.

The way these d-$\mathbb{P}$'s were obtained was, as explained in Sect. 2, by the application of some integrability criterion to an appropriate ansatz. The usual approach was the de-autonomisation of the QRT mapping. By de-autonomisation we mean that we allow the parameters of the QRT mapping, up to 5 in the symmetric case and up to 8 in the asymmetric one, to be functions of the independent variable. The precise form of these functions is obtained through the application of the integrability criterion.

In [21] we have presented a classification of the various forms of the QRT mappings used as a starting point for the de-autonomisation and derivation of d-$\mathbb{P}$'s. We reproduce it below, giving the form of the equations and the corresponding A_1 matrices.

(I) $x_{n+1} + x_{n-1} = f(x_n)$
$$A_1 = \begin{pmatrix} 0 & 0 & 0 \\ 0 & 0 & 0 \\ 0 & 0 & 1 \end{pmatrix}$$

(II) $x_{n+1}x_{n-1} = f(x_n)$
$$A_1 = \begin{pmatrix} 0 & 0 & 0 \\ 0 & 1 & 0 \\ 0 & 0 & 0 \end{pmatrix}$$

(III) $(x_{n+1} + x_n)(x_n + x_{n-1}) = f(x_n)$
$$A_1 = \begin{pmatrix} 0 & 0 & 0 \\ 0 & 0 & 1 \\ 0 & 1 & 0 \end{pmatrix}$$

(IV) $(x_{n+1}x_n - 1)(x_nx_{n-1} - 1) = f(x_n)$
$$A_1 = \begin{pmatrix} 0 & 0 & 0 \\ 0 & 1 & 0 \\ 0 & 0 & -1 \end{pmatrix}$$

(V) $\dfrac{(x_{n+1}+x_n+2z)(x_n+x_{n-1}+2z)}{(x_{n+1}+x_n)(x_n+x_{n-1})} = f(x_n)$
$$A_1 = \begin{pmatrix} 0 & 0 & 1 \\ 0 & 2 & 2z \\ 1 & 2z & 0 \end{pmatrix}$$

(VI) $\dfrac{(x_{n+1}x_n-z^2)(x_nx_{n-1}-z^2)}{(x_{n+1}x_n-1)(x_nx_{n-1}-1)} = f(x_n)$
$$A_1 = \begin{pmatrix} 1 & 0 & 0 \\ 0 & -z^2-1 & 0 \\ 0 & 0 & z^2 \end{pmatrix}$$

(VII) $\dfrac{(x_{n+1}-x_n-z^2)(x_{n-1}-x_n-z^2)+x_nz^2}{x_{n+1}-2x_n+x_{n-1}-2z^2} = f(x_n)$
$$A_1 = \begin{pmatrix} 0 & 0 & 1 \\ 0 & -2 & -2z^2 \\ 1 & -2z^2 & z^4 \end{pmatrix}$$

(VIII) $\quad \frac{(x_{n+1}z^2 - x_n)(x_{n-1}z^2 - x_n) - (z^4 - 1)^2}{(x_{n+1}z^{-2} - x_n)(x_{n-1}z^{-2} - x_n) - (z^{-4} - 1)^2} = f(x_n)$

$$A_1 = \begin{pmatrix} 0 & 0 & z^4 \\ 0 & -z^2(z^4 + 1) & 0 \\ z^4 & 0 & (z^4 - 1)^2 \end{pmatrix}$$

The forms presented above correspond to symmetric mappings but they can be extended to asymmetric ones directly, the A_1 matrix being the same. To these cases one must add the explicitly asymmetric one,

(IX) $\quad x_{n+1} + x_n = f(y_n), \quad y_n y_{n-1} = g(x_n) \qquad\qquad A_1 = \begin{pmatrix} 0 & 0 & 0 \\ 0 & 0 & 1 \\ 0 & 0 & 0 \end{pmatrix}.$

One can recognize in the forms presented above the autonomous limits of, among others, the equations d-$P_{I/II}$ (I), q-P_{III} (II), d-P_{IV} (III), q-P_V (IV), d-P_V (V,VII,IX), q-P_{VI} (VI,VIII). We must stress here that the forms presented above are not the only ones that one may encounter when studying d-$\mathbb{P}$'s. The mappings are given up to homographic transformations, and thus the d-$\mathbb{P}$'s may assume forms different from the ones above.

From the canonical forms above it seems that the vast majority of d-$\mathbb{P}$'s will have a symmetric, in the QRT sense, form. Nothing could be farther from the truth. As we have seen in the previous section, the application of the singularity confinement method quite often leads to terms of the form $(-1)^n$, but also j^n where $j^3 = 1$, i^n and even k^n where $k^5 = 1$. Our initial (and erroneous) tendency was to discard these terms on the ground that "they do not possess a continuous limit". This is just not true. The terms with binary, ternary, etc. symmetry indicate that the equation is better written as a system of two, three, etc. equations. They also introduce one or more extra parameters with obvious consequences when it comes to continuous limit [50].

Let us illustrate this in the case of the d-P_I we already encountered, namely (3.5),

$$x_{n+1} + x_{n-1} + x_n = \frac{\alpha n + \beta + \gamma(-1)^n}{x_n} + a. \tag{4.1}$$

Due to the presence of the $(-1)^n$ term, it is clear that we should distinguish even and odd terms. Let us write the mapping for even and odd indices explicitly

$$x_{2m+1} + x_{2m} + x_{2m-1} = \frac{z_{2m} + \gamma}{x_{2m}} + a \tag{4.2a}$$

$$x_{2m+2} + x_{2m+1} + x_{2m} = \frac{z_{2m+1} - \gamma}{x_{2m+1}} + a, \tag{4.2b}$$

where $z_m = \alpha m + \beta$. Next we introduce two variables, $X_m = x_{2m}$ and $Y_m = x_{2m+1}$, one for each parity. We can now rewrite this system as

$$Y_m + X_m + Y_{m-1} = \frac{Z_m + \gamma}{X_m} + a \tag{4.3a}$$

$$X_{m+1} + X_m + X_m = \frac{Z_m + \alpha - \gamma}{Y_m} + a \tag{4.3b}$$

where $Z_m = 2\alpha m + \beta$.

Thus, in the new variables, the mapping becomes a system of two two-point mappings. The important remark is that this mapping has now one more genuine parameter than the symmetric d-P_I. A careful computation of the continuous limit of this asymmetric d-P_I leads to P_{II}. In fact, setting $X = 1+\epsilon w+\epsilon^2 u$, $Y = 1-\epsilon w+\epsilon^2 u$, $Z = 1-\epsilon^3 m$, $a = 2$ and $\gamma = -\epsilon^3 c/4$, we find a first relation, $u = \frac{1}{4}(w^2-w'+t)$, with $t = \epsilon m$, leading to $w'' = 2w^3+2tw+c$.

In order to deal with the profusion of asymmetric forms, we made the terminology situation even more complicated by introducing the qualifier 'asymmetric' (borrowed from the 'asymmetric QRT' mapping) in front of the names derived from the continuous limit of the *symmetric* forms. Thus the equation we have just examined is called 'asymmetric d-P_I'. (It is a source of wonder to the authors that people still manage to follow their work through this labyrinth of complications.)

Quite expectedly the limit of asymmetric d-P_{II} is P_{III}, that of asymmetric q-P_{III} is P_{VI} (as has been shown by Jimbo and Sakai [51]). The limits of asymmetric d-P_{IV} and q-P_V are given below. We start from the forms

$$(x_{n+1} + y_n)(y_n + x_n) = \frac{(y_n - a)(y_n - b)(y_n - c)(y_n - d)}{(y_n - z - \kappa/2)^2 - e^2}$$

$$(y_n + x_n)(x_n + y_{n-1}) = \frac{(x_n + a)(x_n + b)(x_n + c)(x_n + d)}{(x_n - z)^2 - f^2}, \tag{4.4}$$

with a constraint $a + b + c + d = 0$, and

$$(x_{n+1}y_n - 1)(y_n x_n - 1) = \frac{\lambda rs(y_n - a)(y_n - b)(y_n - c)(y_n - d)}{(y_n - p)(y_n - q)}$$

$$(y_n x_n - 1)(x_n y_{n-1} - 1) = \frac{pq(x_n - 1/a)(x_n - 1/b)(x_n - 1/c)(x_n - 1/d)}{\lambda(x_n - r)(x_n - s)}, \tag{4.5}$$

with a constraint $pq = \lambda abcdrs$. The last two systems are new forms of d-P_{VI}, as can be assessed from their continuous limits. For (4.4) we take $a = 1/2+\epsilon\alpha$, $b = 1/2 - \epsilon\alpha$, $c = -1/2 + \epsilon\beta$, $d = -1/2 - \epsilon\beta$, $e = \epsilon\gamma, f = \epsilon\delta$, $x = w - 1/2$, $z = \zeta + 1/2$, $y = w(\zeta - 1)/(w - \zeta) + 1/2 + \epsilon u$, while for (4.5) we take $a = \theta e^{\epsilon\alpha}$, $b = \theta e^{\epsilon\beta}$, $c = \theta^{-1}e^{\epsilon\gamma}$, $d = \theta^{-1}e^{\epsilon\delta}$, $\lambda = e^\epsilon$, $p = e^{\epsilon(n+\phi)}$, $q = e^{\epsilon(n+\psi)}$, $r = e^{\epsilon(n+\omega)}$, $s = e^{\epsilon(n+\chi)}$, $w = (x - \theta)/(\theta^{-1} - \theta)$, $\zeta = (z - \theta)/(\theta^{-1} - \theta)$, $y = (z(x - \theta^{-1} - \theta) + 1)/(x - z) + \epsilon u$, with $\phi + \psi = 1 + \alpha + \beta + \gamma + \delta + \omega + \chi$ (consequence of the constraint $pq = \lambda abcdrs$). In both cases the limit, when ϵ goes to zero, is

$$\frac{d^2w}{d\zeta^2} = \frac{1}{2}\left(\frac{1}{w} + \frac{1}{w-1} + \frac{1}{w-\zeta}\right)\left(\frac{dw}{d\zeta}\right)^2 - \left(\frac{1}{\zeta} + \frac{1}{\zeta-1} + \frac{1}{w-\zeta}\right)\frac{dw}{d\zeta}$$
$$+ \frac{w(w-1)(w-\zeta)}{2\zeta^2(\zeta-1)^2}\left(A + \frac{B\zeta}{w^2} + \frac{C(\zeta-1)}{(w-1)^2} + \frac{D\zeta(\zeta-1)}{(w-\zeta)^2}\right), \quad (4.6)$$

where $A = 4\gamma^2$, $B = -4\alpha^2$, $C = 4\beta^2$ and $D = 1 - 4\delta^2$ in the case of (4.4), while in the case of (4.5) $A = (\omega - \chi)^2$, $B = -(\alpha - \beta)^2$, $C = (\gamma - \delta)^2$ and $D = 1 - (\phi - \psi)^2$.

One should not draw, based on the above examples, the conclusion that the only 'asymmetric' forms are two-component mappings. Indeed there exist systems which require a higher number of components. Take for instance the mapping

$$x_{n+1}x_{n-1} = a(x_n - 1). \quad (4.7)$$

Its asymmetric form is easily again obtained by the application of singularity confinement. The final result is

$$\log a_n = kn + p + rj^n + sj^{2n} + t(-1)^n, \quad (4.8)$$

where j is a cubic root of unity. It is clear from this expression that the fully asymmetric form of (4.7) has four parameters. We can thus rewrite (4.7) as a system

$$w_{m-1}y_m = q_m(x_m - 1)ad$$

$$x_m z_m = q_m\lambda(y_m - 1)b/d$$

$$y_m u_m = q_m\lambda^2(z_m - 1)cd$$

$$z_m v_m = q_m\lambda^3(u_m - 1)a/d \quad (4.9)$$

$$u_m w_m = q_m\lambda^4(v_m - 1)bd$$

$$v_m x_{m+1} = q_m\lambda^5(w_m - 1)c/d,$$

where $q_m = \lambda^{6m}$. We must point out here that although the mapping is written as a system of six equations, it is still of second order. Four of the equations are just local relations between the variables.

In these last two sections we have focused on the direct derivation of discrete $\mathbb{P}$'s, based on some integrability criterion. While this method is very powerful, it has one drawback. It does not lead to the Lax pair for the equation, nor does it provide any clue for its derivation. There exist, however, as we have already explained, other methods for deriving discrete $\mathbb{P}$'s. One of them is particularly interesting since it is constructive, in the sense that it derives the d-$\mathbb{P}$ together with its Lax pair. This method is based on the deep relation that exists between continuous and discrete, difference equations [52]. Let us start with the Lax pair of a continuous Painlevé equation. It has the general form

$$\psi_\zeta = A\psi \tag{4.10a}$$

$$\psi_z = B\psi, \tag{4.10b}$$

where ζ is the spectral parameter and A, B are matrices depending explicitly on ζ and the dependent as well as the independent variables, w and z. The continuous Painlevé equation is obtained from the compatibility condition $\psi_{\zeta z} = \psi_{z\zeta}$ leading to

$$A_z - B_\zeta + AB - BA = 0 \tag{4.11}$$

In general, the Painlevé equation depends on parameters $(\alpha, \beta, \dots)$ which are associated to the monodromy exponents θ_i appearing explicitly in the Lax pair. The Schlesinger transform relates two solutions ψ and ψ' of the isomonodromy problem for the equation at hand corresponding to different sets of parameters $(\alpha, \beta, \dots)$ and $(\alpha', \beta', \dots)$. The main characteristic of these transforms is that the monodromy exponents, at the singularities of the associated linear problem, related to sets $(\alpha, \beta, \dots)$ and $(\alpha', \beta', \dots)$, differ by integers (or half-integers). The general form of a Schlesinger transformation is

$$\psi' = R\psi, \tag{4.12}$$

where R is another matrix depending on ζ, w, z and the monodromy exponents θ_i. The important remark is that (4.10a) together with (4.12) constitute *the Lax pair of the discrete, difference equation*. The latter is obtained from the compatibility conditions

$$R_\zeta + RA - A'R = 0. \tag{3.8}$$

Thus, the difference equations are intimately related to the continuous ones. They are their *contiguity* relations. In fact this is precisely how Jimbo and Miwa [11] have derived the so-called alternate d-P_I (1.2). Of course, this method, while interesting, should not be considered to be a panacea. First, with this approach one does not have a real control over the equation one derives. One only knows how many parameters are involved. Moreover, since the richest continuous $\mathbb{P}$, namely P_{VI}, has only four parameters, one can obtain, at best, a difference-$\mathbb{P}$ with three parameters. But we know that there exist difference-$\mathbb{P}$'s with many more parameters than that. The relation of these equations to continuous systems is an open question (on which even the authors of the present review have not succeeded in reaching unanimity). Finally the whole class of q-equations lies beyond the scope of this method. As we shall see in the following chapters, q-$\mathbb{P}$'s are usually their own contiguity relations.

Ιδιότητες

5 What Makes Discrete Painlevé Equations Special?

The discrete Painlevé equations have a host of special properties. Most of these are the analogues of the properties of the continuous Painlevé equations. This parallel between continuous and discrete systems has been of tremendous help in the investigation of integrable mappings. In most cases, one had to start from some property of the continuous system, namely, of the continuous $\mathbb{P}$ in the case at hand, and ask how this property could be transposed to the discrete case. (There even exist cases where, having obtained some unexpected result for the discrete system, we have looked for, and found, its continuous analogue which was, to our knowledge, a new result [53]).

Degeneration Cascade

The first property of the discrete Painlevé equations we are going to discuss is that of degeneration through coalescence. In practice, this means that, starting from a d-$\mathbb{P}$ with a given number of parameters, and introducing a special limit of the dependent and/or independent variables, as well as the parameters, one obtains a d-$\mathbb{P}$ with one fewer parameter. The degeneration pattern for the "standard" family of d-$\mathbb{P}$'s takes the form of the cascade:

$$
\begin{array}{ccccc}
q\text{-P}_{\text{VI}} & \longrightarrow & q\text{-P}_{\text{V}} & \longrightarrow & q\text{-P}_{\text{III}} \\
\downarrow & & \downarrow & & \downarrow \\
\text{d-P}_{\text{V}} & \longrightarrow & \text{d-P}_{\text{IV}} & \longrightarrow & \text{d-P}_{\text{II}} & \longrightarrow & \text{d-P}_{\text{I}}.
\end{array}
$$

In what follows, we will present the result for the seven standard forms. The following conventions will be used. The variables and parameters of the 'higher' equation will be given in capital letters (X, Z, P, Q, A, B, C, D), while those of the 'lower' equation will be given in lowercase letters (x, z, p, q, a, b, c, d). The small parameter that will introduce the coalescence limit will be denoted by δ.

In order to illustrate the process, let us work out in full detail the case d-P$_{\text{II}} \to$ d-P$_{\text{I}}$. We start with the equation d-P$_{\text{II}}$

$$
X_{n+1} + X_{n-1} = \frac{Z_n X_n + A}{1 - X_n^2} \tag{5.1}
$$

We set $X = 1 + \delta x$, whereupon the equation becomes

$$
4 + 2\delta(x_{n+1} + x_{n-1} + x_n) + \mathcal{O}(\delta^2) = -\frac{Z_n(1 + \delta x_n) + A}{\delta x_n}. \tag{5.2}
$$

Now, clearly, Z must cancel A up to order δ and this suggests the ansatz $Z = -A - 2\delta^2 z$. Moreover, the constant term in the right-hand side must cancel the 4 of the left-hand side, and we are thus led to $A = 4 + 2\delta a$. Using these values of Z and A we find (at $\delta \to 0$)

$$
x_{n+1} + x_{n-1} + x_n = \frac{z_n}{x_n} + a, \tag{5.3}
$$

i.e., precisely d-P$_{\text{I}}$.

The coalescence d-P_{III} to d-P_{II} requires a more delicate limit since the independent variable of d-P_{III} enters in an exponential way. We start from:

$$X_{n+1}X_{n-1} = \frac{AB(X_n - P_n)(X_n - Q_n)}{(X_n - A)(X_n - B)} \tag{5.4}$$

The ansatz for X is here, too, $X = 1 + \delta x$. For the remaining quantities we set

$$A = 1 + \delta, \quad B = 1 - \delta$$

$$P = 1 + \delta + \delta^2(z + a)/2 + \mathcal{O}(\delta^3), \quad Q = 1 - \delta + \delta^2(z - a)/2 + \mathcal{O}(\delta^3), \tag{5.5}$$

and at the limit $\delta \to 0$, d-P_{III} reduces exactly to d-P_{II}

$$x_{n+1} + x_{n-1} = \frac{z_n x_n + a}{1 - x_n^2}. \tag{5.6}$$

In perfect analogy to the continuous case, d-P_{IV} also reduces to d-P_{II}. Here we start from

$$(X_{n+1} + X_n)(X_n + X_{n-1}) = \frac{(X_n^2 - A^2)(X_n^2 - B^2)}{(X_n - Z_n)^2 - C^2} \tag{5.7}$$

and set $X = 1 + \delta x$. We also set

$$A = 1 + \delta, \quad B = 1 - \delta$$

$$C = \delta - \delta^2 a/2, \quad Z = 1 - \delta^2 z/4. \tag{5.8}$$

The result at $\delta \to 0$ is precisely d-P_{II} given by (5.6).

In the case of q-P_V

$$(X_{n+1}X_n - 1)(X_n X_{n-1} - 1) = \frac{(X_n - A)(X_n - 1/A)(X_n - B)(X_n - 1/B)}{(1 - CX_nQ_n)(1 - X_nQ_n/C)}, \tag{5.9}$$

two different limits exist. In order to obtain d-P_{IV} we set $X = 1 + \delta x$ and

$$A = 1 + \delta a, \quad B = 1 - \delta b,$$

$$C = 1 + \delta c, \quad Q_n = 1 - \delta z_n, \tag{5.10}$$

i.e., $\lambda = 1 - \alpha\delta$, such that $z_n = \alpha n + \beta$. At the limit $\delta \to 0$ we find d-P_{IV} (5.7) in terms of the variable x. The case of the coalescence d-P_V to d-P_{III} requires a different ansatz. Here we set $X = x/\delta$. Moreover we set

$$C = c, \quad Q_n = \frac{q_n}{\delta}, \quad A = \frac{a}{\delta}, \quad B = \frac{b}{\delta}, \tag{5.11}$$

We then find at the limit $\delta \to 0$

$$x_{n+1}x_{n-1} = \frac{(x_n - a)(x_n - b)}{(1 - cx_nq_n)(1 - x_nq_n/c)} \tag{5.12}$$

While this is not exactly the form of d-P_{III} (5.4) it is very easy to reduce it to the latter. We introduce y through $x = y\lambda^n$ (recall $q_n = q_0\lambda^n$) and find with $\mu = 1/\lambda$:

$$y_{n+1}y_{n-1} = \frac{(x_n - a\mu^n)(x_n - b\mu^n)}{(x_n - c)(x_n - 1/c)}, \tag{5.13}$$

that is obviously of the form (5.4).

In the case of d-P_V,

$$\frac{(X_n + X_{n+1} - Z_n - Z_{n+1})(X_n + X_{n-1} - Z_n - Z_{n-1})}{(X_n + X_{n+1})(X_n + X_{n-1})}$$
$$= \frac{(X_n - Z_n - A)(X_n - Z_n + A)(X_n - Z_n - B)(X_n - Z_n + B)}{(X_n - C)(X_n + C)(X_n - D)(X_n + D)}, \tag{5.14}$$

there is only one limit which is d-P_{IV}. We set $X = x$ but $Z = z + 1/\delta$ and

$$A = c + 1/\delta, \quad B = -c + 1/\delta, \quad C = a, \quad D = b. \tag{5.15}$$

At the limit $\delta \to 0$, we find d-P_{IV} (5.7) in terms of the variable x.

Finally, in the case of q-P_{VI},

$$\frac{(X_n X_{n+1} - Q_n Q_{n+1})(X_n X_{n-1} - Q_n Q_{n-1})}{(X_n X_{n+1} - 1)(X_n X_{n-1} - 1)} =$$
$$\frac{(X_n - AQ_n)(X_n - Q_n/A)(X_n - BQ_n)(X_n - Q_n/B)}{(X_n - C)(X_n - 1/C)(X_n - D)(X_n - 1/D)}$$

$$\tag{5.16}$$

there are again two limits, to d-P_V and q-P_V.

If we set

$$X = 1 + \delta x, \quad A = 1 + \delta a, \quad B = 1 + \delta b, \quad C = 1 + \delta c \quad \text{and} \quad D = 1 + \delta d, \tag{5.17}$$

i.e., $\lambda = 1 + \delta a$, so $Q = 1 + \delta z$, where $z_n = \alpha n + \beta$ we recover exactly d-P_V at the limit $\delta \to 0$. On the other hand, with the choice $X = x$, $Q_n = q_n/\delta$, $A = c/\delta$, $B = 1/\delta c$, $C = a$, $D = b$, one recovers at the limit $\delta \to 0$ the equation q-P_V.

It goes without saying that this degeneration through coalescence follows the well-known pattern of the continuous $\mathbb{P}$'s,

$$P_{VI} \to P_V \to \{P_{IV}, P_{III}\} \to P_{II} \to P_I,$$

while accomodating the particularities of the d-$\mathbb{P}$'s. As we shall see in the next section, the situation is in fact much more complicated.

Special Solutions

Another property of the d-$\mathbb{P}$'s is that they possess solutions expressible in terms of special functions. These solutions exist only for particular values of the parameters, and of course they do not capture the full freedom of the solution which is essentially transcendental, even for these special values. Since another article [54] in this volume is devoted to the study of the special solutions of d-$\mathbb{P}$'s, we shall not go into details here.

Miura/Auto-Bäcklund/Schlesinger Transformations

The discrete $\mathbb{P}$'s, just like their continuous brethren, have all kinds of interrelations. We reserve the name of Miura for transformations which relate two different equations. While Miura transformations relate the solutions of two different d-$\mathbb{P}$'s, the auto-Bäcklund transformations are relations that allow one to relate a solution of a given d-$\mathbb{P}$ to a solution of the same d-$\mathbb{P}$ with different values of the parameters. The Schlesinger transformations are just particular auto-Bäcklund transformations. As such they relate solutions of the same equation. The Schlesinger transformations for continuous equations relate solutions corresponding to the same monodromy data except for *integer* differences in the monodromy exponents. In the discrete case the very existence of monodromy exponents to be related to the Schlesinger transformations is not always clear. However, we can use an analogy with the continuous case. If one uses the proper parametrisation of the equation, the Schlesinger transformations can be shown to be associated to elementary changes of the parameters. The discrete case can be analysed in the same spirit. By using the proper parametrisation, one can identify, among the auto-Bäcklund transformations, those which correspond to elementary changes of the parameters and which can thus be dubbed Schlesinger transformations.

Let us illustrate the application of Miura transformations in the case of d-$\mathbb{P}_{II}$. We introduce the system, where $\tilde{z} = (z_n + z_{n+1})/2$ and $\delta = z_{n+1} - z_n$),

$$y_n = (1 + x_n)(1 - x_{n+1}) - \frac{\tilde{z}}{2}. \tag{5.18}$$

$$x_n = \frac{m + y_n - y_{n-1}}{y_n + y_{n-1}} \tag{5.19}$$

Eliminating y from this system, we obtain the d-$\mathbb{P}_{II}$ in the usual form,

$$x_{n+1} + x_{n-1} = \frac{m - \delta/2 + z_n x_n}{1 - x_n^2}, \tag{5.20}$$

while, by eliminating, x we find

$$(y_{n+1} + y_n)(y_n + y_{n-1}) = \frac{4y_n^2 - m^2}{y_n + \tilde{z}/2}, \tag{5.21}$$

which is the discrete form of equation 34 in the Painlevé/Gambier classification, usually denoted by d-$\mathbb{P}_{34}$.

Miura transformations are particularly interesting in the case of d-P$_{\text{I}}$ [55] which has no parameter. Let us illustrate this in the case of the d-P$_{\text{I}}$

$$x_{n+1} + x_{n-1} = \frac{z_n}{x_n} + \frac{a}{x_n^2}. \tag{5.22}$$

The Miura transformation $y_n = x_n x_{n+1}$ can be applied in a straightforward way, leading to

$$(y_n + y_{n-1} - z_n)(y_n + y_{n+1} - z_{n+1}) = \frac{a^2}{y_n}. \tag{5.23}$$

This form was identified in [47] as another form of d-P$_{\text{I}}$. The Miura transformation $y_n = x_{n+1}/x_n$ leads to a more interesting result. We multiply (5.22) by x_n^2 and take the discrete derivative, $i.e.$, subtract it from its up-shift to eliminate the constant a. Using systematically the Miura transformation $y_n = x_{n+1}/x_n$ we obtain a four-point equation for y

$$\frac{y_{n+1}y_n + 1 - y_{n+1}^2 y_n(y_{n+2}y_{n+1} + 1)}{y_n y_{n-1} + 1 - y_n^2 y_{n-1}(y_n y_{n+1} + 1)} = \frac{y_{n+1}z_{n+2} - z_{n+1}}{y_n z_{n+1} - z_n} \frac{1}{y_n y_{n-1}}, \tag{5.24}$$

which satisfies the singularity confinement requirement. The continuous limit is obtained by $y = 1 - \epsilon^3 w$ and $z = -6 + \epsilon^4 t$, leading to

$$ww''' = (w'' - 1)w' + 12w^3. \tag{5.25}$$

Multiplying (5.25) by $(w'' - 1)$, we can rewrite it as

$$\frac{d}{dt}\left(\frac{(w'' - 1)^2 - 24w^2(w' - t)}{w^2}\right) = 0, \tag{5.26}$$

and absorbing the integration constant by a translation of t we obtain

$$(w'' - 1)^2 - 24w^2(w' - t) = 0, \tag{5.27}$$

$i.e.$, Cosgrove's equation SD$_{\text{V}}$ [56], which is a form of modified P$_{\text{I}}$. Thus (5.24) is a discrete form of the second-degree Painlevé equation SD$_{\text{V}}$ in derivative form.

How can one find the auto-Bäcklund (and Schlesinger) transformations for a given d-P? The general principle is the following. First obtain a Miura transformation that transforms the equation into a new one (the 'modified' one). Second, find the invariance of the latter, usually associated to some discrete transformations. Third implement these discrete transformations and return to the initial equation by the inverse of the Miura transformation. In the process we find that the parameters of the initial equation have been modified and thus the chain of transformations indeed defines an auto-Bäcklund transformation. Obtaining the Miura transformation can be facilitated once we remark that all known Miura transformations have the form of a discrete Riccati equation, i.e., a homographic mapping.

An interesting case of auto-Bäcklund/Schlesinger construction is the one concerning q-P_{III}. We start from the form

$$x_{n+1}x_{n-1} = \frac{cd(x_n - a)(x_n - b)}{(x_n - c)(x_n - d)}, \tag{5.28}$$

where c, d are constants and a, b proportional to λ^n. Following the derivation we presented in [57], we introduce

$$u_n = (x_n - c)(x_{n+1} - d). \tag{5.29}$$

This is the first half of the Miura transformation. The second half of the Miura transformation involves a rational expression which must be homographic in both u_n and u_{n-1}. We readily find that

$$x_n = \frac{u_n u_{n-1}/cd - u_n - u_{n-1} + cd - ab}{-u_n/d - u_{n-1}/c + c + d - a - b}. \tag{5.30}$$

Eliminating x_n and x_{n+1} between (5.29), (5.30) and its upshift leads to an equation for u_{n-1}, u_n, u_{n+1}. This equation is, after a change of variables, a discrete form of d-P_V equation, although not all the parameters of a d-P_V are present. Introducing $U = u - cd$, we find that

$$(U_n U_{n+1} - \lambda^2 abcd)(U_n U_{n-1} - abcd) = \\ \frac{cd(U_n + bd)(U_n + \lambda ac)(U_n + ad)(U_n + \lambda bc)}{U_n + cd}. \tag{5.31}$$

In order to define a different Miura transformation we can introduce the quantity

$$w_n = (1/x_n - 1/a)(1/x_{n+1} - 1/\lambda b). \tag{5.32}$$

The second half of the Miura transformation, analogous to (5.30), is

$$x_n = \frac{-aw_{n-1}/\lambda - bw_n\lambda + 1/a + 1/b - 1/c - 1/d}{abw_n w_{n-1} + 1/ab - 1/cd - w_n\lambda - w_{n-1}/\lambda}. \tag{5.33}$$

Again eliminating x leads to an equation for w. Introducing $W_n = w_n - 1/\lambda ab$, we find for this equation

$$(W_n W_{n+1} - \frac{1}{\lambda^2 abcd})(W_n W_{n-1} - \frac{1}{abcd}) = \\ \frac{(W_n + 1/\lambda bd)(W_n + 1/ac)(W_n + 1/\lambda bc)(W_n + 1/ad)}{\lambda abW_n + 1}. \tag{5.34}$$

The quantity $\tilde{W}_n = U/\lambda abcd$ satisfies an equation obtained from (5.31), namely,

$$(\tilde{W}_n\tilde{W}_{n+1} - \frac{1}{\lambda^2 abcd})(\tilde{W}_n\tilde{W}_{n-1} - \frac{1}{abcd}) =$$
$$\frac{(\tilde{W}_n + 1/\lambda ac)(\tilde{W}_n + 1/bd)(\tilde{W}_n + 1/\lambda bc)(\tilde{W}_n + 1/ad)}{\lambda ab\tilde{W}_n + 1}. \qquad (5.35)$$

Equation (5.34) has the same form as (5.35), provided one introduces the parameters

$$\tilde{a} = a\sqrt{\lambda}, \ \tilde{b} = b/\sqrt{\lambda}, \ \tilde{c} = c\sqrt{\lambda}, \ \tilde{d} = d/\sqrt{\lambda}. \qquad (5.36)$$

We define

$$\tilde{w}_n = \tilde{W}_n + 1/\lambda\tilde{a}\tilde{b} = u_n/\lambda abcd \qquad (5.37)$$

and

$$\tilde{x}_n = \frac{-\tilde{a}\tilde{w}_{n-1}/\lambda - \tilde{b}\tilde{w}_n\lambda + 1/\tilde{a} + 1/\tilde{b} - 1/\tilde{c} - 1/\tilde{d}}{\tilde{a}\tilde{b}\tilde{w}_n\tilde{w}_{n-1} + 1/\tilde{a}\tilde{b} - 1/\tilde{c}\tilde{d} - \tilde{w}_n\lambda - \tilde{w}_{n-1}/\lambda}. \qquad (5.38)$$

Given this definition of $\tilde{x}$, and since $\tilde{W}$ satisfies (5.35), it follows that

$$\tilde{w}_n = (1/\tilde{x}_n - 1/\tilde{a})(1/\tilde{x}_{n+1} - 1/\lambda\tilde{b}), \qquad (5.39)$$

and therefore $\tilde{x}$ satisfies d-P$_{\text{III}}$ with parameters $\tilde{a}, \tilde{b}, \tilde{c}, \tilde{d}$. The transformation from x to $\tilde{x}$ defines an auto-Bäcklund transformation for d-P$_{\text{III}}$. In this case this is indeed a Schlesinger transformation, which we denote by S_c^a. (The convention used here is to give explicitly the parameters associated to x_n, rather than x_{n+1}, in (5.29) and (5.32).) The inverse transformation $(S_c^a)^{-1}$ can be obtained by defining w by (5.32), $u = w\lambda abcd$ and finally x through the analogue of (5.30).

In a similar way we can introduce the transformations S_c^b, $S_d^a = (S_c^b)^{-1}$ and $S_d^b = (S_c^a)^{-1}$. They correspond to multiplying the two parameters which appear explicitly by $\sqrt{\lambda}$ while dividing the two others by the same quantity. These are the most elementary Schlesinger transformations. Using them we can construct further Schlesinger transformations that act separately on $\{a,b\}$ or $\{c,d\}$. For instance the product $S_c^a S_d^a$ corresponds to $a \to a\lambda$, $b \to b/\lambda$, $c \to c$, $d \to d$.

The procedure presented here for the construction of the transformations of the solutions of d-$\mathbb{P}$'s may, understandably, appear tedious and the ansatzes introduced somewhat arbitrary. We must reassure the reader that the whole approach will become clearer and less arbitrary once the appropriate bilinear formalism for d-$\mathbb{P}$'s is introduced and associated with their systematic geometric description.

Ταξινόμησις

6 Putting Some Real Order to the d-$\mathbb{P}$ Chaos

The Bilinearisation of the d-$\mathbb{P}$'s
As a first step towards the classification of the d-$\mathbb{P}$'s we shall present their bilinearisation in the framework of the Hirota formalism. While the latter is quite powerful and has been applied to the study of nonlinear evolution equations (both continuous and discrete), curiously its use in the case of $\mathbb{P}$'s has been rather limited, a notable exception being the work of Okamoto [58]. This is rather intriguing since the solutions of the continuous $\mathbb{P}$'s are meromorphic functions of the independent variable and thus should possess simple expressions in terms of entire functions. This is precisely what the Hirota formalism is doing. It introduces a dependent variable transformation, and thus makes possible the expression of the original one in terms of τ-functions, which are assumed to be entire. Our guide to the bilinearisation of d-$\mathbb{P}$'s will be their singularity structure and the property of singularity confinement.

Let us examine two cases which are a perfect illustration of the method, namely d-$\mathrm{P_I}$ and d-$\mathrm{P_{II}}$.

$$x_{n+1} + x_{n-1} = -x_n + \frac{z_n}{x_n} + a \tag{6.1}$$

$$x_{n+1} + x_{n-1} = \frac{z_n x_n + a}{1 - x_n^2} \tag{6.2}$$

respectively, where $z_n = \alpha n + \beta$, and a, α, β are constants. For the needs of the present paper a schematic singularity structure will suffice, the precise balancing can be found in Sect. 1 and is not necessary here. In the case of d-$\mathrm{P_I}$, we have a singularity whenever the x in the denominator happens to vanish. This has as a consequence that both x_{n+1} and x_{n+2} diverge, whereupon x_{n+3} vanishes again and x_{n+4} is finite, *i.e.*, the singularity is indeed confined. Thus the singularity pattern is $\{0, \infty, \infty, 0\}$. In the case of d-$\mathrm{P_{II}}$ a singularity appears whenever x_n in the denominator takes the value $+1$ or -1. Thus we have two singularity patterns which, in this case, turn out to be $\{-1, \infty, +1\}$ and $\{+1, \infty, -1\}$.

How can we use this information in order to express x in terms of τ-functions [59]? Let us start with d-$\mathrm{P_I}$. As a first step, we surmise that there exists a relationship between the singularity patterns of a d-$\mathbb{P}$ and the number of τ-functions necessary to express the original variable. Thus in the case of d-$\mathrm{P_I}$, which has a unique singularity pattern, it is enough to introduce just one τ-function. Since τ-functions are entire, x must be a ratio of products of such functions. Hence, let us assume that x_n contains a τ-function F_n in the numerator, and that F_n passes through zero. Since x_{n+1} and x_{n+2} are infinite, the denominator of x must contain F_{n-1} and F_{n-2}, which ensures that F_n

appears in the denominators of x_{n+1} and x_{n+2}, respectively. Finally since x_{n+3} vanishes, x_n must contain F_{n-3} in the numerator. Thus, the expression of x, dictated by the singularity pattern, is:

$$x_n = \frac{F_n F_{n-3}}{F_{n-1} F_{n-2}}. \tag{6.3}$$

As we shall see below, this expression suffices for the multilinearisation, more precisely the trilinearisation, of d-P$_I$. That the choice (6.3) is a reasonable one also can be seen by the continuous limit of this expression. We know, for d-P$_I$, that the continuous limit is obtained by $x = 1 + \epsilon^2 w$ at $\epsilon \to 0$. Implementing this limit in (6.3) we find that $w = 2\partial_z^2 \log F$, a transformation that is at the base of the (continuous) Hirota bilinear formalism.

In the case of d-P$_{II}$, we have two singularity patterns, and so we expect two τ-functions to appear in the expression of x. Let us start with the pattern $\{-1, \infty, +1\}$. The diverging x may be related to a vanishing τ-function, say F, in the denominator. In order to ensure that x_{n-1} and x_{n+1} are respectively -1 and $+1$, we choose x_n of the form $x(n) = -1 + \frac{F_{n+1}}{F_n} p = 1 + \frac{F_{n-1}}{F_n} q$, where p, q must be expressed in terms of a second τ-function, G. We turn now to the second pattern $\{+1, \infty, -1\}$ related to the vanishing of the τ-function G. We find that in this case, $x_n = 1 + \frac{G_{n+1}}{G_n} r = -1 + \frac{G_{n-1}}{G_n} s$, where r and s are expressed in terms of F. Combining the two expressions in terms of F and G we find, with the appropriate choice of gauge, the following simple expression for x,

$$x(n) = -1 + \frac{F_{n+1} G_{n-1}}{F_n G_n} = 1 - \frac{F_{n-1} G_{n+1}}{F_n G_n}, \tag{6.4}$$

which satisfies both singularity patterns. Thanks to this particular choice of gauge the relative sign is such that the continuous limit of (6.4), obtained by $x = \epsilon w$, is $w = \partial_z \log \frac{F}{G}$, i.e., precisely the expected transformation in the case of P$_{II}$.

Having explained the general procedure we are now ready to perform the bilinearisation. For d-P$_I$ we implement ansatz (6.3), not directly on (6.1), but on its discrete derivative,

$$x_{n+1} - x_{n-2} = \frac{z_n}{x_n} - \frac{z_{n-1}}{x_{n-1}}, \tag{6.5}$$

and obtain the following trilinear equation,

$$F_{n+3} F_{n-2} F_{n-1} - z_n F_{n+1}^2 F_{n-2} = F_{n-3} F_{n+2} F_{n+1} - z_{n-1} F_{n-1}^2 F_{n+2}, \tag{6.6}$$

which we can further regroup into

$$(F_{n-1} F_{n+3} - z_n F_{n+1}^2) F_{n-2} = (F_{n-3} F_{n+1} - z_{n-1} F_{n-1}^2) F_{n+2}. \tag{6.7}$$

This is as far as we can go with just one τ-function. In order to really bilinearise the equation we must introduce an auxiliary τ-function. (This is something one could not guess from the singularity pattern alone and it is in fact the reason why the bilinearisation procedure cannot be entirely automated.) So, introducing the auxiliary τ-function G_n and splitting (6.7) into a system, we find that

$$F_{n-2}F_{n+2} - z_n F_n^2 = G_{n+1}F_{n-1}$$

$$F_{n-2}F_{n+2} - z_{n-1}F_n^2 = G_{n-1}F_{n+1}. \tag{6.8}$$

This constitutes the bilinearisation of the standard d-P_I.

In the case of d-P_{II}, we start from ansatz (6.4) which has already furnished one bilinear equation. Eliminating the denominator, $F_n G_n$, we obtain

$$F_{n+1}G_{n-1} + F_{n-1}G_{n+1} - 2F_n G_n = 0. \tag{6.9}$$

In order to obtain the second equation we rewrite d-P_{II} as $(x_{n+1} + x_{n-1})(1 - x_n)(1 + x_n) = zx_n + a$. We use the two possible definitions of x_n in terms of F and G in order to simplify the expressions $1 - x_n$ and $1 + x_n$. Next, we obtain two equations by using these two definitions for x_{n+1} combined with the alternate definition for x_{n-1}. We thus obtain

$$F_{n+2}F_{n-1}G_{n-1} - F_{n-2}F_{n+1}G_{n+1} = F_n^2 G_n(zx_n + a) \tag{6.10a}$$

and

$$G_{n-2}G_{n+1}F_{n+1} - G_{n+2}G_{n-1}F_{n-1} = G_n^2 F_n(zx_n + a). \tag{6.10b}$$

Finally, we add (6.10a) multiplied by G_{n+2}, and (6.10b) multiplied by F_{n+2}. Up to the use of the upshift of (6.9), a factor $F_{n+1}G_{n+1}$ appears in both sides of the resulting expression. After simplification, the remaining equation is indeed bilinear,

$$F_{n+2}G_{n-2} - F_{n-2}G_{n+2} = z(F_{n+1}G_{n-1} - F_{n-1}G_{n+1}) + 2aF_n G_n \tag{6.11}$$

where a symmetric expression was used for x in the right-hand side, obtained as the arithmetic mean of the two right-hand sides of (6.4). Equations (6.9) and (6.11), taken together, are the bilinear form of d-P_{II}.

Complete results on the bilinearisation of d-P's can be found in [60]. We find that d-P's with more parameters involve a higher number of τ-functions. In that work, we have also examined the bilinearisation of the continuous P's. The latter was first obtained by Hietarinta and Kruskal [61] for the first five equations. Since that of the continuous P_{VI} was missing, we have taken advantage of our bilinearisation of d-P's in order to fill this gap. Let us present here this method which shows the interplay between continuous and discrete integrable systems. Our starting point is the discrete form of P_{VI} discovered by Jimbo and Sakai [51]. The q-P_{VI} equation is written in the form of a system,

$$x_{n+1}x_n = \frac{(y_n - \alpha\tilde{z})(y_n - \beta\tilde{z})}{(y_n - \gamma)(y_n - 1/\gamma)} \tag{6.12a}$$

$$y_n y_{n-1} = \frac{(x_n - az)(x_n - bz)}{(x_n - c)(x_n - 1/c)}, \tag{6.12b}$$

where $z = \lambda^n$, $\tilde{z} = z\sqrt{\lambda}$, with the constraint $ab = \alpha\beta$. The τ-functions are introduced by

$$x_n = c\left(1 + (1-z)^{1/2}\frac{M_n N_{n-1}}{F_n G_n}\right) = \frac{1}{c}\left(1 + (1-z)^{1/2}\frac{M_{n-1}N_n}{F_n G_n}\right) = \frac{H_n K_n}{F_n G_n}$$

$$\frac{1}{x_n} = \frac{1}{az}\left(1 - (1-z)^{1/2}\frac{P_n Q_{n-1}}{H_n K_n}\right) = \frac{1}{bz}\left(1 - (1-z)^{1/2}\frac{P_{n-1}Q_n}{H_n K_n}\right) = \frac{F_n G_n}{H_n K_n}$$

$$y_n = \gamma\left(1 + (1-\tilde{z})^{1/2}\frac{F_{n+1}G_n}{M_n N_n}\right) = \frac{1}{\gamma}\left(1 + (1-\tilde{z})^{1/2}\frac{F_n G_{n+1}}{M_n N_n}\right) = \frac{P_n Q_n}{M_n N_n}$$

$$\tag{6.13}$$

and

$$\frac{1}{y_n} = \frac{1}{\alpha\tilde{z}}\left(1 - (1-\tilde{z})^{1/2}\frac{H_{n+1}K_n}{P_n Q_n}\right)$$

$$= \frac{1}{\beta\tilde{z}}\left(1 - (1-\tilde{z})^{1/2}\frac{H_n K_{n+1}}{P_n Q_n}\right) = \frac{M_n N_n}{P_n Q_n}$$

leading to

$$2F_n G_n + (1-z)^{1/2}(M_n N_{n-1} + M_{n-1}N_n) = \left(c + \frac{1}{c}\right)H_n K_n$$

$$2H_n K_n - (1-z)^{1/2}(P_n Q_{n-1} + P_{n-1}Q_n) = (a+b)zF_n G_n$$

$$2M_n N_n + (1-\tilde{z})^{1/2}(F_{n+1}G_n + F_n G_{n+1}) = \left(\gamma + \frac{1}{\gamma}\right)P_n Q_n$$

$$2P_n Q_n - (1-\tilde{z})^{1/2}(H_{n+1}K_n + H_n K_{n+1}) = (\alpha+\beta)\tilde{z}M_n N_n$$

$$\left(c - \frac{1}{c}\right)F_n G_n + (1-z)^{1/2}\left(cM_n N_{n-1} - \frac{1}{c}M_{n-1}N_n\right) = 0. \tag{6.14}$$

$$\left(\frac{1}{a} - \frac{1}{b}\right)H_n K_n - (1-z)^{1/2}\left(\frac{1}{a}P_n Q_{n-1} - \frac{1}{b}P_{n-1}Q_n\right) = 0$$

$$\left(\gamma - \frac{1}{\gamma}\right)M_n N_n + (1-\tilde{z})^{1/2}\left(\gamma F_{n+1}G_n - \frac{1}{\gamma}F_n G_{n+1}\right) = 0$$

$$\left(\frac{1}{\alpha} - \frac{1}{\beta}\right)P_n Q_n - (1-\tilde{z})^{1/2}\left(\frac{1}{\alpha}H_{n+1}K_n - \frac{1}{\beta}H_n K_{n+1}\right) = 0.$$

We go to the continuous limit by $a = 1 + \epsilon a_1 + \epsilon^2 a_2$, $b = 1 - \epsilon a_1 + \epsilon^2 b_2$, $c = 1 + \epsilon c_1$, $\alpha = 1 + \epsilon\alpha_1 + \epsilon^2\alpha_2$, $\beta = 1 - \epsilon\alpha_1 + \epsilon^2\beta_2$, $\gamma = 1 + \epsilon\gamma_1$. The constraint $ab = \alpha\beta$ translates into $a_2 + b_2 - a_1^2 = \alpha_2 + \beta_2 - \alpha_1^2$. The result is

$$FG + (1 - z)^{1/2}MN = HK$$

$$HK - (1 - z)^{1/2}PQ = zFG$$

$$c_1FG + (1 - z)^{1/2}\left(D_\zeta + c_1\right)M{\cdot}N = 0$$

$$a_1HK + (1 - z)^{1/2}\left(D_\zeta - a_1\right)P{\cdot}Q = 0$$

$$\gamma_1MN + (1 - z)^{1/2}\left(D_\zeta + \gamma_1\right)F{\cdot}G = 0 \qquad (6.15)$$

$$\alpha_1PQ + (1 - z)^{1/2}\left(D_\zeta - \alpha_1\right)H{\cdot}K = 0$$

$$(1 - z)D_\zeta^2 F{\cdot}G - (1 - z)^{1/2}D_\zeta^2 M{\cdot}N + (1 - z)^{1/2}D_\zeta^2 P{\cdot}Q$$
$$= -(a_2 + b_2)zFG - c_1^2 HK + \gamma_1^2(1 - z)^{1/2}PQ$$
$$(1 - z)D_\zeta^2 H{\cdot}K - z(1 - z)^{1/2}D_\zeta^2 M{\cdot}N + (1 - z)^{1/2}D_\zeta^2 P{\cdot}Q$$
$$= -(a_2 + b_2)zFG - c_1^2 zHK - (\alpha_2 + \beta_2)z(1 - z)^{1/2}MN,$$

where $z = e^\zeta$, $w = x = HK/FG$ and, in addition, at the continuous limit, $w = 1 + (1-z)^{1/2}MN/FG$, $1/w = (1 - (1-z)^{1/2}PQ/HK)/z$. We thus obtain the continuous P_{VI},

$$
\begin{aligned}
w_{zz} = {}&\frac{1}{2}\left(\frac{1}{w} + \frac{1}{w - 1} + \frac{1}{w - z}\right)w_z^2 - \left(\frac{1}{z} + \frac{1}{z - 1} - \frac{1}{z - w}\right)w_z \\
&+ \frac{w(w - 1)(w - z)}{z^2(z - 1)^2}\left(\frac{\gamma_1^2}{2} - \frac{\alpha_1^2}{2}\frac{z}{w^2} + \frac{c_1^2}{2}\frac{z - 1}{(w - 1)^2} + \frac{1 - a_1^2}{2}\frac{z(z - 1)}{(w - z)^2}\right),
\end{aligned}
$$

$$(6.16)$$

One last word is in order at this point. Our analysis of d-$\mathbb{P}$'s was based on the assumption that the "right" number of τ-functions is at least equal to the number of different singularity patterns. This led to the introduction of up to 8 τ-functions for the higher d-$\mathbb{P}$'s. This proliferation of τ functions may appear strange and one may be tempted to try to obtain a description involving fewer τ functions. In order to cut short this speculation we must make clear that only *one* τ function exists. This τ function is multidimensional *i.e.*, it involves several parameters, and the various symbols F, G, H etc. that appeared above are just realisations of the same τ function for different values of its parameters. Thus they are all related through Miura transformations, and the use of the different symbols is a mere convenience. The details of the description of the Painlevé equations with a multidimensional τ function will be given in what follows.

The Property of Self-Duality

The Miura/auto-Bäcklund transformations we have discussed in the previous section can be described in a most elegant way, for continuous $\mathbb{P}$'s, through the Hamiltonian formulation of Okamoto [62]. It is indeed known that all

Painlevé's can be obtained from the equations of motion of a polynomial Hamiltonian. Starting with $H(x, p, t)$ and the equations

$$f(t)\frac{dx}{dt} = \frac{\partial H}{\partial p}$$

$$f(t)\frac{dp}{dt} = -\frac{\partial H}{\partial x}, \tag{6.17}$$

where $f(t)$ is a rational function defined once and for all for each Painlevé equation, we obtain, eliminating p, the desired equation for x. One can, naturally, wonder what may be obtained if one eliminates x. The answer is, another Painlevé equation for p. Thus system (1) can be viewed as defining a Miura transformation connecting these two Painlevé equations. (Sometimes the two equations turn out to be the same, in which case this Miura transformation is just an auto-Bäcklund.)

The crucial observation of Okamoto was that the Hamiltonian is related to the τ-functions by

$$f(t)\frac{d}{dt}\log\tau = H. \tag{6.18}$$

The Bäcklund tranformations mentioned above are introduced through birational canonical transformations on H which induce a correspondence between two τ-functions associated to different sets of parameters. In particular, when these transformations in the set of parameters is a translation, the iteration of the Bäcklund tranformation defines a τ-sequence $\{\tau_m, m \in \mathbb{Z}\}$. Okamoto has shown that the τ-sequence of a Painlevé equation obeys the Toda equation,

$$\left(f(t)\frac{d}{dt}\right)^2 \log\tau_m = \frac{\tau_{m-1}\tau_{m+1}}{\tau_m^2}, \tag{6.19}$$

which, in bilinear formalism, can be rewritten

$$(D_z^2 - 2e^{D_m})\tau_m \cdot \tau_m = 0. \tag{6.20}$$

Here we have introduced a new variable, z, to absorb the $f(t)$ factor by $dz = f(t)^{-1}dt$.

While Miura/auto-Bäcklund/Schlesinger transformations have been derived for (almost) all the d-$\mathbb{P}$'s, it has not been possible to obtain the strict equivalent of the Hamiltonian formulation. This has led to a critical examination of the Okamoto formalism which resulted to the conclusion that the Hamiltonian is not the crucial ingredient. The key element is the τ-function. Thus one expects the discrete form of the Toda equation to play a major role.

The first key discovery was that of the property of self-duality. This property first arose as a surprise. In fact, there is no way one could have guessed its existence from experience with continuous Painlevé equations. The first known example is the alternate d-P_{II} equation which we have derived in [52] and studied extensively in [63]. This equation has the form,

$$\frac{z_n}{x_{n+1}x_n + 1} + \frac{z_{n-1}}{x_n x_{n-1} + 1} = -x_n + \frac{1}{x_n} + z_n + a, \tag{6.21}$$

where $z_n = \delta n + z_0$ and a is a parameter. Its bilinearisation was obtained from the singularity structure. Two τ-functions F and G were introduced by

$$x_n = \frac{F_n G_{n-1}}{F_{n-1} G_n}, \tag{6.22}$$

and a third, auxiliary, τ-function E was needed to obtain the system

$$F_{n+1}G_{n-1} + F_{n-1}G_{n+1} = z_n F_n G_n \tag{6.23a}$$

$$F_{n+1}F_{n-1} = F_n^2 + G_n E_n \tag{6.23b}$$

$$G_n E_{n-1} - G_{n-1} E_n = a F_n F_{n-1} \tag{6.23c}$$

The choice of equation (6.23bc) for the introduction of E is more or less arbitrary. One could have equally well replaced (6.23) by:

$$F_{n+1}G_{n-1} + F_{n-1}G_{n+1} = z_n F_n G_n \tag{6.24a}$$

$$G_{n+1}G_{n-1} = G_n^2 + F_n H_n \tag{6.24b}$$

$$H_n F_{n-1} - H_{n-1} F_n = (a + \delta)G_n G_{n-1} \tag{6.24c}$$

where we recall that $z_n = \delta n + z_0$, and thus that $\delta = z_{n+1} - z_n$. By comparing (6.23) and (6.24) one sees that a shift of a to $a + \delta$ is associated to the τ-function transformation $\{E, F, G\} \rightarrow \{F, G, H\}$. In fact, the τ-functions $\ldots, E, F, G, H, \ldots$ constitute a Schlesinger chain, or, in Okamoto's terminology, a τ-sequence.

The Schlesinger transform of (6.21) was presented in [63],

$$\underset{\sim}{x}_n = \frac{1}{x_n} + \frac{a(1 + x_n x_{n-1})}{1 + x_n x_{n-1} - z_{n-1}x_n} \tag{6.25}$$

where $\underset{\sim}{x}_n$ satisfies the alternate d-P$_{\text{II}}$ equation with parameter $a - \delta$. In terms of the τ-functions,

$$\underset{\sim}{x}_n = \frac{E_n F_{n-1}}{E_{n-1} F_n}, \tag{6.26}$$

which is just (6.22) transposed to the case where a is shifted to $a - \delta$, i.e., $\{F, G\} \rightarrow \{E, F\}$. Similarly,

$$\tilde{x}_n = \left(x_n - \frac{(a + \delta)(1 + x_n x_{n-1})}{1 + x_n x_{n-1} - z_{n-1}x_{n-1}} \right)^{-1}, \tag{6.27}$$

satisfying (6.21) with parameter $a + \delta$, and, of course,

$$\tilde{x}_n = \frac{G_n H_{n-1}}{G_{n-1} H_n}. \tag{6.28}$$

We can obtain the discrete equation in parameter-space. Eliminating x_{n-1} between (6.25) and (6.27) we obtain the *dual* equation of alternate d-P$_{II}$, *i.e.*, the equation where parameter a is now the independent variable, since x_n, $\underset{\sim}{x}_n$, $\tilde{x}_n$ are associated to a, $a - \delta$, $a + \delta$ respectively,

$$\frac{a+\delta}{x\tilde{x}-1} + \frac{a}{x\underset{\sim}{x}-1} = x + \frac{1}{x} - a - z. \tag{6.29}$$

Here we have dropped the index n, so $z(\equiv z_n)$ is just a parameter. We remark that (6.29) is essentially alternate d-P$_{II}$ itself. The only, minor, change is that, in order to recover the precise form of (6.21), we must multiply x by i.

A Geometrical Interpretation of d-P's
The discovery of the property of self-duality for alternate-d-P$_{II}$ raised an important question. Is this property exceptional or does it characterise the d-P's in general? We know today that all difference $\mathbb{P}$'s are indeed self-dual and so are, less expectedly, almost all q-$\mathbb{P}$'s. Still, some q-discrete $\mathbb{P}$'s do exist which do not possess this property.

How can self-duality lead us to a classification of the d-P's? We must admit that the link of self-duality to the geometrical interpretation is rather indirect, but that's how intuition works most of the time. Let us see the source of alternate d-P$_{II}$. In [52] we have obtained alternate d-P$_{II}$ as a contiguity relation between solutions of the continuous P$_{III}$,

$$w'' = \frac{w'^2}{w} - \frac{w'}{t} + \frac{1}{t}(\alpha w^2 + \beta) + \gamma w^3 + \frac{\delta}{w}. \tag{6.30}$$

Assuming that $\gamma \neq 0$ and $\delta \neq 0$ one can use scaling of both w and t to obtain $\gamma = 1$ and $\delta = -1$. We have he following relations:

$$w(-\alpha, -\beta) = -w(\alpha, \beta) \tag{6.31}$$

$$w(-\beta, -\alpha) = w^{-1}(\alpha, \beta) \tag{6.32}$$

and

$$w(-\beta - 2, -\alpha - 2) = w(\alpha, \beta)\left(1 + \frac{2+\alpha+\beta}{t(\frac{w'}{w} + w + \frac{1}{w}) - 1 - \beta}\right). \tag{6.33}$$

We assume further that $\alpha \neq \beta$. Using (6.31-33) and the analogue of (6.33) starting from $w(-\beta, -\alpha)$ (which leads to $w(\alpha - 2, \beta - 2)$), we can eliminate w' and obtain a relation between $w(\alpha - 2, \beta - 2)$, $w(\alpha, \beta)$ and $w(\alpha + 2, \beta + 2)$, *i.e.*, a one-dimensional 3-point mapping on the (α, β)-plane. We introduce the independent variable, $z = (\alpha + \beta + 2)/4$, so that the variation of z between two consecutive w's is 1, and the parameters κ and μ by $\mu = (\beta - \alpha - 2)/4, \kappa = -it/2$. We choose $x = i/w$ as the mapping variable and find that

$$\frac{z_n}{x_{n+1}x_n + 1} + \frac{z_{n-1}}{x_n x_{n-1} + 1} = \kappa\left(-x_n + \frac{1}{x_n}\right) + z_n + \mu \qquad (6.34)$$

It is, moreover, straightforward to set $\kappa = 1$ by scaling. Thus one obtains the final form of the alternate d-P_{II} equation. We remark that both the independent variable, z, and the parameter, μ, of alternate d-P_{II} are linear combinations of the parameters α and β of P_{III}. Since α and β play equivalent roles in P_{III}, one expects z and μ to play equivalent roles in alternate d-P_{II}. This is naïvely the origin of self-duality.

But the situation is even deeper than this. Since Okamoto has obtained a formulation for the transformations of the solutions of the continuous $\mathbb{P}$'s, by translations of their parameters, in terms of affine Weyl groups, one expects the same groups to play a role in the case of d-$\mathbb{P}$'s, at least the ones which are contiguities of c-$\mathbb{P}$'s, but, as it turns out, the applicability of the approach is quite general.

Let us present here the description of the asymmetric d-P_{II} and its transformations in a geometrical language based on that of affine Weyl groups. The asymmetric d-P_{II} equation,

$$x_{m+1} + x_m = \frac{y_m z_{m+1/2} + \gamma}{1 - y_m^2}$$

and

$$y_m + y_{m-1} = \frac{x_m z_m + \delta}{1 - x_m^2}, \qquad (6.35)$$

with $z_m = \alpha m + \beta$, is described by the affine Weyl group of $A_3^{(1)}$. That is, the τ functions live on the weight lattice of $A_3^{(1)}$, namely the points of integer coordinates, either all even or all odd. Such points will henceforward be called τ-points. Any τ-point (say, the origin) has 8 nearest-neighbours (NN) at distance $\sqrt{3}$, namely $(\epsilon_1, \epsilon_2, \epsilon_3)$, where $\epsilon_i^2 = 1$, and 6 next-nearest-neighbours (NNN) at distance 2, with one coordinate ± 2 and two vanishing coordinates. We are also interested in next-next-nearest-neighbours (NNNN) at distance $2\sqrt{2}$ with one vanishing coordinate and two coordinates of absolute value 2 (so there are 12 of these).

Nonlinear variables are defined at points of integer coordinates which do not belong to the weight lattice of $A_3^{(1)}$, but rather have two coordinates of one parity and the third one of opposite parity. For instance, at the point $(0,0,1)$ one can define such a variable that we will denote as x_{001}. A site like (n_1, n_2, n_3+1), where we assume that the three n_i have the same parity, is the midpoint of a unique pair of τ-points in NNN position, namely (n_1, n_2, n_3) and $(n_1, n_2, n_3 + 2)$ and of exactly two pairs of τ-points in NNNN position, $(n_1 + 1, n_2 + 1, n_3 + 1)$ and $(n_1 - 1, n_2 - 1, n_3 + 1)$ on the one hand, and $(n_1 + 1, n_2 - 1, n_3 + 1)$ and $(n_1 - 1, n_2 + 1, n_3 + 1)$ on the other. It turns out that the relevant x can be expressed in terms of the τ's associated to these six points in two different ways:

$$x_{n_1,n_2,n_3+1} = \frac{\tau_{n_1-1,n_2-1,n_3+1}\tau_{n_1+1,n_2+1,n_3+1}}{\tau_{n_1,n_2,n_3}\tau_{n_1,n_2,n_3+2}} - 1$$

$$= 1 - \frac{\tau_{n_1-1,n_2+1,n_3+1}\tau_{n_1+1,n_2-1,n_3+1}}{\tau_{n_1,n_2,n_3}\tau_{n_1,n_2,n_3+2}}. \tag{6.36}$$

Of course if the index with a different parity had been the first or the second one, instead of the third one, one should change the indices accordingly. The important point is the overall sign, which depends on the choice of the pair of τ's in NNNN positions, whether the vector joining them has both components of same or opposite parities. Writing that the two expressions for x coincide, we obtain a first bilinear equation relating the τ's,

$$\tau_{n_1-1,n_2-1,n_3+1}\tau_{n_1+1,n_2+1,n_3+1} + \tau_{n_1-1,n_2+1,n_3+1}\tau_{n_1+1,n_2-1,n_3+1} =$$
$$2\tau_{n_1,n_2,n_3}\tau_{n_1,n_2,n_3+2}, \tag{6.37}$$

and, of course two more equations along the two first indices. These equations are autonomous Hirota-Miwa equation [15,64] relating the τ's. Consider now a point with all coordinates half-integer, say $(n_1 + \epsilon_1/2, n_2 + \epsilon_2/2, n_3 + \epsilon_3/2)$, where we again assume that the n_i's have the same parity. This is the mid-point of one pair of τ's in NN position, τ_{n_1,n_2,n_3} and $\tau_{n_1+\epsilon_1,n_2+\epsilon_2,n_3+\epsilon_3}$ and exactly three pairs of τ's at distance $\sqrt{11}$, of the type $\tau_{n_1,n_2,n_3+2\epsilon_3}$ and $\tau_{n_1+\epsilon_1,n_2+\epsilon_2,n_3-\epsilon_3}$. Let us assume that the three quantities,

$$\tau_{n_1,n_2,n_3+2\epsilon_3}\tau_{n_1+\epsilon_1,n_2+\epsilon_2,n_3-\epsilon_3} - \epsilon_1\epsilon_2 z_3\tau_{n_1,n_2,n_3}\tau_{n_1+\epsilon_1,n_2+\epsilon_2,n_3+\epsilon_3} \tag{6.38}$$

and the two others, where the first two coordinates are singularized, are equal. Writing that any two of these quantities are equal leads, for instance, to

$$\tau_{n_1,n_2,n_3+2\epsilon_3}\tau_{n_1+\epsilon_1,n_2+\epsilon_2,n_3-\epsilon_3} - \tau_{n_1,n_2+2,n_3\epsilon_3}\tau_{n_1+\epsilon_1,n_2+\epsilon_2,n_3-\epsilon_3}$$
$$= \epsilon_1(\epsilon_2 z_3 - \epsilon_3 z_2)\tau_{n_1,n_2,n_3}\tau_{n_1+\epsilon_1,n_2+\epsilon_2,n_3+\epsilon_3}. \tag{6.39}$$

These equalities consitute a second set of Hirota Miwa equations, which in this case are nonautonomous. This set is highly overdetermined but internally consistent, and also consistent with the first one, which it implies, provided that $z_i = \alpha n_i + \beta_i$, with a common value for α and three independent arbitrary values for the β's. (In fact dividing by the products of τ's in the right-hand side of (6.39), any of the three quantities (6.38), would allow us to define a new variable on the point of half-integer coordinates, which does appear in an interesting equation we are not going, however, to discuss further here [65]).

We can now proceed to the Miura relation relating three nonlinear variables. Let us consider the variables at the points $(0,0,1)$, $(1,0,1)$ and $(1,0,0)$, namely, $x_{0,0,1}$, etc.). We can thus write

$$x \equiv x_{0,0,1} = \frac{\tau_{1,1,1}\tau_{-1,-1,1}}{\tau_{0,0,0}\tau_{0,0,2}} - 1$$

$$w \equiv x_{1,0,0} = \frac{\tau_{1,1,1}\tau_{1,-1,-1}}{\tau_{0,0,0}\tau_{2,0,0}} - 1. \tag{6.40}$$

$$y \equiv x_{1,0,1} = 1 - \frac{\tau_{2,0,0}\tau_{0,0,2}}{\tau_{1,1,1}\tau_{1,-1,1}}$$

The reason why we chose this particular representation of each variable, rather than the other one will become obvious soon. Indeed, computing $x - w$ we find that

$$x - w = \frac{\tau_{1,1,1}\tau_{-1,-1,1}}{\tau_{0,0,0}\tau_{0,0,2}} - \frac{\tau_{1,1,1}\tau_{1,-1,-1}}{\tau_{0,0,0}\tau_{2,0,0}} = \frac{\tau_{1,1,1}}{\tau_{0,0,0}} \times \frac{\tau_{2,0,0}\tau_{-1,-1,1} - \tau_{0,0,2}\tau_{1,-1,-1}}{\tau_{0,0,2}\tau_{2,0,0}}. \tag{6.41}$$

Using (6.39), or rather, its analogue where the second coordinate is preferred, the numerator on the right-hand side can be rewritten $(z_1 - z_3)\tau_{0,0,0}\tau_{1,-1,1}$, where z_1 and z_3 are in fact computed at y, so $n_1 = n_3 = 1$, and thus we obtain

$$x - w = \frac{z_1 - z_3}{y - 1}. \tag{6.42}$$

This is the Miura transformation relating the three points x, y and w on the vertices of an isosceles right triangle, with y at the right angle. Let us now consider the point $(2,0,1)$ and the associated variable $\overline{x} \equiv x_{2,0,1}$,

$$\overline{x} \equiv x_{2,0,1} = 1 - \frac{\tau_{1,1,1}\tau_{3,-1,1}}{\tau_{2,0,0}\tau_{2,0,2}}. \tag{6.43}$$

Similarly we obtain

$$\overline{x} + w = \frac{z_1 + z_3}{y + 1}. \tag{6.44}$$

Taking the sum of these equations we obtain

$$\overline{x} + x = \frac{z_1 - z_3}{y - 1} + \frac{z_3 + z_1}{y + 1} = \frac{2z_1 y - 2z_3}{y^2 - 1}. \tag{6.45}$$

One must supplement this equation by one relating y at $(1,0,1)$ with $\overline{y}$ at $(3,0,1)$, using for instance, u at $(2,1,1)$ so $\overline{x}$ at $(2,0,1)$ is at the right angle of the isosceles right triangle $\overline{x}\overline{y}u$. One must now compute the z's at $\overline{x}$ so z_2 for $n_2 = 0$ and $\tilde{z}_1 = z_1 + \alpha$ at $n_1 = 2$. Then

$$y + u = \frac{\tilde{z}_1 + z_2}{\overline{x} + 1} \tag{6.46a}$$

$$\overline{y} - u = \frac{\tilde{z}_1 - z_2}{\overline{x} - 1} \tag{6.46b}$$

so

$$\overline{y} + y = \frac{2\tilde{z}_1 \overline{x} - 2z_2}{\overline{x}^2 - 1}. \tag{6.47}$$

Iterating, we see that in the numerator of the right-hand side the coefficient of the independent variable increases by α at each step while the zero-degree terms keep alternating between the values z_3 computed at $n_3 = 1$ and z_2 computed at $n_2 = 0$. This is exactly (6.35).

The Sakai Approach and the Classification of d-$\mathbb{P}$'s

While we have undertaken the classification of d-$\mathbb{P}$'s based on a geometrical description [66], another approach has tackled the same problem and presented a global answer [67]. Our approach has been essentially constructive. Starting from a given d-$\mathbb{P}$ equation, we worked out in detail its geometric description, which turned out to be described by an affine Weyl group. Sakai's approach is complementary. He started from the geometry of affine Weyl groups and obtained d-$\mathbb{P}$ equations in the end. The approach of Sakai also draws its inspiration from the work of Okamoto [68] on continuous $\mathbb{P}$'s. The two key notions are the "space of initial conditions" and the "symmetries under affine Weyl groups". Let us make these notions a little more precise.

The continuous $\mathbb{P}$'s are second-order differential equations. Thus one would expect the space of their initial conditions to be $\mathbb{C} \times \mathbb{C}$ since, for a given value t_0 of the independent variable, the solution is specified by the data of the function and its derivative at this point, with some reserves concerning the points at which the coefficients of the equation become singular. However there exist solutions which diverge at t_0. Thus we must compactify $\mathbb{C}^2$. Once this is done, it may happen that several solutions pass through the point at infinity. We must then separate them. The procedure is by a blowing-up of the space, i.e., through the introduction of local coordinates which make the divergence disappear.

The second idea of Okamoto, pertinent to the work of Sakai, concerns the symmetries of continuous $\mathbb{P}$'s under affine Weyl groups. As Okamoto has shown, the auto-Bäcklund (Schlesinger) transformations of the continuous $\mathbb{P}$'s generate extended affine Weyl groups and he has provided the following correspondences between equations and symmetries: P_{II} - $A_1^{(1)}$, P_{III} - $(2A_1^{(1)})$, P_{IV} - $A_2^{(1)}$, P_V - $A_3^{(1)}$, P_{VI} - $D_4^{(1)}$. (Equation P_I has no parameters and thus no auto-Bäcklund transformation.) We must point out here that Okamoto's methodology was, in spirit, closer to ours, in the sense that he started from a given equation and obtained the space of initial conditions as well as the affine Weyl group corresponding to each equation.

Sakai's approach consisted in studying rational surfaces in connection to extended Weyl groups. Surfaces obtained by successive blow-ups of $\mathbb{P}^2$ or $\mathbb{P}^1 \times \mathbb{P}^1$ have been studied through the connections between Weyl groups and the groups of Cremona isometries on the Picard group of the surfaces. (The Picard group of a rational surface X is the group of isomorphism classes of invertible sheaves on X, and it is isomorphic to the group of equivalent classes of divisors on X. A Cremona isometry is an isomorphism of the Picard group such that a) it preserves the intersection number of any pair of divisors, b) it preserves the canonical divisor K_X, and c) it leaves the set of effective classes of divisors invariant.) In the case where 9 points (for $\mathbb{P}^2$, or 8 points for $\mathbb{P}^1 \times \mathbb{P}^1$) are blown up, if the points are in a generic position, the group of Cremona isometries becomes isomorphic to an extension of the Weyl group of type $E_8^{(1)}$. When the 9 points are not in a generic position, the classification of connections between the group of Cremona isometries and the extended affine Weyl groups was studied in full generality by Sakai. Birational (bimeromorphic) mappings on $\mathbb{P}^2$ (or on $\mathbb{P}^1 \times \mathbb{P}^1$) are obtained by interchanging the procedure of blow-downs. Discrete Painlevé equations are recovered as birational mappings corresponding to translations of affine Weyl groups. We shall not present the work of Sakai in detail (lest we sink into plagiarism). We urge the interested reader to read this excellent piece of work and study it carefully.

The net result of the Sakai approach is a complete classification of the d-P's in terms of affine Weyl groups. Starting from the exceptional Weyl group, $E_8^{(1)}$, he obtained the systems corresponding to the degeneracy pattern below,

$$
\begin{array}{ccccccccccccc}
E_8^c & & & & & & & & & & & & A_1^q \\
\downarrow & & & & & & & & & & & \nearrow & \\
E_8^q & \longrightarrow & E_7^q & \longrightarrow & E_6^q & \longrightarrow & D_5^q & \longrightarrow & A_4^q & \longrightarrow & (A_2+A_1)^q & \longrightarrow & (A_1+A_1)^q & \longrightarrow & A_1^q \\
\downarrow & & \downarrow & & \downarrow & & \downarrow & & \downarrow & & \downarrow & & \downarrow & & \downarrow \\
E_8^\delta & \longrightarrow & E_7^\delta & \longrightarrow & E_6^\delta & \longrightarrow & D_4^c & \longrightarrow & A_3^c & \longrightarrow & & (2A_1)^c & & & A_1^c \\
& & & & & & & \searrow & & \downarrow & & \searrow & & \downarrow \\
& & & & & & & & A_2^c & & \longrightarrow & & A_1^c
\end{array}
$$

In this diagram, we assign to a Weyl group an upper index e if it supports a discrete equation involving elliptic functions, an upper index q if the equation is of q-type, an upper index δ if it is a difference equation not explicitly related to a continuous equation, and an upper index c if it is a difference equation which is explicitly the contiguity relation of one of the (continuous) Painlevé equations, namely P_{VI} for D_4, P_V for A_3, P_{IV} for A_2, (full) P_{III} for $2A_1$ (which means the direct product of twice A_1 in a self-dual way), P_{II} for the A_1 on the last line and finally the one-parameter P_{III} for the A_1 on the next to last line. Neither P_I nor the zero-parameter P_{III} appear here since, having no parameter, they have no contiguity relations, hence no discrete difference equation related to them. For each Weyl group we give below examples of equations that live in them. In what follows, $q_n = q_0\lambda^n$, $\rho_n = q_n/\sqrt{\lambda}$, while $z_n = z_0 + n\delta$ and $\zeta_n = z_n - \delta/2$.

E_8^q

We start from eight constants with the constraint that their product is unity. Let $m_1, m_2, \ldots, m_8$ be the elementary symmetric functions of order 1 to 8 of these eight constants (given the constraint, $m_8 = 1$). Then the mapping is:

$$
\frac{(x_{n+1}\rho_{n+1}^2 - y_n)(x_n\rho_n^2 - y_n) - (\rho_{n+1}^4 - 1)(\rho_n^4 - 1)}{(x_{n+1}/\rho_{n+1}^2 - y_n)(x_n/\rho_n^2 - y_n) - (1 - 1/\rho_{n+1}^4)(1 - 1/\rho_n^4)} =
$$

$$
\frac{y_n^4 - M_1 y_n^3 + M_2 y_n^2 + M_3 y_n + M_4}{y_n^4 - M_7 y_n^3 + M_6 y_n^2 + M_5 y_n + M_4/q_n^8},
$$

$$\frac{(y_{n-1}q_{n-1}^2 - x_n)(y_n q_n^2 - x_n) - (q_{n-1}^4 - 1)(q_n^4 - 1)}{(y_{n-1}/q_{n-1}^2 - x_n)(y_n/q_n^2 - x_n) - (1 - 1/q_{n-1}^4)(1 - 1/q_n^4)} =$$

$$\frac{x_n^4 - N_7 x_n^3 + N_6 x_n^2 + N_5 x_n + N_4}{x_n^4 - N_1 x_n^3 + N_2 x_n^2 + N_3 x_n + N_4/\rho_n^8},$$

where we have introduced the auxiliary quantities $M_1 = m_1 q_n$, $M_2 = m_2 q_n^2 - 3 - q_n^8$, $M_3 = m_7 q_n^7 - m_3 q_n^3 + 2m_1 q_n$, $M_4 = q_n^8 - m_6 q_n^6 + m_4 q_n^4 - m_2 q_n^2 + 1$, $M_5 = m_1/q_n^7 - m_5/q_n^3 + 2m_7/q_n$, $M_6 = m_6/q_n^2 - 3 - 1/q_n^8$, $M_7 = m_7/q_n$ and $N_1 = m_1/\rho_n$, $N_2 = m_2/\rho_n^2 - 3 - 1/\rho_n^8$, $N_3 = m_7/\rho_n^7 - m_3/\rho_n^3 + 2m_1/\rho_n$, $N_4 = \rho_n^8 - m_2 \rho_n^6 + m_4 \rho_n^4 - m_6 \rho_n^2 + 1$, $N_5 = m_1 \rho_n^7 - m_5 \rho_n^3 + 2m_7 \rho_n$, $N_6 = m_6 \rho_n^2 - 3 - \rho_n^8$, $N_7 = m_7 \rho_n$.

$\mathbf{E_8^\delta}$

Here we start from eight constants with the constraint that their sum must be zero. Let $s_2, s_3, \dots, s_8$ be their elementary symmetric functions of order 2 to 8 (from the constraint, $s_1 = 0$). Then the equation is

$$\frac{(y_n - x_{n+1} + 4\zeta_{n+1}^2)(y_n - x_n + 4\zeta_n^2) + 16y_n\zeta_{n+1}\zeta_n}{\zeta_n(y_n - x_{n+1} + 4\zeta_{n+1}^2) + \zeta_{n+1}(y_n - x_n + 4\zeta_n^2)} =$$

$$4\frac{y_n^4 + S_2 y_n^3 + S_4 y_n^2 + S_6 y_n + S_8}{8z_n y_n^3 + S_3 y_n^2 + S_5 y_n + S_7},$$

with $S_2 = s_2 + 28z_n^2$, $S_3 = s_3 + 6z_n s_2 + 56z_n^3$, $S_4 = s_4 + 5z_n s_3 + 15z_n^2 s_2 + 70z_n^4$, $S_5 = s_5 + 4z_n s_4 + 10z_n^2 s_3 + 20z_n^3 s_2 + 56z_n^5$, $S_6 = s_6 + 3z_n s_5 + 6z_n^2 s_4 + 10z_n^3 s_3 + 15z_n^4 s_2 + 28z_n^6$, $S_7 = s_7 + 2z_n s_6 + 3z_n^2 s_5 + 4z_n^3 s_4 + 5z_n^4 s_3 + 6z_n^5 s_2 + 8z_n^7$, $S_8 = s_8 + z_n s_7 + z_n^2 s_6 + z_n^3 s_5 + z_n^4 s_4 + z_n^5 s_3 + z_n^6 s_2 + z_n^8$.

$$\frac{(x_n - y_{n-1} + 4z_{n-1}^2)(x_n - y_n + 4z_n^2) + 16x_n z_{n-1} z_n}{z_n(x_n - y_{n-1} + 4z_{n-1}^2) + z_{n-1}(x_n - y_n + 4z_n^2)} =$$

$$4\frac{x_n^4 + \Sigma_2 x_n^3 + \Sigma_4 x_n^2 + \Sigma_6 x_n + \Sigma_8}{8\zeta_n x_n^3 + \Sigma_3 x_n^2 + \Sigma_5 x_n + \Sigma_7},$$

with $\Sigma_2 = s_2 + 28\zeta_n^2$, $\Sigma_3 = -s_3 + 6\zeta_n s_2 + 56\zeta_n^3$, $\Sigma_4 = s_4 - 5\zeta_n s_3 + 15\zeta_n^2 s_2 + 70\zeta_n^4$, $\Sigma_5 = -s_5 + 4\zeta_n s_4 - 10\zeta_n^2 s_3 + 20\zeta_n^3 s_2 + 56\zeta_n^5$, $\Sigma_6 = s_6 - 3\zeta_n s_5 + 6\zeta_n^2 s_4 - 10\zeta_n^3 s_3 + 15\zeta_n^4 s_2 + 28\zeta_n^6$, $\Sigma_7 = -s_7 + 2\zeta_n s_6 - 3\zeta_n^2 s_5 + 4\zeta_n^3 s_4 - 5\zeta_n^4 s_3 + 6\zeta_n^5 s_2 + 8\zeta_n^7$, $\Sigma_8 = s_8 - \zeta_n s_7 + \zeta_n^2 s_6 - \zeta_n^3 s_5 + \zeta_n^4 s_4 - \zeta_n^5 s_3 + \zeta_n^6 s_2 + \zeta_n^8$.

$\mathbf{E_7^q}$

$$\frac{(y_n x_{n+1} - q_n \rho_{n+1})(x_n y_n - q_n \rho_n)}{(y_n x_{n+1} - 1)(x_n y_n - 1)} =$$

$$\frac{(y_n - aq_n)(y_n - bq_n)(y_n - cq_n)(y_n - dq_n)}{(y_n - p)(y_n - r)(y_n - s)(y_n - t)},$$

$$\frac{(x_n y_n - z_n \rho_n)(x_n y_{n-1} - z_{n-1}\rho_n)}{(x_n y_n - 1)(x_n y_{n-1} - 1)} =$$

$$\frac{(x_n - \rho_n/a)(x_n - \rho_n/b)(x_n - \rho_n/c)(x_n - \rho_n/d)}{(x_n - 1/p)(x_n - 1/r)(x_n - 1/s)(x_n - 1/t)},$$

where a, b, c, d, p, r, s and t are 8 constants satisfying the constraints $abcd = prst = 1$.

$\mathbf{E}_7^\delta$

$$\frac{(y_n + x_{n+1} - z_n - \zeta_{n+1})(x_n + y_n - z_n - \zeta_n)}{(y_n + x_{n+1})(x_n + y_n)} =$$

$$\frac{(y_n - z_n - a)(y_n - z_n - b)(y_n - z_n - c)(y_n - z_n - d)}{(y_n - p)(y_n - r)(y_n - s)(y_n - t)}$$

$$\frac{(x_n + y_n - z_n - \zeta_n)(x_n + y_{n-1} - z_{n-1} - \zeta_n)}{(x_n + y_n)(x_n + y_{n-1})} =$$

$$\frac{(x_n - \zeta_n + a)(x_n - \zeta_n + b)(x_n - \zeta_n + c)(x_n - \zeta_n + d)}{(x_n + p)(x_n + q)(x_n + s)(x_n + t)},$$

where a, b, c, d, p, r, s and t are 8 constants satisfying the constraints $a + b + c + d = p + r + s + t = 0$.

$\mathbf{E}_6^q$

$$(y_n x_{n+1} - 1)(x_n y_n - 1) = \frac{(y_n - p)(y_n - r)(y_n - s)(y_n - t)}{(y_n - aq_n)(y_n - q_n/a)}$$

$$(x_n y_n - 1)(x_n y_{n-1} - 1) = \frac{(x_n - 1/p)(x_n - 1/r)(x_n - 1/s)(x_n - 1/t)}{(x_n - b\rho_n)(x_n - \rho_n/b)},$$

where a, b, p, r, s and t are 6 constants satisfying the constraint $prst = 1$.

$\mathbf{E}_6^\delta$

$$(y_n + x_{n+1})(x_n + y_n) = \frac{(y_n - p)(y_n - r)(y_n - s)(y_n - t)}{(y_n - z_n - a)(y_n - z_n + a)}$$

$$(x_n + y_n)(x_n + y_{n-1}) = \frac{(x_n + p)(x_n + r)(x_n + s)(x_n + t)}{(x_n - \zeta_n - b)(x_n - \zeta_n + b)}$$

where a, b, p, r, s and t are 6 constants satisfying the constraint $p + r + s + t = 0$.

$\mathbf{D}_5^q$

$$x_{n+1} x_n = \frac{(y_n - aq_n)(y_n - q_n/a)}{(y_n - c)(y_n - 1/c)}$$

$$y_n y_{n-1} = \frac{(x_n - b\rho_n)(x_n - \rho_n/b)}{(x_n - d)(x_n - 1/d)},$$

where a, b, c, d are 4 constants.

D$_4^c$

$$\frac{\zeta_{n+1}}{1 - x_{n+1}y_n} + \frac{z_n}{1 - y_n x_n} = z_n + a + \frac{by_n + (1 - y_n^2)(z_n/2 + c)}{(1 + dy_n)(1 + y_n/d)}$$

$$\frac{z_n}{1 - y_n x_n} + \frac{\zeta_n}{1 - x_n y_{n-1}} = \zeta_n + a + \frac{bx_n + (1 - x_n^2)(z_n/2 - c)}{(1 + dx_n)(1 + x_n/d)},$$

where a, b, c and d are 4 constants.

In the same space one has also, in a different direction

$$x_{m+1}x_m = \frac{(y_m - z_m)^2 - a}{y_m^2 - b}$$

$$y_m + y_{m-1} = \frac{\zeta_m - c}{1 + dx_m} + \frac{\zeta_m + c}{1 + x_m/d},$$

where a, b, c and d are 4 constants.

D$_4^q$

$$x_{n+1}x_n = \frac{c(y_n - aq_n)(y_n - q_n/a)}{y_n - 1}$$

$$y_n y_{n-1} = \frac{(x_n - b\rho_n)(x_n - \rho_n/b)}{c(x_n - 1)},$$

where a b, c are 3 constants.

A$_3^c$

$$x_{n+1} + x_n = \frac{y_n z_n - a}{y_n^2 - 1}$$

$$y_n + y_{n-1} = \frac{x_n \zeta_n - b}{x_n^2 - 1},$$

where a and b are 2 constants.

In the same space one has also, in a different direction

$$x_{m+1}x_m = \frac{y_m - z_m}{y_m^2 - a}$$

$$y_m + y_{m-1} = \frac{1}{x_m} + \frac{\zeta_m + b}{1 - x_m}$$

where a and b are 2 constants. .

$(A_2+A_1)^q$

$$x_{n+1}x_n = \frac{aq_ny_n - q_n^2}{y_n(y_n - 1)}$$

$$y_ny_{n-1} = \frac{b\rho_nx_n - \rho_n^2}{x_n(x_n - 1)},$$

where a, b are 2 constants.

$(2A_1)^c$,
which means a self-dual combination of two A_1^c's

$$\frac{\zeta_{n+1}}{1 - x_{n+1}x_n} + \frac{\zeta_n}{1 - x_{n-1}x_n} = x_n + \frac{1}{x_n} + z_n + a,$$

where a is a constant.

A_2^c

$$x_{n+1} + x_n = -y_n + b + \frac{z_n + a}{y_n}$$

$$y_n + y_{n-1} = -x_n + b + \frac{\zeta_n - a}{x_n},$$

where a, b are 2 constants, but b can always be scaled to unity unless it is zero.

$(A_1+A_1)^q$,
which means a *non* self-dual combination of two A_1^q's,

$$x_nx_{n+1} = \frac{1 + y_n/q_n}{y_n(1 + y_n/a)}$$

$$y_ny_{n-1} = a\frac{x_n + 1}{x_n^2}$$

or, in a different direction,

$$(x_{m+1}x_m - 1)(x_{m-1}x_m - 1) = \frac{aq_m^2x_m}{(x_m - q_m)}.$$

We have two different equations corresponding to A_1^c. The one on the next to last line, related to the one-parameter P_{III}, is

$$x_{n+1} + x_{n-1} = \frac{x_nz_n + 1}{x_n^2},$$

while the one on the last line, related to P_{II}, contains the discrete equation

$$\frac{\zeta_{n+1}}{x_{n+1} + x_n} + \frac{\zeta_n}{x_{n-1} + x_n} = x_n^2 + a,$$

where a is a constant which can always be scaled to unity unless it is zero.

Finally we have two different possibilities for equations in A_1^q,

$$x_{n+1}x_{n-1} = \frac{x_n q_n + 1}{x_n^2}$$

and an 'asymmetric' equation, which was shown to be a discrete form of the zero-parameter P_{III},

$$(x_{n-1}y_n - 1)(x_n y_n - 1) = q_n$$

$$(x_n y_n - 1)(x_n y_{n+1} - 1) = q_n x_n^2$$

with $q_n = q_0 \lambda^n$.

One important finding of Sakai is the discovery of the equations related to the group $E_8^{(1)}$. He found that there are three such equations, rather than two as in the case of $E_7^{(1)}$ or $E_6^{(1)}$. The upper index e indicates the third, new, kind of discrete $\mathbb{P}$'s, namely, mappings where the independent variable as well as the parameters enter through the arguments of elliptic functions. Sakai's construction is global. If one wishes to construct explicit examples of the equations associated to specific affine Weyl groups, one has to specify a nonclosed, periodically repeated, pattern in the appropriate space, in order to obtain the corresponding d-$\mathbb{P}$. (A consequence of this last statement is that the potential number of d-$\mathbb{P}$'s is infinite, since any nonclosed periodically repeated pattern in each of the spaces of the affine Weyl groups of the degeneration pattern would lead to a different d-$\mathbb{P}$.) In [69] we have obtained explicit examples of elliptic d-$\mathbb{P}$'s and, despite its lenghty expression, we cannot resist the temptation to present one such example here,

$$\frac{x_{n-1} - \mathrm{sn}^2(\lambda n + \lambda/2 + \phi(n-1) - (-1)^n\psi + \omega(n); m)}{x_{n-1} - \mathrm{sn}^2(\lambda n + \lambda/2 + \phi(n-1) - (-1)^n\psi - \omega(n); m)}$$

$$\times \frac{x_n - \mathrm{sn}^2(-2\lambda n - \phi(n-1) - \phi(n+1) + \omega(n); m)}{x_n - \mathrm{sn}^2(-2\lambda n - \phi(n-1) - \phi(n+1) - \omega(n); m)}$$

$$\times \frac{x_{n+1} - \mathrm{sn}^2(\lambda n - \lambda/2 + \phi(n-1) + (-1)^n\psi + \omega(n); m)}{x_{n+1} - \mathrm{sn}^2(\lambda n - \lambda/2 + \phi(n-1) - (-1)^n\psi - \omega(n); m)} =$$

$$\frac{1 - m^2\mathrm{sn}^2(\lambda n + \alpha(n) + \frac{\omega(n)}{2}; m)\mathrm{sn}^2(-\lambda n + \beta(n) - \frac{\omega(n)}{2}; m)}{1 - m^2\mathrm{sn}^2(\lambda n + \alpha(n) - \frac{\omega(n)}{2}; m)\mathrm{sn}^2(-\lambda n + \beta(n) + \frac{\omega(n)}{2}; m)}$$

$$\times \frac{1 - m^2\mathrm{sn}^2(\lambda n + \alpha(n) - \frac{\omega(n)}{2}; m)\mathrm{sn}^2(\gamma(n) + \frac{\omega(n)}{2}; m)}{1 - m^2\mathrm{sn}^2(\lambda n + \alpha(n) + \frac{\omega(n)}{2}; m)\mathrm{sn}^2(\gamma(n) - \frac{\omega(n)}{2}; m)}$$

$$\times \frac{1 - m^2\mathrm{sn}^2(-\lambda n + \beta(n) + \frac{\omega(n)}{2}; m)\mathrm{sn}^2(\gamma(n) - \frac{\omega(n)}{2}; m)}{1 - m^2\mathrm{sn}^2(-\lambda n + \beta(n) - \frac{\omega(n)}{2}; m)\mathrm{sn}^2(\gamma(n) + \frac{\omega(n)}{2}; m)} \quad (6.48)$$

where we have introduced the auxiliary quantities $\alpha(n) = \lambda/6 + (2\phi(n-1) + \phi(n+1) - (-1)^n\psi)/3$, $\beta(n) = \lambda/6 - (2\phi(n+1) + \phi(n-1) + (-1)^n\psi)/3$,

$\gamma(n) = (\phi(n+1) - \phi(n-1) + 2(-1)^n \psi - \lambda)/3$ and where ψ is a constant, $\phi(n+3) = \phi(n)$, i.e. ϕ has period three and $\omega(n+4) = \omega(n)$, i.e., ω has period four so the whole equation has period 12. The total number of degrees of freedom is 8, including the independent variable.

While the Sakai approach may seem somewhat abstract, it is quite useful for the understanding of various aspects of d-$\mathbb{P}$'s and discrete systems in general. Sakai himself provided the link between the property of singularity confinement and the construction of the space of initial conditions. He has shown that all d-$\mathbb{P}$'s have a maximum of 8 confined singularities, and that they can be described by a maximum of 8 blow-ups. The procedure of blowing-up at each singularity is the one first advocated by Kruskal [70]. According to Kruskal, one must provide a complete description of the dynamics of the mapping and this means lifting the indeterminacies of each of the singularities. This program of complete description of the dynamics of mappings with confined singularities was carried out by Takenawa [71]. He has studied the discrete $\mathbb{P}$'s of the Sakai classification and reconstituted their dynamics through a series of blow-ups and blow-downs. He has used this approach in order to compute the algebraic entropy of these systems, and has shown that all these d-$\mathbb{P}$'s have a degree growth that behaves as n^2, a result previously established, albeit in a nonrigorous way, in [45]. Moreover Takenawa has used this approach on mappings, like the one obtained in [37], which have a positive algebraic entropy while having confined singularities. He was able to obtain analytically the algebraic entropy of such mappings, thus extending and complementing Sakai's approach.

$E\pi\iota\delta\acute{o}\rho\pi\iota o\nu$

7 More Nice Results on d-$\mathbb{P}$'s

This last section will be devoted to topics which either could not fit into the previous sections or would lengthen the presentation there. Still, since they are interesting results we decided not to omit them but regroup them in this section, which, as a result, will appear slightly less well organised than the previous ones.

Limits and Degeneracies of d-$\mathbb{P}$'s
As we have already seen, most d-$\mathbb{P}$'s do contain parameters. One can consider the limit of the d-$\mathbb{P}$ when these parameters take some special values. The usual result of this limiting procedure is again a discrete $\mathbb{P}$ the continuous limit of which is different from that of the initial one.

Let us start from the simplest case, d-$\mathrm{P_I}$, written as

$$x_{n+1} + x_{n-1} + x_n = \frac{z_n}{x_n} + \alpha. \tag{7.1}$$

Parameter α can be scaled to 1, unless it is 0. In this special case, the d-P$_I$ equation reduces to

$$x_{n+1} + x_{n-1} + x_n = \frac{z_n}{x_n}, \tag{7.2}$$

which does not possess any nontrivial continuous limit.

Let us now turn to d-P$_{II}$ and introduce a scaling of all variables and parameters so as to transform the equation to

$$x_{n+1} + x_{n-1} = \frac{z_n x_n + a}{\rho^2 - x_n^2}. \tag{7.3}$$

By taking the limit $\rho \to 0$ we find the equation

$$x_{n+1} + x_{n-1} = \frac{z_n}{x_n} + \frac{a}{x_n^2}, \tag{7.4}$$

which is a well known form of d-P$_I$.

Similarly, one can study limits of q-P$_{III}$, which we shall rewrite here for convenience as

$$x_{n+1} x_{n-1} = \frac{\gamma x_n^2 + \zeta x_n + \mu}{\alpha x_n^2 + \beta x_n + \gamma}. \tag{7.5}$$

The full q-P$_{III}$ corresponds to $\gamma \neq 0$. Without loss of generality, one can set γ to be a constant. We find, by applying the singularity confinement criterion, that α and β can both be set to constants, by redefining x. We further find that $\zeta = \zeta_0 \lambda^n$ and $\mu = \mu_0 \lambda^{2n}$. The limits we consider here correspond to $\gamma = 0$, while keeping the behaviours $\alpha = const.$, $\beta = const.$, $\zeta \propto \lambda^n, \mu \propto \lambda^{2n}$. We first find the equation

$$x_{n+1} x_{n-1} = \frac{\zeta x_n + \mu}{(x_n + \beta) x_n}. \tag{7.6}$$

This is a novel form of q-P$_{II}$. Its continuous limit can be obtained through $x = 1 + \epsilon w$, $\beta = -2 + \epsilon^3 g$, $\zeta = -2\lambda^n$, $\mu = \lambda^{2n}$ where $\lambda = 1 + \epsilon^3/2$, leading to $w'' = 2w^3 + wt + g$. A further limit can be obtained, starting from (7.6), by taking $\beta = 0$, in addition to $\gamma = 0$. In this case we find the equation

$$x_{n+1} x_{n-1} = \frac{\zeta}{x_n} + \frac{\mu}{x_n^2}, \tag{7.7}$$

where, by the gauge $x \to x\lambda^{n/2}$, μ can be taken as to be constant and ζ to be of the form $\zeta_0 \lambda^{n/2}$. Equation (7.7) is a discrete q-P$_I$, as can be seen from the continuous limit obtained by $x = 1 + \epsilon^2 w$, $\zeta = 4\kappa^n$, $\mu = -3$ and $\kappa(\equiv \lambda^{1/2}) = 1 - \epsilon^5/4$, leading to $w'' = 6w^2 + t$. More limiting cases based on equations d-P$_{IV}$ and d-P$_V$ can be found in [47].

The cases associated to the notion that we denoted in [47] by the term of *degeneracy* need some explanation. (The choice of this term is admittedly unfortunate, since it is too similar to the term *degeneration* used for the result of coalescence. However since it is the one we have used in the initial publication on the subject we will, with apologies, retain it here). Here is what we mean by degeneracy. The way we have obtained most d-$\mathbb{P}$'s was to assume a functional form given by the QRT mapping at the autonomous limit, and de-autonomise it, using some integrability criterion. The important assumption at this stage is that the functional form is fixed.

Let us illustrate this with an example based on d-P_{I-II}:

$$x_{n+1} + x_{n-1} = -\frac{\beta x_n^2 + \epsilon x_n + \zeta}{\alpha x_n^2 + \beta x_n + \gamma} \tag{7.8}$$

Suppose that the numerator and denominator in the right-hand side of (7.8) have a common factor. This is what we call degenerate case. This case is of interest only when $\alpha \neq 0$, otherwise the degenerate equation becomes linear. In this case we obtain

$$x_{n+1} + x_{n-1} = \frac{\epsilon}{x_n + \rho}. \tag{7.9}$$

We can translate ρ to zero and de-autonomise (7.9). By Using singularity confinement, we obtain

$$x_{n+1} + x_{n-1} = \frac{z}{x_n} + a \tag{7.10},$$

with again z linear in n and a constant, which is another form of d-P_I. Its continuous limit is obtained through $x = 1 + \epsilon^2 w$, $a = 4$, $z = -2 - \epsilon^5 n$, leading at $\epsilon \to 0$ to $w'' + 2w^2 + t = 0$, with $t = \epsilon n$.

Similarly the degenerate forms of d-P_{III} are obtained when the numerator and the denominator in the right-hand side of (7.5) have a common factor. In this case,

$$x_{n+1} x_{n-1} = \frac{a x_n + b}{c x_n + d}, \tag{7.11}$$

The de-autonomisation of this equation yields $a = a_0 \lambda^n$ and $d = d_0 \lambda^n$. Unless $c = 0$, we can always take $c = 1$, by division, and a proper gauge allows us to take $b = 1$. Equation (7.11) in its nonautonomous form is a novel form of q-discrete P_{II}. The limit $d = 0$ in (7.11) leads to the equation ($c = 1$),

$$x_{n+1} x_{n-1} = a + \frac{1}{x_n}, \tag{7.12}$$

where $a = a_0 \lambda^n$. This is another form of q-P_I. The continuous limit is obtained by $x = x_0(1 + \epsilon^2 w)$, where $x_0^3 = -1/2$, $a = 3x_0^2 \lambda^n$, with $\lambda = 1 - \epsilon^5/3$, leading to $w'' + 3w^2 + t = 0$. An equivalent equation can be obtained from (7.11) by taking $a = 0$,

$$x_{n+1} x_{n-1} = \frac{1}{x_n + d}. \tag{7.13}$$

Equation (7.13) is transformed into (7.12) by taking $x \to 1/x$ and exchanging a and d. Another limit, leading to another q-P_I, is $c = 0$. We find the equation

$$x_{n+1}x_{n-1} = ax_n + b, \qquad (7.14)$$

where a, in this case, is a constant and $b = b_0\lambda^n$, with continuous limit $w'' + 6w^2 + t = 0$ obtained by $x = 1 + \epsilon^2 w$, $a = 2$, $b_0 = -1$ and $\lambda = 1 + \epsilon^5$. An equivalent equation can also be obtained by taking $b = 0$ in (7.11). We find the equation

$$x_{n+1}x_{n-1} = \frac{ax_n}{x_n + d}. \qquad (7.15)$$

Equations (7.14) and (7.15) are related through the transformation $x \to 1/x$, with the appropriate relations of the parameters.

Many more examples, related to equations d-P_{IV} and q-P_V can be found in [47].

On Asymmetric Forms

In the previous sections we have encountered several examples of d-$\mathbb{P}$'s. As we have explained one very simple way to obtain them is to start from a QRT-type mapping,

$$x_{n+1} = \frac{f_1(x_n) - x_{n-1}f_2(x_n)}{f_4(x_n) - x_{n-1}f_3(x_n)} \qquad (7.16)$$

and de-autonomise it by the application of an integrability criterion. The result of this approach is, as we have seen, quite often a mapping where a degree of freedom with binary, ternary or higher periodicity may exist. This suggests that the mapping can be better written as a system of more than one equation. (As we have remarked the mapping remains of the second-order, since all but two of the relations involve dependent variables which are all defined at for the same value of the independent variable, while the last two involve two values of the independent variable, $n - 1$ and n for one of them, n and $n + 1$ for the other. In the symmetric case, of course, all three values appear in the unique equation.)

This approach does still not exhaust all the possibilities. The d-$\mathbb{P}$'s obtained in this way are very slightly asymmetric. One could thus wonder whether more complicated systems would exist. The answer is, quite expectedly, indeed they can.

First as we have seen in the classification of the canonical QRT forms, there exist mappings of the form

$$x_n + x_{n-1} = f(y_n) \qquad (7.17a)$$

$$y_n y_{n+1} = g(x_n). \qquad (7.17b)$$

Several examples of such mappings are known, for instance [72],

$$x_{m+1}x_m = \frac{(y_m - z_m)^2 - a}{y_m^2 - 1} \tag{7.18a}$$

$$y_m + y_{m-1} = \frac{\zeta_m - c}{1 + dx_m} + \frac{\zeta_m + c}{1 + x_m/d}, \tag{7.18b}$$

with $\zeta_n = z_{n-1/2}$, which is a discrete P_V, obtained as a contiguity relation [73] for the solutions of the continuous P_{VI}. Many more equations of this form are known.

Another possibility also exists. Some equations can be written in a such form that the right-hand sides are quite different, without any possibility of further simplification. In [74] we presented the mapping

$$x_n x_{n+1} = \frac{1 + y_n/q_n}{y_n(1 + y_n/d)} \tag{7.19a}$$

$$y_n y_{n-1} = d\frac{x_n + 1}{x_n^2}, \tag{7.19b}$$

which was to be a discrete form of the one-parameter P_{III}. Similarly in [48] we introduced the system of equations

$$(x_n y_n - 1)(x_{n-1}y_n - 1) = q_n \tag{7.20a}$$

$$(x_n y_n - 1)(x_n y_{n+1} - 1) = q_n x_n^2, \tag{7.20b}$$

which was shown to be a q-discrete form of the zero-parameter P_{III}.

More equations of this kind do indeed exist, but, curiously they have not yet been studied in detail. One way to approach this problem would be to start from an asymmetric QRT mapping,

$$x_{n+1} = \frac{f_1(y_n) - x_n f_2(y_n)}{f_4(y_n) - x_n f_3(y_n)} \tag{7.21a}$$

$$y_{n-1} = \frac{g_1(x_n) - y_n g_2(x_n)}{g_4(x_n) - y_n g_3(x_n)}, \tag{7.21b}$$

and investigate explicitly the integrability of forms which are strongly asymmetric.

A remark is in order here concerning the possible forms of d-$\mathbb{P}$'s. Sometimes one obtains equations which are written in a form which is very far from a QRT one. For instance while studying the contiguities of the solutions of P_V [52] we obtained the system

$$x_{n+1} = -x_n(y_n - 1)\left(1 - \frac{x_n(y_n - 1)^2}{sy_n}\right) \tag{7.22a}$$

$$x_{n-1} = x_n \left(1 - \frac{1}{y_n} + \frac{z_n}{y_n(x_n + z_n)} - \frac{(z_n y_n + x_n(y_n - 1))^3}{sy_n^2(z_n + x_n)^2} \right) \tag{7.22b}$$

$$y_{n-1} = -\frac{(z_n y_n + x_n(y_n - 1))^2}{sy_n(z_n + x_n) - (z_n y_n + x_n(y_n - 1))^2} \tag{7.22c}$$

$$y_{n+1} = -\frac{(sy_n - x_n(y_n - 1)^2)^2}{(y_n - 1)(sy_n(x_n(y_n - 1) - z_n - 1) - x_n^2(y_n - 1)^3)}, \tag{7.22d}$$

which looks already awesome even though we do not present it here in its full generality. As was shown in [52], the continuous limit of this mapping is the P_{IV} equation. We speculate that equations of the above form result from complicated evolution paths in the corresponding geometry.

Moreover it is interesting to know that one can also look for d-$\mathbb{P}$'s of the form $x_{n+1} = f(x_n, y_n)$, $y_{n+1} = g(x_n, y_n)$, with the necessary reserve that the backward evolution, $x_{n-1} = h(x_n, y_n)$, $y_{n-1} = k(x_n, y_n)$ be equally defined. We point out here that this evolution is different from the one defined in the QRT mapping. In the latter the variables are staggered, in the sense that y_{n+1} is not defined in terms of x_n and y_n but rather in terms of y_n and x_{n+1}.

Discrete Systems from Contiguities of Continuous $\mathbb{P}$'s by Limiting Procedures

As we have seen in the previous sections, many difference $\mathbb{P}$'s are just the contiguity relations of the solutions of the continuous $\mathbb{P}$'s. While the independent variable is now what was previously a parameter of the continuous $\mathbb{P}$, the continuous independent variable t does not disappear but survives as a parameter of the d-$\mathbb{P}$. Moreover it is not a parameter the value of which can be modified through a Schlesinger transformation. Given the relation to the continuous Painlevé equation, there exist some values of t which play a special role. In particular the values which correspond to the fixed singularities of the continuous Painlevé equations, namely $(0,1,\infty)$ for P_{VI}, $(0,\infty)$ for P_V and P_{III}, and only ∞ for P_{IV} and P_{II}, are expected to play a role.

Let us now present some of these special limits, detailed analysis of which may be found in [75]. Here we shall limit ourselves to the equations obtained from the contiguity relations of the continuous P_{VI}. Two such equations have been presented in [73].

Let us start with the system,

$$x_n + x_{n-1} = \frac{z_n + a}{1 + y_n/t} + \frac{z_n - a}{1 + ty_n} \tag{7.23a}$$

$$y_n y_{n+1} = \frac{(x_n - z_{n+1/2})^2 - p^2}{x_n^2 - c^2}, \tag{7.23b}$$

where $z_n = \alpha n + \beta$, even if n is not an integer. This equation was first derived in [72], where we gave its Lax pair and showed that it was a discrete form of P_V. If we take $t \to 0$ in (7.23), or $t \to \infty$, since (7.23) is clearly invariant under $t \to 1/t$, $a \to -a$, we find that the first equation reduces to

$x_n + x_{n-1} = z_n - a$, i.e., a linear equation. Thus, at this limit, system (7.23) is trivially linearised. Next we take $t \to 1$. In this case (7.23a) reduces to $x_n + x_{n-1} = 2z_n/(1 + y_n)$ which can be solved for y. We can thus obtain a single equation for x,

$$\left(\frac{x_n + x_{n+1} - 2z_{n+1}}{x_n + x_{n+1}} \right) \left(\frac{x_n + x_{n-1} - 2z_n}{x_n + x_{n-1}} \right) = \frac{(x_n - z_{n+1/2})^2 - p^2}{x_n^2 - c^2}. \quad (7.24)$$

This is an equation belonging to the family of d-P$_V$ we presented in [49] and it is in fact linearisable, as we shall show below.

The two limits obtained above were performed in a straightforward way. However it is possible to obtain more limiting cases if we renormalise the variables appropriately. Let us examine again the $t \to 0$ case. We introduce $Y = ty$ and take $tp = k$. The limit of (7.23a) now becomes $x_n + x_{n-1} = (z_n - a)/(1 + Y_n)$ and we can solve this equation for Y. Substituting Y into the limit of (7.23b) we find:

$$\left(\frac{x_n + x_{n+1} - z_{n+1} + a}{x_n + x_{n+1}} \right) \left(\frac{x_n + x_{n-1} - z_n + a}{x_n + x_{n-1}} \right) = \frac{k^2}{x_n^2 - c^2} \quad (7.25)$$

This equation is the Miura of the equation we call the alternate d-P$_{II}$ [63]. In the case $t \to 1$ we introduce $t = 1 + \epsilon$, $x = X/\epsilon$, $y = -1 - \epsilon Y$, $c = 1/\epsilon$, $p^2 = 1/\epsilon^2 - 2q/\epsilon$ and obtain, at the limit $\epsilon \to 0$, the mapping:

$$X_n + X_{n-1} = 2\frac{Y_n z_n + a}{1 - Y_n^2} \quad (7.26a)$$

$$Y_n + Y_{n+1} = 2\frac{X_n z_n + q}{1 - X_n^2}, \quad (7.26b)$$

which is known as the asymmetric d-P$_{II}$ equation and constitutes a discrete form of P$_{III}$.

The second equation coming from P$_{VI}$ is the one we presented in [73] together with its Lax pair.

$$\frac{z_{n+1/2}}{1 - x_n x_{n+1}} + \frac{z_{n-1/2}}{1 - x_n x_{n-1}} = a + z_n + \frac{b(t^2 - 1)x_n + t(1 - x_n^2)(z_n/2 + (-1)^n c)}{(t + x_n)(1 + tx_n)}. \quad (7.27)$$

It also has P$_V$ as a continuous limit. Again the limit $t \to 0$ (or $t \to \infty$) is straightforward. The equation reduces to $z_{n+1/2}/(1 - x_n x_{n+1}) + z_{n-1/2}/(1 - x_n x_{n-1}) = a - b + z_n$ which is linear for the quantity $1/(1 - x_n x_{n+1})$. Thus this limit is trivially linearisable. The limit $t \to 1$ is more interesting. We obtain

$$\frac{z_{n+1/2}}{1 - x_n x_{n+1}} + \frac{z_{n-1/2}}{1 - x_n x_{n-1}} = a + z_n + \frac{1 - x_n}{1 + x_n}(z_n/2 + (-1)^n c). \quad (7.28)$$

This equation is a new discrete system, which we expect to be integrable, since it is a limit of an integrable one. Moreover the iteration of some initial data leads to a linear degree growth. Thus we surmise that this system is linearisable. Its linearisation will be given below.

Just as in the case of (7.23) we can take the limits $t \to 0$ (or $t \to \infty$) and $t \to 1$ in a subtler way. For the $t \to 0$ case we put $x = 2t^{-1}/(X-1)$, $z = 4 + t^2 Z$, $a = -2 + At^2/2$, $b = 2 + At^2/2$ and $c = -Ct^2/2$. At the limit we find:

$$X_{n+1} + X_{n-1} = \frac{Z_n X_n + A + (-1)^n C}{1 - X_n^2}, \tag{7.29}$$

i.e., the asymmetric d-P_{II} equation.

In the case of the limit $t \to 1$ we can obtain an equation somewhat richer than (7.28). It suffices to assume that, while t goes to 1, b diverges so as to keep the quantity $b(t^2 - 1)$ finite. We thus find the limit

$$\frac{z_{n+1/2}}{1 - x_n x_{n+1}} + \frac{z_{n-1/2}}{1 - x_n x_{n-1}} = a + z_n + \frac{k x_n}{(1 + x_n)^2} + \frac{1 - x_n}{1 + x_n}(z_n/2 + (-1)^n c). \tag{7.30}$$

This equation is different from (7.28). As a matter of fact (7.30) is a new discrete Painlevé equation. An equivalent form of this equation can be obtained if we take $x = (1 - X)/(1 + X)$. We find, in this new variable,

$$\frac{z_{n+1/2}}{X_n + X_{n+1}} + \frac{z_{n-1/2}}{X_n + X_{n-1}} = \frac{k}{2} + 2\frac{a + X_n(-z_n/2 + (-1)^n c)}{1 - X_n^2}, \tag{7.31}$$

Next we introduce the new auxiliary variable Y through $k(Y_n - 1)/4 = z_{n+1/2}/(X_n + X_{n+1})$. Equation (7.31) can now be written as a system

$$Y_n + Y_{n-1} = \frac{A + (Z_n + (-1)^n C)X_n}{X_n^2 - 1} \tag{7.32a}$$

$$X_n + X_{n+1} = \frac{Z_{n+1/2}}{Y_n - 1}, \tag{7.32b}$$

with the appropriate redefinitions of a, c, and z. We can recognize (7.32a) as part of the asymmetric d-P_{II} equation.

Let us now present the linearisation of the two equations we obtained above. We start from (7.28), which we rewrite as

$$\frac{2\zeta_{n+1/2}}{1 - x_n x_{n+1}} + \frac{2\zeta_{n-1/2}}{1 - x_n x_{n-1}} = a + \zeta_{n-1/2} + \zeta_{n+1/2} + \frac{1 - x_n}{1 + x_n}Z_n \tag{7.33}$$

by introducing two new functions, ζ and Z, and we have just used the property that the quantity appearing on the right-hand side is the half-sum of the two quantities appearing at the numerators of the left-hand side. With this parametrisation, the linearisation condition is $Z_n + Z_{n+1} = \zeta_{n-1/2} + \zeta_{n+3/2}$. Thus only one free function remains. A most convenient way to parametrise this linearisability condition is to introduce a free function ϕ and express ζ and Z as $\zeta_{n+1/2} = \phi_n + \phi_{n+1}$ and $Z_n = \phi_{n-1} + \phi_{n+1}$.

In order to linearise (7.33) we first introduce the auxiliary variable w by $w_n = 2\zeta_{n+1/2}/(1 - x_n x_{n+1}) - \zeta_{n+1/2}$. Equation (7.33) becomes

$$w_n + w_{n-1} - a = \frac{1 - x_n}{1 + x_n} Z_n, \qquad (7.34)$$

which can be solved for x_n in terms of w_n, w_{n-1}. Substituting this expression for x back into the definition of w we find the equation

$$((w_{n-1} + w_n - a)\zeta_{n+1/2} - Z_n w_n)((w_{n+1} + w_n - a)\zeta_{n+1/2} - Z_{n+1} w_n) = (w_n^2 - \zeta_{n+1/2}^2)Z_n Z_{n+1}. \qquad (7.35)$$

We can formally introduce a parameter p into the equation and rewrite it as

$$((w_{n-1} + w_n - a)\zeta_{n+1/2} - Z_n w_n)((w_{n+1} + w_n - a)\zeta_{n+1/2} - Z_{n+1} w_n) = (w_n^2 - p\zeta_{n+1/2}^2)Z_n Z_{n+1}. \qquad (7.36)$$

If we eliminate p between (7.36) and its upshift we find a four-point equation relating four w's (from $n-1$ to $n+2$). On the other hand consider the linear equation

$$A_n w_{n+1} + B_n(a - w_n) + A_{n+1} w_{n-1} + r(Z_n w_{n+1} + (Z_n + Z_{n+1})(w_n - a) + Z_{n+1} w_{n-1}) = 0, \qquad (7.37)$$

where r is an arbitrary constant and A and B have simple expressions in terms of ϕ: $A_n = \phi_n^2(\phi_{n-1} + \phi_{n+1})$, $B_n = (\phi_n + \phi_{n+1})\phi_{n-1}\phi_{n+2} + (\phi_{n-1} + \phi_{n+2})\phi_n\phi_{n+1}$. Eliminating r between (7.37) and its upshift leads to exactly the same four-point equation as the one obtained from (7.35). The way to integrate (7.35) becomes clear. Start from two initial conditions for w, say w_{-1} and w_0. Use (7.35) to compute w_1, so (7.36) is satisfied with $p = 1$ for these values of the w's. Substitute all three values of w into (7.37) and set r so that the latter is satisfied. From then on, integrate the linear equation (7.37) for this value of r to obtain all the w's. This ensures that the four-point equation we referred to above is always satisfied for these values of the w's and, thus, so is (7.36) for a constant value of p, namely 1. So indeed (7.35) is satisfied for all values of n. Using the definition of x in terms of w we finally obtain the solution of (7.28).

The other case to be linearised is (7.24). We start by rewriting it in a more convenient form as

$$\frac{x_n + x_{n+1} - z_n - z_{n+1}}{x_n + x_{n+1}} \frac{x_n + x_{n-1} - z_n - z_{n-1}}{x_n + x_{n-1}} = \frac{(x_n - z_n)^2 + m + q}{x_n^2 - m + q}, \qquad (7.38)$$

where m and q are constants. Next we compute the discrete derivative of (7.38) obtained by using the fact that q is a constant, i.e., eliminating q between (7.38) and its upshift. We do not give here the lengthy equation, relating four x's (from $n-1$ to $n+2$) that results, but it can readily be obtained using a symbolic-manipulation program. Next we write the *linear* equation

$$z_n x_n = m - k + \left(k + \frac{z_n^2}{2}\right)\left(\frac{x_n + x_{n+1}}{z_n + z_{n+1}} + \frac{x_n + x_{n-1}}{z_n + z_{n-1}}\right), \tag{7.39}$$

where a free constant k appears. It turns out that if we compute the discrete derivative of (7.39) using the fact that k is a constant, *i.e.*, eliminating k between (7.39) and its up-shift, the four-point equation thus obtained is *identical* to the one we found above starting from (7.38). The way to integrate the latter is similar to the way (7.28) was integrated. Start from two initial conditions for x, say x_{-1} and x_0 and two constants m and q. Use (7.38) to compute x_1, substitute all three values of x into (7.39) and fix k so that the latter is satisfied. From then on, integrate (7.39), which is linear, for this value of k to obtain all the x's.

Mappings Obtained for Special Values of q

One interesting result is the recent discovery of second-order autonomous mappings which are not explicitly of QRT form but still possess an invariant. The departure from the QRT type manifests itself in the fact that the invariant, instead of being biquadratic is biquartic. The first such example encountered was the mapping [76]

$$(x_n x_{n+1} - 1)(x_n x_{n-1} - 1) = \frac{(x-a)(x-1/a)(x^2-1)}{p^2 x^2 - 1}. \tag{7.40}$$

The conserved quantity is

$$K = \frac{\left((x_n - x_{n-1})^2 - p^2(x_n x_{n-1} - 1)^2\right)\left((x_n + x_{n-1} - a - 1/a)^2 - p^2(x_n x_{n-1} - 1)^2\right)}{(x_n x_{n-1} - 1)^2}, \tag{7.41}$$

which isquartic in x_n and x_{n-1} separately.

We shall not elaborate here on the possible existence of a transformation which could bring (7.40) into QRT form. We believe that such a transformation exists but must be prohibitively complicated to derive. There exists however a way to explain the existence of a mapping like (7.40) and to link it to the discrete Painlevé equations. Let us start with the full asymmetric q-P$_V$

$$(x_n y_n - 1)(x_n y_{n-1} - 1) = \frac{(x_n - a)(x_n - b)(x_n - c)(x_n - d)}{(pq^n x_n - 1)(rq^n x_n - 1)} \tag{7.42a}$$

$$(x_{n+1} y_n - 1)(x_n y_n - 1) = \frac{(y_n - 1/a)(y_n - 1/b)(y_n - 1/c)(y_n - 1/d)}{(sq^n y_n - 1)(tq^n y_n - 1)}, \tag{7.42b}$$

where the parameters satisfy the constraints $abcd = 1$ and $st = qpr$. Next we try to obtain autonomous reductions of this mapping but, instead of the trivial choice $q = 1$ we take $q = -1$. In order for the mapping to be indeed autonomous we must set $p + r = 0$, $s + t = 0$. Then the constraint becomes $s^2 = -p^2$, or $s = ip$. Next we introduce the scalings $x \to x\sqrt{i}$, $y \to y/\sqrt{i}$, $p \to p/\sqrt{i}$, $a \to a\sqrt{i}$, $b \to b\sqrt{i}$, $c \to c\sqrt{i}$, $d \to d\sqrt{i}$ so that we now have $abcd = -1$ intead of 1. We thus find the mapping

$$(x_n y_n - 1)(x_n y_{n-1} - 1) = \frac{(x_n - a)(x_n - b)(x_n - c)(x_n - d)}{p^2 x_n^2 - 1} \tag{7.43a}$$

$$(x_{n+1} y_n - 1)(x_n y_n - 1) = \frac{(y_n - 1/a)(y_n - 1/b)(y_n - 1/c)(y_n - 1/d)}{p^2 y_n^2 - 1}.$$
$$\tag{7.43b}$$

This autonomous mapping is in fact an asymmetric extension of (7.40). It turns out that it also has a biquartic invariant,

$$K = $$
$$\frac{(x(x - s_1) + y(y - s_{-1}) - (p(xy - 1))^2)^2 - 4(x(x - s_1) + s_2)(y(y - s_{-1}) - s_2)}{(xy - 1)^2}.$$
$$\tag{7.44}$$

where $s_1 = a + b + c + d$, $s_{-1} = 1/a + 1/b + 1/c + 1/d$, $s_2 = (ab + ac + ad + bc + bd + cd)/2$. By invariance, we mean that K has the same value whether $\{x, y\}$ stands for $\{x_n, y_n\}$ or for $\{x_n, y_{n-1}\}$.

We can obtain the symmetric reduction of (7.43) to precisely (7.40). We identify $y_{n-1} = X_{2n-1}$, $x_n = X_{2n}$, $y_n = X_{2n+1}$, etc. and demand that (7.43b) be just the upshift of (7.43a). The root $1/a$ in the right-hand side of (7.43b) should coincide with one of the roots in the right-hand side of (7.43a). So unless $a = \pm 1$, without loss of generality one may assume that $1/a = b$ and thus $d = -1/c$. Then, the root $1/c$ in the right-hand side of (7.43b) can only coincide with the root c in the right-hand side of (7.43a), so $c = -d = \pm 1$. Had we taken $a = \pm 1$ we would have found the same result, up to a renaming of the parameters.

From this construction we see clearly that (7.40) as well as its asymmetric form (7.43) are just special, artificially autonomised, cases of (nonautonomous) discrete Painlevé equations. Once this construction has been obtained for q-P_V, it is quite easy to extend it to other families of discrete Painlevé equations and try to obtain autonomous mappings with quartic invariants.

As an illustration we start from the q-P_V which was introduced in [77],

$$y_n y_{n-1} = \frac{(x_n - aq^n)(x_n - bq^n)}{1 - px_n} \tag{7.45a}$$

$$x_{n+1} x_n = \frac{(y_n - cq^n)(y_n - dq^n)}{1 - ry_n} \tag{7.45b}$$

with the constraint $cd = qab$. We try again to obtain an autonomous reduction with $q = -1$. This imposes $a + b = c + d = 0$, and $c^2 = -a^2$. One could leave the mapping in this form, but for reasons that will appear shortly, we introduce a change of variables, $x \to x/\sqrt{pr}$, $y \to iy/\sqrt{pr}$, $r \to r/i$. Then the mapping becomes

$$y_n y_{n-1} = \frac{x_n^2 - t^2}{s x_n - 1} \tag{7.46a}$$

$$x_{n+1} x_n = \frac{y_n^2 - t^2}{y_n/s - 1} \tag{7.46b}$$

where $s = \sqrt{p/r}$ and $t^2 = pra^2$. The autonomous mapping resulting from this construction is indeed integrable since it is just a subcase of an integrable, nonautonomous discrete Painlevé equation. Moreover, it possesses a biquartic invariant, and again this mapping is *not* of QRT type,

$$K \equiv$$
$$\frac{x^2 y^2 (sy - x/s)^2 + 2xy(x^2 - y^2)(sy - x/s) + 2t^2 xy(sy + x/s) + (x^2 + y^2 - t^2)^2}{x^2 y^2},$$
$$\tag{7.47}$$

where again K is invariant whether $\{x, y\}$ stands for $\{x_n, y_n\}$ or for $\{x_n, y_{n-1}\}$.

With this choice of variables, we again identify $y_{n-1} = X_{2n-1}$, $x_n = X_{2n}$, $y_n = X_{2n+1}$ - but, for simplicity, we denote the new variable by x rather than X - leads, in the special case of $s = 1/s$, to a symmetric, one-component, form of this mapping,

$$x_{n+1} x_{n-1} = \frac{x_n^2 - t^2}{x_n - 1}. \tag{7.48}$$

The invariant for (7.48) can be simply obtained from (7.47),

$$K =$$
$$\frac{x_n^2 x_{n-1}^2 (x_n - x_{n-1})^2 - 2x_n x_{n-1}(x_n + x_{n-1})((x_n - x_{n-1})^2 - t^2) + (x_n^2 + x_{n-1}^2 - t^2)^2}{x_n^2 x_{n-1}^2}.$$
$$\tag{7.49}$$

The integration of mappings like (7.40) was recently presented in [30]. Quite expectedly the solution is still given in terms of elliptic functions.

Discrete $\mathbb{P}$'s for Noncommuting Variables

Integrable equations involving noncommuting variables can be of particular interest since they can have an applicability in quantum field theories. Discrete systems present an additional difficulty. Indeed, most commutation rules one can introduce for the quantification of a discrete system are incompatible with the evolution induced by the mapping. By this we mean that if we assume that the mapping variables obey some commutation rules at some iteration, there is no guarantee whatsoever that, at the next step, the mapping variables will obey the *same* commutation rule. In particular, the Heisenberg commutation relation $[x, y] = 1$ is not necessarily the proper commutation rule for all the mappings we shall examine here, although it *does* work for a certain class). On the other hand there is no deep physical reason, such as the Hamiltonian structure in the case of the Heisenberg rule, why one should choose one rule over another. "Exotic" commutation rules have been introduced in [78,79] and used for the quantization of mappings.

Below we shall summarise our findings on d-ℙ's. As we have already explained, the integrability properties of the d-ℙ's are reflected in the existence of Lax pairs,

$$h\frac{d\Phi_n}{dh} = L_n\Phi_n$$

$$\Phi_{n+1} = M_n\Phi_n, \tag{7.50}$$

the d-ℙ's being obtained from the compatibility

$$h\frac{dM_n}{dh} = L_{n+1}M_n - M_nL_n. \tag{7.51}$$

In the quantized case, the ordering is important and must be respected throughout. Let us illustrate this point in the case of d-P_I. (In what follows we will use the notation, $\bar{x} \equiv x_{n+1}$, $x \equiv x_n$ and $\underline{x} \equiv x_{n-1}$).

In [46] we presented a 3×3 matrix realisation of the Lax pairs,

$$L = \begin{pmatrix} \lambda_1 & x & 1 \\ h & \lambda_2 & c - x - \underline{x} \\ h\underline{x} & h & \lambda_3 \end{pmatrix} \qquad M = \begin{pmatrix} (\lambda_1 - \lambda_2)x^{-1} & 1 & 0 \\ 0 & 0 & 1 \\ h & 0 & 0 \end{pmatrix}, \tag{7.52}$$

where $\lambda_1 = const.$, $\lambda_2 = \lambda + \frac{n}{2}$, $\lambda_2 = \lambda + \frac{n+1}{2}$. Using (7.51), one finds that

$$\bar{x} + x + \underline{x} = c + (\lambda_1 - \lambda_2)x^{-1}, \tag{7.53}$$

i.e., the usual form of d-P_I without any quantum corrections. Moreover the expressions for L and M are the straightforward transcriptions of the classical ones. No ordering ambiguity appears. That this need not be always the case can be seen in the 2×2 Lax pair of Fokas *et al.* [80] for the same equation d-P_I. Starting from

$$M = \begin{pmatrix} 2\mu\bar{x}^{-\frac{1}{2}} & -\bar{x}^{-\frac{1}{2}}x^{\frac{1}{2}} \\ 1 & 0 \end{pmatrix} \qquad L = \begin{pmatrix} -\mu(2x - c) & (x + \bar{x} - c)x^{\frac{1}{2}} \\ -(x + \underline{x} - c)x^{\frac{1}{2}} & \mu(2x - c) \end{pmatrix} \tag{7.54}$$

and the compatibility condition, $\frac{dM_n}{d\mu} = L_{n+1}M_n - M_nL_n$ - notice the different definition of the spectral parameter! -, one has to make in (7.54) specific choices for the order of the x, $\underline{x}$ and $\bar{x}$ terms. The order which was actually used in (7.54) is in fact the one leading to the correct d-P_I. Indeed one finds, using the compatibility relation (7.51),

$$X - \bar{X} + 1 = \bar{x}^{\frac{1}{2}}x\bar{x}^{\frac{1}{2}} - x^{\frac{1}{2}}\bar{x}x^{\frac{1}{2}}, \tag{7.55}$$

where $X = -4\mu^2 x + x^{\frac{1}{2}}(\overline{x} + x + \underline{x} - c)x^{\frac{1}{2}}$. Now, it can be shown that if $[x, \overline{x}] = 1$, then the right-hand side of (7.55) vanishes, whereupon the latter is integrated to $X = n + \kappa$. Multiplying by $x^{-\frac{1}{2}}$ from right and left we obtain (7.53).

While in the last case the choice of ordering was important, for the remaining cases of known Lax pairs for d-$\mathbb{P}$'s the situation is quite simple: No ordering ambiguities exist and one obtains the quantum analogue of the corresponding discrete Painlevé equation in a straightforward way. Thus for the second d-$\mathrm{P_I}$ we have

$$L = \begin{pmatrix} hx + \lambda_1 & h + y \\ h^2 + h(c_2 - y + c_1 x - x^2) & h(c_1 - x) + \lambda_2 \end{pmatrix} \quad M = \begin{pmatrix} (\lambda_1 - \lambda_2)y^{-1} & 1 \\ h & 0 \end{pmatrix},$$

(7.56)

where $\lambda_1 = const.$ and $\lambda_2 = n + c$, leading to

$$x + \overline{x} = c + (\lambda_1 - \lambda_2)y^{-1} \tag{7.57a}$$

$$y + \underline{y} = c_2 + x c_1 - x^2. \tag{7.574b}$$

Similarly, for d-$\mathrm{P_{II}}$ one finds that

$$L = \begin{pmatrix} \lambda_1 & x & 1 & 0 \\ 0 & \lambda_2 & c - \underline{x} & 1 \\ h & 0 & \lambda_3 & c - x \\ h\underline{x} & h & 0 & \lambda_4 \end{pmatrix}$$

and

$$M = \begin{pmatrix} (\lambda_1 - \lambda_2)x^{-1} & 1 & 0 & 0 \\ 0 & 0 & 1 & 0 \\ 0 & 0 & (\lambda_3 - \lambda_4)(c - x)^{-1} & 1 \\ h & 0 & 0 & 0 \end{pmatrix}, \tag{7.58}$$

where $\lambda_1 = const.$, $\lambda_3 = const.$, $\lambda_2 = \frac{n-1}{2} + \lambda$ and $\lambda_4 = \frac{n}{2} + \lambda$ leading to

$$\overline{x} + \underline{x} = c + (\lambda_1 - \lambda_2)x^{-1} + (\lambda_3 - \lambda_4)(c - x)^{-1}. \tag{7.59}$$

Finally for d-$\mathrm{P_{III}}$ we recall that the isospectral problem is of q-difference type rather than a differential one

$$\Phi_n(\rho h) = L_n(h)\Phi_n(h), \quad \Phi_{n+1}(h) = M_n(h)\Phi_n(h). \tag{7.60}$$

The compatibility condition, in the case of (7.60), is

$$M_n(\rho h)L_n(h) = L_{n+1}(h)M_n(h). \tag{7.61}$$

Here, using $x\overline{x} = q\overline{x}x$,

$$L = \begin{pmatrix} \lambda_1 & \lambda_1 + \kappa x^{-1} & \kappa x^{-1} & 0 \\ 0 & \lambda_2 & \lambda_2 + \underline{x} & x \\ hx & 0 & \lambda_3 & \lambda_3 + x \\ h(\lambda_4 + \alpha\kappa x^{-1}) & h\alpha\kappa x^{-1} & 0 & \lambda_4 \end{pmatrix}$$

and

$$M = \begin{pmatrix} \Lambda_1(\lambda_4 + \alpha\kappa x^{-1})^{-1}\,(\lambda_1 x + \kappa)(\lambda_2 x + \kappa)^{-1} & 0 & 0 & 0 \\ 0 & 0 & 1 & 0 \\ 0 & 0 & \Lambda_2(x + \rho\lambda_2)^{-1}\,(x + \lambda_3)(x + \lambda_4)^{-1} & 0 \\ h & 0 & 0 & 0 \end{pmatrix},$$
(7.62)

with $\alpha^2 = \rho$, $\lambda_1 = const.$, $\lambda_3 = const.$, $\lambda_2 = \lambda\alpha^{n-1}$, $\lambda_4 = \lambda\alpha^n$, $\Lambda_1 = \alpha\lambda_1 - \lambda_4$, $\Lambda_2 = \lambda_3 - \rho\lambda_2$, $\kappa = C\alpha^n$.

The quantized mapping is obtained by the application of (7.61),

$$\overline{x}(\kappa + \lambda_1 x)(\kappa + \lambda_2 x)^{-1}\underline{x} = \alpha\kappa(x + \lambda_3)(x + \lambda_4)^{-1}.$$
(7.63)

We can remark at this point that (7.63) does not depend explicitly on the quantum parameter q. This should be related to specific way (7.63) was written. If we compare (7.63) to (3.2), where f_2 is set to zero, - which is the appropriate choice for d-P$_{\mathrm{III}}$ –, we see that the ordering of the variables is different. Had we written (7.63) exactly as (3.2) or, equivalently, in the form $\overline{x}\underline{x} = W(x)$, explicit q-dependence would have appeared.

Since many of the difference $\mathbb{P}$'s can be obtained as contiguity of continuous $\mathbb{P}$'s, and thus possess a Lax pair almost "by construction", it would be interesting to investigate the possibility for a systematic quantization starting from the Lax pairs of the continuous $\mathbb{P}$'s and their Schlesinger transformations.

Delay-Differential Extensions of $\mathbb{P}$'s

Another interesting extension of $\mathbb{P}$'s is to the semi-discrete domain of delay-differential equations. We have examined in [40] a particular class of discrete/continuous systems, differential-delay equations, where the dependent function appears at a given time t and also at previous times $t - \tau, t - 2\tau, \ldots$, where τ is the delay. Our approach treats hystero-differential equations as differential-difference systems. The $u(t + k\tau)$ for various k's are treated as different functions of the continuous variable, indexed by k, $u(t+k\tau) = u_k(t)$. A detailed analysis of a particular class of such delay systems, *i.e.*, equations of the form

$$F(u_k, u_{k-1}, u'_k, u'_{k-1}) = 0$$
(7.64)

which are bi-Riccati, has led to the discovery of a new class of transcendents, the delay-Painlevé equations (D-$\mathbb{P}$'s). We have, for example, with $u = u(t)$ and $\overline{u} = u(t + \tau)$, a particular form of D-P$_{\mathrm{I}}$,

$$u' + \overline{u}' = (u - \overline{u})^2 + k(u + \overline{u}) + \lambda t.$$
(7.65)

Our results indicate that the D-$\mathbb{P}$'s are objects that may go beyond the Painlevé transcendents.

First-order, three-point D-$\mathbb{P}$'s have also been identified although our investigation in this case is still in an initial phase. An example of D-P$_{\mathrm{I}}$ may be written

$$\frac{u'}{u} = \frac{\overline{u}}{\underline{u}} + \lambda t \qquad (7.66)$$

The general form that contains all integrable cases is the mapping,

$$\overline{u} = \frac{f_1(u, u') - f_2(u, u')\underline{u}}{f_4(u, u') - f_3(u, u')\underline{u}}, \qquad (7.67)$$

where the f_i are 'Riccati-like' objects, $f_i = \alpha_i u' + \beta_i u^2 + \gamma_i u + \delta_i$, with $\alpha, \beta, \gamma, \delta$ functions of t. Unfortunately the domain has not attracted much attention since the first exploratory investigations.

Ultra-discrete $\mathbb{P}$'s
The last extension of $\mathbb{P}$'s we shall present here is to the ultra-discrete domain. This name is used to designate systems where the *dependent* variables as well as the independent ones, assume only discrete values. In this respect ultra-discrete systems are generalised cellular automata. The name of ultra-discrete is reserved for systems obtained from discrete ones through a specific limiting procedure introduced in [81] by the Tokyo-Kyoto group.

Before introducing the ultra-discrete limit let us first consider the question of nonlinearity. How simple can a nonlinear system be and still be *genuinely* nonlinear. The nonlinearities to which we are we are accustomed, involving powers, are not necessarily the simplest. It turns out (admittedly with hindsight) that the simplest nonlinear function of x one can think of is $|x|$. It is indeed linear for *both* $x > 0$ and $x < 0$ and the nonlinearity comes only from the different determinations. Thus one would expect the equations involving nonlinearities only in terms of absolute values to be the simplest. The ultra-discrete limit does just that, *i.e.*, it converts a given (discrete) nonlinear equation to one where only absolute-value nonlinearities appear. The key relation is the following limit,

$$\lim_{\epsilon \to 0^+} \epsilon \log(1 + e^{x/\epsilon}) = \max(0, x) = (x + |x|)/2 . \qquad (7.68)$$

Other equivalent expressions exist for this limi, and the notation that is often used is the truncated power function $(x)_+ \equiv \max(0, x)$. It is easy to show that $\lim_{\epsilon \to 0^+} \epsilon \log(e^{x/\epsilon} + e^{y/\epsilon}) = \max(x, y)$, and the extension to n terms in the argument of the logarithm is straightforward.

Two remarks are in order at this point. First, since the function $(x)_+$ takes only integer values when the argument is integer, the ultra-discrete equations can describe generalised cellular automata, provided one restricts the initial conditions to integer values. This approach has already been used in order to introduce cellular automata (and generalised cellular automata) related to many interesting evolution equations [82]. Second, the necessary condition for the procedure to be applicable is that the dependent variables be positive, since we are taking a logarithm and we require that the result assume values in $\mathbb{Z}$. This means that only some solutions of the discrete equations will survive in the ultra-discretisation.

As an illustration of the method and a natural introduction to ultra-discrete Painlevé equations, let us consider the following discrete Toda system,

$$u_n^{t+1} - 2u_n^t + u_n^{t-1} =$$
$$\log(1+\delta^2(e^{u_{n+1}^t}-1))-2\log(1+\delta^2(e^{u_n^t}-1))+\log(1+\delta^2(e^{u_{n-1}^t}-1)), \qquad (7.69)$$

which is the integrable discretisation of the continuous Toda system,

$$\frac{d^2 r_n}{dt^2} = e^{r_{n+1}} - 2e^{r_n} + e^{r_{n-1}}. \qquad (7.70)$$

For the ultra-discrete limit one introduces w by $\delta = e^{-L/2\epsilon}$, $w_n^t = \epsilon u_n^t - L$, and takes the limit $\epsilon \to 0$. Thus the ultra-discrete limit of (7.69) becomes simply

$$w_n^{t+1} - 2w_n^t + w_n^{t-1} = (w_{n+1}^t)_+ - 2(w_n^t)_+ + (w_{n-1}^t)_+ . \qquad (7.71)$$

Equation (7.71) is the cellular automaton analogue of the Toda system (7.69).

Let us now restrict ourselves to a simple periodic case with period 2, $i.e.$, $r_{n+2} = r_n$ and similarly $w_{n+2} = w_n$. Calling $r_0 = x$ and $r_1 = y$, we have, from (7.70), the equations $\ddot{x} = 2e^y - 2e^x$ and $\ddot{y} = 2e^x - 2e^y$ leading to $\ddot{x} + \ddot{y} = 0$. Thus $x + y = \mu t + \nu$, and we obtain, after some elementary manipulations,

$$\ddot{x} = ae^{\mu t}e^{-x} - 2e^x. \qquad (7.72)$$

Equation (7.72) is a special form of the Painlevé P_{III} equation. Indeed, setting $v = e^{x-\mu t/2}$, we find that

$$\ddot{v} = \frac{\dot{v}^2}{v} + e^{\mu t/2}(a - 2v^2). \qquad (7.73)$$

The same periodic reduction can be performed on the ultra-discrete Toda equation (4.13). We introduce $w_0^t = X^t$, $w_1^t = Y^t$ and have, in perfect analogy to the continuous case, $X^{t+1} - 2X^t + X^{t-1} = 2(Y^t)_+ - 2(X^t)_+$ and $Y^{t+1} - 2Y^t + Y^{t-1} = 2(X^t)_+ - 2(Y^t)_+$. Again, $\Delta_t^2(X^t + Y^t) = 0$ and we can take $X^t + Y^t = mt + p$, where m, t, p take integer values. We find thus that X obeys the ultra-discrete equation,

$$X^{t+1} - 2X^t + X^{t-1} = 2(mt + p - X^t)_+ - 2(X^t)_+ . \qquad (7.74)$$

This is the ultra-discrete analogue of the special form (7.73) of the Painlevé P_{III} equation [85].

In order to construct the ultra-discrete analogues of the other Painlevé equations we must start with the discrete form that allows the ultra-discrete limit to be taken. The general procedure is to start with an equation for x, introduce X by $x = e^{X/\epsilon}$, and then take the limit $\epsilon \to 0$. Clearly the substitution $x = e^{X/\epsilon}$ requires x to be positive. This is a stringent requirement that limits the exploitable form of the d-$\mathbb{P}$'s to multiplicative ones. Fortunately many such forms are known for the discrete Painlevé transcendents. For instance for d-$\mathrm{P_I}$ there are the multiplicative, q-forms:

d-$\mathrm{P_{I-1}}$

$$x_{n+1}x_{n-1} = \frac{\lambda^n}{x_n} + \frac{1}{x_n^2}$$

d-$\mathrm{P_{I-2}}$

$$x_{n+1}x_{n-1} = \lambda^n + \frac{1}{x_n}$$

d-$\mathrm{P_{I-3}}$

$$x_{n+1}x_{n-1} = \lambda^n x_n + 1$$

From them it is straightforward to obtain the canonical forms of the ultra-discrete $\mathrm{P_I}$:

u-$\mathrm{P_{I-1}}$
$$X_{n+1} + X_{n-1} + 2X_n = (X_n + n)_+$$

u-$\mathrm{P_{I-2}}$
$$X_{n+1} + X_{n-1} + X_n = (X_n + n)_+$$

u-$\mathrm{P_{I-3}}$
$$X_{n+1} + X_{n-1} = (X_n + n)_+ \; .$$

Ultra-discrete forms have been derived for all Painlevé equations [83]. Moreover we have shown that their properties are in perfect parallel to those of their discrete and continuous analogues (degeneration through coalescence, existence of special solutions, auto-Bäcklund and Schlesinger transformations).

$E\pi i\lambda o\gamma o\varsigma$

8 Epilogue

Faithful readers of our works must have already noticed that while we love introductions we hate conclusions. Following our general practice we shall keep this one as short as possible.

Our plan was to convince our readers that a) discrete systems are the most fundamental ones and b) that discrete $\mathbb{P}$'s are fascinating objects with all kinds of interesting properties. A thorough monitoring of the activity in the domain will test whether or not we have succeeded.

Acknowledgments

Writing a review on a topic like the one treated here is an enterprise which requires a massive dose of *chutzpah*. It was from the outset clear that the end-product could be a) incomplete, b) non rigourous (from the mathematicians' point of view) and c) terribly biased (because of the authors stubbornness). And, *quelle horreur*, some incorrect statements might even creep in (along with the inevitable misprints). Still, because we had the effrontery to try, we went ahead and wrote it. It might even turn out to be useful to some people who may wish to learn more about discrete Painlevé equations.

Our work on d-$\mathbb{P}$'s, and discrete integrable systems in general, was an occasion to meet many people, work with them and, in most cases to establish solid ties of friendship. We do not provide the full list of names here lest these acknowledgments resemble the title page of an experimental high-energy physics article. Our collaborators have tried to transmit to us some of their knowledge and increase our scientific culture. We are grateful to them for their efforts (although we are still not always convinced that they have borne fruits). In any case we assume full responsability for the (absence of) quality of the present review. And since we are convinced that most people read the acknowledgments before the main body of the article (we do!) these words may act also as a warning – whether this *caveat* will spur the interest of the potential reader or, on the contrary, discourage him/her is a risk we are willing to take. In any case, (hand)writing this review was really most enjoyable for the first author, and typing it was a real chore for the second. Well, $\alpha v \tau \acute{\alpha}\ \acute{\epsilon} \chi \epsilon \iota\ \eta\ \zeta \omega \acute{\eta}$.

References

1. M. D. Kruskal, private communication, *circa* 1985.
2. *The Painlevé Property: One Century Later*, edited by R. Conte, CRM Series in Mathematical Physics, Springer-Verlag, New-York (1999).
3. B. Grammaticos and A. Ramani, Integrability – and How to Detect It, Lect. Notes Phys. 638, 31 (2004).
4. R. Mills, *Space, Time and Quanta: an introduction to contemporary physics* W H Freeman & Co, (1994).
5. A. Einstein, *Physics and Reality* in *Essay in Physics*, Philosophical Library, New York, 1950.
   R. P. Feynman, Intl. J. Theor. Phys. 21 (1982) 467.
   E. Witten, *Reflections on the fate of spacetime*, Physics Today, April 1996, p 24.
6. B. Grammaticos and B. Dorizzi, J. Math. Comp. in Sim. 37 (1994) 341.
7. E. Laguerre, J. Math. Pures Appl. 1 (1885) 135.
8. P. Painlevé, Acta Math. 25 (1902) 1.
9. J.A. Shohat, Duke Math. J. 5 (1939) 401.
10. E. Brézin and V.A. Kazakov, Phys. Lett. 236B (1990) 144.

11. M. Jimbo and T. Miwa, Physica 2D (1981) 407.
12. V. Periwal and D. Shevitz, Phys. Rev. Lett. 64 (1990) 1326.
13. F. W. Nijhoff and V. Papageorgiou, Phys. Lett. A 153 (1991) 337.
14. V. Papageorgiou, F. W. Nijhoff and H. Capel, Phys. Lett. A 147 (1990) 106.
15. R. Hirota, J. Phys. Soc. Japan 50 (1981) 3785.
16. J. Satsuma, A. Ramani and B. Grammaticos, Phys. Lett. A 174 (1993) 387.
17. G. R. W. Quispel, J.A.G. Roberts and C.J. Thompson, Physica D34 (1989) 183.
18. A. P. Veselov, Comm. Math. Phys. 145 (1992) 181.
19. B. Grammaticos, F. W. Nijhoff and A. Ramani, Discrete Painlevé equations, pp 413-516 in reference [2].
20. R. J. Baxter, Exactly Solved Models in Statistical Mechanics, Associated Press, London (1982), p 471.
21. A. Ramani, S. Carstea, B. Grammaticos and Y. Ohta, Physica A 305 (2002) 437.
22. A. Iatrou and J.A.G. Roberts, J. Phys. A 34 (2001) 6617.
23. B. Grammaticos, A. Ramani and V. Papageorgiou, Phys. Rev. Lett. 67 (1991) 1825.
24. A. Ramani, B. Grammaticos and J. Hietarinta, Phys. Rev. Lett. 67 (1991) 1829.
25. M. J. Ablowitz, A. Ramani and H. Segur, Lett. Nuovo Cim. 23 (1978) 333.
26. A. Ramani, B. Grammaticos and S. Tremblay, J. Phys. A 33 (2000) 3045.
27. A. Ramani, B. Grammaticos and T. Bountis, Phys. Rep. 180 (1989) 159.
28. B. Grammaticos, A. Ramani and V. Papageorgiou, CRM Lecture Notes 9 (1996) 303.
29. B. Grammaticos, A. Ramani and K. M. Tamizhmani, Jour. Phys. A 27 (1994) 559.
30. C. M. Viallet, A. Ramani and B. Grammaticos, Phys. Lett. A 322 (2004) 186.
31. J. Satsuma, private communication (1992).
32. R. Conte and M. Musette, Phys. Lett. A 223 (1996) 439.
33. V. I. Arnold, Bol. Soc. Bras. Mat. 21 (1990) 1.
34. A. P. Veselov, Comm. Math. Phys. 145 (1992) 181.
35. G. Falqui and C.-M. Viallet, Comm. Math. Phys. 154 (1993) 111.
36. J. Hietarinta and C. Viallet, Phys. Rev. Lett. 81 (1998) 325.
37. T. Takenawa, J. Phys. A 34 (2001) L95.
38. A. Ramani, B. Grammaticos, S. Lafortune and Y. Ohta, J. Phys. A 33 (2000) L287.
39. N. Yanagihara, Arch. Ration. Mech. Anal. 91 (1985) 169.
40. B. Grammaticos, A. Ramani and I. Moreira, Physica A 196 (1993) 574.
41. M. J. Ablowitz, R. Halburd and B. Herbst, Nonlinearity 13 (2000) 889.
42. E. Hille, Ordinary Differential Equations in the Complex Domain, J. Wiley and Sons, New York (1976).
43. G. Valiron, Bull. Soc. Math. France 59 (1931) 17.
44. B. Grammaticos, T. Tamizhmani, A. Ramani and K. M. Tamizhmani, J. Phys A 34 (2001) 3811.
45. Y. Ohta, K. M. Tamizhmani, B. Grammaticos and A. Ramani, Phys. Lett. A 262 (1999) 152.
46. V. G. Papageorgiou, F. W. Nijhoff, B. Grammaticos and A. Ramani, Phys. Lett. A164 (1992) 57.
47. A. Ramani and B. Grammaticos, Physica A 228 (1996) 160.

48. B. Grammaticos, T. Tamizhmani, A. Ramani, A. S. Carstea and K. M. Tamizhmani, J. Phys. Soc. Japan 71 (2002) 443.
49. B. Grammaticos and A. Ramani, Phys. Lett. A 257 (1999) 288.
50. A. Ramani, B. Grammaticos and Y. Ohta, *The Painlevé of discrete equations and other stories*, invited talk at the "Atelier sur la théorie des fonctions spéciales non linéaires: les transcendants de Painlevé", Montréal 1996.
51. M. Jimbo and H. Sakai, Lett. Math. Phys. 38 (1996) 145.
52. A. Fokas, B. Grammaticos and A. Ramani, J. Math. Anal. Appl. 180 (1993) 342.
53. A. Ramani, B. Grammaticos and S. Lafortune, Lett. Math. Phys. 46 (1998) 131.
54. K. M. Tamizhmani, T. Tamizhmani, B. Grammaticos, A. Ramani, Special Solutions for Discrete Painlevé Equations, Lect. Notes Phys. 644, 321 (2004).
55. B. Grammaticos, Y. Ohta, A. Ramani, J. Satsuma and K. M. Tamizhmani, Lett. Math. Phys. 39 (1997) 179.
56. C. Cosgrove and G. Scoufis, Stud. Appl. Math. 88 (1993) 25.
57. M. Jimbo, H. Sakai, A. Ramani and B. Grammaticos, Phys. Lett. A 217 (1996) 111.
58. K. Okamoto, Physica D 2 (1981) 525.
59. A. Ramani, B. Grammaticos and J. Satsuma, Jour. Phys. A 28 (1995) 4655.
60. Y. Ohta, A. Ramani, B. Grammaticos and K. M. Tamizhmani, Phys. Lett. A 216 (1996) 255.
61. J. Hietarinta and M. D. Kruskal, *Hirota forms for the six Painlevé equations from singularity analysis*, NATO ASI series B278, Plenum 1992, p 175.
62. K. Okamoto, *The Painlevé equations and Dynkin diagrams*, NATO ASI series B278, Plenum 1992, p 299.
63. F. W. Nijhoff, J. Satsuma, K. Kajiwara, B. Grammaticos and A. Ramani, Inverse Probl. 12 (1996) 697.
64. T. Miwa, Proc. Japan Acad. 58 (1982) 9.
65. T. Tokihiro, B. Grammaticos and A. Ramani, J. Phys. A 35 (2002) 5943.
66. A. Ramani and B. Grammaticos, *The Grand Scheme for discrete Painlevé equations*, Lecture at the Toda symposium (1996).
    A. Ramani, Y. Ohta, J. Satsuma and B. Grammaticos, Comm. Math. Phys. 192 (1998) 67.
    A. Ramani, B. Grammaticos and Y. Ohta, Comm. Math. Phys. 217 (2001) 315.
    A. Ramani, B. Grammaticos and Y. Ohta, J. Phys. A 34 (2001) 2505.
    Y. Ohta, A. Ramani and B. Grammaticos, J. Phys. A 34 (2001) 10523.
67. H. Sakai, Commun. Math. Phys. 220 (2001) 165.
68. K. Okamoto, Japan J. Math. 5 (1979) 1.
69. Y. Ohta, A. Ramani and B. Grammaticos, J. Phys. A 35 (2002) L653.
70. M. D. Kruskal, private communication, circa 1997.
71. T. Takenawa, J. Phys. A 34 (2001) 10533.
72. B. Grammaticos, Y. Ohta, A. Ramani and H. Sakai, J. Phys. A 31 (1998) 3545.
73. F. W. Nijhoff, A. Ramani, B. Grammaticos and Y. Ohta, Stud. Appl. Math. 106 (2000) 261.
74. A. Ramani, B. Grammaticos, T. Tamizhmani and K. M. Tamizhmani, J. Phys. A 33 (2000) 579.
75. A. Ramani, B. Grammaticos, Y. Ohta and B. Grammaticos, Nonlinearity 13 (2000) 1073.

76. K. Kimura, H. Yahagi, R. Hirota, A, Ramani, B. Grammaticos and Y. Ohta, J. Phys. A 35 (2002) 9205.

77. M. D. Kruskal, K. M. Tamizhmani, B. Grammaticos and A. Ramani, Reg. Chaot. Dyn. 5 (2000) 273.

78. B. Grammaticos, A. Ramani, V. Papageorgiou and F. W. Nijhoff, Jour. Phys. A 25 (1992) 6419

79. A. Ramani, K. M. Tamizhmani, B. Grammaticos and T. Tamizhmani, *The extension of integrable mappings to non-commuting variables*, to appear in J. Nonlinear Math. Phys.

80. A. R. Its, A. V. Kitaev and A. Fokas, Russ. Math. Surv. 45 (1990) 155.

81. T. Tokihiro, D. Takahashi, J. Matsukidaira and J. Satsuma, Phys. Rev. Lett. 76 (1996) 3247.

82. B. Grammaticos, Y. Ohta, A. Ramani, D. Takahashi and K. M. Tamizhmani, Phys. Lett. A 226 (1997) 53.

83. B. Grammaticos, Y. Ohta, A. Ramani and D. Takahashi, Physica D 114 (1998) 185.

Special Solutions
for Discrete Painlevé Equations

K. M. Tamizhmani[1], T. Tamizhmani[2], B. Grammaticos[3], and A. Ramani[4]

[1] Departement of Mathematics, Pondicherry University, Kalapet, Pondicherry 605014, India, tamizh@yahoo.com
[2] Department of Mathematics, Kanchi Mamunivar Centre for Postgraduate Studies, Pondicherry 605008, India, arasi55@yahoo.com
[3] GMPIB, Université Paris VII, Tour 24-14, 5^{e}étage, case 7021, 75251 Paris, France, grammati@paris7.jussieu.fr
[4] CPT, Ecole Polytechnique, CNRS, UMR 7644, 91128 Palaiseau, France, ramani@cpht.polytechnique.fr

Abstract. We construct special solutions for the discrete Painlevé equations. We start with a review of the corresponding solutions in the case of the continuous Painlevé equations and then proceed to construct the solutions in the discrete case. We show how, starting from an elementary, seed solution, one can use the auto-Bäcklund transformations in order to build iteratively 'higher' solutions. Using the bilinear formalism we show that the τ-functions for these 'higher' solutions can be cast into the form of Casorati determinants.

Prologue

Solutions are what differential equations are all about. When one formulates a problem in terms of a differential system what one is interested in are the solutions of the system, their properties and, whenever possible, their exact dependence on the independent variable, the parameters of the equation and the initial conditions. Painlevé equations occupy a very special position among differential equations [1]. These second-order equations were, indeed, proposed as the first example of equations which introduce new functions. The name Painlevé transcendents was coined especially in order to denote the solutions of the Painlevé equations, which in general cannot be expressed in terms of elementary functions. Although the existence of functions defined by the solutions of the Painlevé equations was guaranteed by the Painlevé property [2] (absence of initial-condition-dependent, multivaluedness-inducing singularities), it was not clear how to obtain them. The construction of the solutions of the Painlevé equations had to wait three-quarters of a century. It was finally obtained in [3] following the development of spectral methods like the Inverse Scattering Transform for the study of the solution of integrable evolution equations.

Unfortunately the solution of the Painlevé equations by means of linear integro-differential equations is not quite explicit and, although one can, in

K.M. Tamizhmani, T. Tamizhmani, B. Grammaticos, and A. Ramani, Special Solutions for Discrete Painlevé Equations, Lect. Notes Phys. **644**, 323–382 (2004)
http://www.springerlink.com/ © Springer-Verlag Berlin Heidelberg 2004

principle deduce all the properties of the transcendents, one cannot easily form a qualitative idea of their behaviour. Fortunately, while the *general* solution of a Painlevé equation cannot be expressed in terms of elementary functions, there exist simple solutions which can be obtained through the solution of Riccati (or even linear) first-order equations. These solutions are not only one-degree-of-freedom ones but, moreover, they exist provided the parameters of the Painlevé equation satisfy some constraint. The method for the derivation of such solutions will be reviewed in what follows. Obtaining a solution of a Painlevé equation in terms of an elementary function allows's one to study it easily, tabulate its values and graph its behaviour. This often suffices for a rough, qualitative idea of how the solution behaves.

Another, quite different, approach is that of numerical simulations. Starting form a given Painlevé equation, one introduces a difference scheme which approximates the equation and uses it to construct iteratively the solution starting from given initial data. This is where discrete Painlevé equations enter the scene [4], they constitute *integrable* difference schemes for the continuous Painlevé equations (although, as explained in [5], they were not introduced in this way). Discrete Painlevé equations are perfect simulators for continuous Painlevé equations since they preserve their essential character, integrability. Moreover, the property of singularity confinement [6] provides the natural answer to one of the major problems of numerical analysis, the treatment of divergences.

However the study of discrete Painlevé equations has made clear that these objects are not only interesting for their own sake but are, in fact, more fundamental than their continuous counterparts. They possess various regimes [7] in which they simulate not only the continuous Painlevé equations but other systems as well. This has led to a series of works which has tremendously increased our understanding of these particularly rich systems. Quite expectedly, the properties of the continuous Painlevé equations find themselves reflected in the properties of their discrete analogues.

In this expository article we shall focus on the existence of elementary solutions for discrete Painlevé equations. As in the continuous case, these solutions are obtained through some linearisation procedure, reduction to Riccati or first-order linear equation. Moreover they exist only when the coefficients of the equation satisfy some constraint. Such solutions are all the more important since the solution of the inverse problem has not yet been performed for the vast majority of the discrete Painlevé equations.

1 What Is a Discrete Painlevé Equation?

By a discrete Painlevé equation (d-$\mathbb{P}$) we mean a nonautonomous, integrable mapping which becomes, at the continuous limit, some (continuous) Painlevé equation. Moreover we reserve the name of discrete Painlevé equation to second-order mappings, in parallel with what is customary in the continuous case. As explained in [8], numerous methods exist for the derivation of d-$\mathbb{P}$'s:

i) The methods related to some inverse problem. The discrete AKNS method, the methods of orthogonal polynomials, of discrete dressing, of non-isospectral deformations, etc. belong to this class.

ii) The methods based on some reduction. Similarity reduction of integrable lattices is the foremost among them, but this class contains the methods based on limits, coalescences and degeneracies of d-$\mathbb{P}$'s as well as stationary reductions of nonautonomous differential-difference equations.

iii) The contiguity relations approach. Discrete $\mathbb{P}$'s can be obtained from the auto-Bäcklund, Miura and Schlesinger transformations of both continuous and discrete Painlevé equations.

iv) The direct, constructive, approach. Two methods fall under this heading. One is the construction of discrete Painlevé equations from the geometry of some affine Weyl group. The other is the method of de-autonomisation using the singularity confinement approach.

In [9] we have focused on the constructive method which uses as starting point the QRT [10] mapping. The latter is known under two canonical forms called symmetric and asymmetric, and which involve one and two dependent functions respectively. We have thus, for the symmetric QRT mapping the form:

$$x_{n+1}x_{n-1}f_3(x_n) - (x_{n+1} + x_{n-1})f_2(x_n) + f_1(x_n) = 0, \qquad (1.1)$$

where the $f_i's$ are specific quartic polynomials involving 5 parameters. For the asymmetric form we have the system:

$$x_{n+1}x_n f_3(y_n) - (x_{n+1} + x_n)f_2(y_n) + f_1(y_n) = 0, \qquad (1.2a)$$

$$y_n y_{n-1} g_3(x_n) - (y_n + y_{n-1})g_2(x_n) + g_1(x_n) = 0, \qquad (1.2b)$$

where the f_i's, and g_i's are specific quartic polynomials involving 8 parameters. The *rationale* for the choice of the QRT mapping as starting point is that its solutions are given in terms of elliptic functions. Since the autonomous limits of continuous Painlevé equations are equations solvable in terms of elliptic functions [11], it is quite reasonable to start from the QRT mapping which has the latter property and de-autonomise it. The de-autonomisation procedure consists of assuming that the parameters of the mapping are functions of the dependent variable. We determine their precise form, compatible with integrability, by the application of an integrability criterion. Depending on the mapping from which we start, we may obtain a symmetric or asymmetric d-$\mathbb{P}$. In the latter case it turns out sometimes that the system can be written in terms of more than two dependent variables. In these cases, the equations are still of second-order since all but two equations are local, rational relations of the dependent variables. We shall not go here neither into the details of the derivation of the d-$\mathbb{P}$'s nor into their classification. These subjects are covered in the article of Grammaticos and Ramani [12]. It suffices to say that our current knowledge of d-$\mathbb{P}$'s is quite detailed, and that the various approaches have resulted in an identification of three kinds of discrete Painlevé equations.

The first kind of d-$\mathbb{P}$ is that of difference equations. The oldest example known is that obtained in [13] by Shohat in 1939. In his study of orthogonal polynomials he obtained the (integrable) recursion relation:

$$x_{n+1} + x_n + x_{n-1} = \frac{\alpha n + \beta + \gamma(-1)^n}{x_n} + 1 \qquad (1.3)$$

This equation was eventually shown [14] to be a discrete analogue of Painlevé I, when $\gamma = 0$. As a matter of fact, due to the presence of the $(-1)^n$ term, this equation has a more natural form where even- and odd-index terms are distinguished. Setting $X_m = x_{2m}$ and $Y_m = x_{2m+1}$, we find that,

$$X_{m+1} + X_m + Y_m = \frac{Z_m + C}{Y_m} + 1$$

$$Y_m + Y_{m-1} + X_m = \frac{Z_m - C}{X_m} + 1, \qquad (1.4)$$

where $Z_m = \alpha(2m+1/2)+\beta$ and $C = \alpha/2-\gamma$ is a genuine parameter. So, due to the even-odd degree of freedom the equation becomes a one-parameter d-$\mathbb{P}_{\mathrm{II}}$ equation when γ is non-zero [15]. This example shows that the distinction between "symmetric" and "asymmetric" forms is rather artificial. We will keep it only for tradition's sake and for lack of better terminology. Equation (1.3) is a difference equation: the independent variable enters the equation in an additive way, $\alpha n + \beta$.

The second kind of d-$\mathbb{P}$ is that of multiplicative, q-equations. They were discovered by two of the authors in collaboration with J. Hietarinta [4] in 1991. The first such equation obtained was the q-discrete analogue of $\mathbb{P}_{\mathrm{III}}$:

$$x_{n+1}x_{n-1} = \frac{\gamma_{e,o}x_n^2 + \zeta_{e,o}\lambda^n x_n + \mu\lambda^{2n}}{\alpha x_n^2 n + \beta_{e,o}x_n + \gamma_{o,e}}. \qquad (1.5)$$

Again we remark the presence of parity-dependent terms, $\gamma_{e,o}$, $\zeta_{e,o}$ and $\beta_{e,o}$, which suggests an asymmetric form for this equation and introduce three more degrees of freedom with respect to the symmetric case, $\gamma_e = \gamma_o, \zeta_e = \zeta_o$ and $\beta_e = \beta_o$. This asymmetric equation was shown by Jimbo and Sakai [16] to be a discrete analogue of $\mathbb{P}_{\mathrm{VI}}$. Equation (1.5) is a multiplicative equation since the independent variable enters through a term λ^n. This means that if x_n denotes the value of the dependent variable at a point z, i.e., $x_n = x(z)$, then $x_{n\pm1}$ are the values of x at points λz and z/λ, respectively.

The third kind of d-$\mathbb{P}$ was discovered by Sakai [17] in 1999 (in his PhD thesis). He derived the classification of d-$\mathbb{P}$'s in terms of affine Weyl groups and found that equations related to the group $E_8^{(1)}$ could be not only of difference and multiplicative kinds but also of a third kind where the independent variable and the parameters of the equation enter through the arguments of elliptic functions. These elliptic discrete Painlevé equations have very complicated expressions as one can judge from the example below derived recently by two of the authors in collaboration with Y. Ohta [18]:

$$\frac{x_{n-1} - \text{sn}^2(\lambda n + \lambda/2 + \phi(n-1) - (-1)^n\psi + \omega(n); m)}{x_{n-1} - \text{sn}^2(\lambda n + \lambda/2 + \phi(n-1) - (-1)^n\psi - \omega(n); m)}$$

$$\times \frac{x_n - \text{sn}^2(-2\lambda n - \phi(n-1) - \phi(n+1) + \omega(n); m)}{x_n - \text{sn}^2(-2\lambda n - \phi(n-1) - \phi(n+1) - \omega(n); m)}$$

$$\times \frac{x_{n+1} - \text{sn}^2(\lambda n - \lambda/2 + \phi(n-1) + (-1)^n\psi + \omega(n); m)}{x_{n+1} - \text{sn}^2(\lambda n - \lambda/2 + \phi(n-1) - (-1)^n\psi - \omega(n); m)} =$$

$$\frac{1 - m^2\text{sn}^2(\lambda n + \alpha(n) + \frac{\omega(n)}{2}; m)\text{sn}^2(-\lambda n + \beta(n) - \frac{\omega(n)}{2}; m)}{1 - m^2\text{sn}^2(\lambda n + \alpha(n) - \frac{\omega(n)}{2}; m)\text{sn}^2(-\lambda n + \beta(n) + \frac{\omega(n)}{2}; m)}$$

$$\times \frac{1 - m^2\text{sn}^2(\lambda n + \alpha(n) - \frac{\omega(n)}{2}; m)\text{sn}^2(\gamma(n) + \frac{\omega(n)}{2}; m)}{1 - m^2\text{sn}^2(\lambda n + \alpha(n) + \frac{\omega(n)}{2}; m)\text{sn}^2(\gamma(n) - \frac{\omega(n)}{2}; m)}$$

$$\times \frac{1 - m^2\text{sn}^2(-\lambda n + \beta(n) + \frac{\omega(n)}{2}; m)\text{sn}^2(\gamma(n) - \frac{\omega(n)}{2}; m)}{1 - m^2\text{sn}^2(-\lambda n + \beta(n) - \frac{\omega(n)}{2}; m)\text{sn}^2(\gamma(n) + \frac{\omega(n)}{2}; m)} \quad (1.6)$$

where we have introduced the auxiliary quantities $\alpha(n) = \lambda/6 + (2\phi(n-1) + \phi(n+1) - (-1)^n\psi)/3$, $\beta(n) = \lambda/6 - (2\phi(n+1) + \phi(n-1) + (-1)^n\psi)/3$, $\gamma(n) = (\phi(n+1) - \phi(n-1) + 2(-1)^n\psi - \lambda)/3$ and where ψ is a constant, $\phi(n+3) = \phi(n)$, i.e., ϕ has period three and $\omega(n+4) = \omega(n)$ i.e. ω has period four so the whole equation has period 12. The total number of degrees of freedom is 8, including the independent variable.

The domain of elliptic d-$\mathbb{P}$'s is essentially unexplored. The possible forms of these equations are far from known. In any case the solutions of elliptic d-$\mathbb{P}$'s will not be the object of the present course.

One thing that must be stated clearly is that the naming of d-$\mathbb{P}$'s is not without some arbitrariness. The convention that has been adopted was to name a d-$\mathbb{P}$ according to its continuous limit. Thus equation (1.3) for $\gamma = 0$ is referred to as d-$\mathbb{P}_{\text{I}}$. (Even this convention is not applied consistently since equation (1.4) is often referred to as "asymmetric d-$\mathbb{P}_{\text{I}}$"). The naming of d-$\mathbb{P}$'s after their continuous limit is inadequate. First, since there are only 6 continuous Painlevé equations and several dozen known d-$\mathbb{P}$'s, it is natural to have many discrete equations with the same limit. (In [19] we have compiled a, certainly non-exhaustive, list of 15 different d-$\mathbb{P}_{\text{I}}$'s). Moreover, since the richest continuous Painlevé equation is $\mathbb{P}_{\text{VI}}$, which contains only four parameters, it is natural that all d-$\mathbb{P}$'s with more than four parameters (they can have up to 7) possess $\mathbb{P}_{\text{VI}}$ as their continuous limit . One other difficulty derives from the fact that a given d-$\mathbb{P}$ may have more than one continuous limit. In [19] we have shown that the mapping

$$x_{n+1}x_{n-1} = \frac{\alpha\lambda^n}{x_n} + \frac{1}{x_n^2}, \quad (1.7)$$

which was known to be a q-discrete form of P_I, may also become, at a different continuous limit, the zero-parameter P_{III} equation. The situation can be made even worse if one takes into account the fact that some d-$\mathbb{P}$'s do not seem to possess any nontrivial continuous limit. In [5] we have given the example of such an equation:

$$x_{n+1} + x_n + x_{n-1} = \frac{\alpha n + \beta}{x_n}. \tag{1.8}$$

These arguments show that the names we shall attribute to the equations we shall examine in what follows must be considered *cum grano salis*. Still the relation to continuous Painlevé equations will be a useful guide throughout. (We are talking here about the relation in terms of the continuous limit. The other relation some d-$\mathbb{P}$'s bear to continuous Painlevé equations, namely that of the former being contiguity relations of the latter, is an exact one and the consequences drawn from this relationship are expected to be of general validity).

2 Finding Special-Function Solutions

2.1 The Continuous Painlevé Equations and Their Special Solutions

Before proceeding to the derivation of the solutions of discrete Painlevé equations, let us, as a reminder (and for pedagogical reasons as well), present the derivation of the solutions of the continuous Painlevé equations in terms of special functions, and the way they organise themselves in the degeneration through the coalescence cascade of the equations.

The general form of a continuous Painlevé equations is

$$w'' = f(w', w, z), \tag{2.1.1}$$

where f is polynomial in w', rational in w and analytic in z. In order to find a solution of (2.1.1) in terms of special functions we assume that w is a solution of a Riccati equation,

$$w' = Aw^2 + Bw + C, \tag{2.1.2}$$

where A, B and C are functions of z to be determined. Substituting (2.1.2) into (2.1.1) yields an over-determined system which allows the determination of A, B and C and fixes the parameters of (2.1.1). Equation (2.1.2) is subsequently linearised by the transformation

$$w = -\frac{u'}{Au}. \tag{2.1.3}$$

In the case of the Painlevé equations, the end result is an equation of the hypergeometric family. The special solutions of the continuous Painlevé equations have been studied in detail in the monograph of Gromak and Lukashevich [20]. However this is an almost inaccessible reference since the book is currently out of print and moreover it is in Russian. We prefer to retrace these results here so as to present a consistent choice of normalisations, which may be of help to people interested in the matter. Our aim here is not to just present the solutions of continuous Painlevé equations is terms of special functions, but also to insert them into the coalescence framework [21]. We start with the P_{VI} equation,

$$W'' = \frac{W'^2}{2}\left(\frac{1}{W} + \frac{1}{W-1} + \frac{1}{W-Z}\right) - W'\left(\frac{1}{Z} + \frac{1}{Z-1} + \frac{1}{W-Z}\right)$$
$$+ \frac{W(W-1)(W-Z)}{2Z^2(Z-1)^2}\left(A - \frac{BZ}{W^2} + C\frac{Z-1}{(W-1)^2} - \frac{(D-1)Z(Z-1)}{(W-Z)^2}\right). \quad (2.1.4)$$

When one requires the existence of a solution given by a Riccati equation (2.1.2), the result is

$$W' = \frac{P}{Z(Z-1)}W^2 + \frac{QZ+M}{Z(Z-1)}W + \frac{N}{Z-1} \quad (2.1.5)$$

where the parameters P, Q, M and N are related by

$$P + Q + M + N = 0, \quad (2.1.6)$$

and their relation to those of P_{VI} is

$$A = P^2, \quad B = N^2, \quad C = (Q+N)^2, \quad D = (P+Q-1)^2. \quad (2.1.7)$$

The condition for the existence of (2.1.5) is obtained when one eliminates N, P and Q from equation (2.1.7),

$$\epsilon_1\sqrt{A} + \epsilon_2\sqrt{B} + \epsilon_3\sqrt{C} + \epsilon_4\sqrt{D} = 1, \quad (2.1.8)$$

where the ϵ_i's are arbitrary signs. The linearisation of (2.1.5) is obtained in a straightforward way by

$$W = -\frac{Z(Z-1)U'}{PU}, \quad (2.1.9)$$

and the transformation $\zeta = (1-Z)^{-1}$ converts (2.1.5) to a hypergeometric equation [22],

$$\zeta(1-\zeta)\frac{d^2U}{d\zeta^2} + (Q - (1-N-P)\zeta)\frac{dU}{d\zeta} - NPU = 0. \quad (2.1.10)$$

Before proceeding to the first coalescence, we introduce the following convention of notation. The variables and parameters of the "higher" equation will be represented by upper-case letters, while those of the "lower" equation will be represented by low-case ones. The small parameter will be denoted by δ, and the coalescence corresponds to the limit $\delta \to 0$.

Going from P_{VI} to P_V we set $W = w$, $Z = 1 + \delta z$, $A = a$, $B = b$, $C = d/\delta^2 + c/\delta$, $D = d/\delta^2$. One obtains thus P_V,

$$w'' = w'^2\left(\frac{1}{2w} + \frac{1}{w-1}\right) - \frac{w'}{z} + \frac{(w-1)^2}{2z^2}\left(aw - \frac{b}{w}\right) + \frac{cw}{2z} - \frac{dw(w+1)}{2(w-1)}. \quad (2.1.11)$$

This coalescence limit is compatible with the linearisable case provided $N = n$, $P = p$, $Q = q/\delta$ (and $M = -q/\delta - p - n$). Using (2.1.6), the Riccati equation becomes

$$w' = \frac{pw^2}{z} + \frac{(qz - p - n)w}{z} + \frac{n}{z}. \quad (2.1.12)$$

The parameter constraints are transformed into

$$a = p^2, \quad b = n^2, \quad c = 2q(n + 1 - p) \quad \text{and} \quad d = q^2, \quad (2.1.13)$$

and the condition is obtained readily,

$$\epsilon_1\sqrt{a} + \epsilon_2\sqrt{b} + \epsilon_3\frac{c}{2\sqrt{d}} = 1, \quad (2.1.14)$$

for some choice of the signs ϵ_i's. Equation (2.1.12) is linearised by a Cole-Hopf transformation $w = -(z/p)u'/u$ to a confluent hypergeometric equation,

$$u'' - \left(q - \frac{1+n+p}{z}\right)u' + \frac{npu}{z^2} = 0, \quad (2.1.15)$$

which can be transformed either to a Kummer or a Whittaker equation.

From P_V we can obtain two coalescence limits to P_{IV} and P_{III}. In the first case we set $W = \delta w$, $Z = 1 + 2\delta z$, $A = 1/4\delta^4$, $B = b$, $C = -1/2\delta^4$, and $D = 1/4\delta^4 + a/\delta^2$ and obtain P_{IV} in the form

$$w'' = \frac{w'^2}{2w} + \frac{3w^3}{2} + 4zw^2 + 2w(z^2 + a) - \frac{2b}{w}. \quad (2.1.16)$$

This limit is compatible with the Riccati equation (2.1.12), which becomes

$$w' = w^2 + 2zw + 2n, \quad (2.1.17)$$

provided that $P = Q = 1/2\delta^2$, $N = n$. The coefficients of P_{IV} (2.1.16), in the linearisable case are given by

$$a = n + 1, \quad b = n^2, \quad (2.1.18)$$

with the obvious relation

$$a + \epsilon\sqrt{b} = 1 \quad (2.1.19)$$

for some choice of the sign ϵ. The Riccati equation (2.1.17) linearises by $w = -u'/u$ to the Hermite equation

$$u'' - 2zu' + 2nu = 0. \quad (2.1.20)$$

For the second limit, to P$_{III}$, we set $W = 1 + \delta w$, $Z = z$, $A = b/\delta^2 + a/\delta$, $B = b/\delta^2$, $C = c\delta$, $D = d\delta^2$ and obtain P$_{III}$ in the noncanonical form

$$w'' = \frac{w'^2}{w} - \frac{w'}{z} + \frac{bw^3}{z^2} + \frac{aw^2}{2z^2} + \frac{c}{2z} - \frac{d}{w}. \qquad (2.1.21)$$

For the linearisation of (2.1.18) we obtain the Riccati equation

$$w' = \frac{nw^2}{z} + \frac{pw}{z} + q \qquad (2.1.22)$$

from the limit of (2.1.12) by $N = n/\delta$, $P = n/\delta + p$, and $Q = q\delta$. The parameters of the Riccati equation are related to those of P$_{III}$ by

$$a = 2np, \quad b = n^2, \quad c = 2q(1 - p), \quad d = q^2 \qquad (2.1.23)$$

corresponding to the linearisability condition

$$\epsilon_1 \frac{a}{2\sqrt{b}} + \epsilon_2 \frac{c}{2\sqrt{d}} = 1 \qquad (2.1.24)$$

for some choice of the signs. The Riccati equation (2.1.22) is linearised through the Cole-Hopf transformation $w = -(z/n)u'/u$ leading to the equation

$$zu'' + (1 - p)u' + nqu = 0. \qquad (2.1.25)$$

The solution of the latter is given in terms of the Bessel function $\mathcal{C}$ as $u = z^{p/2}\mathcal{C}_p(2\sqrt{nqz})$.

Both P$_{IV}$ and P$_{III}$ go to P$_{II}$ by coalescence. In the first case we set $W = 2/\delta^3 + w/\delta$, $Z = -2/\delta^3 + \delta z$, $A = 2/\delta^6 + a$, $B = 4/\delta^{12}$ and obtain:

$$w'' = 2w^3 + 8wz + 4a. \qquad (2.1.26)$$

The Riccati equation (2.1.17) becomes

$$w' = w^2 + 4z, \qquad (2.1.27)$$

provided we take $N = 2/\delta^6$ with the linearisability condition

$$a = 1. \qquad (2.1.28)$$

The linearisation of (2.1.27) is straightforward, $w = -u'/u$, and leads to the Airy equation

$$u'' + 4zu = 0. \qquad (2.1.29)$$

In the second case we start from the noncanonical form of P$_{III}$ (2.1.21). It turns out that this does not make any difference at the level of P$_{II}$, apart from some unimportant coefficients. We set $W = 1 + \delta w$, $Z = 1 + \delta^2 z$, $A = -\delta^{-6}$, $B = \delta^{-6}/4 + b\delta^{-3}$, $C = \delta^{-6}$, and $D = \delta^{-6}/4$ and find, at the limit $\delta \to 0$, P$_{II}$ in the form

$$w'' = \frac{1}{2}w^3 + \frac{1}{2}wz + b. \tag{2.1.30}$$

The limit of the Riccati equation (2.1.22) is obtained by $Q = \delta^{-3}/2$, $N = \delta^{-3}/2$, and $P = -\delta^{-3}$

$$w' = \frac{1}{2}(w^2 + z). \tag{2.1.31}$$

The linearisability condition is simply

$$2\epsilon b = 1, \tag{2.1.32}$$

and the linearisation of the Riccati equation (2.1.31), with $w = -2u'/u$, leads again to the Airy equation,

$$u'' + \frac{z}{4}u = 0 \tag{2.1.33}$$

Equation P_{II} degenerates to P_I through the appropriate limit. However this coalescence does not present any interest for our purpose since it is incompatible with the existence of a Riccati equation. Indeed P_I does not possess any particular solution.

2.2 Special Function Solutions for Symmetric Discrete Painlevé Equations

In this section we shall concentrate on d-$\mathbb{P}$'s which are given in a symmetric QRT form,

$$x_{n+1} = \frac{f_1(x_n, n) - x_{n-1}f_2(x_n, n)}{f_4(x_n, n) - x_{n-1}f_3(x_n, n)} \tag{2.2.1}$$

where the f_i's are polynomials in x_n of degree four at maximum. The solution of (2.2.1) in terms of special functions proceeds by the introduction of a discrete Riccati equation,

$$x_{n+1} = \frac{A_n x_n + B_n}{C_n x_n + D_n}, \tag{2.2.2}$$

where A_n, B_n, C_n and D_n are functions of n to be determined by substituting (2.2.2) into (2.2.1). As in the continuous case, this fixes the parameters of the d-$\mathbb{P}$. The linearisation of (2.2.2) is again obtained by a Cole-Hopf transformation,

$$x_n = \left(\frac{D_n}{C_n}\right)\frac{y_{n+1} - y_n}{y_n}, \tag{2.2.3}$$

leading to the linear equation

$$\frac{D_{n+1}}{C_{n+1}}y_{n+2} - \left(\frac{D_{n+1}}{C_{n+1}} + \frac{A_n}{C_n}\right)y_{n+1} + \left(\frac{A_n}{C_n} - \frac{B_n}{D_n}\right)y_n = 0. \tag{2.2.4}$$

Equation (2.2.4) turns out to be, in all cases concerning discrete Painlevé equations, the discrete analogue of the hypergeometric equation or of one of its degenerate forms.

Let us present our results following the degeneration cascade of the d-$\mathbb{P}$'s depicted in the diagram,

$$
\begin{array}{ccccc}
q\text{-}\mathrm{P_{VI}} & \longrightarrow & q\text{-}\mathrm{P_V} & \longrightarrow & q\text{-}\mathrm{P_{III}} \\
\downarrow & & \downarrow & & \downarrow \\
\delta\text{-}\mathrm{P_V} & \longrightarrow & \delta\text{-}\mathrm{P_{IV}} & \longrightarrow & \delta\text{-}\mathrm{P_{II}} & \longrightarrow & \delta\text{-}\mathrm{P_I}
\end{array}
$$

The convention introduced in the case of continuous $\mathbb{P}$'s, namely that the variables and parameters of the equation higher in the degeneration cascade will be represented by upper-case letters while those of the "lower" equation will be denoted by low-case letters, will be used also here.

We start with q-$\mathrm{P_{VI}}$,

$$
\frac{(x_n x_{n+1} - z_n z_{n+1})(x_n x_{n-1} - z_n z_{n-1})}{(x_n x_{n+1} - 1)(x_n x_{n-1} - 1)} =
$$
$$
\frac{(x_n - a z_n)(x_n - z_n/a)(x_n - b z_n)(x_n - z_n/b)}{(x_n - c)(x_n - 1/c)(x_n - d)(x_n - 1/d)}, \qquad (2.2.5)
$$

where $z_n = z_0 \lambda^n$ and a, b, c, d are free constants. The continuous $\mathrm{P_{VI}}$ equation has solutions in terms of hypergeometric functions for some special values of the parameters. The same is true for q-$\mathrm{P_{VI}}$. The simplest way to obtain these special solutions is to use the splitting technique [23]. We separate equation (2.2.5) in two discrete Riccati, homographic, equations in the following way,

$$
\frac{x_n x_{n+1} - z_n z_{n+1}}{x_n x_{n+1} - 1} = \frac{(x_n - a z_n)(x_n - z_n/a)}{(x_n - c)(x_n - d)} \qquad (2.2.6a)
$$

$$
\frac{x_n x_{n-1} - z_n z_{n-1}}{x_n x_{n-1} - 1} = \frac{(x_n - b z_n)(x_n - z_n/b)}{(x_n - 1/c)(x_n - 1/d)}. \qquad (2.2.6b)
$$

The two equations of system (2.2.6) are indeed homographic and compatible provided the condition $ab = \lambda cd$ is satisfied. The linearisation of the discrete Riccati equation is obtained by a Cole-Hopf transformation, $x_n = P_n/Q_n$, resulting in the linear equation,

$$
Q_{n+1}(a z_n - d)(a z_n - c)((a+b)z_{n-1} - c - d)
$$
$$
+ a Q_n \left((a+b)z_{n-1}((ab-1)z_n^2 + \lambda - cd) - (c+d)((ab - 1/\lambda)z_n^2 + 1 - cd) \right)
$$
$$
- Q_{n-1}(a - d z_n)(a - c z_n)((a+b)z_n - c - d) = 0. \qquad (2.2.7)
$$

Equation (2.2.7) has the hypergeometric equation as continuous limit. This limit is simpler to obtain if we start from the discrete Riccati equation (2.2.6) and implement the continuous limit by: $\lambda = e^\epsilon$, $a = -e^{\epsilon\alpha}$, $b = e^{\epsilon\beta}$, $c = -e^{\epsilon\gamma}$, $d = e^{\epsilon\delta}$ and moreover $z = (1 + \sqrt{\zeta})/(1 - \sqrt{\zeta})$ and the transformation

$x = (\sqrt{\zeta} + w)/(\sqrt{\zeta} - w)$. In the limit $\epsilon \to 0$ we are led to a continuous Riccati equation which can be linearized through the Cole-Hopf $w = \zeta - \frac{\zeta(1-\zeta)}{\gamma u} \frac{du}{d\zeta}$, leading to

$$\zeta(1 - \zeta)\frac{d^2u}{d\zeta^2} + (\beta - \delta - (\beta + \gamma + 1)\zeta)\frac{du}{d\zeta} - \beta\gamma u = 0, \qquad (2.2.8)$$

i.e., precisely the Gauss hypergeometric equation.

From q-P$_{VI}$ we can obtain two different degenerations through coalescence: to d-P$_V$ and to q-P$_V$. Let us start with d-P$_V$. If we set $X = 1 + \delta x$, $A = 1 + \delta a$, $B = 1 + \delta b$, $C = 1 + \delta c$, $D = 1 + \delta d$, *i.e.*, $\lambda = 1 + \alpha\delta$, so $Z = 1 + \delta z$, where $z = \alpha n + \beta$, we recover exactly d-P$_V$ at the limit $\delta \to 0$,

$$\frac{(x_n + x_{n+1} - z_n - z_{n+1})(x_n + x_{n-1} - z_n - z_{n-1})}{(x_n + x_{n+1})(x_n + x_{n-1})}$$
$$= \frac{(x_n - z_n - a)(x_n - z_n + a)(x_n - z_n - b)(x_n - z_n + b)}{(x_n - c)(x_n + c)(x_n - d)(x_n + d)}, \qquad (2.2.9)$$

where a, b, c, d are constants and $z_n = \alpha n + \beta$. The linearisation of (2.2.9) can be obtained simply from the coalescence limit applied to (2.2.6a). We find thus the mapping:

$$\frac{x_n + x_{n+1} - z_n - z_{n+1}}{x_n + x_{n+1}} = \frac{(x_n - z_n - a)(x_n - z_n + a)}{(x_n - c)(x_n - d)}, \qquad (2.2.10)$$

provided condition (obtained as well from the coalescence of the condition for q-P$_{VI}$)

$$a + b = c + d + \alpha \qquad (2.2.11)$$

is satisfied, in which case (2.2.10) is indeed homographic:

$$x_{n+1} = \frac{((z_{n+1} - a)(z_{n+1} - b) - cd)x_n + cd(z_n + z_{n+1})}{x_n(z_n + z_{n+1}) - (z_n + a)(z_n + b)}. \qquad (2.2.12)$$

The discrete P$_V$ has only one degeneration, to d-P$_{IV}$. We set $X = x$, $Z = z + 1/\delta$ and

$$A = c + 1/\delta, \quad B = -c + 1/\delta, \quad C = a, \quad D = b. \qquad (2.2.13)$$

At the limit $\delta \to 0$ we find d-P$_{IV}$ in terms of the variable x.

The second coalescence of q-P$_{VI}$ leads to q-P$_V$ through the choice $X = x$, $Z_n = z_n/\delta$, $A = c/\delta$, $B = 1/\delta c$, $C = a$, $D = b$. At the limit $\delta \to 0$ one recovers the equation q-P$_V$, written below in slightly different notation,

$$(x_{n+1}x_n - 1)(x_nx_{n-1} - 1) = \frac{(x_n - u)(x_n - 1/u)(x_n - v)(x_n - 1/v)}{(x_n/p - 1)(x_n/q - 1)}, \qquad (2.2.14)$$

where u and v are constants and p_n, q_n are proportional to λ^n.

Again we seek a factorisation of the equation into

$$x_{n+1}x_n - 1 = \frac{(x_n - u)(x_n - v)}{uv(x_n/p - 1)} \tag{2.2.15a}$$

$$x_n x_{n-1} - 1 = \frac{uv(x_n - 1/u)(x_n - 1/v)}{x_n/q - 1}. \tag{2.2.15b}$$

Equation (2.2.15a) can be rewritten as a homographic mapping (discrete Riccati equation),

$$x_{n+1} = \frac{x_n - u - v + uv/p}{uv(x_n/p - 1)}, \tag{2.2.16}$$

and by up-shifting (2.2.15b) and solving for x_{n+1} we obtain the same homographic mapping, provided that

$$uvq\lambda = p. \tag{2.2.17}$$

If this condition is satisfied, then q-P$_V$ possesses solutions that are obtained through the linearisation of the discrete Riccati equation (2.2.16). The linearisation of the latter was given in [10], where we showed that x can be expressed in terms of discrete confluent hypergeometric functions. Indeed, setting $x = R/S$, we find that $R_n = p(S_n - S_{n+1})$ and S obeys the discrete confluent hypergeometric equation,

$$S_{n+2} + (\frac{1}{\lambda uv} - 1)S_{n+1} + \frac{1}{\lambda}(\frac{1}{u} - \frac{1}{p})(\frac{1}{v} - \frac{1}{p})S_n = 0, \tag{2.2.18}$$

where $p \propto \lambda^n$. The continuous limit of (2.2.16) should coincide with the Riccati equation obtained for P$_V$ in Sect. 2.1. As a matter of fact, implementing the continuous limit by $\lambda = 1 + \epsilon$, $u = 1 + \epsilon\nu$, $v = -1 - \epsilon\rho$, $p = (1/\epsilon + \mu)/z$, $q = (-1/\epsilon + \mu)/z$, one does not obtain, at $\epsilon \to 0$, the same Riccati equation. This is due to the fact that one has also to transform the dependent variable. Thus $x = (1+w)/(1-w)$ where w is the variable that becomes that of P$_V$ in the continuous limit. Using this transformation, one obtains P$_V$ with $a = \rho^2$, $b = \nu^2$, $c = -8\mu$ and $d = 4$. Linearisation constraint (2.2.17) becomes, at the limit, $\rho + \nu + 1 = 2\mu$ which is consistent with the continuous condition (2.1.14). A computation of the continuous limit of the Riccati equation for W yields

$$w' = \frac{\rho w^2}{z} + \frac{(2z - \rho + \nu)w}{z} - \frac{\nu}{z}, \tag{2.2.19}$$

which is exactly (2.1.12), with $p = \rho$, $n = -\nu$ and $q = 2$, which is in accordance with (2.1.13).

We proceed now to the coalescence q-P$_V \to$d-P$_{IV}$, using the convention of upper/lower-case characters as explained above. Setting $X = 1 + \delta x$, $U = 1 + \delta a$, $V = 1 + \delta b$, $P = 1 + \delta(z + c)$ and $Q = 1 + \delta(z - c)$, i.e., $\Lambda = 1 + \delta a$, such that $z = \alpha n + \beta$, we obtain, at $\delta \to 0$, the discrete P$_{IV}$,

$$(x_{n+1} + x_n)(x_n + x_{n-1}) = \frac{(x_n^2 - a^2)(x_n^2 - b^2)}{(x_n - z)^2 - c^2}. \tag{2.2.20}$$

The linearisation is again obtained by factorisation,

$$x_{n+1} + x_n = \frac{(x_n - a)(x_n - b)}{x_n - z + c} \tag{2.2.21a}$$

$$x_n + x_{n-1} = \frac{(x_n + a)(x_n + b)}{x_n - z - c}, \tag{2.2.21b}$$

and the two equations are compatible if the following constraint is satisfed,

$$a + b + \alpha = 2c, \tag{2.2.22}$$

which is exactly what would result from the coalescence limit of (2.2.17). Equation (2.2.21a) is indeed the discrete Riccati equation one obtains from q-P_V or d-P_V through coalescence. Moreover its continuous limit is the Riccati equation for the "linearisable" solutions of P_{IV}. Indeed the continuous limit of (2.2.21a), obtained through $c = 1/\epsilon$, $b = 2/\epsilon$, $a = \nu\epsilon$, with $x = w$ and z_n becoming the continuous variable, z, is

$$w' = w^2 - 2zw - 2\nu, \tag{2.2.23}$$

i.e., the equation we obtained by linearising P_{IV}. Equation (2.2.21a) has been shown to be solvable in terms of the discrete analogues of Hermite functions [25].

The second coalescence one can obtain from q-P_V is that to q-P_{III}. It is based on the limit $\delta \to 0$, where $X = x/\delta$, $\Lambda = \lambda$, $U = a/\delta$, $V = \delta/b$, $P = p/\delta$, $Q = q/\delta$ leading to

$$x_{n+1}x_{n-1} = \frac{(x_n - a)(x_n - b)}{(x_n/p - 1)(x_n/q - 1)}, \tag{2.2.24}$$

where $p = p_0\lambda^n$, $q = q_0\lambda^n$. The linearisation is again given by a factorisation

$$x_{n+1} = \frac{b}{a}\frac{(x_n - a)}{(x_n/p - 1)} \tag{2.2.25a}$$

$$x_{n-1} = \frac{a}{b}\frac{(x_n - b)}{(x_n/q - 1)}, \tag{2.2.25b}$$

and the compatibility condition of (2.2.25a) and (2.2.25b) is

$$bp = aq\lambda, \tag{2.2.26}$$

in which case (2.2.25) is solved in terms of discrete Bessel functions. Again, (2.2.26) is exactly the limit of (2.2.17) under the coalescence procedure. In perfect analogy to the q-$P_V \to$ d-P_{IV} case, we can show that (2.2.25a) is the coalescence limit of (2.2.16). As far as the continuous limit is concerned, we must take into account the noncanonical character of (2.1.21). Setting $x = w$, $\lambda = 1 + \epsilon$, $b = -a + \epsilon c$, $p_0 = -1$, $q_0 = 1 + \epsilon d$ we find for the continuous limit of q-P$_{\text{III}}$ the (again noncanonical) equation,

$$w'' = \frac{w'^2}{w} - \frac{w'}{z} + w^3 + \frac{dw^2}{z} - \frac{c}{z^2} - \frac{a^2}{wz^2}, \tag{2.2.27}$$

where $z = n\epsilon$. The continuous limit of (2.2.25a) is the Riccati equation,

$$w' = -w^2 - \frac{c}{a}\frac{w}{z} - \frac{a}{z}, \tag{2.2.28}$$

which, with $w = u'/u$, is linearised to $zu'' + cu'/a + au = 0$. The latter equation is solvable, as expected, in terms of Bessel functions, $u = z^{\frac{a-c}{2a}}C_{1-c/a}(2\sqrt{az})$.

Two coalescence limits remain to be considered, those of d-P$_{\text{IV}}$ and q-P$_{\text{III}}$ to d-P$_{\text{II}}$. In the first case we set $X = 1 + \delta x$, $A = 1 + \delta$, $B = -1 + \delta$, $Z = 1 - \delta^2 z/4$, $C = \delta - \delta^2 a/4$ and find

$$x_{n+1} + x_{n-1} = \frac{z_n x_n + a}{1 - x_n^2} \tag{2.2.29}$$

Starting from (2.2.22) we implement the coalescence limit and find that the linearisability condition of d-P$_{\text{II}}$ is

$$a = \frac{\alpha}{2}, \tag{2.2.30}$$

while the homographic mapping (2.2.21) reduces to

$$x_{n+1} + 1 = \frac{a + z_n}{2(1 - x_n)} \tag{2.2.31a}$$

$$x_{n-1} - 1 = \frac{a - z_n}{2(1 + x_n)}. \tag{2.2.32b}$$

This leads indeed to the linearisation of d-P$_{\text{II}}$. It can be shown that the continuous limit of (2.2.31), obtained by $x = \epsilon w$, $a = 2\epsilon^3$, while the discrete variable z_n is related to the continuous variable z through $z_n = 2 + 4\epsilon^2 z$, coincides with the Riccati equation (2.1.27) from P$_{\text{II}}$.

In a similar way, one can work out the coalescence q-P$_{\text{III}} \rightarrow$ d $-$ P$_{\text{II}}$. We start by transforming (2.2.29) by $x = y/z$, where $z = \lambda^n$, to

$$y_{n+1}y_{n-1} = \frac{p_0 q_0 (y_n - a/z_n)(y_n - b/z_n)}{(y_n - p_0)(y_n - q_0)}. \tag{2.2.32}$$

Next we introduce $Y = 1 + \delta x$, $P_0 = 1 + \delta$, $Q_0 = 1 - \delta$, $A = 1 + \delta + \delta^2 a/2$, $B = 1 - \delta - \delta^2 a/2$ and $\Lambda = 1 - \delta^2 a/2$ leading to $Z = 1 - z\delta^2/2$, and we find in the limit precisely d-P$_{\text{II}}$ in the form (2.2.29). Again the linearisability condition, resulting from the limit of (2.2.26), is identical to (2.2.30). The discrete Riccati equation is also identical to (2.2.31). As we have shown in [26], the solution of the latter (and thus of d-P$_{\text{II}}$ for $a = \alpha/2$) is given in terms of discrete Airy functions.

2.3 The Case of Asymmetric Discrete Painlevé Equations

The degeneration cascade we have presented above does not, and by far, exhaust all d-$\mathbb{P}$'s. As a matter of fact, the majority of the d-$\mathbb{P}$'s have a natural form which is asymmetric, with two and sometimes more components. In [8] we have presented their organisation in a degeneration cascade, first obtained by Sakai. Before proceeding to the solutions of the d-$\mathbb{P}$'s corresponding to these asymmetric forms, let us give the general method for their derivation [27]. In analogy to the symmetric case, we seek a solution that satisfies a discrete Riccati equation in one of the variables

$$x_n = \frac{ax_{n-1} + b}{cx_{n-1} + d}, \tag{2.3.1a}$$

while the two variables are related through a homographic transformation,

$$y_n = \frac{fx_n + g}{hx_n + k}. \tag{2.3.1b}$$

As we have already explained, the asymmetric discrete Painlevé equations we shall examine are organised in a coalescence cascade where a given 'higher' equation leads to one (or more) 'lower' ones through a limiting procedure involving the dependent and independent variables as well as the parameters.

a. The Asymmetric q-P$_{\mathrm{V}}$. We start with the asymmetric q-P$_{\mathrm{V}}$ equation which by definition the system:

$$(y_n x_n - 1)(y_n x_{n-1} - 1) = \frac{(y_n - u)(y_n - v)(y_n - w)(y_n - s)}{(1 - py_n/z_n)(1 - y_n/pz_n)} \tag{2.3.2a}$$

$$(y_n x_n - 1)(y_{n+1} x_n - 1) = \frac{(x_n - 1/u)(x_n - 1/v)(x_n - 1/w)(x_n - 1/s)}{(1 - rx_n/z_{n+1/2})(1 - x_n/rz_{n+1/2})}, \tag{2.3.2b}$$

with the constraint $uvws = 1$, and where $z_n = z_0\lambda^n$ and $z_{n+1/2} = z_0\lambda^{n+1/2}$. For future convenience we introduce the parameter $\mu = \lambda^{1/2}$. The linearisation of (2.3.2) can be obtained most simply by the splitting procedure *i.e.*, splitting each of the equations of the system in two parts and requiring that the resulting system:

$$y_n x_n - 1 = -\frac{(y_n - u)(y_n - v)}{uv(1 - py_n/z_n)}$$

$$y_n x_{n-1} - 1 = -\frac{(y_n - w)(y - s)}{(1 - y_n/pz_n)ws}$$

$$y_n x_n - 1 = -\frac{uv(x_n - 1/u)(x_n - 1/v)}{(1 - rx_n/z_{n+1/2})} \tag{2.3.3}$$

$$y_{n+1}x_n - 1 = -\frac{ws(x_n - 1/w)(x_n - 1/s)}{(1 - x_n/rz_{n+1/2})},$$

be compatible. The condition for compatibility is

$$r = \mu uvp, \qquad (2.3.4)$$

which is precisely the linearisability condition. Indeed, when (2.3.4) is satisfied, we can obtain from (2.3.3) a homographic mapping for x of the form

$$x_n =$$

$$\frac{x_{n-1}sw(puv - (u+v)z_n + pz_n^2) - (p^2uv - sw)z_n + p(u+v-s-w)z_n^2}{x_{n-1}z_n(p^2 - 1) + puv(sw - pz_n(s+w) + z_n^2)}.$$

$$(2.3.5)$$

This discrete Riccati equation can be linearised by a Cole-Hopf transformation. The resulting equation has one more parameter than the q-hypergeometric equation, just as asymmetric q-P$_\text{V}$ has one extra parameter compared to P$_\text{VI}$. However, as we have shown in [28], asymmetric q-P$_\text{V}$ does become P$_\text{VI}$ at the continuous limit. Indeed setting $u = \theta e^{\epsilon a}$, $v = \theta^{-1}e^{\epsilon b}$, $w = \theta e^{-\epsilon a}$, $s = \theta^{-1}e^{-\epsilon b}$, $\lambda = e^\epsilon$, $p = e^{\epsilon c}$, $r = e^{\epsilon d}$, $\omega = (x - \theta)/(\theta^{-1} - \theta)$, $\zeta = (z - \theta)/(\theta^{-1} - \theta)$, $y = (z(x - \theta^{-1} - \theta) + 1)/(x - z) + \epsilon\psi$, where the constraint $uvws = 1$ has been implemented, we obtain after eliminating ψ from two first-order equations,

$$\frac{d^2\omega}{d\zeta^2} = \frac{1}{2}\left(\frac{1}{\omega} + \frac{1}{\omega - 1} + \frac{1}{\omega - \zeta}\right)\left(\frac{d\omega}{d\zeta}\right)^2 - \left(\frac{1}{\zeta} + \frac{1}{\zeta - 1} + \frac{1}{\omega - \zeta}\right)\frac{d\omega}{d\zeta}$$
$$+ \frac{\omega(\omega - 1)(\omega - \zeta)}{2\zeta^2(\zeta - 1)^2}\left(A + \frac{B\zeta}{\omega^2} + \frac{C(\zeta - 1)}{(\omega - 1)^2} + \frac{D\zeta(\zeta - 1)}{(\omega - \zeta)^2}\right), \quad (2.3.6)$$

i.e., precisely P$_\text{VI}$, where $A = 4c^2$, $B = -4b^2$, $C = 4a^2$ and $D = 1 - 4d^2$.

Similarly the continuous limit of the discrete Riccati equation (2.3.5), is the Riccati equation:

$$\zeta(1 - \zeta)\frac{d\omega}{d\zeta} = 2d\omega^2 + (2(a+b)\zeta - 2a - 2c)\omega - 2b\zeta, \qquad (2.3.7)$$

where the linearisation condition is now $d = a+b+c+1/2$, and the Cole-Hopf transformation $\omega = \zeta - \frac{\zeta(1-\zeta)}{2cG}\frac{dG}{d\zeta}$ linearises the equation to

$$\zeta(1 - \zeta)\frac{d^2G}{d\zeta^2} + \left(2a + 2c + 1 - (2c + 2d + 1)\zeta\right)\frac{dG}{d\zeta} - 4cdG = 0, \qquad (2.3.8)$$

i.e., the Gauss hypergeometric equation in canonical form.

The coalescence procedure applied to asymmetric q-P$_\text{V}$ allows one to obtain either asymmetric d-P$_\text{V}$ or asymmetric q-P$_\text{III}$. Let us study the first limit.

b. The Asymmetric d-P_{IV}. In order to obtain asymmetric d-P_{IV} starting from asymmetric q-P_V we introduce the following transformation, $X = 1 + \delta x$, $Y = 1 + \delta y$, $Z = 1 + \delta z$, $\lambda = 1 + \delta a$, $U = 1 + \delta u$, $V = 1 + \delta v$, $W = 1 + \delta w$, $S = 1 + \delta s$, $P = 1 + \delta p$, $R = 1 + \delta r$, where now $z = \alpha n + \beta$. At the limit $\delta \to 0$, we obtain the system,

$$(y_n + x_n)(y_n + x_{n-1}) = \frac{(y_n - u)(y_n - v)(y_n - w)(y_n - s)}{(y_n + p - z_n)(y_n - p - z_n)} \qquad (2.3.9a)$$

$$(y_n + x_n)(y_{n+1} + x_n) = \frac{(x_n + u)(x_n + v)(x_n + w)(x_n + s)}{(x_n + r - z_{n+1/2})(x_n - r - z_{n+1/2})} \qquad (2.3.9b)$$

with the constraint $u + v + w + s = 0$, and where $z_{n+1/2} = z_n + \alpha/2$. Instead of performing the linearisation splitting from the start, we use the coalescence limit on the asymmetric q-P_V. We thus find the system,

$$y_n + x_n = \frac{(y_n - u)(y_n - v)}{(y_n + p - z_n)}$$

$$y_n + x_{n-1} = \frac{(y_n - w)(y_n - s)}{(y_n - p - z)}$$

$$y_n + x_n = \frac{(x_n + u)(x_n + v)}{(x_n + r - z_{n+1/2})} \qquad (2.3.10)$$

$$y_{n+1} + x_n = \frac{(x_n + w)(x_n + s)}{(x_n - r - z_{n+1/2})}$$

and the compatibility-linearisability condition is

$$r = u + v + p + \alpha/2. \qquad (2.3.11)$$

The discrete Riccati equation is

$$x_n = $$
$$\frac{x_{n-1}\big((z_n + p)(z_n - p - u - v) + uv\big) + (z_n - p - u - v)(sw - uv) - 2puv}{2px_{n-1} + (z_n - p)(z_n + p + u + v) + sw}.$$
$$\qquad (2.3.12)$$

Its linearisation leads again to a discrete linear equation with more parameters than the hypergeometric. As in the case of asymmetric q-P_V, the continuous limit can be easily obtained. For asymmetric d-P_{IV} it leads to P_{VI} [28]. We set $u = 1/2 + \epsilon a$, $v = -1/2 + \epsilon b$, $w = 1/2 - \epsilon a$, $s = -1/2 - \epsilon b$, $p = \epsilon c$, $r = \epsilon d$, $x = w - 1/2$, $z = \zeta - 1/2$, $y = \omega(\zeta - 1)/(\omega - \zeta) + 1/2 + \epsilon \psi$, and after again eliminating ψ in two first-order equations, at $\epsilon \to 0$ we recover equation (2.5) with $A = 4c^2$, $B = -4a^2$, $C = 4b^2$ and $D = 1 - 4d^2$. The same approach on the Riccati equation, where the linearisation condition is now

$d = a + b + c + 1/2$, leads to a continuous equation linearised with the same Cole-Hopf transformation as (2.3.8) to:

$$\zeta(1 - \zeta)\frac{d^2G}{d\zeta^2} + \left(d - a - (2c + 2d + 1)\zeta\right)\frac{dG}{d\zeta} - 4cdG = 0, \qquad (2.3.13)$$

again the hypergeometric equation.

c. The Asymmetric q-P_{III}. The asymmetric q-P_V has another coalescence limit to the asymmetric q-P_{III} equation. Putting: $X = x/\delta$, $Y = y/\delta$, $Z = z/\delta$, $U = u/\delta$, $V = v\delta$, $W = w/\delta$, $S = s\delta$, $P = p$, $R = r$, at $\delta \to 0$ we find the mapping

$$x_n x_{n-1} = \frac{(y_n - u)(y_n - w)}{(1 - y_n p/z_n)(1 - y_n/pz_n)} \qquad (2.3.14a)$$

$$y_n y_{n+1} = \frac{(x_n - 1/v)(x_n - 1/s)}{(1 - x_n r/z_{n+1/2})(1 - x_n/rz_{n+1/2})}, \qquad (2.3.14b)$$

with the obvious condition $uvws = 1$. Equations (2.3.14) can be written in canonical form by introducing a gauge $y \to zy$, $x \to \zeta x$. We thus obtain

$$x_n x_{n-1} = \frac{(y_n - u/z_n)(y_n - w/z_n)}{(1 - y_n p)(1 - y_n/p)} \qquad (2.3.15a)$$

$$y_n y_{n+1} = \frac{(x_n - 1/z_{n+1/2}v)(x_n - 1/z_{n+1/2}s)}{(1 - x_n r)(1 - x_n/r)}. \qquad (2.3.15b)$$

Equation (2.3.15) was studied by Jimbo and Sakai [16], who have shown that it is a q-discrete form of P_{VI}. Thus this equation is often referred to as the q-P_{VI} equation. Its linearisation was also obtained by Jimbo and Sakai. Provided $r = \mu uvp$ is satisfied we can obtain for x the discrete Riccati equation,

$$x_n = \frac{x_{n-1}(u - pz_n) + p(u - w)}{x_{n-1}z_n(p^2 - 1) + p(pw - z_n)}. \qquad (2.3.16)$$

The equation can be linearised by the Cole-Hopf transformation $x = H/G$, leading to

$$G_{n+1} + (2\lambda pz_n - \lambda u - p^2 w)G_n + \lambda p(z_n - pu)(pz_n - w)G_{n-1} = 0. \quad (2.3.17)$$

Jimbo and Sakai, who first obtained this mapping, identified it as the equation for the q-hypergeometric $_2\phi_1$.

d. The Discrete P_V. From the diagram in Sect. 2.2 we can see that the asymmetric d-P_{IV} and the asymmetric q-P_{III} become the same equation in the coalescence limit. This equation was first identified in [29], where we showed that it is a discrete form of P_V. Let us first examine the degeneration asymmetric d-$P_{IV} \to$ d-P_V. We set $X = k+x$, $Y = -k+(y+z)\delta$, $Z = -k+z\delta$,

$$U = k + r + u\delta, \; V = -k + v\delta, \; W = k - r + w\delta, \; P = p\delta, \; R = r, \; S = -k + s\delta,$$

and from (2.3.9) we obtain, at the limit $\delta \to 0$,

$$x_n x_{n-1} = \frac{(y_n + z_n - v)(y_n + z_n - s)}{(y_n + p)(y_n - p)} \qquad (2.3.18a)$$

$$y_n + y_{n+1} = -\frac{z_{n+1/2} + u}{x_n/c + 1} - \frac{z_{n+1/2} + w}{x_n c + 1} \qquad (2.3.18b)$$

where the constraint $u + v + w + s = 0$ is still satisfied, and we have moreover set $4k^2 - r^2 = 1$, $c = 2k + r$.

The linearisation of d-P$_V$ can be obtained from the direct splitting of (2.3.18), but also from the degeneration of the linearisation of asymmetric d-P$_{IV}$. The result is the system,

$$x_n = -\frac{c(y + z_n - v)}{y_n + p}$$

$$x_{n-1} = -\frac{y_n + z_n - s}{c(y_n - p)}$$

$$y_n = -\frac{z_{n+1/2} + u}{x_n/c + 1} - p \qquad (2.3.19)$$

$$y_{n+1} = -\frac{z_{n+1/2} + w}{x_n c + 1} + p,$$

under the linearisation constraint

$$u + v + p + \alpha/2 = 0. \qquad (2.3.20).$$

Recall that $z_n = \alpha n + \beta$. A discrete Riccati equation is easily obtained from (2.3.19):

$$x_n = \frac{x_{n-1} c^2 (v - p - z_n) + c(v - s)}{2 x_{n-1} c p + p + s - z_n}. \qquad (2.3.21)$$

The linearisation of (2.3.21) can be obtained through a Cole-Hopf transformation $x = H/G$, leading to

$$G_{n+1} + \left((c^2 + 1) z_n + (c^2 - 1) p - c^2 v - s + \alpha\right) G_n + c^2 (z_n - v - p)(z_n - s + p) G_{n-1} = 0 \qquad (2.3.22)$$

This equation can be transformed by a gauge transformation, $G = \Phi F$, with $\Phi_n = (v + p - z_n) \Phi_{n-1}$,

$$(z_{n+1} - v - p) F_{n+1} - \left((c^2 + 1) z_n + (c^2 - 1) p - c^2 v - s + \alpha\right) F_n + c^2 (z_n - s + p) F_{n-1} = 0. \qquad (2.3.23)$$

It can be easily shown that (2.3.23) is just one of the Gauss relations for contiguous hypergeometric functions. In fact, equation (2.3.23) is satisfied by $F(1 + (z - v - p)/\alpha, (s - v)/\alpha; 1 + (s - v - 2p)/\alpha; 1 - 1/c^2)$ [22].

The relation of the special function solutions of d-P_V to the hypergeometric equation is not at all astonishing. Indeed in [30] we showed that (2.3.18) can be obtained from the Schlesinger transformations of P_{VI}. This means that the dependent variable of the discrete equation coincides, under the proper choice, with that of the continuous equation. Thus the result that the special solutions of the discrete P_V obey the contiguity relations of the function that appears in the special solutions of P_{VI}, namely the hypergeometric function makes perfect sense.

As we explained above d-P_V, in the form of equation (2.3.18), can be obtained as a degeneration of asymmetric q-P_{III}. This is in fact how this equation was first obtained. We shall not go into these details. It is a mere (and straightforward) verification to show that the linearisation of the d-P_V equations obtained from that of the q-P_{III} through the coalescence procedure gives the same result as the one obtained above.

The discrete P_V has two possible degenerations to P_{IV} and to asymmetric P_{II}. Let us start with the first degeneration.

e. The Discrete P_{IV}. This degeneration was first obtained in [29]. Starting from d-P_V, equation (2.3.18), we introduce the coalescence: $X = x/\delta$, $Y = y$, $U = u/\delta^2$, $V = -u/\delta^2$, $W = w$, $S = s$, $P = p$, $C = -\delta$. At the limit $\delta \to 0$ we obtain:

$$x_n x_{n-1} = u \frac{(y_n + z_n - s)}{(y_n + p)(y_n - p)} \tag{2.3.24a}$$

$$y_n + y_{n+1} = \frac{u}{x_n} + \frac{z_{n+1/2} + w}{x_n - 1}, \tag{2.3.24b}$$

which was shown in [29] to become P_{IV} at the continuous limit. From the linearisation equations for d-P_V we obtain simply

$$x_n = \frac{u}{y_n + p}$$

$$x_{n-1} = \frac{y_n + z_n - s}{y_n - p}$$

$$y_n + p = \frac{u}{x_n} \tag{2.3.25}$$

$$y_{n+1} - p = \frac{z_{n+1/2} + w}{x_n - 1},$$

and the linearisability condition $p = s + w - \alpha/2$. The discrete Riccati equation for x now becomes

$$x_n = \frac{u(x_{n-1} - 1)}{2px_{n-1} + z_n - s - p}. \tag{2.3.26}$$

The linearisation of this equation by a Cole-Hopf transformation, $x = H/G$, results in the linear equation,

$$G_{n+1} - (z_n + \alpha + u - s - p)G_n + u(z_n + p - s)G_{n-1} \tag{2.3.27}$$

This equation is, up to a trivial gauge transformation, the recurrence relation (with respect to the second parameter) for the Kummer confluent hypergeometric U function [22], which is quite reasonable since d-P_{IV} is related to the continuous P_V equation [31].

f. The Asymmetric d-P_{II}, Discrete P_{III}, Equation. The other degeneration of d-P_V is towards the asymmetric d-P_{II}, which was shown in [23] to be a discrete form of the P_{III} equation. Starting from d-P_V, equation (2.3.18), we set $X = 1 + \delta x$, $Y = y$, $Z = \delta z$, $U = -\delta w = -W$, $V = 1 + \delta v = -S$, $C = -1 - \delta$, $P = 1$, and we obtain at $\delta \to 0$,

$$x_n + x_{n-1} = \frac{z_n + s}{y_n - 1} + \frac{z_n - s}{y_n + 1} = \frac{2z_n y_n + 2s}{y_n^2 - 1} \qquad (2.3.28a)$$

$$y_n + y_{n+1} = \frac{z_{n+1/2} - w}{x_n - 1} + \frac{z_{n+1/2} + w}{x_n + 1} = \frac{2z_{n+1/2} x_n - 2w}{x_n^2 - 1}. \qquad (2.3.28b)$$

The linearisation can be obtained from that of d-P_V or by direct splitting of (2.3.28) to

$$x_n - 1 = \frac{z_n + s}{y_n - 1}$$

$$x_{n-1} + 1 = \frac{z_n - s}{y_n + 1}$$

$$y_n - 1 = \frac{z_{n+1/2} - w}{x_n - 1} \qquad (2.3.29)$$

$$y_{n+1} + 1 = \frac{z_{n+1/2} + w}{x_n + 1}.$$

The linearisability/compatibility condition is $s + w = \alpha/2$, and leads to the discrete Riccati equation,

$$x_n = \frac{(2 - z_n - s)x_{n-1} + 2 - 2z_n}{2x_{n-1} + 2 - z_n + s}. \qquad (2.3.30)$$

The linearisation is again obtained by $x = H/G$ and results in

$$G_{n+1} + (2z_n + \alpha - 4)G_n + (z_n^2 - s^2)G_{n-1} = 0. \qquad (2.3.31)$$

This equation, just like (2.3.27), is a recurrence relation of the Kummer U function [22] with respect to its first parameter, and up to a simple gauge transformation. As a matter of fact, the discrete P_{III} equation can also be obtained [31] from the Schlesinger transformations of the continuous P_V, and it is intimately related to the discrete P_{IV} equation (2.3.24). The two equations share the same 'Grand Scheme' [32].

g. The Asymmetric d-P$_I$ Equation. This equation was studied in great detail in [15] where we showed its relation to the continuous P$_{IV}$. This asymmetric equation is just another form of d-P$_{II}$. In the coalescence cascade we presented in Sect. 2.2, it can be obtained as a degeneration of both d-P$_{IV}$ and asymmetric d-P$_{II}$. Let us show how the first limit can be obtained. We start from (2.3.24) and set $X = 1 + \delta x/2$, $Y = 1 + \delta y$, $S = 1 + \delta^2 s/2$, $W = \delta^2 w/2$, $P = 1$, $Z = \delta^2 z/2$. At the limit $\delta \to 0$, we find

$$x_n + x_{n-1} = -y_n + u + \frac{z_n - s}{y_n} \qquad (2.3.32a)$$

$$y_n + y_{n+1} = -x_n + u + \frac{z_{n+1/2} + w}{x_n}. \qquad (2.3.32b)$$

The linearisation splitting is:

$$x_n = -y_n + u$$

$$x_{n-1} = \frac{z_n - s}{y_n}$$

$$y_n = -x_n + u \qquad (2.3.33)$$

$$y_{n+1} = \frac{\eta_n + w}{x_n},$$

and the condition is $s + w = \alpha/2$. Using (2.3.33) we can obtain a discrete Riccati equation for x,

$$x_n = u + \frac{s - z_n}{x_{n-1}}, \qquad (2.3.34)$$

which linearises, by $x = H/G$, to:

$$G_{n+1} - uG_n + (z_n - s)G_{n-1} = 0, \qquad (2.3.35)$$

i.e., a discrete analogue of the Airy equation which is nothing but a recurrence relation of the parabolic cylinder equation, a fact that is expected given the relation of asymmetric d-P$_I$ to P$_{IV}$.

The asymmetric d-P$_I$ equation can be also obtained from the asymmetric d-P$_{II}$ through a coalescence limit. This procedure is essentially the same as the one introduced in [4] for the degeneration of the *symmetric* d-P$_{II}$ to d-P$_I$. The linear equation resulting form this coalescence is, of course, the same discrete Airy equation as in (2.3.35).

3 Solutions by Direct Linearisation

In the previous subsection we have presented solutions of the continuous and discrete Painlevé equations which are expressed in terms of special functions. These solutions are obtained by the assumption that the Painlevé equation is satisfied by the solution of some Riccati equation. However there exists another class of solutions which will be the object of this subsection [33, 34]. Let us present the argument schematically, in the simplest case, that of the fundamental solution. Suppose that for some value of the parameters, a linearisability condition is satisfied and the solution is furnished by a Riccati equation,

$$w' = \alpha w^2 + \beta w + \gamma, \tag{3.0.1}$$

where the prime denotes the derivative with respect to the independent variable t. Now it turns out that when a further constraint is satisfied, the coefficient α vanishes, whereupon the Riccati equation becomes linear. Equivalently, when γ vanishes, (3.0.1) becomes linear for the quantity $1/w$. In both cases the solution w is obtained by a linear first-order equation and involves one integration constant. When $\alpha\gamma \neq 0$, we may still reduce (3.0.1) to a first-order equation if we are able, by inspection or any other means, to obtain an explicit special solution, ϕ. In this case setting $w = \phi + 1/\psi$, we obtain for ψ the first-order inhomogeneous linear equation

$$\psi' + (2\alpha\phi + \beta)\psi + \alpha = 0. \tag{3.0.2}$$

The integration of (3.0.2) introduces an integration constant and thus $w = \phi + 1/\psi$ is indeed the general solution of (3.0.1). These special solutions will be the object of the present study. It is clear from the argument above that the minimal number of parameters for such a solution to exist is two, thus P_{II}, and a *fortiori* P_I, are excluded, to say nothing of the one-parameter P_{III} [35].

While our arguments have been presented for the continuous $\mathbb{P}$'s, they can be transposed, *mutatis mutandis*, to the discrete case for both difference and q-discrete Painlevé equations.

3.1 Continuous Painlevé Equations

In this section we shall present the special solutions of the continuous Painlevé equations. In what follows, the notation ϵ_i will be reserved for a free sign, ± 1.

a) The P_{III} Equation
The Painlevé III equation, in the following normalisation,

$$w'' = \frac{w'^2}{w} - \frac{w'}{t} + w^3 + \frac{1}{t}(\alpha w^2 + \beta) - \frac{1}{w}, \tag{3.1.1}$$

does possess solutions obtained from those of a linear equation whenever the condition [21],

$$\epsilon_1 \alpha + \epsilon_2 \beta = 2, \tag{3.1.2}$$

is satisfied. These solutions are obtained by the Riccati equation:

$$w' = \epsilon_1 w^2 + \frac{\alpha \epsilon_1 - 1}{t} w + \epsilon_2. \qquad (3.1.3)$$

Given the structure of (3.1.3), it is clear that it cannot be reduced to a first-order linear equation. In this case, one can find a special solution with the help of the rational solutions of P_{III}. If $\alpha = -\epsilon_3 \beta$ equation (3.1.1) has an elementary, rational, solution $w^2 = \epsilon_3$. Demanding that this solution satisfy the Riccati equation (3.1.3), we obtain the conditions $\alpha = \epsilon_1$, $\beta = \epsilon_2$ and $\epsilon_3 = -\epsilon_1 \epsilon_2$. The Riccati equation now reduces to $w' = \epsilon_1(w^2 - \epsilon_3)$, which we can integrate by quadrature in an elementary way. We obtain

$$w = \epsilon_2 \tanh(t - t_0), \qquad (3.1.4a)$$

for $\epsilon_3 = 1$, i.e., $\epsilon_2 = -\epsilon_1$, and

$$w = \epsilon_2 \tan(t - t_0), \qquad (3.1.4b)$$

for $\epsilon_3 = -1$ (i.e. $\epsilon_2 = \epsilon_1$). This one-parameter solution was obtained in [36] (see also [20]). Using the auto-Bäcklund transformations of P_{III} one can construct higher solutions involving tangents. For instance, for $\alpha = 1$ and $\beta = -3$ we find the solution, $w = 1/\tanh(t - t_0) - 1/t$.

b) The P_{IV} Equation
The Painlevé IV equation,

$$w'' = \frac{w'^2}{2w} + \frac{3w^3}{2} + 4tw^2 + 2w(t^2 + \alpha) - \frac{2\beta^2}{w}, \qquad (3.1.5)$$

has linearisable solutions whenever constraint

$$\epsilon_1 \alpha + \epsilon_2 \beta = 1 \qquad (3.1.6)$$

is satisfied [21]. They are given by the solutions of the Riccati equation,

$$w' = \epsilon_1(w^2 + 2tw) - 2\epsilon_2 \beta. \qquad (3.1.7)$$

Clearly, if $\beta = 0$, in which case $\alpha = \epsilon_1$, the Riccati equation becomes a linear equation for $u = 1/w$:

$$u' = -\epsilon_1(2tu + 1). \qquad (3.1.8)$$

The integration of (3.1.8) is straightforward,

$$u = \left(c - \epsilon_1 \int e^{\epsilon_1 t^2} dt \right) e^{-\epsilon_1 t^2}, \qquad (3.1.9)$$

with c an integration constant, i.e., u, or equivalently w, can be expressed in terms of the Error function (of t for $\epsilon_1 = -1$ and of it for $\epsilon_1 = 1$) [37]. One remark is in order here. If one applies the auto-Bäcklund transformation of P_{IV},

$$\tilde{w} = \frac{w' - w^2 - 2tw + 2\beta}{2w}, \tag{3.1.10}$$

one obtains a solution of (3.1.5) with parameters $\tilde{\alpha} = (3\beta - \alpha - 1)/2$ and $\tilde{\beta} = \epsilon(\beta + \alpha - 1)/2$. Starting from $\beta = 0$ and $\alpha = \epsilon_1 = -1$ we find that $\tilde{\alpha} = 0$ and $\tilde{\beta} = -\epsilon$, satisfying the condition $\tilde{\alpha} - \epsilon\tilde{\beta} = 1$. Thus, while (3.1.7) in this case is not linear either in w or in $1/w$, the transformed solution is again an elementary one of the linearisable class. It goes without saying that repeated application of (3.1.10) will lead to higher solutions involving Error functions.

c) The P_V Equation
We start from

$$w'' = w'^2 \left(\frac{1}{2w} + \frac{1}{w-1} \right) - \frac{w'}{t} + \frac{(w-1)^2}{2t^2} \left(\alpha^2 w - \frac{\beta^2}{w} \right) + \frac{\gamma w}{t} - \frac{w(w+1)}{2(w-1)}. \tag{3.1.11}$$

The linearisablity condition, with the notation we chose above, is just [21]:

$$\epsilon_1 \alpha + \epsilon_2 \beta + \epsilon_3 \gamma = 1, \tag{3.1.12}$$

and the associated Riccati equation is

$$w' = \alpha \epsilon_1 \frac{w(w-1)}{t} + \epsilon_3 w + \epsilon_2 \beta \frac{w-1}{t}. \tag{3.1.13}$$

When $\alpha = 0$, or, equivalently, $\beta = 0$ and $w \to 1/w$, equation (3.1.13) becomes linear,

$$w' = (\epsilon_3 + \epsilon_2 \frac{\beta}{t}) w - \epsilon_2 \frac{\beta}{t}. \tag{3.1.14}$$

Its solution,

$$w = \left(c - \epsilon_2 \beta \int e^{\epsilon_3 t} t^{\epsilon_2 \beta - 1} dt \right) e^{\epsilon_3 t} t^{-\epsilon_2 \beta}, \tag{3.1.15}$$

involves the incomplete Gamma function [20].

Just as in the case of P_{IV}, the auto-Bäcklund transformation of P_V generates solutions belonging to the linearisable class and which involve the incomplete Gamma function.

d) The P_{VI} Equation
The most general of the second-order Painlevé equations is

$$w'' = \frac{w'^2}{2} \left(\frac{1}{w} + \frac{1}{w-1} + \frac{1}{w-t} \right) - w' \left(\frac{1}{t} + \frac{1}{t-1} + \frac{1}{w-t} \right)$$
$$+ \frac{w(w-1)(w-t)}{2t^2(t-1)^2} \left(\alpha^2 - \beta^2 \frac{t}{w^2} + \gamma^2 \frac{t-1}{(w-1)^2} - (\delta^2 - 1) \frac{t(t-1)}{(w-t)^2} \right). \tag{3.1.16}$$

The linearisability condition in this case is, [21],

$$\epsilon_1 \alpha + \epsilon_2 \beta + \epsilon_3 \gamma + \epsilon_4 \delta = 1, \tag{3.1.17}$$

and the Riccati equation is:

$$w' = \frac{\epsilon_1 \alpha}{t(t-1)} w^2 + \frac{(\epsilon_3 \gamma + \epsilon_2 \beta)t - (\epsilon_1 \alpha + \epsilon_3 \gamma)}{t(t-1)} w - \frac{\epsilon_2 \beta}{t-1}. \tag{3.1.18}$$

Equation (3.1.18) reduces to a linear one, when $\alpha = 0$,

$$w' = \frac{(\epsilon_3 \gamma + \epsilon_2 \beta)t - \epsilon_3 \gamma}{t(t-1)} w - \frac{\epsilon_2 \beta}{t-1}. \tag{3.1.19}$$

The integration of (3.1.19) leads to a solution expressed in terms of the incomplete Beta function [20],

$$w = \left(c - \epsilon_2 \beta \int t^{-\epsilon_3 \gamma}(t-1)^{-\epsilon_2 \beta - 1} dt \right) t^{\epsilon_3 \gamma}(t-1)^{\epsilon_2 \beta}. \tag{3.1.20}$$

A similar result could have been obtained by setting $\beta = 0$, in which case a linear equation would have been obtained for $u = 1/w$.

Thus all the Painlevé equations with at least two parameters possess solutions involving special functions which for P_{IV}, P_V and P_{VI} are expressed as integrals.

3.2 Symmetric Discrete Painlevé Equations

Let us now turn to the discrete Painlevé equations which are the main subject of this course. We start with discrete $\mathbb{P}$'s of the form:

$$x_{n+1} = \frac{f_1(x_n, n) - x_{n-1} f_2(x_n, n)}{f_4(x_n, n) - x_{n-1} f_3(x_n, n)}. \tag{3.2.1}$$

As we have seen, the special solutions of the d-$\mathbb{P}$'s are obtained, provided that some linearisability constraint is satisfied by the discrete Riccati equation,

$$x_{n+1} = -\frac{\alpha x_n + \beta}{\gamma x_n + \delta}, \tag{3.2.2}$$

where α, β, γ and δ are functions of the independent variable n. Now, it may happen that, when some further constraint is satisfied, γ or β vanishes, in which case (3.2.2) is transformed to a linear equation for x_n or $1/x_n$. The integration of such a linear equation is straightforward. Starting from

$$\delta_n x_{n+1} + \alpha_n x_n + \beta_n = 0, \tag{3.2.3}$$

we first obtain a solution, ξ_n, of the homogeneous equation

$$\delta_n x_{n+1} + \alpha_n x_n = 0. \tag{3.2.4}$$

Formally, this solution is given by $\xi_n = A \prod_{k=0}^{n-1}(-\alpha_k/\delta_k)$ where A a is constant. Next, using the standard "variation of constant" procedure, *i.e.*, considering A as dependent on n, we obtain the solution of the full equation. We find

$$A_{n+1} - A_n = \frac{\beta_n}{\alpha_n \prod_{k=0}^{n-1}(-\alpha_k/\delta_k)} \tag{3.2.5}$$

Thus, formally, we have $A_n = \sum_n \beta_n/(\alpha_n \prod_{k=0}^{n-1}(-\alpha_k/\delta_k)) + c$, where c is the integration constant. In this way the solution of the Painlevé equation can be expressed in terms of a discrete quadrature.

For the discrete ℙ's that we shall examine here, it turns out that the homogeneous part of equation (3.2.3) can be solved explicitly. For all cases of difference equations we find an expression of the form,

$$\frac{\xi_{n+1}}{\xi_n} = \omega \frac{\prod_i(n + \rho_i)}{\prod_i(n + \sigma_i)}, \tag{3.2.6}$$

where the number of terms in the product may vary from case to case. Given the form of (3.2.6), the solution is expressed as a product of Gamma functions:

$$\xi_n = \xi_0 \omega^n \prod_i \frac{\Gamma(n + \rho_i)}{\Gamma(\rho_i)} \prod_i \frac{\Gamma(\sigma_i)}{\Gamma(n + \sigma_i)}. \tag{3.2.7}$$

In the case of q-equations,

$$\frac{\xi_{n+1}}{\xi_n} = \omega \frac{\prod_i(\lambda^{n+\rho_i} - 1)}{\prod_i(\lambda^{n+\sigma_i} - 1)}, \tag{3.2.8}$$

and the solution is expressed in terms of q-Gamma functions:

$$\xi_n = \xi_0 \omega^n \prod_i \frac{\Gamma_\lambda(n + \rho_i)}{\Gamma_\lambda(\rho_i)} \prod_i \frac{\Gamma_\lambda(\sigma_i)}{\Gamma_\lambda(n + \sigma_i)}. \tag{3.2.9}$$

However it does not seem possible, even given this explicit form of ξ, to perform the last quadrature in closed form.

Now it may also turn out that there is no possibility to set either β or γ to zero. In this case we can still proceed, provided we can find one special solution, η_n, of the Riccati equation (3.2.2). In this case, setting $x = \eta + 1/y$, we find that y satisfies the linear, inhomogeneous equation,

$$(\gamma_n \eta_{n+1} + \alpha_n)y_{n+1} + (\gamma_n \eta_n + \delta_n)y_n + \gamma_n = 0. \tag{3.2.10}$$

Just as for the continuous ℙ's, this last case will apply to the discrete ℙ$_{\mathrm{III}}$.

a) The Discrete ℙ$_{\mathrm{III}}$
The form of the q-ℙ$_{\mathrm{III}}$ we are going to work with is

$$x_{n+1}x_{n-1} = \frac{(x_n - a)(x_n - b)}{(1 - x_n z_n/c)(1 - x_n z_n/d)}, \tag{3.2.11}$$

where a, b, c and d are constants and $z_n = \lambda^n$. The linearisability condition is

$$ad = bc\lambda \tag{3.2.12}$$

and leads to the homographic mapping

$$x_{n+1} = \frac{d}{\lambda} \frac{a - x_n}{c - x_n z_n} \tag{3.2.13}$$

As we showed in [21], equation (3.2.13) leads to solutions of q-$\mathrm{P}_{\mathrm{III}}$ in terms of discrete Bessel functions. Given the form of (3.2.13), it is not possible to reduce the discrete Riccati equation directly to a linear equation. On the other hand it is straightforward to obtain one special solution of (3.2.13). We introduce $\mu \equiv \sqrt{\lambda}$ and find that $x_n = k/\sqrt{z_n}$, with $k = \sqrt{ac}$, satisfies (3.2.13), provided that condition $c\mu + d = 0$ is satisfied. Using this particular solution we can obtain a linear first-order equation by putting $x_n = k/\sqrt{z_n} + 1/y_n$. We find for y the linear mapping

$$y_{n+1}(\sqrt{az_n/c} + 1) + \mu y_n(\sqrt{az_n/c} - 1) + \mu z_n/c = 0. \tag{3.2.14}$$

The solution of the homogeneous part of this mapping is:

$$\eta_n = A\sqrt{z_n} \prod^{n-1} \tanh \frac{1}{4} \ln\left(\frac{c}{az_k}\right). \tag{3.2.15}$$

where A is a constant. The general solution of (3.2.14) can be obtained through the variation of the constant A:

$$A_{n+1} - A_n = \frac{\sqrt{z_n}}{(\sqrt{acz_n} - c) \prod^{n-1} \tanh \frac{1}{4} \ln\left(\frac{c}{az_k}\right)} \tag{3.2.16}$$

The formal integration of (3.2.16) by a discrete quadrature introduces one integration constant.

One remark is in order at this point concerning the comparison of the solution of q-$\mathrm{P}_{\mathrm{III}}$ obtained through (3.2.14) and the explicit solution of $\mathrm{P}_{\mathrm{III}}$ exhibited in Sect. 3.1. We expect the two solutions to be equivalent at the continuous limit up to the allowed transformations of the dependent and independent variables. As a matter of fact the continuous limit of (3.2.11) is not the canonical form of $\mathrm{P}_{\mathrm{III}}$ (3.1.1) but

$$w'' = \frac{w'^2}{w} - \frac{w'}{t} + w^3 + \frac{\alpha w^2}{t} + \frac{\beta}{t^2} - \frac{1}{wt^2} \tag{3.2.17}$$

obtained by $\lambda = 1 + \epsilon$, $(z \to t)$, $x = w$, $a = \epsilon$, $b = -\epsilon + \epsilon^2\beta$, $c = 1/\epsilon$, $d = -1/\epsilon - \alpha$. (Note that (3.2.17) can be transformed to (3.1.1) for ω and s through the change of variables, $s = \sqrt{t}$, $\omega = w\sqrt{t}$). The condition for special

solution (3.2.15-16) to exist is $\alpha = \beta = 1/2$. In this case the continuous limit
of the Ricccati equation (3.2.2) is just

$$x' = x^2 - \frac{x}{2t} - \frac{1}{t}.$$
(3.2.18)

The solution to this equation is

$$x = -\frac{\tanh(2\sqrt{t} - \phi)}{\sqrt{t}},$$
(3.2.19)

(where ϕ is the integration constant) which should be compared to (3.1.4a).
Let us now show how we can obtain this solution from the continuous limit
of (3.2.15-16). We set $\theta = \eta/\sqrt{z}$ and find from (3.2.15) that

$$\frac{\theta_{n+1}}{\theta_n} = \frac{1 - \epsilon\sqrt{z}}{1 + \epsilon\sqrt{z}}.$$
(3.2.20)

In order to find the continuous limit of (3.2.20), we must take into account
that $z_{n+1} = z_n + \epsilon z_n$ and thus, if $\theta_n \to g$ at the continuous limit, $\theta_{n+1} = g + \epsilon z g' + \mathcal{O}(\epsilon^2)$. At the limit $\epsilon \to 0$ we find, from (3.2.20),

$$\frac{g'}{g} = -\frac{2}{\sqrt{t}},$$
(3.2.21)

and after an integration, $g = Ae^{-4\sqrt{t}}$. In order to obtain the solution of the
nonhomogeneous equation (3.2.14), we must solve $A_{n+1} - A_n \to -\epsilon\sqrt{z}e^{4\sqrt{z}}$,
or, taking the limit,

$$A' = -\frac{e^{4\sqrt{t}}}{\sqrt{t}}.$$
(3.2.22)

Integrating (3.2.22) we find that $A = -e^{4\sqrt{t}}/2 + k$, where k is the integration constant. So, finally, in this limit y becomes $\sqrt{t}(-1/2 + ke^{-4\sqrt{t}})$.
Since the limit of the particular solution is $1/\sqrt{t}$, the full solution is $w = 1/\sqrt{t}(1 + 1/(-1/2 + ke^{-4\sqrt{t}})) = -\tanh(2\sqrt{t} - \phi/\sqrt{t})$, with $e^{2\phi} = -2k$. Thus
the continuous limit of special solution (3.2.15-16) is precisely (3.2.19).

b) The Discrete P_{IV}
Here we shall consider the d-P_{IV},

$$(x_{n+1} + x_n)(x_n + x_{n-1}) = \frac{(x_n^2 - a^2)(x_n^2 - b^2)}{(x_n - z_n)^2 - c^2},$$
(3.2.23)

where a, b, c are constants and $z_n = \delta n + z_0$. The linearisability condition is

$$2c - a - b = \delta,$$
(3.2.24)

and the corresponding homographic mapping is:

$$x_{n+1} = \frac{x_n(a+b-c-z_n) - ab}{-x_n + c + z_n}. \tag{3.2.25}$$

This mapping can obviously be made linear for $y \equiv 1/x$ provided we take $ab = 0$. Taking, for instance, $b = 0$ and implementing (3.2.24) we obtain

$$y_{n+1} = \frac{y_n(c+z_n) - 1}{c - z_{n+1}}. \tag{3.2.26}$$

The homogeneous part of this equation can be solved simply in terms of Gamma functions whereupon the general solution of (3.2.26) is given in terms of a discrete quadrature. This special solution is precisely the one discovered by Bassom and Clarkson [37] who have shown that it is the discrete equivalent of the Error function solution of continuous P_{IV}.

c) The q-Discrete P_V

The q-P_V we are now going to study is given by

$$(x_{n+1}x_n - 1)(x_n x_{n-1} - 1) = \frac{(x_n - a)(x_n - 1/a)(x_n - b)(x_n - 1/b)}{(1 - x_n z_n/c)(1 - x_n z_n/d)}, \tag{3.2.27}$$

where a, b, c, d are constants and $z_n = \lambda^n$. The linearisability condition is:

$$abd = \lambda c, \tag{3.2.28}$$

while the discrete Riccati equation, which leads to solutions expressed in terms of the discrete confluent hypergeometric function, is

$$x_{n+1} = \frac{x_n - a - b + abz_n/c}{ab(x_n z_n/c - 1)}. \tag{3.2.29}$$

A direct linearisation of (3.2.29) seems impossible, but this is only due to the fact that x is not the appropriate variable. As we remarked already in [21], the variable that becomes that of P_V in the continuous limit is related to x by a homographic transformation. We introduce the same homographic transformation here, $x = (1+y)/(1-y)$. We readily obtain for y the discrete Riccati equation,

$$y_{n+1} = \frac{y_n(-2abz_n - c(ab - a - b - 1) + c(a-1)(b-1)}{y_n c(a+1)(b+1) + 2abz_n - c(ab + a + b - 1)}. \tag{3.2.30}$$

Clearly (3.2.30) becomes linear in y if $a = -1$ or $b = -1$, and also linear in $1/y$ if $a = 1$ or $b = 1$. Let us take for instance $b = -1$. We find that

$$y_{n+1} = \frac{-y_n a(z_n + c) + c(a-1)}{az_n - c}. \tag{3.2.31}$$

The solution to the homogeneous part of (3.2.31) can be expressed in terms of q-Gamma functions, and a final quadrature leads to the general solution of (3.2.31) which is just a q-discrete analogue of the incomplete Gamma function. Indeed, setting $y = w$, $\lambda = 1 + \epsilon$, $a = 1 + \epsilon\beta$, $c = 2/\epsilon$ we obtain (2.1.12) as the continuous limit at $\epsilon \to 0$ of (3.2.29). In the same limit the linear equation (3.2.31) becomes (3.1.14), with $\alpha = 0$, $\epsilon_3 = 1$, $\epsilon_2 = 1$, which is precisely the equation whose solution is the incomplete Gamma function.

d) The (difference) Discrete P_V

This equation was introduced much more recently than its q-discrete analogue [38]. Its form was given as,

$$\frac{(x_n + x_{n+1} - z_n - z_{n+1})(x_n + x_{n-1} - z_n - z_{n-1})}{(x_n + x_{n+1})(x_n + x_{n-1})}$$
$$= \frac{(x_n - z_n - a)(x_n - z_n + a)(x_n - z_n - b)(x_n - z_n + b)}{(x_n - c)(x_n + c)(x_n - d)(x_n + d)}, \qquad (3.2.32)$$

where a, b, c and d are constants and $z_n = \delta n + z_0$. However it is not difficult to show, using the standard factorisation technique, that (3.2.32) has solutions given by the mapping:

$$\frac{x_n + x_{n+1} - z_n - z_{n+1}}{x_n + x_{n+1}} = \frac{(x_n - z_n - a)(x_n - z_n - b)}{(x_n - c)(x_n - d)}, \qquad (3.2.33)$$

provided condition

$$a + b = c + d + \delta \qquad (3.2.34)$$

holds, in which case (3.2.33) is indeed homographic,

$$x_{n+1} = \frac{((z_{n+1} - a)(z_{n+1} - b) - cd)x_n + cd(z_n + z_{n+1})}{x_n(z_n + z_{n+1}) - (z_n + a)(z_n + b)}. \qquad (3.2.35)$$

Clearly, (3.2.35) can be directly linearised for $1/x$, provided that $cd = 0$. Let us take $c = 0$ and introduce $y = 1 - z/x$, since the latter is a more convenient variable for the continuous limit. We thus obtain

$$y_{n+1} = \frac{-z_{n+1}(z_n + a)(z_n + b)y_n + ab(z_n + z_{n+1})}{z_n(z_{n+1} - a)(z_{n+1} - b)}. \qquad (3.2.36)$$

The solution of the homogeneous part of the linear equation can be given in terms of Gamma functions, whereupon the solution of (3.2.36) is reduced to a discrete quadrature. The solution thus obtained is expected to be the discrete analogue of the incomplete Gamma function. This can be assessed by the continuous limit obtained by taking $y = w$, $\delta = \epsilon$, $b = 1/\epsilon$, $a = \epsilon\beta$, $c = \epsilon\alpha$ and $d = 1/\epsilon - \epsilon\gamma$ and $z = \sqrt{t}$, whereupon, at the limit $\epsilon \to 0$ (3.2.33), becomes P_V. We find that under constraints $\alpha = 0$ and $\gamma + \beta = 1$, the linear equation (3.2.36) becomes (3.1.14).

e) The q-Discrete P_{VI}

The q-discrete analogue of P_{VI} was introduced in [38]. Its form is:

$$\frac{(x_n x_{n+1} - z_n z_{n+1})(x_n x_{n-1} - z_n z_{n-1})}{(x_n x_{n+1} - 1)(x_n x_{n-1} - 1)} =$$

$$\frac{(x_n - az_n)(x_n - z_n/a)(x_n - bz_n)(x_n - z_n/b)}{(x_n - c)(x_n - 1/c)(x_n - d)(x_n - 1/d)} \qquad (3.2.37)$$

where a, b, c, d are constants and $z_n = \lambda^n$. Its linearisable solutions were also presented in [38]. It turns out that when condition

$$ab = cd\lambda \qquad (3.2.38)$$

is satisfied, the mapping

$$\frac{x_n x_{n+1} - z_n z_{n+1}}{x_n x_{n+1} - 1} = \frac{(x_n - az_n)(x_n - bz_n)}{(x_n - c)(x_n - d)} \qquad (3.2.39)$$

becomes homographic and has solutions which are given by a discrete equation having the Gauss hypergeometric function as its continuous limit. However just as in the q-P_V case, this mapping is not convenient for direct linearisation. Again the clue to the proper variable to choose is furnished by the continuous limit. We thus introduce a change of the dependent variable,

$$x = \frac{y(1-z) - 1 - z}{y(1-z) + 1 + z}, \qquad (3.2.40)$$

and obtain a homographic mapping for y which can be directly linearised provided $(d-1)(c-1) = 0$. Let us choose $d = 1$ and let us use condition (3.2.38). We thus obtain the linear mapping,

$$y_{n+1} =$$
$$\left(\frac{z_{n+1}+1}{z_{n+1}-1}\right) \frac{\lambda(1-z_n)(bz_n - 1)(az_n - 1)y_n + (z_n + 1)(z_{n+1}z_n - 1)(ab + \lambda)}{(z_n + 1)(z_{n+1} - a)(z_{n+1} - b)}.$$
$$(3.2.41)$$

The homogeneous equation for y can be solved in terms of q-Gamma functions, and the full equation is then solved through a discrete quadrature. For the continuous limit we must take $\lambda = e^\epsilon$, $y = w$, $a = -e^{\epsilon\delta}$, $b = e^{\epsilon\gamma}$, $c = -e^{\epsilon\beta}$, $d = e^{\epsilon\alpha}$ and $z = (\sqrt{t} - 1)/(\sqrt{t} + 1)$, whereupon (3.2.37) becomes the continuous P_{VI} (3.1.16). Under the constraints (3.2.38), which in the continuous limit are just $\gamma + \delta - \alpha - \beta = 1$, and $d = 1$, i.e., $\alpha = 0$, the continuous limit of (3.2.41) is just (3.1.19), with $\epsilon_2 = -1$ and $\epsilon_3 = 1$. Thus the solution of (3.2.41) can be considered to be a q-discrete analogue of the incomplete Beta function.

3.3 Asymmetric Discrete Painlevé Equations

For asymmetric d-$\mathbb{P}$'s, the procedure is quite similar. In the results below we shall lump together the equations according to their common symmetry group. Since we are going to deal with both difference- and q-discrete equations, we shall introduce the notations $z_n = \delta(n-n_0)$, $z_{n+1/2} = \delta(n-1/2-n_0)$ $q_n = q_0\lambda^n$, $\rho_n = q_0\lambda^{n-1/2}$, and we will use $\mu = \sqrt{\lambda}$.

Equations with Geometry Described by $A_3^{(1)}$

We start with two equations whose geometry is described by the affine Weyl group of $A_3^{(1)}$. These equations are contiguity relations of the continuous P_V [24]. The first equation we shall examine is the discrete P_{IV} [29],

$$x_{n+1}x_n = \frac{y_n + z_n}{y_n^2 - a^2} \tag{3.3.1a}$$

$$y_n + y_{n-1} = \frac{1}{x_n} + \frac{z_{n+1/2} + r}{x_n - 1} \tag{3.3.1b}$$

where a and r are two constants. The linearisability constraint in this case is $a = r + \delta/2$.

The discrete Riccati equation is

$$x_{n+1} = \frac{1 + (z_n - a)x_n}{1 - 2ax_n}, \tag{3.3.2}$$

where y is given by:

$$y_n = \frac{1}{x_n} - a. \tag{3.3.3}$$

The discrete Riccati equation can be directly linearised if $a = 0$. In this case we obtain for x the equation,

$$x_{n+1} = z_n x_n + 1. \tag{3.3.4}$$

Introducing $t = 1/\delta$, we can show that (3.3.4) is the contiguity relation of the incomplete Gamma function, namely,

$$x_n = e^t t^{1-n} \gamma(n, t). \tag{3.3.5}$$

This is another kind of discrete Error function, different from and simplerthan the one introduced by Bassom and Clarkson [37].

We now turn to a second equation, known as the asymmetric d-P_{II}, and which is in fact a discrete form of P_{III} [23],

$$x_{n+1} + x_n = 2\frac{y_n z_n - p}{y_n^2 - 1} \tag{3.3.6a}$$

$$y_n + y_{n-1} = 2\frac{x_n z_{n+1/2} - r}{x_n^2 - 1}, \tag{3.3.6b}$$

where p and r are constants. The linearisability constraint,

$$p - r + \delta/2 = 0, \tag{3.3.7}$$

leads to the homographic equation for x,

$$x_{n+1} = \frac{x_n(2 + p - z_n) + 2 - 2z_n}{2x_n + 2 - p - z_n}, \tag{3.3.8}$$

with y given by

$$y_n = \frac{z_{n+1/2} + r}{x_n + 1} - 1. \tag{3.3.9}$$

It turns out that there is no possibility of obtaining a direct linearisation of the Riccati equation, a situation which exists also in the case of continuous P_{III} and the standard q-P_{III} equations. In these two cases, special solutions were obtained [33] using one particular solution, which happens to be rational, of the Riccati equation whereupon the reduction to a first order linear equation is straightforward. This is not possible here due to the fact that the asymmetric d-P_{II} and discrete P_{III} equation do not possess solutions in the linearisable class which are rational functions of the independent variable.

Equations with Geometry Described by $A_2^{(1)} \oplus A_1^{(1)}$

The equation we will here is:

$$x_{n+1}x_n = \frac{q_n^2 - aq_n y_n}{y_n(y_n - 1)} \tag{3.3.10a}$$

$$y_n y_{n-1} = \frac{\rho_n^2 - b\rho_n x_n}{x_n(x_n - 1)}, \tag{3.3.10b}$$

where a and b are constants, and, as was shown in [39], is a discrete form of P_{III}. The linearisability constraint is

$$b = a\mu, \tag{3.3.11}$$

and the homographic system is

$$x_{n+1} = \frac{q_n(x_n + a^2)}{a(x_n + aq_n)}, \tag{3.3.12}$$

with y given by

$$y_n = -\frac{aq_n}{x_n}. \tag{3.3.13}$$

No direct linearisation is expected here, but one can still obtain a special solution. It turns out that when a and b satisfy an extra constraint, $ab = 1$, which, given (3.3.11) means that $a = 1/\sigma^2$, $b = \sigma^2$, where $\lambda = \sigma^8$, one special solution does exist, $x_n = \pm Q_n/\sigma^3$, $y_n = \mp \sigma Q_n$, with $Q_n = \sqrt{q_n}$. From this elementary solution one can derive

$$x_n = \frac{Q_n}{\sigma^3} \frac{Z_n + 1}{Z_n - 1} \qquad (3.3.14a)$$

$$y_n = \sigma Q_n \frac{1 - Z_n}{1 + Z_n}, \qquad (3.3.14b)$$

provided that Z satisfies the equation

$$Z_{n+1} = Z_n \frac{\sigma Q_n + 1}{\sigma Q_n - 1}. \qquad (3.3.14c)$$

It is interesting to observe that, at the continuous limit obtained by $x = w/\epsilon$, this solution becomes a well-known solution of P_{III} for w. For this limit we choose $\lambda = 1 + \epsilon$, $q_0 = a^2/\epsilon^2$ and introduce the continuous independent variable $t = e^{-n\epsilon/2}$. We find that $Q = a/(\epsilon t)$ and $Z = e^{-4t/a}$ leading to a solution for w in terms of the hyperbolic tangent. The solution to P_{III} in terms of a tangent function could have been obtained through a different choice of signs for a and b, $a = -1/\sigma^2$, and $b = -\sigma^2$.

Equations with Geometry Described by $A_4^{(1)}$

The only equation associated to this affine Weyl group that we will discuss here is the discrete q-P_V [40]:

$$x_{n+1}x_n = \frac{(y_n - pq_n)(y_n - q_n/p)}{1 - y_n/a} \qquad (3.3.15a)$$

$$y_n y_{n-1} = \frac{(x_n - r\rho_n)(x_n - \rho_n/r)}{1 - ax_n}, \qquad (3.3.15b)$$

where a, p, r are constants. The linearisability constraint is

$$pr = \mu a^2 \qquad (3.3.16)$$

The homographic system is now

$$x_{n+1} = \frac{px_n + a^2(p^2 - 1)q_n}{apx_n + a^3(ap - q_n)}, \qquad (3.3.17)$$

where y is given by

$$y_n = -\frac{x_n - r\rho_n}{a^2}. \qquad (3.3.18)$$

The extra condition $p = \pm 1$ allows one to linearise directly the discrete Riccati equation in terms of $1/x$,

$$\frac{1}{x_{n+1}} = \frac{a^3(a \mp q_n)}{x_n} + a. \qquad (3.3.19)$$

Observe that the choice $r^2 = 1$ would lead to an equation linear in $1/y$ by interchanging $n + 1$ with $n - 1$. Thus we have essentially only one special solution.

The solution of the homogeneous equation $Z_{n+1} = a^3(a \mp q_n)Z_n$ is just a q-Gamma function, $Z_n = Z_0(\mp a^4)^n \Gamma_\lambda(n-n_1)/\Gamma_\lambda(-n_1)$, where $\lambda^{n_1} = \pm a/q_0$. We expect the solutions of the inhomogeneous equation (3.3.19) to be some form of the incomplete q-Gamma function, which, however, does not seem to be well-known.

Equations with Geometry Described by $D_4^{(1)}$

The equation we shall consider here is one obtained from the auto-Bäcklund transformation of the continuous Painlevé VI [30], although is was first identified in the degeneration through coalescence of the asymmetric q-P$_{\mathrm{III}}$ [29]. It is another discrete form of P$_\mathrm{V}$,

$$x_{n+1}x_n = \frac{(y_n - z_n)^2 - p^2}{y_n^2 - a^2} \tag{3.3.20a}$$

$$y_n + y_{n-1} = \frac{z_{n+1/2} - r}{1 - bx_n} + \frac{z_{n+1/2} + r}{1 - x_n/b}, \tag{3.3.20b}$$

where a, b, p and r are constants. The linearisability constraint can be obtained from the relation to P$_\mathrm{VI}$,

$$a + p + r + \delta/2 = 0. \tag{3.3.21}$$

In this case the discrete Riccati equation is

$$x_{n+1} = \frac{b^2(z_n + a - p)x_n + 2bp}{2abx_n + z_n + p - a}, \tag{3.3.22}$$

where y is given by

$$y_n = \frac{z_n + p + abx_n}{1 - bx_n}. \tag{3.3.23}$$

Let us first make a remark about the $b = 1$ case. For $b = 1$ one obtains from (3.3.22) and (3.3.22) a linear equation for y. However, as we showed in [41], the full, discrete system (3.3.20) for $b = 1$ is not a d-$\mathbb{P}$ anymore but rather a linearisable equation.

We turn now to the special solutions of the genuine d-P$_\mathrm{V}$, $b \neq 1$. The extra constraint, $a = 0$, leads to the linear equation,

$$\frac{z_n + p}{2bp}x_{n+1} = \frac{b(z_n - p)}{2p}x_n + 1. \tag{3.3.24}$$

Similarly, for $p = 0$, we can obtain an equation that is linear in $1/x$,

$$\frac{b(z_n + a)}{2ax_{n+1}} = \frac{z_n - a}{x_n} + 1. \tag{3.3.25}$$

In an analogous way, the constraint $r = -a$, which from (3.3.21) entails $p + \delta/2 = 0$, leads to an equation linear in terms of a simple homographic transformation of y.

All these linear equations can be shown to be contiguity relations of the incomplete Beta function. Let us introduce the quantity $U_t(\mu, \nu) = t^{-\mu}(1-t)^{-\nu}B_t(\mu, \nu)$ where $B_t(\mu, \nu) = \int_0^t \tau^{\mu-1}(1-\tau)^{\nu-1}d\tau$ is usually called the incomplete Beta function. From its properties, we can easily show that U obeys the contiguity relation,

$$\mu U_t(\mu, \nu) - t(\mu + \nu)U_t(\mu + 1, \nu) = 1, \tag{3.3.26}$$

which defines a recursion along the line $\nu = cnst$. By inspection we find for the solution of (3.3.24),

$$x_n = -\frac{2p\delta}{b}U_{1/b^2}(n - n_0 - p/\delta, 2p/\delta), \tag{3.3.27}$$

and, for the solution of (3.3.25),

$$x_n = -\frac{\delta}{U_{2a/\delta}(n - n_0 - a/\delta, 2a/\delta)}. \tag{3.3.28}$$

Equations with Geometry Described by $D_5^{(1)}$

The equation associated to this group is the asymmetric q-P_{III} which was shown by Jimbo and Sakai [16] to be a discrete form of P_{VI},

$$x_{n+1}x_n = \frac{(y_n - pq_n)(y_n - q_n/p)}{(y_n - a)(y_n - 1/a)} \tag{3.3.29a}$$

$$y_n y_{n-1} = \frac{(x_n - r\rho_n)(x_n - \rho_n/r)}{(x_n - b)(x_n - 1/b)}, \tag{3.3.29b}$$

where a, b, p, r are constants. The linearisability constraint is

$$ab\mu = pr, \tag{3.3.30}$$

and the homographic system is

$$x_{n+1} = \frac{p(apq_n - 1)x_n - ab(p^2 - 1)q_n}{bp(a^2 - 1)x_n + ab^2(q_n - ap)}, \tag{3.3.31}$$

with y given by,

$$y_n = \frac{x_n - r\rho_n}{a(x_n - b)} \tag{3.3.32}$$

The linearisation of the Riccati equation (3.3.31) leads to an equation identified by Jimbo and Sakai [16] as that of the q-hypergeometric function $_2\phi_1$. Taking $a = \pm 1$ for x leads to the linear equation,

$$b^2(q_n \mp p)x_{n+1} - p(pq_n \mp 1)x_n + (p^2 - 1)bq_n = 0. \tag{3.3.33}$$

Similarly, the choice $p = 1$ leads to an equation linear in $1/x$, while the choice $b = \pm 1$ and $r = \pm\sqrt{\mu}$ leads to linear equations involving y

and $1/y$ respectively. From the relation of the solution of (3.3.31) to the q-hypergeometric function $_2\phi_1$, we expect equation (3.3.33) to be related to the incomplete q-Beta function. In any case, it can be shown that at the continuous limit, $\lambda \to 1 + \epsilon$, $q \to t$, $b = -1 + \epsilon(\mu - 1)/2$, $p = 1 - \epsilon\nu/2$, equation (3.3.33) becomes

$$t(1 - t)\frac{dx}{dt} + (t(1 - \mu - \nu) + mu - 1)x - \nu t = 0, \qquad (3.3.34)$$

with solution $x = \nu t^{1-\mu}(1 - t)^{-\nu}B_t(\mu, \nu)$.

Equations with Geometry Described by $E_6^{(1)}$

Two well-known equations are associated to this affine Weyl group, a difference equation and a q-equation [28]. The former is known as the asymmetric d-$\mathrm{P_{IV}}$ equation, with $\mathrm{P_{VI}}$ as continuous limit,

$$(y_n + x_{n+1})(x_n + y_n) = \frac{(y_n - p)(y_n - r)(y_n - s)(y_n - t)}{(y_n - z_n - a)(y_n - z_n + a)}, \qquad (3.3.35a)$$

$$(x_n + y_n)(x_n + y_{n-1}) = \frac{(x_n + p)(x_n + r)(x_n + s)(x_n + t)}{(x_n - z_{n+1/2} - b)(x_n - z_{n+1/2} + b)} \qquad (3.3.35b)$$

where a, b, p, r, s and t are constants satisfying the constraint $p+r+s+t = 0$. The linearisability constraint,

$$a + b = p + r + \delta/2, \qquad (3.3.36)$$

leads to the homographic system,

$$x_{n+1} = \frac{y_n(\zeta_{n+1} - b) + pr}{y_n - z_n - a}, \qquad (3.3.37a)$$

where one should substitute the value of y_n,

$$y_n = \frac{x_n(z_n - a) + st}{x - z_{n+1/2} - b}. \qquad (3.3.37b)$$

From (3.3.37) we can obtain several examples where the discrete Riccati equation reduces to a first-order linear equation. First we have the case where $a = 0$. The linear equation, for x, is

$$x_{n+1} = \frac{x_n(z_n - p)(z_n - r) + z_n(st - pr) - (p + r)(pr + st)}{(z_n - s)(z_n - t)}. \qquad (3.3.38)$$

In an analogous way, taking $b = \delta/2$, one finds a similar linear equation for the variable y. For the second family of linear equations, the proper variable is not x or y but an auxiliary variable obtained through homographic transformation of it. These solutions are obtained with the extra constraint

$t = p$, or analogously any of the choices $s = p$, $s = r$, $t = r$. We introduce the variable X by $x = 1/X - p$, and obtain for X the equation

$$X_{n+1} = \frac{X_n(z_n + a - p)(z_n + p + b - \delta/2) - 2a}{(z_n - a - p)(z_n + p - b + \delta/2)}. \qquad (3.3.39)$$

The integration of these linear equations proceeds along the lines described at the beginning of this section. We expect the solutions of these linear equations to be discrete analogues of the incomplete Beta function.

The second equation associated with $E_6^{(1)}$ is the asymmetric q-P$_V$ equation,

$$(y_n x_{n+1} - 1)(x_n y_n - 1) = q_n^2 \frac{(y_n - p)(y_n - r)(y_n - s)(y_n - t)}{(y_n - a q_n)(y_n - q_n/a)} \qquad (3.3.40a)$$

$$(x_n y_n - 1)(x_n y_{n-1} - 1) = \rho_n^2 \frac{(x_n - 1/p)(x_n - 1/r)(x_n - 1/s)(x_n - 1/t)}{(x_n - b\rho_n)(x_n - \rho_n/b)}, \qquad (3.3.40b)$$

where a, b, p, r, s and t are constants satisfying the constraint $prst = 1$. Despite the fact that this equation has 5 genuine parameters, its continuous limit is just P$_{VI}$, as was shown in [28]. The linearisability constraint in this case is

$$ab = pr\mu, \qquad (3.3.41)$$

and the homographic system becomes

$$x_{n+1} = \frac{1 + aq_n(y_n - p - r)/pr}{y_n - aq_n}, \qquad (3.3.42a)$$

where one should substitute the value of y_n,

$$y_n = \frac{1 + b\rho_n(x_n st - s - t)}{x_n - b\rho_n}. \qquad (3.3.42b)$$

Let us point out that while the linearisation of (3.3.42) leads to an equation with more parameters than the discrete hypergeometric function, its continuous limit is the Gauss hypergeometric equation, which is expected since the continuous limit of (3.3.40) is P$_{VI}$. Again we obtain a special solution of asymmetric q-P$_V$ by a first-order linear q-discrete equation in several cases. Taking $a = 1$ we find the linear mapping,

$$x_{n+1} = \frac{x_n(q_n - r)(q_n - p)st + q_n^2(p + r - s - t) + q_n(st - pr)}{pr(q_n - s)(q_n - t)}. \qquad (3.3.43)$$

Similarly, a linear mapping for y can be obtained when $b = \mu$. As in the case of asymmetric d-P$_{IV}$, we obtain more instances of direct linearisation for variables obtained by a homographic transformation from x or y. These solutions are obtained with the extra constraint $t = p$ or analogously any of the choices $s = p$, $s = r$, $t = r$. We introduce the variable X through $x = 1/X + 1/p$ and obtain the equation,

$$X_{n+1} = \frac{X_n(aq_n - p)(bpq_n - \mu) + prq_n(bps - a\mu)}{p(aq_n - r)(bsq_n - \mu)} \tag{3.3.44}$$

The solutions of these equations have one more parameter than the incomplete q-Beta function.

Equations with Geometry Described by $E_7^{(1)}$

The first equation, whose geometry is described by the $E_7^{(1)}$ affine Weyl group, is the asymmetric d-P_V,

$$\frac{(y_n + x_{n+1} - z_n - \zeta_{n+1})(x_n + y_n - z_n - z_{n+1/2})}{(y_n + x_{n+1})(x_n + y_n)} =$$
$$\frac{(y_n - z_n - a)(y_n - z_n - b)(y_n - z_n - c)(y_n - z_n - d)}{(y_n - p)(y_n - r)(y_n - s)(y_n - t)} \tag{3.3.45a}$$

$$\frac{(x_n + y_n - z_n - z_{n+1/2})(x_n + y_{n-1} - z_{n-1} - z_{n+1/2})}{(x_n + y_n)(x_n + y_{n-1})} =$$
$$\frac{(x_n - z_{n+1/2} + a)(x_n - z_{n+1/2} + b)(x_n - z_{n+1/2} + c)(x_n - z_{n+1/2} + d)}{(x_n + p)(x_n + r)(x_n + s)(x_n + t)}, \tag{3.3.45b}$$

where a, b, c, d, p, r, s and t are constants satisfying the constraints $a + b + c + d = p + r + s + t = 0$. This equation, just like the asymmetric q-P_V above, has more parameters than P_{VI}. However at the continuous limit these extra parameters do not survive, and thus at the limit we obtain just P_{VI} [38]. The linearisability constraint is

$$a + b = p + r + \delta/2, \tag{3.3.46}$$

or equivalently $s + t = c + d + \delta/2$. The homographic system now becomes

$$x_{n+1} = \frac{y_n(\zeta_{n+1}^2 - (a + b)\zeta_{n+1} + ab - pr) + pr(z_n + \zeta_{n+1})}{y_n(z_n + \zeta_{n+1}) - z_n^2 - z_n(a + b) - ab + pr}, \tag{3.3.47a}$$

where one should substitute the value of y_n

$$y_n = \frac{x_n(z_n^2 + (c + d)z_n + cd - st) + st(z_n + \zeta_n)}{x_n(z_n + \zeta_n) - z_{n+1/2}^2 + z_{n+1/2}(c + d) - cd + st}. \tag{3.3.47b}$$

Again, the linearisation of (3.3.47) leads to an equation which goes beyond the discrete hypergeometric function . The direct linearisation can be obtained, provided one extra constraint is satisfied together with (3.3.46). This constraint can be any one of $t = p$, $s = p$, $s = r$, $t = r$, $a = c$, $b = c$, $a = d$, or $b = d$. For instance if we take $t = p$ and introduce the auxiliary variable X by $x = 1/X - p$, we find that

$$X_{n+1} = \frac{X_n(z_n + a - p)(z_n + b - p)(z_{n+1/2} + p - c)(z_{n+1/2} + p - d) + Q_n}{(z_n + a - p)(z_n + d - p)(\zeta_{n+1} + p - a)(\zeta_{n+1} + p - b)},$$
(3.3.48)

where Q_n is a polynomial quadratic in z, $Q_n = -2z_n^2(a+b+p+r)+2z_n(-ab+cd+4p(p+r))+\delta(p^2+(ab+cd)/2)$.

The second equation we shall examine is a q-equation, which, in the symmetric case, is a discrete form of P_{VI} [38]. It can be written as

$$\frac{(y_n x_{n+1} - q_n \rho_{n+1})(x_n y_n - q_n \rho_n)}{(y_n x_{n+1} - 1)(x_n y_n - 1)} =$$
$$\frac{(y_n - aq_n)(y_n - bq_n)(y_n - cq_n)(y_n - dq_n)}{(y_n - p)(y_n - r)(y_n - s)(y_n - t)}$$
(3.3.49a)

$$\frac{(x_n y_n - q_n \rho_n)(x_n y_{n-1} - q_{n-1}\rho_n)}{(x_n y_n - 1)(x_n y_{n-1} - 1)} =$$
$$\frac{(x_n - \rho_n/a)(x_n - \rho_n/b)(x_n - \rho_n/c)(x_n - \rho_n/d)}{(x_n - 1/p)(x_n - 1/r), (x_n - 1/s)(x_n - 1/t)}$$
(3.3.49b)

where a, b, c, d, p, r, s and t are constants satisfying the constraints $abcd = prst = 1$. The condition for the existence of a linearisable solution is

$$ab = pr\mu,$$
(3.3.50)

or, equivalently, $st = cd\mu$. Whenever this linearisability constraint is satisfied the solution is given by the homographic system,

$$x_{n+1} = \frac{y_n(1 - q_n \rho_{n+1}) + q_n(\rho_{n+1}(r + p) - a - b)}{y_n(r + p - q_n(a + b)) + pr(q_n \rho_{n+1} - 1)},$$
(3.3.51a)

where one should substitute the value of y_n,

$$y_n = \frac{x_n st(q_n \rho_n - 1) + q_n(c + d - \rho_n(s + t))}{x_n(q_n(c + d) - s - t) + 1 - q_n \rho_n}.$$
(3.3.52b)

Just as in the case of asymmetric d-P_V above, the direct linearisation can be obtained provided one extra constraint is satisfied together with (3.3.50), and this constraint can be any of $t = p$, $s = p$, $s = r$, $t = r$, $a = c$, $b = c$, $a = d$, or $b = d$. For instance we can take $t = p$ and introduce again an auxiliary variable X by $x = 1/X + 1/p$. We find that

$$X_{n+1} = \frac{X_n(aq_n - p)(bq_n - p)(pq_n - c\mu)(pq_n - d\mu) + Q_n}{(p\rho_{n+1} - a)(p\rho_{n+1} - b)(cq_n - p)(dq_n - p)},$$
(3.3.52)

where Q_n is a polynomial cubic in q, $Q_n = \mu p^2 q_n^3(1/a + 1/b - 1/c - 1/d) + pq_n^2(\mu^2(p/s + 1) - p/r - 1) + \mu p^3 q_n(r(c + d) - s(a + b)) - \mu p^4(s - r)$.

3.4 Other Types of Solutions for d-$\mathbb{P}$'s

The continuous Painlevé equations possess another type of special solution which exists whenever one particular constraint is satisfied among their parameters. This constraint is, in general, different from the linearisability constraint. The solutions obtained in this case have no degree of freedom, they are rational functions of the independent variable. The same kind of solution does exist for discrete Painlevé equations. Let us illustrate such solutions in the case of the symmetric d-P_{II},

$$x_{n+1} + x_{n-1} = \frac{2x_n z_n + 2m}{x_n^2 - 1}, \tag{3.4.1}$$

where $z_n = \alpha n + \beta$. Equation (3.4.1), obviously, has the solution $x_n = 0$ for $m = 0$. It is also straightforward to construct higher-degree rational solutions. We find, for example, for $m = -\alpha$,

$$x_n = \frac{\alpha}{1 + z_n}, \tag{3.4.2}$$

and for $m = -2\alpha$,

$$x_n = \frac{2\alpha}{1 + z_n} \frac{\alpha^2 z_n - (1 + z_n)^3}{\alpha^2(z_n + 3) - (1 + z_n)^3}. \tag{3.4.3}$$

Observe that higher-degree rational solutions exist for all values of m which are integral multiples of α, the step in z_n. We shall return to the symmetric d-P_{II} in the following section.

Another type of solution is known as "molecule" solutions. The name comes from the semi-infinite or finite Toda lattice equation which is also known under the name of Toda molecule equation, which has the form,

$$\frac{d^2}{dt^2} \log V_n = V_{n+1} + V_{n-1} - 2V_n, \tag{3.4.4}$$

with $V_0 = 0$. It admits the solution

$$V_n = \frac{\tau_{n-1}\tau_{n+1}}{\tau_n^2}, \tag{3.4.5}$$

where the τ-function is given by

$$\tau_N = \begin{vmatrix} f & \frac{d}{dt}f & \cdots & (\frac{d}{dt})^{N-1}f \\ \frac{df}{dt} & (\frac{d}{dt})^2 f & \cdots & (\frac{d}{dt})^N f \\ \vdots & \vdots & \ddots & \vdots \\ (\frac{d}{dt})^{N-1}f & (\frac{d}{dt})^N f & \cdots & (\frac{d}{dt})^{2N-2}f \end{vmatrix}, \tag{3.4.6}$$

where f is an arbitrary function of t. The crucial point is that the size of the determinants entering into the expression of a solution at site n depend on n. A molecule-type solution, obviously, can exist only for discrete equations.

Let us illustrate such a molecule-type solution in the case of asymmetric d-P_{II} (3.3.6). This equation is in fact a contiguity relation for the confluent hypergeometric equation, which, in special cases, can be a rational function of its argument. So this equation does possess another type of interesting solution, namely solutions where x_n is a rational expression of δ but of degree depending on n called molecule-type solutions in Hirota's terminology [42]). Indeed, for the special value $p = -\delta/4$, which implies that $r = \delta/4$ under the constraint (3.3.7), the ansatz

$$x_n = \frac{\delta}{8}\frac{v_n}{v_{n-1}} + \frac{\delta}{2}(n - k - \frac{1}{4}) - 1 \tag{3.4.7}$$

leads to the equation for v,

$$v_{n+1} + 8(n - k + \frac{1}{2} - \frac{2}{\delta})v_n + 16(n - k + \frac{1}{4})(n - k - \frac{1}{4})v_{n-1} = 0. \tag{3.4.8}$$

One solution can be given in terms of Hermite polynomials. Indeed, eliminating H_{m+1} and H_{m-1} from the usual recursion relation for Hermite polynomials of argument t,

$$H_{m+1} - 2tH_m + 2mH_{m-1} = 0 \tag{3.4.9}$$

and its up- and down-shifts, one obtains

$$H_{m+2} + (4m + 2 - 4t^2)H_m + 4m(m - 1)H_{m-2} = 0, \tag{3.4.10}$$

This equation is just (3.4.8) for $m = 2(n - k) + 1/2$, which is an integer provided that $k \pm 1/4$ is an integer and $t^2 = 4/\delta$. One solution of (3.4.8) is thus $v_n = H_m(2/\sqrt{\delta})$, leading for x to an expression that is rational in δ (since only polynomials of the same parity enter the equation) but of degree dependent on n.

4 From Elementary to Higher-Order Solutions

4.1 Auto-Bäcklund and Schlesinger Transformations

In the previous section we have presented solutions of Painlevé equations, both continuous and discrete, in terms of "classical" transcendents. What must be made clear at the outset is that these solutions are not isolated. They are just the first, elementary, members of large families with an infinite number of members. The importance of the elementary solutions just obtained lies in the fact that they are the "seeds" for the calculation of the higher solutions. The way one obtains these higher solutions is through the use of auto-Bäcklund and Schlesinger transformations of the equation.

Let illustrate this aproach with two selected examples. We start with the case of continuousP_{II}:

$$w'' = 2w^3 + tw + \mu \tag{4.1.1}$$

For $\mu = 1/2$, a special solution exists which can be expressed in terms of Airy functions. Setting $w = -u'/u$, we find that u satisfies the equation

$$u'' + \frac{t}{2}u = 0. \tag{4.1.2}$$

Next we use the auto-Bäcklund trasformation for P_{II} [36],

$$w(\mu + 1) = -w(\mu) - \frac{2\mu + 1}{2w^2(\mu) + 2w'(\mu) + t}, \tag{4.1.3}$$

and construct the higher Airy-type solutions. For $\mu = 1/2$ in (4.1.3), we obtain

$$w(3/2) = \frac{u'}{u} - \frac{1}{2(u'/u)^2 + t}. \tag{4.1.4}$$

Similarly, for $\mu = 5/2$, we find that

$$w(5/2) = -w(3/2) - \frac{(2(u'/u)^2 + t)^2}{4(u'/u)^3 + 2tu'/u - 1} \tag{4.1.5}$$

Higher solutions can be obtained through the repeated application of the auto-Bäcklund transformation.

We turn now to the discrete P_{II}

$$x_{n+1} + x_{n-1} = \frac{2z_n x_n + 2m}{x_n^2 - 1}, \tag{4.1.6}$$

where $z_n = \alpha n + \beta$. For the special value, $m = -\alpha/2$, d-P_{II} possesses a special solution which can be expressed in terms of a discrete Airy function. Setting $x_n = Q_{n+1}/Q_n - 1$, we find that Q satisfies the discrete equation,

$$Q_{n+1} - 2Q_n + (m - z_{n-1})Q_{n-1} = 0. \tag{4.1.7}$$

The auto-Bäcklund transformation of d-P_{II} is [26]

$$x_n(m - \alpha) = -x_n(m) + \frac{(2m - \alpha)(1 - x_n(m))}{(x_{n+1}(m) + 1)(x_n(m) - 1) - z_n - m}. \tag{4.1.8}$$

Using this transformation, we can construct a solution for $m = -3\alpha/2$. We find that

$$x_n(-3\alpha/2) = -x_n(-\alpha/2) + \frac{2\alpha Q_{n+1}(Q_{n+1} - 2Q_n)}{2Q_{n+1}^2 + Q_{n+1}Q_n(\alpha - 4) - (2z_n + \alpha)Q_n^2}. \tag{4.1.9}$$

Similarly we can construct higher Airy-type solutions for d-P_{II}.

The way to obtain higher solutions by means of auto-Bäcklund transformations, though quite systematic, does have a drawback. If one wishes to compute the Nth higher solution, one must start from the basic, elementary, one and compute the higher ones successively till one reaches the Nth. Thus no closed form expression can be given for these solutions; one has to content himself with recursion for their derivation. Fortunately an alternative approach does exist.

4.2 The Bilinear Formalism for d-$\mathbb{P}$s

A formalism which is particularly convenient for the study of the special solutions of d-$\mathbb{P}$'s is the Hirota bilinear formalism. This is not in the least astonishing since it is the very same formalism which, in the context of continuous integrable evolution equations, has made possible the systematic derivation of multi-soliton solutions.

The application of the bilinear formalism to discrete Painlevé equations was (first) introduced in [43]. As in the continuous case, one must propose an ansatz in order to express the dependent variable in terms of τ-functions. The success of the bilinearisation depends largely on the adequate choice of this ansatz. Fortunately there exists a precious guide, the singularity structure. This works equally well in the continuous as well as in the discrete case, but we shall concentrate here on the latter.

The simplest way to illustrate this is through an example. Let us start from the symmetric d-$\mathbb{P}_{\mathrm{II}}$ equation,

$$x_{n+1} + x_{n-1} = \frac{zx_n + a}{1 - x_n^2}, \qquad (4.2.1)$$

with $z = \alpha n + \beta$. Obviously, a singularity appears whenever the x in the denominator takes the value $+1$ or -1. A detailed study of these singularities leads to the singularity patterns $\{-1, \infty, +1\}$ and $\{+1, \infty, -1\}$. The assumption made in [43] was that there exists a relationship between the number of singularity patterns of a d-$\mathbb{P}$ and the minimum number of the τ-functions necessary for its description. In the case at hand, this means that we must introduce two τ-functions, F and G. Since the τ-functions are entire, x must involve ratios of such functions. With the appropriate choice of gauge, the ansatz for x turns out to be

$$x_n = -1 + \frac{F_{n+1}G_{n-1}}{F_n G_n} = 1 - \frac{F_{n-1}G_{n+1}}{F_n G_n}. \qquad (4.2.2)$$

Equating the two rightmost sides of this relation leads to a first bilinear equation for d-$\mathbb{P}_{\mathrm{II}}$,

$$F_{n+1}G_{n-1} + F_{n-1}G_{n+1} - 2F_n G_n = 0, \qquad (4.2.3)$$

and substituting the ansatz into (4.2.1) yields the second equation,

$$F_{n+2}G_{n-2} - F_{n-2}G_{n+2} = z(F_{n+1}G_{n-1} - F_{n-1}G_{n+1}) + 2aF_n G_n. \qquad (4.2.4)$$

However the assumption that the number of τ-functions is related to the number of singularity patterns should not be interpreted as a strict equality. Quite often the number of τ-functions is larger than that of the singularity patterns. This is due to the fact that auxiliary τ-functions are necessary for the bilinearisation. Let us illustrate this in the case of the "alternate" d-$\mathbb{P}_{\mathrm{II}}$ equation,

$$\frac{z_n}{x_{n+1}x_n + 1} + \frac{z_{n-1}}{x_n x_{n-1} + 1} = -x_n + \frac{1}{x_n} + z_n + \mu. \tag{4.2.5}$$

The singularity analysis of (4.2.5) results into two singularity patterns, $\{0, \infty\}$ and $\{\infty, 0\}$. This suggests the introduction of two τ-functions and the substitution:

$$x_n = \frac{F_n G_{n-1}}{F_{n-1} G_n}. \tag{4.2.6}$$

From the forms that appear in the denominators of the left hand side of (4.2.5) it becomes clear that the only hope for a simplification is when there exists some relation between numerator and denominator which leads to the introduction of a first, bilinear condition,

$$F_{n+1}G_{n-1} + F_{n-1}G_{n+1} = z_n F_n G_n. \tag{4.2.7}$$

However, even with the use of (4.2.7), the equation we obtain is still quadrilinear. In order to simplify it further we observe that if $G_n = 0$, then $F_{n+1}F_{n-1} = F_n^2$. Thus, when $G_n \neq 0$, we can extend this relation by the introduction of a third, auxiliary, τ-function, E

$$F_{n+1}F_{n-1} = F_n^2 + G_n E_n \tag{4.2.8}$$

Using (4.2.8) and (4.2.7) we can simplify the expression of alternate d-P_{II}. We finally obtain

$$G_n E_{n-1} - G_{n-1} E_n = \mu F_n F_{n-1}. \tag{4.2.9}$$

Equations (4.2.9), (4.2.8) and (4.2.7) constitute the bilinearization of alternate d-P_{II}. Similarly we could have considered what happens when $F_n = 0$. In this case we could have introduced another τ-function H by

$$G_{n+1}G_{n-1} = G_n^2 + F_n H_n, \tag{4.2.10}$$

and obtained a third equation

$$F_{n-1}H_n - F_n H_{n-1} = (\mu + \Delta z)G_n G_{n-1}, \tag{4.2.11}$$

where $\Delta z = z_{n+1} - z_n$. By comparing (4.2.9) and (4.2.11) one sees that a shift from μ to $\mu + \Delta z$ is associated to the transformation of the τ-functions $\{E, F, G\} \to \{F, G, H\}$. In fact, the τ-functions E, F, G and H constitute a Schlesinger chain in the sense of Okamoto [44].
    Let us now use the bilinear form of (4.2.5) in order to construct the elementary solution in terms of discrete Airy functions. We start with $x_n = F_n G_{n-1}/F_{n-1} G_n$ and require that $F_n = 1$ for all n. In this case (4.2.7) becomes

$$G_{n-1} + G_{n+1} = z_n G_n, \tag{4.2.12}$$

and it is not difficult to recognize in (4.2.12) a discrete form of the Airy equation. Moreover $x_n = G_{n-1}/G_n$ defines a Cole-Hopf-like transformation analogous to the one used in the continuous case. Substituting $F = 1$ in (4.2.8) we find that $E = 0$, and from (4.2.9) we obtain $\mu = 0$.

One interesting interpretation of this solution of the alternate d-P_{II} can be obtained by the relation of this equation to the continuous P_{III}, which we have investigated in detail in [45]. As we have seen in Sect. 2 the special solutions of P_{III} can be expressed in terms of Bessel functions. From the Riccati equation $w' = -w^2 - (1 + \alpha)w/t + 1$, setting $w = u'/u$ we obtain,

$$u'' + \frac{1 + \alpha}{t} u' - u = 0. \tag{4.2.13}$$

Equation (4.2.13) is solved in terms of a Bessel function, C_ν, $u = t^\nu C_\nu(it)$, where $\nu = -\alpha/2$. Using the well-known property of the Bessel functions, $C_\nu'/C_\nu = -\nu/t + C_{\nu-1}/C_\nu$ we can express w simply as $w = iC_{\nu-1}/C_\nu$. As shown in [45] the variable of alternate d-P_{II} is related to that of P_{III} by $x = i/w = C_\nu/C_{\nu-1}$. A straightforward calculation shows that x satisfies (4.2.5), provided that $\mu = 0$. We can easily show that $\mu = 0$ is precisely the condition for the existence of the special solution of P_{III}. We can wonder at this point what this special solution of alternate d-P_{II} may be. The answer is furnished by P_{III} itself. The Bessel functions C_ν satisfy the recursion relation,

$$C_{\nu+1} + C_{\nu-1} = \frac{2\nu}{t} C_\nu. \tag{4.2.14}$$

Equation (4.2.14), considered as an equation for ν, is nothing but the discrete form of the Airy equation, which can be confirmed by the continuous limit in ν. Thus the special solution of alternate d-P_{II} is expressed in terms of the discrete Airy function, as expected.

4.3 The Casorati Determinant Solutions

Why is the bilinear formalism useful for the construction of solutions of d-$\mathbb{P}$'s? In the beginning of this section we have pointed out the difficulty in constructing higher solutions using the auto-Bäcklund and Schlesinger transformations of the discrete Painlevé equations. In order to construct the Nth solution one has to calculate all the previous ones. However a small miracle, one of those which abound in the domain of integrability, occurs. The τ-function for higher solutions of the d-$\mathbb{P}$'s can be expressed as a Casorati determinant. Following the analysis of Sect. 2, we expect four different kinds of Casorati determinant solutions to exist, but it turns out that in some case not all four types are present. The first one involves the discrete special functions obtained as elementary solutions. These special functions appear as entries in the Casorati determinant. At the continuous limit, they become the corresponding solutions of the continuous Painlevé equations. We call such solutions "lattice-type solutions". The second one is that of the rational solutions. They are expressed in two ways; in the continuous limit, one reduces to the Hankel determinant representation of rational solutions for the corresponding continuous equation, while the other reduces to the Devisme polynomial representation [42]. The two other ones are what we call "molecule-type solution" (in terms of either special function or polynomials). This type of solution is specific to discrete equations: the size of the determinant in the solutions depends on the lattice site and thus the determinant structure cannot have any continuous limit. The names "lattice" and "molecule" come from the two types of the Toda equations on an infinite lattice and on a semi-infinite or finite lattice. The former is sometimes referred to as the "Toda lattice", and the latter as the "Toda molecule". The Toda lattice equation admits Casorati determinant solutions, and the size of the determinant is the number of solitons in superposition. For the Toda molecule equation, the size of the determinant is just the lattice site [42].

Let us first examine the lattice-type solutions for the alternate d-P_{II} equation. We show that they are expressed by the Casorati determinant whose entries are given by the discrete Airy function defined by (4.2.12). We here present only the results. The details of the derivation can be found in [45].

We consider the following τ-function,

$$\tau_N^n = \begin{vmatrix} G_n & G_{n+1} & \cdots & G_{n+N-1} \\ G_{n+1} & G_{n+2} & \cdots & G_{n+N} \\ \vdots & \vdots & \ddots & \vdots \\ G_{n+N-1} & G_{n+N} & \cdots & G_{n+2N-2} \end{vmatrix}, \tag{4.3.1}$$

where G_n satisfies (4.2.12),

$$G_{n+1} + G_{n-1} = (an + b)G_n. \tag{4.3.2}$$

We can then show that

$$X_n^N = \frac{\tau_{N+1}^{n-1}\tau_N^{n+1}}{\tau_{N+1}^n \tau_N^n} \tag{4.3.3}$$

gives the solution of alternate d-P_{II}, in the form

$$\frac{an + aN + b}{X_{n+1}^N X_n^N + 1} + \frac{a(n-1) + aN + b}{X_n^N X_{n-1}^N + 1} = -X_n^N + \frac{1}{X_n^N} + (an + 2aN + b). \tag{4.3.4}$$

Note that the size of the τ-function is related to the parameter N in this equation. The elementary solution is included as the special case $N = 0$, if we define $\tau_0^n = 1$. The continuous limit is obtained as follows. Setting $a = \epsilon^3$, $b = 2$, $n\epsilon = t$, $X_n^N = 1 - \epsilon u$, we find that, when $\epsilon \to 0$, the above result reduces to

$$\tau_N = \begin{vmatrix} Ai & \left(\dfrac{d}{dt}\right) Ai & \cdots & \left(\dfrac{d}{dt}\right)^{N-1} Ai \\ \left(\dfrac{d}{dt}\right) Ai & \left(\dfrac{d}{dt}\right)^{2} Ai & \cdots & \left(\dfrac{d}{dt}\right)^{N} Ai \\ \vdots & \vdots & \ddots & \vdots \\ \left(\dfrac{d}{dt}\right)^{N-1} Ai & \left(\dfrac{d}{dt}\right)^{N} Ai & \cdots & \left(\dfrac{d}{dt}\right)^{2N-2} Ai \end{vmatrix}, \tag{4.3.5}$$

where Ai is an Airy function satisfying $\frac{d^2}{dt^2} Ai = t\, Ai$. The continuous dependent variable is

$$u = \frac{d}{dt} \log \left(\frac{\tau_{N+1}}{\tau_N} \right), \tag{4.3.6}$$

and satisfies the P_{II} equation,

$$\frac{d^2}{dt^2} u = 2u^3 - 2tu + (2N + 1), \tag{4.3.7}$$

as expected.

We next turn to the molecule-type solutions of alternate d-P_{II} We start from the Toeplitz determinant,

$$\tau_N^n = \begin{vmatrix} f_n & f_{n-1} & \cdots & f_{n-N+1} \\ f_{n+1} & f_n & \cdots & f_{n-N+2} \\ \vdots & \vdots & \ddots & \vdots \\ f_{n+N-1} & f_{n+N-2} & \cdots & f_n \end{vmatrix}, \tag{4.3.8}$$

where the f_n's satisfy

$$f_{n+1} - f_{n-1} = anf_n, \tag{4.3.9}$$

where a a constant. We can show that

$$X_N^n = -\frac{\tau_{N+1}^n \tau_N^{n+1}}{\tau_{N+1}^{n+1} \tau_N^n} \tag{4.3.10}$$

satisfies the alternate d-P_{II},

$$\frac{a(N+1)}{X_{N+1}^n X_N^n + 1} + \frac{aN}{X_N^n X_{N-1}^n + 1} = -X_N^n + \frac{1}{X_N^n} + (aN + an + a). \tag{4.3.11}$$

Observe that in this case the size of the determinant is the lattice site, and hence the determinant structure of these solutions cannot survive under the continuous limit.

One can wonder whether the solutions expressed as Casorati-determinants have the properties one would expect for d-$\mathbb{P}$'s. We have seen above how, at the continuous limit, the lattice solutions of the alternate d-P_{II} become the solutions of P_{II}. Another property is that of degeneration through coalescence.

In this case, the example of P_{II} is a bad one, since this equation degenerates to P_I which has no special solutions. Instead we shall present an example based on P_{III}. In what follows we shall present both the continuous P_{III} and the q-P_{III} cases. We start from P_{III}, which is given here in the convenient form

$$W'' = \frac{W'^2}{W} + e^{2Z}(W^3 - \frac{1}{W}) + e^Z(AW^2 + B). \tag{4.3.12}$$

The higher linearisation condition which allows the solution of P_{III} to be expressed as a ratio of two τ-functions is

$$A + B = 2(2N + 1), \tag{4.3.13}$$

for integer N. Setting $A = -2\nu + 2N$ and $B = 2\nu + 2N + 2$, we find that W can be expressed as [47]

$$W = e^{-Z}\left(N + \nu + \frac{d}{dZ}\ln\left(\frac{\tau_N^{\nu+1}}{\tau_{N+1}^{\nu}}\right)\right), \tag{4.3.14}$$

where τ_N^{ν} is just the $(N \times N)$ Wronski determinant,

$$\tau_N^{\nu} = \begin{vmatrix} J_\nu & \frac{d}{dZ}J_\nu & \cdots & \frac{d^N}{dZ^N}J_\nu \\ \frac{d}{dZ}J_\nu & \ddots & & \vdots \\ \vdots & & & \\ \frac{d^N}{dZ^N}J_\nu & \cdots & & \frac{d^{2N}}{dZ^{2N}}J_\nu \end{vmatrix}. \tag{4.3.15}$$

Here, the J_ν's are Bessel functions or, more precisely, $J_\nu(e^Z)$ is solution of the equation

$$\frac{d^2}{dZ^2}J_\nu + (e^{2Z} - \nu^2)J_\nu = 0. \tag{4.3.16}$$

In order to proceed to the coalescence limit, we introduce the following transformation of the independent variable $e^Z = 1/\delta^3 + z/\delta$ together with $\nu = 1/\delta^3$. Thus, if $\delta \to 0$, then $\nu \to \infty$ and equation (4.3.16) becomes

$$\frac{d^2 J}{dz^2} + 2zJ = 0, \tag{4.3.17}$$

i.e., the Airy equation. Simultaneously, we have, for the dependent variable: $W = 1 + \delta w$, and by taking $A = -2/\delta^3 + 2N$, $B = 2/\delta^3 + 2N + 2$, we obtain for w equation P_{II} in the form

$$w'' = 2w^3 + 4zw + 4N + 2. \tag{4.3.18}$$

The only subtle point remaining is to prove that the limit of (4.3.14) leads to the Wronskian solution of (4.3.18). As a matter of fact, (4.3.14) is expressed in terms of τ^ν and $\tau^{\nu+1}$ which involve J^ν and $J^{\nu+1}$ respectively. These Bessel functions of argument e^Z are related by $J^{\nu+1} = e^{-Z}(\nu J^\nu - \frac{d}{dZ}J^\nu)$. Introducing the transformation of independent variable from Z to z, we find that $J^{\nu+1} = J^\nu + \mathcal{O}(\delta)$. Thus, at the limit $\delta \to 0$, τ^ν and $\tau^{\nu+1}$ are expressed in terms of the same solution of (4.3.17). In conclusion, the coalescence of the Wronskian solutions of P_{III} leads to the Wronskian solutions of P_{II} [48],

$$w = \frac{d}{dz}\ln\frac{\tau_N}{\tau_{N+1}}, \qquad (4.3.19)$$

for all integer N, where the τ-functions are given by expression (4.3.15), and where the entries are in terms of the solution J of the Airy equation (4.3.17).

We turn now to the case of the coalescence from q-P_{III} to d-P_{II}. We start from q-P_{III} in the form

$$X_{n+1}X_{n-1} = \frac{PQ(X_n - AZ)(X_n - BZ)}{(X_n - P)(X_n - Q)}, \qquad (4.3.20)$$

where $Z = \Lambda^n$ and A, B, P and Q are constants.

The higher linearisability condition is

$$AQ = BP\Lambda^{1+2N}, \qquad (4.3.21)$$

where N is an integer. For $N = 0$ we set:

$$X_n = P + \frac{J_{n+1}}{J_n}, \qquad (4.3.22)$$

and find for J the equation

$$J_{n+2} + (P - Q)J_{n+1} + Q(AZ - P)J_n = 0. \qquad (4.3.23)$$

The function J is characterised by one parameter, ν, which can be related to the parameters of q-P_{III} by $P/Q = -\Lambda^\nu$. In fact a simple expression for ν exists for any value of N,

$$\frac{AP}{BQ} = \Lambda^{1+2\nu}. \qquad (4.3.24)$$

Equation (4.3.23) is a discrete form of the Bessel equation. One can easily derive the contiguity relations for the discrete Bessel functions $J_n^{(\nu)}$. We find:

$$J_n^{(\nu+1)} = \frac{1}{\sqrt{Z}}(J_n^{(\nu)} + \frac{1}{P}J_{n+1}^{(\nu)}) \qquad (4.3.25a)$$

$$J_n^{(\nu-1)} = \frac{1}{\sqrt{Z}}(J_n^{(\nu)} - \frac{1}{Q}J_{n+1}^{(\nu)}), \qquad (4.3.25b)$$

where $\nu + 1$ and $\nu - 1$ are associated to values of parameters $A\sqrt{\Lambda}$, $B/\sqrt{\Lambda}$, $P\sqrt{\Lambda}$, $Q/\sqrt{\Lambda}$ and $A/\sqrt{\Lambda}$, $B\sqrt{\Lambda}$, $P/\sqrt{\Lambda}$, $Q\sqrt{\Lambda}$, respectively.

For generic N, the form of X is given in terms of τ-functions as [49],

$$X_n = P + \frac{\tau_{n+1}^{(\nu,N+1)}\tau_n^{(\nu+1,N)}}{\tau_n^{(\nu,N+1)}\tau_{n+1}^{(\nu+1,N)}}, \tag{4.3.26}$$

where $\tau_n^{(\nu,N)}$ is given by the $N \times N$ Casorati determinant of discrete Bessel functions,

$$\tau_n^{(\nu,N)} = \begin{vmatrix} J_n^{(\nu)} & J_{n+1}^{(\nu)} & \cdots & J_{n+N-1}^{(\nu)} \\ J_{n+2}^{(\nu)} & \ddots & & \vdots \\ \vdots & & & \\ J_{n+2N-2}^{(\nu)} & \cdots & & J_{n+3N-3}^{(\nu)} \end{vmatrix}. \tag{4.3.27}$$

We now proceed to the coalescence limit and introduce $X = 1 + \delta x$, $P = 1-\delta$, $Q = 1+\delta$, $A = 1-\delta-\delta^2 a/2$, $B = 1+\delta+\delta^2 a/2$ and $\Lambda = 1+\delta^2\alpha/2$, leading to $Z = 1 + z\delta^2/2$. Moreover, since $P/Q = -\Lambda^\nu$, we obtain for ν at a leading order, $\nu = 2i\pi/(\alpha\delta^2)$, which means that ν diverges at the limit $\delta \to 0$. We thus obtain d-P$_{\text{II}}$, where parameter a, resulting from the limit of (4.3.21), is

$$a = \alpha(N + 1/2). \tag{4.3.28}$$

The coalescence of the discrete Bessel equation (4.3.23) first leads to:

$$J_{n+2} - 2\delta J_{n+1} + \delta^2(z - a)J_n/2 = 0, \tag{4.3.29}$$

and we absorb the δ factors through a gauge transformation, $J_n = \delta^n K_n$. We thus find for K the equation

$$K_{n+2} - 2K_{n+1} + (z - a)K_n/2 = 0, \tag{4.3.30}$$

which is precisely the discrete form of the Airy equation. In the coalescence process, ν disappears from the limit equation. Observe that, during the limiting process, the solution K_n could a priori have retained a memory of the value of ν, by the value of $J^{(\nu)}$ from which it came. However this is not the case. In fact, since $J_n = \delta^n K_n$, $J_{n+1}^{(\nu)}$ becomes negligible in the contiguity relation (4.3.25). Moreover, since Z goes to 1 at lowest order, the K_n's are indeed independent of the index ν. By taking the limit of (4.3.26), we obtain the Casorati solution of d-P$_{\text{II}}$. Starting from (4.3.27) and expressing J_n in terms of K_n we obtain τ-functions of the same form, where J_n is replaced by K_n, and a global factor that is a power of δ. However it turns out that when we compute the ratio appearing in (4.3.26), all the δ's but one drop out. We thus have

$$X_n = 1 - \delta + \delta \frac{\tau_{n+1}^{(N+1)}\tau_n^{(N)}}{\tau_n^{(N+1)}\tau_{n+1}^{(N)}}, \tag{4.3.31}$$

and, since $X = 1 + \delta x$, we see that a solution of d-P$_{\text{II}}$ is given by

$$x_n = -1 + \frac{\tau_{n+1}^{(N+1)}\tau_n^{(N)}}{\tau_n^{(N+1)}\tau_{n+1}^{(N)}}, \tag{4.3.32}$$

for all integers N.

We turn now to the study of rational solutions. We shall concentrate on the example of the "standard" d-P_{II} for which Kajiwara and Ohta [46] have derived Casorati solutions. These authors have shown that the Casorati determinant,

$$
\tau_N^n = \begin{vmatrix} L_N^{(n)} & L_{N+1}^{(n)} & \cdots & L_{2N-1}^{(n)} \\ L_{N-2}^{(n)} & L_{N-1}^{(n)} & \cdots & L_{2N-3}^{(n)} \\ \vdots & \vdots & \ddots & \vdots \\ L_{N+2}^{(n)} & L_{N+3}^{(n)} & \cdots & L_1^{(n)} \end{vmatrix},
\tag{4.3.33}
$$

whose elements are Laguerre polynomials, satisfies the d-P_{II} equation. We recall the definition of the Laguerre polynomials $L_k^{(n)}(t)$ in terms of their generating function,

$$
\sum_{k=0}^{\infty} L_k^{(n)}(t)\lambda^k = (1 - \lambda)^{1-n} e^{\frac{\lambda t}{1-\lambda}},
\tag{4.3.34}
$$

and $L_k^{(n)}(t) = 0$ for $k < 0$. Thus $L_k^{(n)}(t)$ is polynomial in n of degree k. The nonlinear variable x is related to τ by

$$
x_n^N = \frac{\tau_{N+1}^{n+1} \tau_N^{n-1}}{\tau_{N+1}^n \tau_N^n} - 1,
\tag{4.3.35}
$$

and satisfies the d-P_{II},

$$
x_{n+1}^N + x_{n-1}^N = \frac{2}{t} \frac{(n+1)x_n^N - (N+1)}{1 - (x_n^N)^2}.
\tag{4.3.36}
$$

What is really interesting at this point is to link the continuous limit of this result to the one already obtained for the rational solutions of P_{II}. This limit was obtained by Kajiwara and Ohta, who showed that the rational solution of d-P_{II} becomes the rational solution of P_{II}. For completeness, let us mention the result concerning this latter solution. First one introduces the Devisme polynomials, $p_k(z, s)$, by the generating function,

$$
\sum_{k=0}^{\infty} p_k(z, s)\eta^k = e^{z\eta + s\eta^2 + \eta^3/3},
\tag{4.3.37}
$$

and $p_k(z, s)$ for $k < 0$. Then the Casorati determinant

$$
\tau_N = \begin{vmatrix} p_N(z, s) & p_{N+1}(z, s) & \cdots & p_{2N-1}(z, s) \\ p_{N-2}(z, s) & p_{N-1}(z, s) & \cdots & p_{2N-3}(z, s) \\ \vdots & \vdots & \ddots & \vdots \\ p_{-N+2}(z, s) & p_{-N+3}(z, s) & \cdots & p_1(z, s) \end{vmatrix}
\tag{4.3.38}
$$

can be shown to be independent of s. Introducing the nonlinear variable

$$v = \frac{d}{dz} \log \frac{\tau_{N+1}}{\tau_N}, \tag{4.3.39}$$

one can show that it satisfies the P_{II} equation,

$$v'' = 2v^3 - 4zv + 4(N+1), \tag{4.3.40}$$

where the prime denotes the derivative with respect to z.

5 Bonus Track:
Special Solutions of Ultra-discrete Painlevé Equations

In this section we shall present the special solutions of another brand of Painlevé equations, ultra-discrete $\mathbb{P}$'s [50]. These equations are the cellular-automaton-analogues of the Painlevé equations where the dependent variable takes discrete values. The systematic derivation [51] of these systems became possible thanks to the ultra-discretisation procedure introduced by the Tokyo-Kyoto group [52]. This method is based on a transformation of the dependent variable followed by a limiting procedure. We start with the transformation which relates the variable of the discrete system x to that of the ultra-discrete X, by $x = e^{X/\epsilon}$. An essential requirement is that the variable of the discrete equation assume only positive values. In practice this means that the ultra-discretisation will isolate the positive solutions of the discrete system. In order to obtain a cellular automaton, one takes the limit $\epsilon \to +0$. The cornerstone of the procedure is the identity, $\lim_{\epsilon \to +0} \epsilon \log(e^{a/\epsilon} + e^{b/\epsilon}) = \max(a, b)$. Thus, if a and b are integers, the result of the operation will be an integer.

Let us show how this method works for the P_I equation. In order to ensure the positivity requirement, the discrete forms considered were the multiplicative ones, $i.e.$, the q-Painlevé equations. Thus, for example, we start from the three expressions of q-P_I,

$$x_n^\sigma x_{n+1} x_{n-1} = z x_n + 1, \tag{5.1}$$

where $\sigma = 0$, 1 or 2 and $z = \lambda^n$. From (5.1), with $\lambda = e^{1/\epsilon}$, we obtain the ultra-discrete forms,

$$X_{n+1} + X_{n-1} + \sigma X_n = \max(0, X_n + n). \tag{5.2}$$

In a series of works [50], [51], [53] we showed that the properties of the ultra-discrete $\mathbb{P}$'s follow those of the d-$\mathbb{P}$'s, which parallel the ones of the continuous $\mathbb{P}$'s. Thus the question of the existence of special solutions [53], like the ones which have been the object of the present course, is quite natural.

Let us start with the ultra-discrete P_{II} of first kind,

$$X_{n+1} + X_{n-1} = \max(0, n - X_n) - \max(0, X_n + n - a), \tag{5.3}$$

which is derived from the multiplicative d-P_{II},

$$x_{n+1}x_{n-1} = \frac{x_n + z}{x_n(1 + \alpha z x_n)},$$ (5.4)

where $z = \lambda^n$, and α is a parameter, by setting $x = e^{X/\epsilon}$, $\lambda = e^{1/\epsilon}$, $\alpha = e^{-a/\epsilon}$ and taking the limit $\epsilon \to +0$. By the transformation of the variable,

$$X_n = \tau_{n-1} - \tau_{n-2},$$ (5.5)

we obtain the ultra-discrete "bilinear" equation,

$$\tau_n + \max(\tau_{n-2}, \tau_{n-1} + n - a) = \tau_{n-3} + \max(\tau_{n-1}, \tau_{n-2} + n).$$ (5.6)

We remark that in the ultra-discrete world, "max" and "+" should be regarded as "addition" and "multiplication", respectively, because $e^{A/\epsilon} + e^{B/\epsilon}$ and $e^{A/\epsilon}e^{B/\epsilon}$ go to $\max(A, B)$ and $A + B$, respectively, under the operation $\lim_{\epsilon \to +0} \epsilon \log$. Thus the above equation is indeed a bilinear form in τ. Now (5.3) is invariant under the transformation $a \to -a$, $n \to n - a$, $X \to -X$, thus we can assume $a \geq 0$ without loss of generality. The u-P_{II} (5.3) admits rational solutions for $a = 4m$, where m is a nonnegative integer. The τ-function for the rational solution is

$$\tau_n = \sum_{j=0}^{m-1} \max(0, n - 3j),$$ (5.7)

which is also expressed as

$$\tau_n = \max_{0 \leq j \leq m} \left(jn - \frac{3}{2}j(j - 1)\right).$$ (5.8)

Next we consider the special solution for the u-P_{II} of second kind,

$$X_{n+1} + X_{n-1} - X_n = \max(0, n - X_n) - \max(0, X_n + n - a).$$ (5.9)

This is derived from another multiplicative d-P_{II},

$$x_{n+1}x_{n-1} = \frac{x_n + z}{1 + \alpha z x_n},$$ (5.10)

$z = \lambda^n$, and α is a parameter, by setting $x = e^{X/\epsilon}$, $\lambda = e^{1/\epsilon}$, $\alpha = e^{-a/\epsilon}$ and taking the limit $\epsilon \to 0$. The ultra-discrete bilinear form of (5.9) is

$$\tau_n + \max(\tau_{n-3}, \tau_{n-1} + n - a) = \tau_{n-4} + \max(\tau_{n-1}, \tau_{n-3} + n),$$ (5.11)

where $X_n = \tau_{n-1} - \tau_{n-3}$. For $a = 6m$, with m a nonnegative integer, there exist rational solutions of (5.9). The τ-function for the solution is given by

$$\tau_n = \sum_{j=0}^{m-1} \max(0, n - 4j). \tag{5.12}$$

For this τ-function, X_n gives a multistep solution with the elementary pattern of two successive jumps at $n = 4j - 2$ ($1 \leq j \leq m$) followed by two steps with constant value.

Let us proceed to the rational solution of ultra-discrete P_{III}. We consider only a degenerate case, namely, the case in which the u-P_{III} is decomposed into two parts. The u-P_{III},

$$X_{n+1} + X_{n-1} - 2X_n =$$
$$\max(0, n - X_n) - \max(0, X_n + n - a) + \max(0, n - X_n + b) - \max(0, X_n + n + b), \tag{5.13}$$

is derived from the d-P_{III},

$$x_{n+1}x_{n-1} = \frac{(x_n + z)(x_n + \beta z)}{(1 + \alpha z x_n)(1 + \beta z x_n)} \tag{5.14}$$

, where $z = \lambda^n$ and α, β are parameters, by setting $x = e^{X/\epsilon}$, $\lambda = e^{1/\epsilon}$, $\alpha = e^{-a/\epsilon}$, $\beta = e^{b/\epsilon}$ and taking the limit $\epsilon \to +0$. Now decomposing (5.13) in the following two equations,

$$X_{n+1} + X_{n-1} = \max(0, n - X_n) - \max(0, X_n + n - a), \tag{5.15a}$$

$$2X_n = \max(0, X_n + n + b) - \max(0, n - X_n + b) \tag{5.15b}$$

we obtain bilinear equations by $X_n = \tau_{n-1} - \tau_{n-2}$,

$$\tau_n + \max(\tau_{n-2}, \tau_{n-1} + n - a) = \tau_{n-3} + \max(\tau_{n-1}, \tau_{n-2} + n), \tag{5.16a}$$

$$\tau_{n-1} + \max(\tau_{n-1}, \tau_{n-2} + n + b) = \tau_{n-2} + \max(\tau_{n-2}, \tau_{n-1} + n + b). \tag{5.16b}$$

The first equation, (5.15a) or (5.16a), is nothing but the bilinear form of the u-P_{II} of first kind, therefore what we have to do is to prove that solution (5.7) simultaneously satisfies the second bilinear equation (5.16b). Another bilinear equation can be easily obtained,

$$\max(\tau_{n-q_1}, \tau_{n-p_1} + n) - \tau_{n-p_1} = \max(\tau_{n-q_2}, \tau_{n-p_2} + n) - \tau_{n-p_2},$$
$$0 \leq p_i \leq q_i + k, q_i \geq 0, \tag{5.17}$$

which yields (5.16b) by setting $p_1 = q_2 = b + 1$, $p_2 = q_1 = b + 2$ and $k = 3$. Hence we have proved that the same τ as in the rational solution of u-P_{II} yields the solution for the u-P_{III} (5.13) with $a = 4m$, where m is a nonnegative integer, and $b \geq -1$.

Since ultra-discrete analogues have also been established for the higher Painlevé equations, it is certainly possible to obtain "multistep" solutions. However, due to the reluctance of the authors to embark upon a derivation which presents considerable computational difficulties, none has been derived yet.

Epilogue

In this course we have reviewed the special solutions of Painlevé equations with an emphasis on the solutions of the discrete $\mathbb{P}$'s. We have shown that, for both the continuous and the discrete equations, the principle for the construction of the basic, "elementary", solution is the same. One requires that the Painlevé equation, be satisfied by the solution of a first-order Riccati, or even linear, equation. This is, of course, possible only for some special values of the parameters of the Painlevé equation and the solution thus obtained does not exist for arbitrary initial conditions. After all one expects the general solution of the Painlevé equation to be transcendental.

Several open problems remains. The Casorati determinant solutions have been computed for only a few selected d-$\mathbb{P}$'s. While for the special-function type solutions, the situation is bad, it is even worse for the rational solutions. For the former one knows at least what will be the elements of the determinant. They are the special functions obtained at the first stage, that of the calculation of the lower, elementary, solution. However for the rational solutions, there is no way to know beforehand what special polynomials will appear as elements of the Casorati determinant. So a good deal of experimentation is needed, unless one is able to link the d-$\mathbb{P}$ to some other system which can provide a useful guide.

Another point which would require additional study is that of the characterisation of the discrete special functions which appear as solutions of the linear second-order mappings obtained from the linearisation of the discrete Riccati equations. For some of them, it is possible to provide a description either as contiguities of continuous special functions or, in the case of q-mappings, as degenerate forms of the q-hypergeometric function. In all cases, of course, the continuous limit of the linear equation is useful, be it only for attributing a name to the discrete special functions. However for the higher-order linear mappings and the related special functions the question is still open.

References

1. P. Painlevé, Acta Math. 25 (1902) 1.
2. B. Gambier, Acta Math. 33 (1910) 1.
3. M. J. Ablowitz and H. Segur, Phys. Rev. Lett. 38 (1977) 1103.
4. A. Ramani, B. Grammaticos and J. Hietarinta, Phys. Rev. Lett. 67 (1991) 1829.
5. B. Grammaticos, B. Dorizzi, J. Math. Comp. in Sim. 37 (1994) 341.
6. B. Grammaticos, A. Ramani and V. Papageorgiou, Phys. Rev. Lett. 67 (1991) 1825.
7. B. Grammaticos, A. Ramani and O. Gerard, *Integrability and Chaos in Discrete and Ultradiscrete Systems*, in " Nonlinear Dynamics: Integrability and Chaos in Dynamical Systems", Narosa (2000) 163.
8. B. Grammaticos, A. Ramani, Reg. and Chaot. Dyn., 5 (2000) 53.

9. B. Grammaticos, F. Nijhoff and A. Ramani, *Discrete Painlevé equations*, in "The Painlevé Property: One Century Later", CRM series in Mathematical Physics, Springer (1999) 413.
10. G. R. W. Quispel, J. A. G. Roberts and C. J. Thompson, Physica D 34 (1989) 183.
11. E. L. Ince, *Ordinary Differential Equations*, Dover, London, (1956).
12. B. Grammaticos and A. Ramani, Integrability – and How to Detect it, Lect. Notes Phys. 638, 31 (2004).
13. J. A. Shohat, Duke Math. J. 5 (1939) 401.
14. E. Brézin and V. A. Kazakov, Phys. Lett. 236B (1990) 144.
15. B. Grammaticos and A. Ramani, J. Phys. A 31 (1998) 5787.
16. M. Jimbo and H. Sakai, Lett. Math. Phys. 38 (1996) 145.
17. H. Sakai, Commun. Math. Phys. 220 (2001) 165.
18. B. Grammaticos, T. Tamizhmani, A. Ramani, A. S. Carstea and K. M. Tamizhmani, J. Phys. Soc. Japan 71 (2002) 443.
19. Y. Ohta, A. Ramani and B. Grammaticos, J. Phys. A 35 (2002) L653.
20. V. A. Gromak and N. A. Lukashevich, *The analytic solutions of the Painlevé equations*, (Universitetskoye Publishers, Minsk 1990), in Russian.
21. K. M. Tamizhmani, A. Ramani, B. Grammaticos, and K. Kajiwara, J. Phys. A 31 (1998) 5799.
22. A. Abramowitz and I. Stegun, *Handbook of Mathematical Functions*, Dover (1965).
23. B. Grammaticos, F. W. Nijhoff, V. Papageorgiou, A. Ramani and J. Satsuma, Phys. Lett. A 185 (1994) 446.
24. K. M. Tamizhmani, A. Ramani, B. Grammaticos, and Y. Ohta, Lett. Math. Phys. 38 (1996) 289.
25. K. M. Tamizhmani, B. Grammaticos, and A. Ramani, Lett. Math. Phys. 29 (1993) 49.
26. A. Ramani and B. Grammaticos, J. Phys. A 25 (1992) L633.
27. T. Tamizhmani , K. M. Tamizhmani, B. Grammaticos, A. Ramani, J. Phys. A 32 (1999) 4553.
28. A. Ramani, B. Grammaticos and Y. Ohta, *The Painlevé of discrete equations and other stories*, invited talk at the "Atelier sur la théorie des fonctions spéciales non linéaires: les transcendants de Painlevé", Montréal 1996.
29. B. Grammaticos, Y. Ohta, A. Ramani and H. Sakai, J. Phys. A 31 (1998) 3545.
30. F. W. Nijhoff, A. Ramani, B. Grammaticos and Y. Ohta, Stud. Appl. Math. 106 (2000) 261.
31. T. Tokihiro, B. Grammaticos and A. Ramani, J. Phys. A 35 (2002) 5943.
32. A. Ramani, Y. Ohta, J. Satsuma and B. Grammaticos, Comm. Math. Phys. 192 (1998) 67.
33. T. Tamizhmani, B. Grammaticos, A. Ramani and K. M. Tamizhmani, Physica A 295 (2001) 359.
34. T. Tamizhmani, B. Grammaticos, A. Ramani and K. M. Tamizhmani, Physica A 315 (2002) 569.
35. A. Ramani, B. Grammaticos, T. Tamizhmani and K. M. Tamizhmani, J. Phys. A 33 (2000) 579.
36. A. S. Fokas and M. J. Ablowitz, J. Math. Phys. 23 (1982) 2033.
37. A. Bassom and P. A. Clarkson, Phys. Lett. A 194 (1994) 358.
38. B. Grammaticos and A. Ramani, Phys. Lett. A 257 (1999) 288.

39. M. D. Kruskal, K. M. Tamizhmani, B. Grammaticos and A. Ramani, Reg. Chaot. Dyn. 5 (2000) 273.
40. A. Ramani, B. Grammaticos and Y. Ohta, J. Phys. A 34 (2001) 2505.
41. A. Ramani, B. Grammaticos, Y. Ohta, B. Grammaticos, Nonlinearity 13 (2000) 1073.
42. K. Kajiwara, *The discrete Painlevé II equations and classical special functions*, in "Symmetries and Integrability of Difference Equations", London Math. Soc. Lect. Note Ser. 255, Cambridge Univ. Press (1999) 217.
43. A. Ramani, B. Grammaticos and J. Satsuma, J. Phys. A 28 (1995) 4655.
44. K. Okamoto, Japan J. Math 5 (1979) 1.
45. F. Nijhoff, J. Satsuma, K. Kajiwara, B. Grammaticos and A. Ramani, Inverse Problems 12 (1996) 697.
46. K. Kajiwara and Y. Ohta, J. Math. Phys. 37 (1996) 4693.
47. K. Okamoto, Math. Ann. 275 (1986) 221.
48. K. Kajiwara, Y. Ohta and J. Satsuma, J. Math. Phys 35, (1995) 4162.
49. K. Kajiwara, Y. Ohta, J. Satsuma, B. Grammaticos and A. Ramani, J. Phys. A 27 (1994) 915.
50. B. Grammaticos, Y. Ohta, A. Ramani, D. Takahashi and K. M. Tamizhmani, Phys. Lett. A 226 (1997) 53.
51. B. Grammaticos, Y. Ohta, A. Ramani and D. Takahashi, Physica D 114 (1998) 185.
52. T. Tokihiro, D. Takahashi, J. Matsukidaira and J. Satsuma, Phys. Rev. Lett. 76 (1996) 3247.
53. D. Takahashi, T. Tokihiro, B. Grammaticos, Y. Ohta and A. Ramani, J. Phys. A 30 (1997) 7953.

Ultradiscrete Systems (Cellular Automata)

Tetsuji Tokihiro

Graduate School of Mathematical Sciences, University of Tokyo, 3-8-1 Komaba, Meguro-ku, Tokyo 153-8914, Japan, toki@ms.u-tokyo.ac.jp

Abstract. Ultradiscretization is a limiting procedure which allows one to obtain a cellular automaton (CA) from continuous equations. Using this method, we can construct *integrable* CAs from integrable partial difference equations. In this course, we focus on a typical integrable CA, called a Box and Ball system (BBS), and review its peculiar features. Since a BBS is an ultradiscrete limit of the discrete KP equation and discrete Toda equation, we can obtain explicit solutions and conserved quantities for the BBS. Furthermore the BBS is also regarded as a limit (crystallization) of an integrable lattice model. Recent topics, and a periodic BBS in particular are also reviewed.

1 Introduction

A cellular automaton (CA) is a discrete dynamical system consisting of a regular array of cells [1]. Each cell takes only a finite number of states and is updated in discrete time steps. Although the updating rules are simple, CAs often exhibit very complicated time evolution patterns which resemble natural phenomena such as chemical reactions, turbulent flows, nonlinear dispersive waves and solitons. As an example, let us consider a one dimensional CA, each cell of which is in either an active or an inactive state. We assume the updating rule at the next time step to be:

1. An active cell always becomes inactive.
2. An inactive cell becomes active if and only if one of the two adjacent cells is active.

Figure 1 shows a time evolution pattern of this CA. A black (white) cell denotes an active (inactive) cell. Clearly the pattern shows self-similarity and, in an appropriate limit, coincides with a fractal pattern (a Sierpinski gasket). From this example, we see that CAs provide models for natural phenomena which are not easily modeled by continuous systems described by differential and/or difference equations.

On the other hand, some CAs show behaviour similar to that explained in general by partial differential equations. In 1985 Wolfram listed up 20 important problems in the research of CAs. [2]. The 9th problem asks "What is the correspondence between cellular automata and continuous systems ?". In a comment on this problem, he pointed out the similarity between time

T. Tokihiro, Ultradiscrete Systems (Cellular Automata), Lect. Notes Phys. **644**, 383–424 (2004)
http://www.springerlink.com/ © Springer-Verlag Berlin Heidelberg 2004

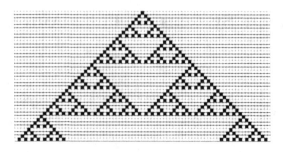

Fig. 1. A time evolution pattern of a cellular automaton

evolution patterns of CAs and behaviour of continuous systems described by differential equations, and stated that discretization of time and spatial valuables would correspond to an approximation in numerical calculation, but that the meaning of a discretization of physical quantities is not clear. He also mentioned "Explicit example of cellular automaton approximations to partial differential equations would be variable".

Concerning this problem, a systematic method to construct CAs from soliton equations (nonlinear integrable equations) was proposed in [3,4]. The method is called ultradiscretization[1] and is based on a limiting procedure through which an equation turns into a piecewise linear equation so that it can be closed under a finite number of discrete values. Hence, in the ultra-discretization, the discretization of physical quantities is understood as an approximation of the equations for the quantities by piecewise linear equations. Since the limiting process is continuous, if the solutions or conserved quantities of the system are stable under the limiting procedure they are naturally transformed into those of the CAs, and the features of the original systems are preserved in the CAs. For soliton equations there exists a stable transformation parameter (a spectral parameter) and through ultradiscretization we can construct CAs which have soliton solutions and an infinite number of conserved quantities. However, it is a rather difficult problem to find such a parameter in a general partial differential equation, and when we construct a CA through ultradiscretization its time evolution patterns often do not reflect the original system. Hence the mathematical structure of ultradiscrete systems is well elucidated only for the integrable CAs, *i.e.*, the CAs obtained from soliton equations.

In this chapter, we review recent developments in ultradiscrete systems. Since, as mentioned above, there remain many problems for general ultradiscrete systems we focus on the integrable CAs, in particular, on so called box-ball systems (BBSs) whose mathematical structure has been clarified with respect to soliton equations and two-dimensional integral lattice models.

---

[1] This name was given by B. Grammaticos at the previous CIMPA meeting, Pondicherry, 1996.

2 Box-Ball System

Solitonic behaviour is widely observed in filter type CAs [5] and these filter type CAs are sometimes called *soliton cellular automata* [6–8]. One of the typical CAs exhibiting solitonic behaviour is the so called box and ball system (BBS). The BBS is a dynamical system of balls in an array of boxes [9] which is a reinterpretation of the filter type CA proposed by Takahashi and Satsuma [10]. Let us consider a one-dimensional array of an infinite number of boxes. At the initial time step $t = 0$, all but a finite number of the boxes are empty, and each of the remaining boxes contains a ball as shown in Fig. 2.

Fig. 2. A state of BBS

As a CA, a vacant box corresponds, say, to an inactive cell and a filled box to an active cell. The time evolution rule for this system from time t to $t + 1$ is given as follows.

1. Move every ball only once.
2. Move the leftmost ball to its nearest right empty box.
3. Move the leftmost ball of the remaining balls to its nearest right empty box.
4. Repeat the above procedure until all the balls have moved.

An example of the time evolution is shown in Fig. 3. In this example, we clearly see the solitonic behavior of balls which is a general property of the BBS. The following facts hold for general time evolution patterns of the BBS.

1. The speed of a sequence of consecutive balls is proportional to its length.
2. Two such sequences do not change their lengths after collision.
3. The phase of a sequence, however, shifts after collision.

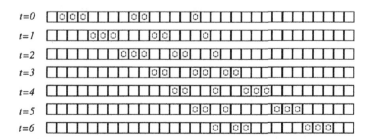

Fig. 3. Time evolution of the BBS

Thus a 'soliton' in the BBS is a sequence of consecutive balls and the 'length' of the sequence corresponds to the 'amplitude' of a KdV soliton. We can prove that every state of this BBS consists of only solitons in the sense that after finitely many time steps it becomes a state consisting of only freely moving solitons [11].

The reason why the BBS has a solitonic nature is well understood by the idea of ultradiscretization. The BBS is regarded as an ultradiscrete limit of the discrete KP equation (dKP eq.) or discrete Toda equation (dToda eq.) [12]. In this limit a soliton solution to the dKP eq. turns into one for the BBS and a conserved quantity for the dToda eq. into that for the BBS.

Although the original BBS is a reinterpretation of the soliton CA by Takahashi and Satsuma, it can be extended to BBSs with many kinds of balls, larger box capacity and carrying carts [13]. All these BBSs can be regarded as ultradiscrete limits of some integrable equations and the solutions and conserved quantities are obtained in concrete forms [14,15]. Furthermore the BBS can be constructed from two dimensional integrable lattice models through so called *crystallization* [16]. Hence the BBS is not only a limit of classical integrable systems (integrable partial differential equations) but also of quantum integrable systems. Using the idea of crystallization, we can regard the BBS as an $A_M^{(1)}$ crystal lattice [17,18] (Fig. 11). Hence we may say that the BBS is an *integrable* CA [19,20].

In the next section, we explain the relation of the BBS to classical integrable systems through ultradiscretization.

3 Ultradiscretization

Ultradiscretization is a limiting procedure which transforms a (continuous valued) discrete equations to a piecewise linear equation. By this method the equation becomes closed under a finite number of values and it can be regarded as a CA. In the next two subsections, we show how this method is used to obtain the BBS from the dKP equation and the dToda equation.

3.1 BBS as an Ultradiscrete Limit of the Discrete KP Equation

The aim of the present subsection is to relate the BBS to one of the most important integrable discrete equation, the discrete KP equation. The discrete KP equation is often called the Hirota-Miwa equation because it was first proposed by Hirota [21] as a discrete generalized Toda equation and Miwa showed that it is essentially equivalent to a generating formula of the KP hierarchy [22,23].

The KP hierarchy is the $\infty \times \infty$ simultaneous nonlinear partial differential equations with ∞ independent variables $t := (t_1 \equiv x, t_2, t_3, \dots)$ and ∞ dependent variables $u := (u_1, u_2, u_3, \dots)$ [24]:

$$\frac{\partial u_i(t)}{\partial t_j} = F_{ij}(u, u_x, u_{xx}, \dots) \quad (1 \le i, \ 2 \le j)$$

where F_{ij} are polynomials in u and their differentials with respect to x. The KP hierarchy is integrable as simultaneous partial differential equations, *i.e.*,

$$\frac{\partial^2 u_i}{\partial t_j \partial t_k} = \frac{\partial^2 u_i}{\partial t_k \partial t_j} \quad (^\forall i, \ j, \ k).$$

The theory of the KP hierarchy (Sato theory) was developed by Sato [25] and Date-Jimbo-Kashiwara-Miwa [26, 27] in the early 1980s. The Sato theory clarifies the structure of the space of solutions, the transformation groups on that space and the relation to linear equations (Lax equations). Some important results of the theory are,

(1) The KP hierarchy is equivalent to the bilinear identity for a single dependent variable $\tau(t)$:

$$\text{Res}_{\lambda = \infty} \left[\tau(t - \epsilon(\lambda))\tau(t' + \epsilon(\lambda))e^{\xi(t - t'; \lambda)} \right] = 0 \quad (^\forall t, t'), \tag{1}$$

where

$$\epsilon(\lambda) := \left(\frac{1}{\lambda}, \frac{1}{2\lambda^2}, \frac{1}{3\lambda^3}, \cdots \right), \quad \xi(t - t'; \lambda) := \sum_{j=1}^{\infty} (t_j - t'_j)\lambda^j.$$

(2) The existence of the boson-fermion correspondence ι , that is, the isomorphism between the ring of formal power series of infinite variables, $\mathbb{C}[t]$, and the fermion Fock space, $\mathcal{F}$, on which infinite numbers of creation and annihilation operators act,

$$\iota: \quad |\phi\rangle \in \mathcal{F} \quad \xrightarrow{\sim} \quad \langle \text{vac}|e^{H(t)}|\phi\rangle \in \mathbb{C}[t].$$

Here $|\text{vac}\rangle$ denotes the vacuum state of the Fock space $\mathcal{F}$ and $H(t) := \sum_{k=1}^{\infty} t_k H_k$ is a linear operator with the boson operators H_k constructed from fermion creation and annihilation operators.

(3) On $\mathcal{F}$ acts the infinite-dimensional transformation group GL_∞ which is, roughly speaking, the group of linear transformation of one particle states in $\mathcal{F}$. The necessary and sufficient condition for $\tau(t) \in \mathbb{C}[t]$ to satisfy the bilinear identity (1) is that the state $|\phi\rangle := \iota^{-1}\tau(t)$ is on an orbit of the vacuum, that is,

$$^\exists g \in GL_\infty \quad |\phi\rangle = g|\text{vac}\rangle.$$

From these results, we see that the bilinear identity (1) is the key equation for the KP hierarchy and that all the solutions are expressed concretely with fermions.

Now, let a, b, c be arbitrary but different complex numbers. Substituting

$$t = (\ell+1)\epsilon(a) + (m+1)\epsilon(b) + (n+1)\epsilon(c), \quad t' = \ell\epsilon(a) + m\epsilon(b) + n\epsilon(c),$$

into (1), we obtain

$$(a-b)\tau_{\ell+1,m+1,n}\tau_{\ell,m,n+1} + (b-c)\tau_{\ell,m+1,n+1}\tau_{\ell+1,m,n}$$
$$+(c-a)\tau_{\ell+1,m,n+1}\tau_{\ell,m+1,n} = 0, \qquad (2)$$

where $\tau_{\ell,m,n} := \tau(\ell\epsilon(a) + m\epsilon(b) + n\epsilon(c))$. Equation (2) is the celebrated discrete KP equation and reduces to the bilinear form of the KP (Kadomtsev-Petviashvili) equation by taking an appropriate continuous limit.

More generally, using three series of arbitrary complex numbers $\{a(i)\}_{i=-\infty}^{+\infty}$, $\{b(j)\}_{j=-\infty}^{+\infty}$ and $\{c(k)\}_{k=-\infty}^{+\infty}$, we can construct the non-autonomous discrete KP (ndKP) equation:

$$(b(m) - c(n)) \, \tau(\ell, m-1, n-1)\tau(\ell-1, m, n)$$
$$+(c(n) - a(\ell)) \, \tau(\ell-1, m, n-1)\tau(\ell, m-1, n)$$
$$+(a(\ell) - b(m)) \, \tau(\ell-1, m-1, n)\tau(\ell, m, n-1) = 0$$

$$(3)$$

by the substitution

$$t = \sum_i^{\ell+1} \epsilon(a(i)) + \sum_j^{m+1} \epsilon(b(j)) + \sum_k^{n+1} \epsilon(c(k)),$$

$$t' = \sum_i^{\ell} \epsilon(a(i)) + \sum_j^{m} \epsilon(b(j)) + \sum_k^{n} \epsilon(c(k))$$

where the symbols $\sum_i^{\ell}$ etc. denote the convention:

$$\sum_i^{\ell} \equiv \begin{cases} \sum_{i=1}^{\ell} & \text{for } \ell \geq 1 \\ 0 & \text{for } \ell = 0 \\ -\sum_{i=\ell+1}^{0} & \text{for } \ell \leq -1 \end{cases} \qquad (4)$$

and

$$\tau(\ell, m, n) := \tau \left(\sum_i^{\ell} \epsilon(a(i)) + \sum_j^{m} \epsilon(b(j)) + \sum_k^{n} \epsilon(c(k)) \right).$$

The ndKP equation is used to construct generalized BBSs through ultradiscretization.

Since the dKP equation is obtained from (1) by a simple transformation of independent variables, its solutions are immediately constructed in concrete forms. For example, its one-soliton solution has arbitrary parameters p, q ($p \neq q$) and γ and is given as

$$\tau_{\ell,m,n} = \langle vac \,|1 + \gamma \psi^*(t, q) \psi(t, p)|\, vac \rangle$$

$$= 1 + \frac{\gamma}{p - q} \left(\frac{a - q}{a - p}\right)^\ell \left(\frac{b - q}{b - p}\right)^m \left(\frac{c - q}{c - p}\right)^n. \tag{5}$$

Here $\psi^*(t, q)$ and $\psi(t, p)$ are time dependent fermionic field operators[2]:

$$\psi^*(t, q) = \frac{1}{(q - a)^\ell (q - b)^m (q - c)^n} \psi^*(q),$$

$$\psi(t, p) = (p - a)^\ell (p - b)^m (p - c)^n \psi(p),$$

which satisfy

$$\langle vac \,|\psi(p_1)\psi(p_2) \cdots \psi(p_N)\psi^*(q_N)\psi^*(q_{N-1}) \cdots \psi^*(q_1)|\, vac \rangle$$

$$= \det \left[\frac{1}{p_i - q_j}\right]_{1 \le i, j \le N}.$$

Similarly its two-soliton solution has parameters p_i, q_i ($p_i \neq q_i$), γ_i ($i = 1, 2$),

$$\tau_{\ell,m,n} = \left\langle vac \left| \prod_{i=1}^{2} (1 + \gamma_i \psi^*(t, q_i)\psi(t, p_i)) \right| vac \right\rangle$$

$$= 1 + \frac{\gamma_1}{p_1 - q_1} \left(\frac{a - q_1}{a - p_1}\right)^\ell \left(\frac{b - q_1}{b - p_1}\right)^m \left(\frac{c - q_1}{c - p_1}\right)^n + (1 \to 2)$$

$$+ \frac{\gamma_1 \gamma_2 (p_1 - p_2)(q_2 - q_1)}{(p_1 - q_1)(p_1 - q_2)(p_2 - q_1)(p_2 - q_2)} \left(\frac{(a - q_1)(a - q_2)}{(a - p_1)(a - p_2)}\right)^\ell$$

$$\times \left(\frac{(b - q_1)(b - q_2)}{(b - p_1)(b - p_2)}\right)^m \left(\frac{(c - q_1)(c - q_2)}{(c - p_1)(c - p_2)}\right)^n. \tag{6}$$

The dKP equation is a partial difference equation with three independent variables. If we impose a constraint on τ, we can obtain a partial difference equation with only two independent variables. (This procedure to construct a lower dimensional equation is called *reduction*.) When we put $a = 0$, $b = 1 - \delta$, $c = 1$ and impose a condition

$$\tau_{\ell+1,m+1,n} = \tau_{\ell,m,n}, \tag{7}$$

the dKP equation turns into the discrete KdV (dKdV) equation

$$\tau_{n+1}^{t+1}\tau_n^{t-1} = (1 - \delta)\tau_n^t \tau_{n+1}^t + \delta \tau_{n+1}^{t-1}\tau_n^{t+1}. \tag{8}$$

Here we put $\tau_n^t := \tau_{t,0,n}$. The dKdV equation also becomes a bilinear form of the KdV (Korteweg-de Vries) equation in an appropriate continuous limit.

---

[2] See the chapter by Willox and Satsuma in this volume for details.

For the soliton solutions, the constraint (7) corresponds to the condition $p_i + q_i = 1 - \delta$ for the parameters q_i, p_i. Hence the soliton solutions to the dKdV equation are equal to those to the dKP equation with a condition on the parameters. In the continuous limit, they naturally become the soliton solutions to the KdV equation.

Now we ultradiscretize the dKdV equation and obtain ultradiscrete KdV equation. Here an *ultradiscrete* equation means a piecewise linear equation in general. For that purpose, we define a new parameter ε by $\delta = e^{-1/\varepsilon}$. Since a solution to the dKdV eq. (8) also depends on ε, we express τ_n^t as $\tau_n^t(\varepsilon)$ and define the new dependent variable $\rho_n^t(\varepsilon)$ by

$$\tau_n^t(\varepsilon) = e^{\rho_n^t(\varepsilon)/\varepsilon}.$$

Now suppose that *there exists a one parameter family of solutions* $\tau_n^t(\varepsilon)$ *to the dKdV eq. (8) such that it has the limit*:

$$\lim_{\varepsilon \to +0} \varepsilon \log \tau_n^t(\varepsilon) = \lim_{\varepsilon \to +0} \rho_n^t(\varepsilon) =: \rho_n^t,$$

then using the identity for $a, b \in \mathbb{R}$

$$\lim_{\varepsilon \to +0} \varepsilon \log \left[e^{a/\varepsilon} \cdot e^{b/\varepsilon} \right] = a + b, \quad \lim_{\varepsilon \to +0} \varepsilon \log \left[e^{a/\varepsilon} + e^{b/\varepsilon} \right] = \max[a, b], \quad (9)$$

we obtain the ultradiscrete KdV (udKdV) equation

$$\rho_{n+1}^{t+1} + \rho_n^{t-1} = \max \left[\rho_n^t + \rho_{n+1}^t, \rho_{n+1}^{t-1} + \rho_n^{t+1} - 1 \right]. \quad (10)$$

The udKdV equation (10) coincides with (8) except for the contribution of the parameter δ if we put $\times \to +$, $+ \to \max$. Furthermore we introduce a new dependent variable

$$u_n^t := \rho_n^t - \rho_n^{t+1} - \rho_{n-1}^t + \rho_{n-1}^{t+1} \quad (11)$$

with the boundary condition $\lim_{n \to -\infty} u_n^t = \lim_{t \to +\infty} u_n^t = 0$. Noticing the fact $\min[a, b] = -\max[-a, -b]$, we have

$$u_n^{t+1} = \min \left[1 - u_n^t, \sum_{k=-\infty}^{n-1} u_k^t - \sum_{k=-\infty}^{n-1} u_k^{t+1} \right] \quad (12)$$

Equation (12) is also a piecewise linear equation and is closed under $u_n^t \in \{0, 1\}$. Hence we suppose that u_n^t takes only the values 0 or 1 and we can rewrite (12) as

$$u_n^{t+1} = \begin{cases} 1 & u_n^t = 0 \text{ and } \displaystyle\sum_{k=-\infty}^{n-1} u_k^t - \sum_{k=-\infty}^{n-1} u_k^{t+1} \geq 1 \\ 0 & \text{otherwise} \end{cases} \quad (13)$$

However, if we regard u_n^t as the number of balls in the nth box at time step t, (13) is equivalent to the time evolution rule of the BBS. Thus, in summary, we find that the generating formula of the KP hierarchy (1) gives the dKP equation (2) by a change of independent variables and the dKdV equation (8) as its reduction, and the ultradiscrete limit of the dKdV equation gives the BBS.

As is obvious by the construction, a solution to the BBS is obtained from solutions to the dKdV equation through the limit $\varepsilon \to +0$. For a one-soliton solution, we take $p = e^{-P/\varepsilon}$, $\gamma = -e^{\theta/\varepsilon}$ ($P \in \mathbb{Z}_+$, $\theta \in \mathbb{Z}$) and take the limit $\varepsilon \to +0$. Noticing the constraint $p + q = 1 - e^{-1/\varepsilon}$, we have

$$\rho_n^t = \max[0, \theta + tP - n],$$

which gives a one-soliton solution with length P as is easily seen from (11). Similarly a two-soliton solution to the BBS is given as

$$\rho_n^t = \max[0, \xi_1, \xi_2, \xi_1 + \xi_2 + A_{12}]$$

where $\xi_i := \theta_i + tP_i - n$ ($i = 1, 2$), and $A_{12} := -2\min[P_1, P_2]$. The expression for a general N-soliton solution is also given in a similar manner [3].

$$\rho_n^t = \max_{\mu_i = 0, 1} \left[\sum_{i=1}^{M} \mu_i \xi_i - \sum_{i>j} \mu_i \mu_j A_{ij} \right]. \tag{14}$$

As we have seen in this subsection, the BBS can be derived from the KP hierarchy through ultradiscretization, and solitons in the BBS correspond to the soliton solutions of the KP hierarchy. This is the reason why the BBS has a solitonic nature.

3.2 BBS as Ultradiscrete Limit of the Discrete Toda Equation

One of the most important integrable dynamical systems is the Toda lattice in which particles interact with repulsive exponential forces [28]. Its equation of motion is given by

$$\frac{d^2 q_n(t)}{dt^2} = e^{-(q_n(t) - q_{n-1}(t))} - e^{-(q_{n+1}(t) - q_n(t))} \qquad (n = 1, 2, \ldots, N), \tag{15}$$

with some boundary condition such as $q_0(t) = q_{N+1}(t) = 0$. This Toda equation (15) has N independent and involutive conserved quantities, and is integrable in the sense of Liouville. By putting $V_n(t) = e^{-(q_{n+1}(t) - q_n(t))}$ and $I_n(t) = \dot{q}_n(t)$, (15) can be rewritten as the equation of nonlinear LC circuits

$$\begin{cases} \dfrac{dV_n(t)}{dt} = V_n(t)\,(I_n(t) - I_{n+1}(t)) \\[2mm] \dfrac{dI_n(t)}{dt} = V_{n-1}(t) - V_n(t) \end{cases} \tag{16}$$

The Toda equation is derived from the two-component KP hierarchy or so-called Toda lattice hierarchy whose mathematical structure is essentially equivalent to the KP hierarchy. Therefore, as in the previous subsection, it can be discretized preserving its solutions into the discrete Toda (dToda) equation[3]

$$
\begin{cases}
I_n^{t+1} = I_n^t + V_n^t - V_{n-1}^{t+1} \\
V_n^{t+1} = \dfrac{I_{n+1}^t V_n^t}{I_n^{t+1}}
\end{cases}
\tag{17}
$$

When we impose the boundary condition $V_0^t = V_N^t = 0$, (17) is often called the discrete Toda molecule equation and by introducing

$$
M_t := \begin{pmatrix} I_i^t & 1 & & \\ & I_2^t & \ddots & \\ & & \ddots & 1 \\ & & & I_N^t \end{pmatrix} \qquad R_t := \begin{pmatrix} 1 & & & \\ V_1^t & 1 & & \\ & \ddots & \ddots & \\ & & V_{N-1}^t & 1 \end{pmatrix}
$$

it is rewritten as

$$
R_{t+1} M_{t+1} = M_t R_t.
\tag{18}
$$

We define $N \times N$ matrix L_t by $L_t := R_t M_t$. Then, from (18), we have the Lax form of the dToda (molecule) equation as

$$
L_{t+1} M_t = M_t L_t.
\tag{19}
$$

Since

$$
C_{t+1}(\lambda) := \det(\lambda E - L_{t+1}) = \det(\lambda E - L_t) = C_t(\lambda)
\tag{20}
$$

(E denotes the $N \times N$ unit matrix), the polynomial $C_t(\lambda)$ of order N does not depend on time step t. Hence, denoting the coefficients of λ^k ($k = 0, 1, 2, ..., N - 1$) by $C^{(k)}$, these are N independent conserved quantities of the dToda equation.

In a state of the BBS, a set of consecutive filled boxes is located between sets of vacant boxes and vice versa. We denote by Q_1^t the length of the leftmost set of filled boxes at time step t, by W_1^t that of the right adjacent set of vacant boxes, by Q_2^t that of the next set of filled boxes, and so on (Fig. 4). Then we find that they satisfy

$$
\begin{cases}
Q_n^{t+1} = \min \left[W_n^t, \ \sum_{j=1}^n Q_j^t - \sum_{j=1}^{n-1} Q_j^t \right] \\
W_n^{t+1} = Q_{n+1}^t + W_n^t - Q_n^{t+1}
\end{cases}
\qquad (n = 1, 2, \dots, N - 1), \tag{21}
$$

---

[3] Historically the dToda equation was first derived by Hirota under the condition that the discretization preserves the soliton solutions [21].

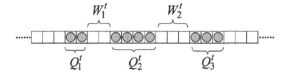

Fig. 4. Variables Q_n^t and W_n^t for udToda eq.

with the boundary condition

$$W_0^t = W_N^t = +\infty. \tag{22}$$

This equation (21) is also an ultradiscrete equation for the BBS. We suppose that the dToda equation (17) has a parameter ε and rewrite the dependent variables as

$$I_n^t = \exp[-Q_n^t(\varepsilon)/\varepsilon], \quad V_n^t = \exp[-W_n^t(\varepsilon)/\varepsilon].$$

When the limit

$$Q_n^t := \lim_{\varepsilon \to +0} Q_n^t(\varepsilon), \quad W_n^t := \lim_{\varepsilon \to +0} W_n^t(\varepsilon)$$

exists, an argument similar to that in the previous section shows that the above Q_n^t, W_n^t satisfy (21) with the boundary condition (22), which means that the BBS is also to be regarded as an ultradiscrete limit of the dToda (molecule) equation.

Using ultradiscrete limits, we can construct conserved quantities of the BBS from those of the dToda equation, *i.e.*, $C^{(k)}$ ($k = 0, 1, \ldots, N-1$). By defining $uC^{(k)} := - \lim_{\varepsilon \to +0} -\varepsilon \log C^{(k)}$, we obtain [29]

$$uC^{(N-1)} = \min \left(\min_{1 \leq i \leq N} Q_i^t, \ \min_{1 \leq j \leq N-1} W_j^t \right)$$

$$uC^{(N-2)} = \min \left(\min_{1 \leq i_1 < i_2 \leq N} (Q_{i_1}^t + Q_{i_2}^t), \right.$$

$$\left. \min_{i_1 \notin \{j_1, j_1+1\}} (Q_{i_1}^t + W_{j_1}^t), \ \min_{1 \leq j_1 < j_2 \leq N} (W_{j_1}^t + W_{j_2}^t) \right)$$

$$\cdots$$

$$uC^{(0)} = \sum_{i=1}^{N} Q_i^t.$$

These N conserved quantities are independent and of course each of them corresponds to a conserved quantity of the dToda equation. Although they have rather complicated expressions, there is an alternative way to calculate the same set of conserved quantities. Denoting a vacant box by 0 and a filled box by 1, we obtain the $0, 1$ sequence corresponding to a state of the BBS. Then the explicit algorithm to construct the conserved quantities is given as follows [30].

1. Let p_1 be the number of 10s in the sequence.
2. Eliminate all the 10s in the original sequence and let p_2 be the number of 10s in the new sequence.
3. Repeat the above procedure until all the 1s are eliminated.
4. Then the weakly decreasing positive integer sequence $\{p_1, p_2, p_3, \dots\}$ consists of the conserved quantities.

For example, for the state

(#) $\dots 0011101110010001111000110100000 \dots$

we have $p_1 = 6$, and eliminating 10s, we obtain a new sequence

$\dots 0011110001110010000 \dots$

and $p_2 = 3$. In a similar manner, we have $p_3 = 2$, $p_4 = 2$, $p_5 = 1$. To see that these $\{p_j\}$ are conserved, we evolve (#) by one time step

(#′) $\dots 000001000110111000011100101111100 \dots$.

By applying the above algorithm again, we find the same integer sequence $\{p_j\}$ $(j = 1, 2, \dots, 5)$.

The sequence $p_1\ p_2\ \dots$ is a weakly decreasing positive integer sequence and we can associate a Young diagram to it by regarding p_j as the number of squares in the jth column of the Young diagram. For example the Young diagram associates to (#) is shown in Fig. 5.

$$p_1\ p_2\ p_3\ p_4 p_5$$

Fig. 5. Young diagram corresponding to the conserved quantities of (#)

When we denote by L_j the length of the jth row of the Young diagram, the weakly decreasing integer sequence $\{L_1, L_2, \dots\}$ is another expression for the conserved quantities of the BBS. When $t \to +\infty$, the state of the BBS consists of solitons which are arranged according to the order of their lengths and move freely. We can prove that the length of the jth largest soliton among these freely moving solitons coincides with L_j [11]. Hence, for a given initial state, we can find the solitons which constitute that state after sufficiently many time steps by constructing the corresponding Young diagram.

Young diagrams and Young tableaux are used in combinatorics, representation theory and the geometry of varieties [31]. Though the Young diagram that appeared in the conserved quantities of BBS does not have a direct connection to such mathematical subjects, generalized BBSs given in the subsequent section show combinatoric aspects related to the representations of quantum algebras. As a consequence, the BBS posesses links to two-dimensional integrable lattice models.

4 Generalization of BBS

4.1 BBS Scattering Rule and Yang-Baxter Relation

As has been mentioned in the previous section, the BBS is just a reinterpretation of the soliton CA proposed by Takahashi and Satsuma. The merit of this reinterpretation is that we can introduce new degrees of freedom–'box capacity' and 'species of balls'. Furthermore we can introduce a 'capacity of carrying cart' [13]. In this section, we consider the BBS with M kinds of balls. We leave the capacity of the boxes to be one.

For simplicity, we distinguish the species of balls by integer indices $1, 2, \ldots, M$. An example of a state of the BBS with M kinds of balls is shown in Fig. 6. Accordingly the time evolution rule is changed as follows

Fig. 6. A state of a BBS with $M(= 3)$ kinds of balls

1. Move every ball only once in one time step.
2. Move the leftmost ball with index 1 to the nearest right vacant box.
3. Move the leftmost ball with index 1 among the rest to its nearest right vacant box.
4. Repeat this procedure until all the balls with index 1 have been moved.
5. Do the same procedure (2)–(4) for the balls with index 2.
6. Repeat this procedure successively until all the balls have been moved.

An example of a time evolution pattern is shown in Fig. 7. After several trials, one may notice that a soliton in this BBS should be defined by a sequence of consecutive balls with indices arranged in weakly increasing order. We also define that the length or the amplitude of a soliton is the number of balls which constitute the soliton. Hereafter we denote a ball with index j by a number 'j' and a vacant box by '.' for simplicity. Then, for example, a state

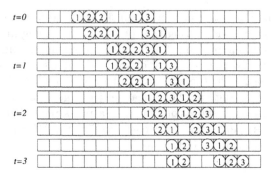

Fig. 7. An example of time evolution pattern of BBS with $M(=3)$ kinds of balls

of the BBS is expressed as

$$....1256334...45123...2132444...$$

In this example, there are seven solitons–'1256', '334', '45', '123', '2', '13' and '2444'. Each soliton has the length (or amplitude) 4,3,2,3,1,2 and 4 respectively. With these definitions, the solitonic features of the BBS are described as follows [14].

1. The number of solitons does not change in time evolution.
2. Suppose that solitons are far enough apart from each other at an initial state. Then their length is not altered after their collisions. Furthermore, the final results do not depend on the initial arrangements of solitons, in the sense that the ingredients of each soliton are not altered by changing the initial distance between the solitons, as long as they are well separated.

We will not give the proofs of the above features but examine the feature of soliton scattering in detail, for it has a deep relation to the so-called R matrices of two-dimensional integrable lattice models.

Figures 8a–c show three examples of two-soliton scatterings:

$$
\begin{array}{llll}
\text{(a)} & \text{'1135'} + \text{'24'} & \longrightarrow & \text{'35'} + \text{'1124'} \\
\text{(b)} & \text{'23455'} + \text{'124'} & \longrightarrow & \text{'235'} + \text{'12445'} \\
\text{(c)} & \text{'11555'} + \text{'234'} & \longrightarrow & \text{'555'} + \text{'11234'}.
\end{array}
$$

As suggested by these examples, two soliton scattering is governed by a deterministic rule (the BBS scattering rule). Consider two solitons '$i_1 i_2 \cdots i_m$' and '$j_1 j_2 \cdots j_n$' that collide and transform into '$j'_1 j'_2 \cdots j'_n$' and '$i'_1 i'_2 \cdots i'_m$' $(m > n)$. Defining $\Lambda := \{i_1, i_2, \ldots, i_m\}$, $\Lambda' := \{i'_1, i'_2, \ldots, i'_m\}$, $\Sigma := \{j_1, j_2, \ldots, j_n\}$ and $\Sigma' := \{j'_1, j'_2, \ldots, j'_n\}$, we clearly have $\Lambda + \Sigma = \Lambda' + \Sigma'$. Furthermore the set Λ' is always a subset of Σ and hence Λ is always a subset of Σ'. The set Λ' is now determined by:

```
(a)   . 1 1 3 5 .  .  . 2 4 . .
        . . 1 1 3 5 . 2 4 . .
            . . 1 3 5 1 2 4 . .
                . . 3 5 . 1 1 2 4 .
                    . . 3 5 . . . 1 1 2 4 . .


(b)   . 2 3 4 5 5 .  . 1 2 4 . .
        . . 2 3 4 5 5 1 2 4 . .
            . . 2 3 5 1 2 4 4 5 . .
                . . 2 3 5 . . 1 2 4 4 5 . .
                    . . 2 3 5 . . . . 1 2 4 4 5 . .


(c)   . 1 1 5 5 5 .  . 2 3 4 . .
        . . 1 1 5 5 5 2 3 4 . .
            . . 5 5 5 1 1 2 3 4 . .
                . . 5 5 5 . . 1 1 2 3 4 . .
                    . . 5 5 5 . . . . 1 1 2 3 4 . .
```

Fig. 8. Examples of two-soliton scattering in BBS

1. If there is no integer in Σ greater than i_1, choose the smallest integer in Σ; otherwise choose the smallest integer among those in Σ greater than i_1. Denote this integer by k_1.
2. Next, if there is no integer in $\Sigma \setminus \{k_1\}$ greater than i_2, choose the smallest integer in $\Sigma \setminus \{k_1\}$; otherwise choose the smallest integer among those in $\Sigma \setminus \{k_1\}$ greater than i_2. Denote it by k_2.
3. Repeat the procedure until k_M has been chosen among $\Sigma \setminus \{k_1, k_2, \cdots, k_{M-1}\}$. Then $\Lambda' = \{k_1, k_2, \cdots, k_M\}$.

For example, in the case

$$1123455 + 156 \quad \longrightarrow \quad 112 + 1345556$$

$\Sigma = \{1, 1, 2, 3, 4, 5, 5\}$ and $\Lambda = \{1, 4, 4, 6\}$. For $i_1 = 1$, the smallest integer in Σ which is greater than i_1 is 2. Hence we obtain $k_1 = 2$. For $i_2 = 5$, no integer in $\Sigma \setminus \{k_1\} = \{1, 1, 3, 4, 5, 5\}$ is greater than i_2 and we choose $k_2 = \min[\Sigma \setminus \{k_1\}] = 1$. Similarly we have $k_3 = 1$ and $\Lambda' = \{2, 1, 1\}$ which coincides with the result. The scattering rule for this example is depicted in Fig. 9.

Next we consider three-soliton scattering. Examples of three-soliton interaction patterns are shown in Figs. 10a–c. As should be the case for genuine solitons, the final (emerging) sequences do not depend upon the order of the collisions. The initial states of Figs. 10 are all composed of '11334', '255' and '23' solitons, the order of collisions however differs because of the different initial arrangements of the solitons. Thus, in Fig. 10a we see the sequence of interactions

(1) $11334 + 255 \longrightarrow 113 + 23455$
(2) $23455 + 23 \longrightarrow 34 + 22355$
(3) $113 + 34 \longrightarrow 11 + 334,$

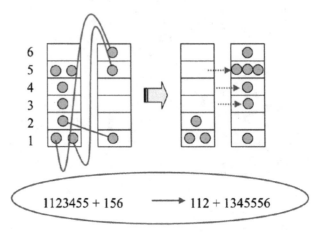

Fig. 9. Schematic representation of the BBS scattering rule

```
(a)  ..11334..255....23...
        ...113..23455.23....
        ...113....2342355....
        ...113....34..22355....
        ...113...34.....22355...
        ...113..34........22355...
        ...11..334.........22355...


(b)  ......255.23....
        ........25523.....
        ...........55223.....
        11334.........55.223.....
        ....11334......55..223.....
        .....11334...55...223.....
        ....113..33455.223.....
        ....11.....333422355.....
        ...11.......334..22355....
        ...11........334....22355....


(c)  ...11334....255.23...
        ...11334..25523.....
        .....113..3422355.....
        ....11.334....22355....
        ...11..334......22355......
```

Fig. 10. Three soliton interactions in a BBS

while in Fig. 10b we have

(1) $255 + 23 \longrightarrow 55 + 223$
(2) $11334 + 55 \longrightarrow 11 + 33455$
(3) $33455 + 223 \longrightarrow 334 + 22355.$

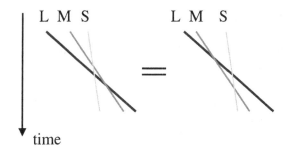

Fig. 11. Yang-Baxter relation for the BBS. Three-soliton scatterings do not depend on the order of collisions

Figure 10c shows three solitons colliding simultaneously. In all three cases the final patterns are the same, *i.e.*, '11', '334' and '22355'. This feature of three-soliton scatterings implies that the N ($N \geq 3$) soliton scattering rules are governed by those for two-soliton scatterings. Scattering properties of this type (in the exact solution of actual one-dimensional problems) were first established by Yang [32] and Baxter [33] and are now commonly referred to as Yang-Baxter relations [34] (Fig. 11). Hence we may regard the soliton scattering in a BBS as an analogue of the Yang-Baxter relation on the level of cellular automata. To satisfy the Yang-Baxter relation is usually understood as a sufficient condition of integrability for the R-matrices in one-dimensional quantum models and two-dimensional lattice models. The BBS scattering rule is essentially equivalent to the action of the combinatorial R-matrix on the symmetric tensor product representation of the quantum algebra $U_q(A_M^{(1)})$ at $q \to 0$ when the states of the BBS are identified with the classical crystal of $U_q(A_M^{(1)})$ [35]. This fact suggests a connection between the BBS and quantum integarable models, however before establishing the direct connection between them, we consider further extensions of the BBSs in the next subsection.

4.2 Extensions of BBSs and Non-autonomous Discrete KP Equation

In this subsection we introduce an extended box and ball system which may very well be the most general BBS imaginable. Its soliton behaviour arises from a reduction of the non-autonomous discrete KP equation (3). It will be explained however that this correspondence entails a lot more complications than in the simpler cases discussed before.

In the previous subsection we discussed the properties of a BBS with M species of balls. As was mentioned at the time, a natural extension of such a BBS is obtained by introducing varying *box capacities*, e.g., the box at site n can accommodate up to $\theta_n \geq 1$ balls and this *capacity* may be location dependent (*i.e.*, θ_n may be a function of n). In practice this means that we change the BBS evolution rule for the case of M species of balls in such a way that one does not move a ball to the first *empty* box on the right, but rather to the first box on the right which has a vacant spot[4]. An example of such a BBS is shown in Fig. 12.

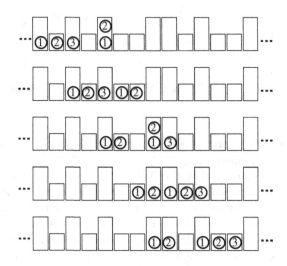

Fig. 12. Example of a BBS with varying box capacities

Besides species of balls or box capacities, there is a third degree of freedom available, one that is directly connected to the time evolution itself. Namely, we can introduce a *carrier* with capacity κ_t which will transport the balls to their new locations. This carrier comes in from the left, proceeds to the right and depending on its remaining capacity (*i.e.* $\kappa_t - \#$ of balls present in the carrier) picks up the balls in a box and drops them into the first available spot ; the 'unloading' proceeds according to the general rule explained above. As can be seen in Fig. 13, while the carrier goes through the array of boxes the process of successively loading and unloading induces the motion of balls over the boxes and hence induces the time evolution of the system. We shall denote such a general box and ball system by its 3 relevant parameters (M, θ_n, κ_t), where (as before) the variables $n, t \in \mathbb{Z}$ play the role of space and time coordinates respectively.

The original BBS corresponds to the choice $(M = 1, \forall \theta_n = 1, \forall \kappa_t = \infty)$ whereas the case $(M \geq 1, \forall \theta_n = 1, \forall \kappa_t = \infty)$ is of course the one we studied in the previous subsection. Other special cases such as $(M = 1, \forall \theta_n = \theta, \forall \kappa_t = \kappa)$ with $\kappa > \theta$ and $(M, \theta_n, \forall \kappa_t = \infty)$ have been treated in [13] and [15] respectively. Remark that, just as the box capacity θ_n can depend on the site

---

[4] In case a box should contain more than one ball of the same species, one may pick any one of them when determining which ball is the 'leftmost' one.

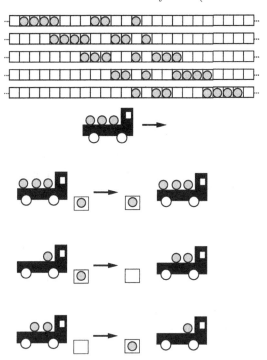

Fig. 13. Example of a BBS with a (capacity 3) carrier and the different loading/unloading actions depending on the remaining capacities of boxes and carrier

n, the carrier capacity κ_t may also vary in time. It is the soliton scattering in this general case (M, θ_n, κ_t) we now wish to elaborate upon.

First, let us introduce some notations. Let b_n^t denote the balls contained in the nth box (with capacity θ_n) at time t and let v_n^t stand for the balls present in the carrier (with capacity κ_t) at that same time t. A more detailed description of the contents of both boxes and carrier is provided by the variables $u_{n,j}^t$ and $v_{n,j}^t$ (for $1 \leq j \leq M+1$) : these represent the number of balls of the $(M+1-j)$th species[5] present in b_n^t and v_n^t, i.e. :

$$b_n^t \equiv (u_{n,M+1}^t, u_{n,M}^t, \cdots, u_{n,1}^t), \tag{23}$$

$$v_n^t \equiv (v_{n,M+1}^t, v_{n,M}^t, \cdots, v_{n,1}^t). \tag{24}$$

Using this notation, the evolution in a general BBS (M, θ_n, κ_t) can be expressed in the following way $(1 \leq j \leq M)$,

---

[5] This notation allows for a ball of species '0', which actually stands for the state where there are no balls at all in a particular box. Although somewhat peculiar, this notation proves useful when discussing the relation with exactly solvable lattice models. Note also that the defining equation (26) remains valid for such '0'th species balls (i.e., for $j = M+1$).

$$u_{n,j}^{t+1} - v_{n,j}^t = \max[X_1 - \theta_n, X_2 - \theta_n, \cdots, X_{j-1} - \theta_n, X_j - \kappa_t, \cdots, X_M - \kappa_t, 0]$$
$$- \max[X_1 - \theta_n, X_2 - \theta_n, \cdots, X_j - \theta_n,$$
$$X_{j+1} - \kappa_t, \cdots, X_M - \kappa_t, 0], \qquad (25)$$

$$v_{n+1,j}^t = u_{n,j}^t + v_{n,j}^t - u_{n,j}^{t+1}, \qquad (26)$$

where $X_\ell = X_{n;\ell}^t := \sum_{i=\ell}^{M} u_{n,i}^t + \sum_{i=1}^{\ell} v_{n,i}^t$.

What we want to show next is that the defining equations (25) and (26) are actually the ultradiscrete limits of a special reduction of the ndKP equation (3) written here in the form :

$$(b_n - c_j)\tau\tau_{tnj} + (c_j - a_t)\tau_{tn}\tau_j + (a_t - b_n)\tau_n\tau_{tj} = 0 \qquad (27)$$

for a tau-function $\tau(t,n,j)$ and for $a(t), b(n)$ and $c(j)$ which are arbitrary functions of t, n and j respectively, after performing the change of variables $\{\ell, m, n\}$ to $\{-(t+1), n, j\}$ and $a(\ell)$ to $a(t)$. Here we used the incremental notation, *i.e.* $\tau \equiv \tau(t,n,j)$, $\tau_t \equiv \tau(t+1,n,j)$, $\tau_{tn} \equiv \tau(t+1,n+1,j)$, etc.

The reduction we want to impose on (27) consists of assigning special values to the lattice parameters, $c(1) = 1, c(2) = c(3) = \cdots = c(M+1) = 0$ and of imposing the (periodicity) constraint,

$$\tau(t, n, j + M + 1) = \tau(t, n, j), \qquad (28)$$

on the tau-functions. As it turns out, the constraint (28) is analogous to what one would call an M-reduction of the KP hierarchy, restricting the infinite dimensional Lie algebra $gl(\infty)$ to the finite dimensional algebra $A_M^{(1)}$.

In this reduction – the details of which we omit here – it can be shown, following a line of reasoning very similar to the case $(M = 1, \forall\theta_n = 1, \forall\kappa_t = \infty)$ treated in Sect. 3.1, that the CA defined by (25) and (26) is obtained by ultradiscretizing the fields

$$U_{n,j}^t \equiv \frac{\tau(t, n+1, j)\tau(t, n, j+1)}{\tau(t, n, j)\tau(t, n+1, j+1)},$$
$$(29)$$
$$V_{n,j}^t \equiv \frac{\tau(t+1, n, j+1)\tau(t, n, j)}{\tau(t+1, n, j)\tau(t, n, j+1)}$$

(defined for $1 \le j \le M$). In particular one finds that in the parametrization $a_t = 1 + \delta_t$, $b_n = 1 + \gamma_n$ with $\delta_t = \exp[-\kappa_t/\varepsilon]$ and $\gamma_n = \exp[-\theta_n/\varepsilon]$, the ultradiscrete limits,

$$u_{n,j}^t = \lim_{\varepsilon\to+0} \varepsilon \log U_{n,j}^t, \qquad (30)$$

$$v_{n,j}^t = \lim_{\varepsilon\to+0} \varepsilon \log V_{n,j}^t, \qquad (31)$$

yield solutions $\{u^t_{n,j}\}$ and $\{v^t_{n,j}\}$ for the system (25) and (26) (for $1 \leq j \leq M$). The values for $u^t_{n,M+1}$ and $v^t_{n,M+1}$ are specified by the requirements $\sum_{j=1}^{M+1} u^t_{n,j} = \theta_n$ and $\sum_{j=1}^{M+1} v^t_{n,j} = \kappa_t$, which fix the numbers of vacant spots in a specific box and in the carrier. Because of the link to an $A^{(1)}_M$ reduced KP equation we shall refer to the BBS described by (25) and (26) as an $A^{(1)}_M$ automaton.

Let us consider the soliton solutions for this automaton. As before, if the limit

$$Y^t_{n,j} := \lim_{\varepsilon \to +0} \varepsilon \log \tau(t, n, j)$$

for the $A^{(1)}_M$ reduced tau-functions exists, then applying the ultradiscrete limit to the expressions (29) produces solutions of the system (25) and (26) in the form $(1 \leq j \leq M)$:

$$u^t_{n,j} = Y^t_{n+1,j} + Y^t_{n,j+1} - Y^t_{n,j} - Y^t_{n+1,j+1}, \tag{32}$$
$$v^t_{n,j} = Y^{t+1}_{n,j+1} + Y^t_{n,j} - Y^{t+1}_{n,j} - Y^t_{n,j+1}. \tag{33}$$

Hence we shall call the quantity $Y^t_{n,j}$ an N-soliton solution to the $A^{(1)}_M$ automaton if it is the ultradiscrete limit in ε of a one-parameter family of appropriately chosen soliton solutions $\tau(t, n, j)$ for (27); the main task is the identification of these corresponding KP-type soliton solutions.

Using the simple correspondence between (27) and the generic non-autonomous discrete KP equation (3), it is clear that the former admits N-soliton solutions corresponding to some elements in GL_∞, where the time evolutions of the fermion operators $\psi(p)$ and $\psi^*(q)$ are now given by the expressions, $\boldsymbol{x} = (t, n, j)$,

$$\psi(\boldsymbol{x}, p) = \left[\prod_{t'}^t (a_{t'} - p) \prod_{n'}^n \frac{1}{b_{n'} - p} \prod_{j'=1}^j \frac{1}{p - c_{j'}} \right] \psi(p), \tag{34}$$

$$\psi^*(\boldsymbol{x}, q) = \left[\prod_{t'}^t \frac{1}{a_{t'} - q} \prod_{n'}^n (b_{n'} - q) \prod_{j'=1}^j (q - c_{j'}) \right] \psi^*(q), \tag{35}$$

$$\tag{36}$$

for the multiplication convention similar to (4).

The $A^{(1)}_M$ reduction (28) of such an N-soliton is obtained by imposing the constraint :

$$\left(\frac{q_k}{p_k} \right)^M \left(\frac{1 - q_k}{1 - p_k} \right) = 1 \quad (k = 1, 2, \cdots, N). \tag{37}$$

Note that, for a given p_k, there are M q_k's which satisfy (37) and $q_k \neq p_k$.

However, the soliton solutions we need to consider here – as opposed to the $M = 1$ case, *i.e.*, the CA (14) – are far more complicated than one might think at first. In fact, because we allow for an arbitrary number of species $M \geq 1$, we need to start from soliton solutions generated by elements :

$$g(\boldsymbol{x}) = \prod_{k=1}^{N} \left(1 + \psi(\boldsymbol{x}, p_k) \phi^*(\boldsymbol{x}, p_i) \right)$$

of $\overline{GL}(\infty)$, with

$$\phi^*(\boldsymbol{x}, p) \equiv \sum_{\ell=1}^{M} c_\ell(p) \psi^*(\boldsymbol{x}, q_\ell),$$

for carefully chosen values of the parameters c_ℓ and for the M roots q_ℓ of the constraint (37) for a given p_k.[6]

Finally, from these solutions for the $A_M^{(1)}$ reduction of (27), the N-soliton solutions for the $A_M^{(1)}$ automaton are found to have the form,

$$Y_{n+1,j+1}^{t+1} = \max_{\boldsymbol{\mu}} \left[\sum_{i=1}^{N} \mu_i K^{(i)}(t, n, j) - A(\boldsymbol{\mu}; j) \right];$$

$\boldsymbol{\mu} = (\mu_1, \mu_2, ..., \mu_N)$ $(\mu_i = 0, 1)$ and $\max_{\boldsymbol{\mu}}[\cdots]$ denotes the maximum among the 2^N values obtained by setting $\mu_i = 0$ or 1 for $i = 1, 2, \cdots, N$. In this expression there appear the functions $K^{(i)}(t, n, j) = K_0^{(i)} - \sum_{j'=1}^{j} \ell_{j'}^{(i)} - \sum_{t'}^{t} \min[\kappa_{t'}, L^{(i)}] + \sum_{n'}^{n} \min[\theta_{n'}, L^{(i)}]$, with sums taken as in the summation convention (4). The (non negative) integers $L^{(i)}$, $\ell_j^{(i)}$ $(1 \leq i \leq N, 1 \leq j \leq M)$ appearing in these functions satisfy $L^{(i)} = \sum_{j=1}^{M} \ell_j^{(i)}$, $L^{(1)} \geq L^{(2)} \geq \cdots \geq L^{(N)}$,

$\ell_j^{(1)} \geq \ell_j^{(2)} \geq \cdots \geq \ell_j^{(N)}$, $(j = 1, 2, \cdots, M)$, $K_0^{(i)}$ being an arbitrary integer. Restricting the set of $\{\mu_i\}$ as in $\mu_i = 1$ for $i = i_1, i_2, \cdots, i_p$ and $\mu_i = 0$ otherwise, the phase factor $A(\boldsymbol{\mu}; j)$ takes the following simple form,

$$A(\boldsymbol{\mu}; j) = \sum_{k=1}^{p} (k - 1) L^{(i_k)}$$

$$+ \sum_{k=1}^{p} \left(X^{(i_k)}(j + k - 1) - X^{(i_k)}(j) \right),$$

with $X^{(i)}(j) = \sum_{j'=1}^{j} \ell_{j'}^{(i)}$ and $\ell_{j+M}^{(i)} = \ell_j^{(i)}$.

---

[6] Hence, these solutions should actually be interpreted as $(N \times M)$-soliton solutions, where the constituent solitons are themselves "degenerate" M-soliton solutions. This can be easily seen from their generating elements, $1 + \psi(\boldsymbol{x}, p) \phi^*(\boldsymbol{x}, p) \equiv \prod_{\ell=0}^{M-1} (1 + c_\ell(p) \psi(\boldsymbol{x}, p) \psi^*(\boldsymbol{x}, q_\ell))$.

5 From Integrable Lattice Model to BBS

5.1 Two-Dimensional Integrable Lattice Models and R-Matrices

It is well known that some of the two dimensional lattice models in physics are solvable in the sense that their partition functions and correlation functions can be obtained exactly [33]. In most cases, the transfer matrices of such solvable lattice models are equivalent to the unitary operators (time evolution operators) of some quantum spin chains and they are then called quantum integrable lattices. The mathematical reasoning behind the solvability of a quantum integrable system relies on the fact that the system has a symmetry corresponding to a quantum algebra; the R-matrix, which gives the Boltzmann weight for a local configuration of the lattice, is regarded as the intertwiner of two tensor representations of the quantum algebra. One of the simplest examples of such a system is the celebrated 6-vertex model which has the symmetry of the quantum affine algebra $U_q(A_2^{(1)})$.

A state of the 6-vertex model is expressed by a square lattice with two kinds of links. An example is shown in Fig. 14, where the two types of links

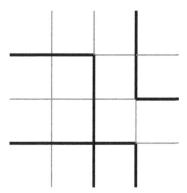

Fig. 14. A part in the state of 6 vertex model

are distinguished by thick and thin lines. The total energy of the system is the sum of the local energies determined by the local configuration of the 4 links around each vertex, and only 6 configurations out of $2^4 = 16$ total configurations have finite energy, *i.e.*, are realized in the model. These finite energy configurations are classified into three types (a)–(c) as shown in Fig. 15, and their energies are denoted by E_a, E_b and E_c respectively. Then, denoting the inverse temperature by $\beta := \dfrac{1}{k_B T}$ where k_B is the Boltzmann constant and T is the temperature, the partition function $Z(\beta)$ is defined by

$$Z(\beta) := \sum_{p \in P} \text{Prob.}(p), \quad \text{Prob.}(p) := \prod_{v \in V} w_v(p : \beta).$$

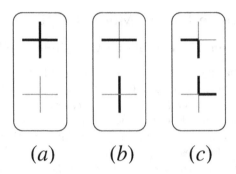

$$(a) \qquad (b) \qquad (c)$$

Fig. 15. Allowed six vertexes. They are classified into three types (a)–(c)

Here P denotes the set of all finite energy vertex configurations, V denotes the set of all vertices, and the Boltzmann weight $w_v(p : \beta)$ at the vertex v for the configuration p is given by the local energy $E_{v(p)}$ $(v(p) \in \{a,\ b,\ c\})$ as

$$w_v(p : \beta) := e^{-\beta E_{v(p)}}.$$

(Subscript $v(p)$ denotes one of the three types of vertices at v in the configuration p.) In Fig. 14, there are 2 vertices of type (a), 4 of type (b) and 3 of type (c), and hence the contribution of this state p to the partition function is given by

$$\text{Prob.}(p) = a^2 b^4 c^3,$$

where we used the abbreviation, $a := e^{-\beta E_a}$ etc.

To calculate the partition function $Z(\beta)$, we usually construct the transfer matrix of the system, and the transfer matrix is, roughly speaking, obtained from the R-matrix[7]. The R-matrix is, taking equivalence between a linear map and a matrix into consideration, an isomorphism between two tensor products of vector spaces,

$$R : \quad V_1 \otimes V_2 \quad \longrightarrow \quad V_2 \otimes V_1$$

where V_1 (V_2) denotes the vector space of a vertical (horizontal) link and the isomorphism is uniquely determined by the Boltzmann weight $w_v(p : \beta)$. In the 6-vertex model, both the vertical and horizontal links have two states and hence $dim(V_1) = dim(V_2) = 2$. Let e_1 and e_2 be a basis of V_1. The basisvector e_1 (e_2) corresponds to the thick (thin) link. Similarly let f_1 and f_2 be a basis of V_2, where f_1 and f_2 correspond to the thick and thin link, respectively. Then the R-matrix of the 6-vertex model is the linear map, $R : V_1 \otimes V_2 \to V_2 \otimes V_1$,

---

[7] More precisely, the transfer matrix is the trace of the matrix product of the so-called L-matrices over an auxiliary degree of freedom, and the L-matrix is essentially the product of an R-matrix and a permutation matrix (in most cases).

$$R(e_1 \otimes f_1) = af_1 \otimes e_1$$
$$R(e_1 \otimes f_2) = bf_2 \otimes e_1 + cf_1 \otimes e_2$$
$$R(e_2 \otimes f_1) = bf_1 \otimes e_2 + cf_2 \otimes e_1$$
$$R(e_2 \otimes f_2) = af_2 \otimes e_2.$$

The R-matrix is of course expressed as a '4 by 4 matrix' $R_{i,j}^{k,l}$ ($i, j, k, l \in \{1, 2\}$) by the relation

$$R(e_i \otimes f_j) = \sum_{k,l} R_{i,j}^{k,l} f_k \otimes e_l.$$

It is well known that a sufficient condition for a two-dimensional lattice model to be integrable is that the R-matrices satisfy the Yang-Baxter relation. The Yang-Baxter relation is the following identity, an equality of the two linear maps $V_1 \otimes V_2 \otimes V_3 \to V_3 \otimes V_2 \otimes V_1$,

$$R_{23}R_{13}R_{12} = R_{12}R_{13}R_{23}, \tag{38}$$

where R_{ij} acts on $V_i \otimes V_j$ as the R-matrix and acts as the identity matrix on the third vector space[8]. For example,

$$R_{12}(V_1 \otimes V_2 \otimes V_3) = R(V_1 \otimes V_2) \otimes V_3 \subseteq V_2 \otimes V_1 \otimes V_3$$
$$R_{13}(V_2 \otimes V_1 \otimes V_3) = V_2 \otimes R(V_1 \otimes V_3) \subseteq V_2 \otimes V_3 \otimes V_1$$

and the R-matrix for $V_1 \otimes V_3$ does not necessarily coincide with that for $V_1 \otimes V_2$. (We often consider the case $V_2 \not\cong V_3$.) In the BBS with M kinds of balls, the vector space of solitons with length L is given as

$$V^{(L)} := \text{span} \left\{ \boxed{k_1}\boxed{k_2}\boxed{\cdots}\boxed{k_L}, \quad 1 \le k_1 \le k_2 \le \ldots \le k_L \le M \right\},$$
$$\forall j \; k_j \in \mathbb{Z}_+. \tag{39}$$

Hence the Yang-Baxter relation of soliton scattering in a BBS is given by the isomorphism determined by the BBS scattering rule:

$$V^{(L_1)} \otimes V^{(L_2)} \otimes V^{(L_3)} \longrightarrow V^{(L_3)} \otimes V^{(L_2)} \otimes V^{(L_1)} \quad (L_1 > L_2 > L_3).$$

Turning to the 6-vertex model, when the ratio of the Boltzmann weights $a(:= e^{-\beta E_A})$, b, c is given with arbitrary complex parameters q, ζ by

$$a : b : c = 1 - q^2 \zeta^2 : (1 - \zeta^2)q : (1 - q^2)\zeta, \tag{40}$$

a direct calculation shows that the R-matrices, $R(\zeta, q)$, satisfy the Yang-Baxter relation:

---

[8] In the Yang-Baxter relation of integrable lattice models, we have to take into account the spectral parameter dependence of the R-matrices, however we will omit this point here. (See the case of the 6-vertex model below as an example.)

$$R_{12}(\zeta_1/\zeta_2, q)R_{13}(\zeta_1/\zeta_3, q)R_{23}(\zeta_2/\zeta_3, q) =$$
$$R_{23}(\zeta_2/\zeta_3, q)R_{13}(\zeta_1/\zeta_3, q)R_{12}(\zeta_1/\zeta_2, q)$$

where ζ_i ($i = 1, 2, 3$) are arbitrary nonzero complex numbers and $V_1 \cong V_2 \cong V_3$. (Parameter ζ is a specral parameter and q stands for the temperature of the model.)

From the above explanation, we see that we can construct an integrable lattice model if we know R-matrices which satisfy the Yang-Baxter relation. The origin of the R-matrix in the framework of the representation theory of quantum groups is elucidated by Drinfel'd and Jimbo [36]. For example, for the 6-vertex model, the relevant quantum group is the quantum affine algebra $U_q(\hat{sl}_2)$. A quantum group U naturally carries a Hopf algebra structure, and we can define the tensor product of two representations using the coproduct Δ. Hence for given U-modules V_1, V_2 we can define the action of $^\forall x \in U$ on $V_1 \otimes V_2$ and $V_2 \otimes V_1$[9]. Suppose there exists an intertwiner for $V_1 \otimes V_2$ and $V_2 \otimes V_1$, that is, a linear operator $R: V_1 \otimes V_2 \longrightarrow V_2 \otimes V_1$ commuting with the action of U,

$$R\Delta(x) = \Delta(x)R \qquad ^\forall x \in U.$$

Then, if the tensor products $V_i \otimes V_j \otimes V_k$ are irreducible, the Yang-Baxter relation follows automatically[10]. Thus once we find an intertwiner of irreducible tensor product representations of some quantum group, we can immediately construct an integrable two-dimensional lattice model. In the next subsection, we shall see that general BBSs result from integrable lattice models by this fact and the idea of 'crystallization' [16].

5.2 Crystallization and BBS

In a lattice model at finite temperature, various configurations are allowed in general for a given boundary condition. However, at zero temperature, the system takes only the lowest energy state and its configuration is uniquely determined by a boundary condition. For example, the zero temperature limit of the 6-vertex model is realized by $q \to 0$ in (40). In this limit, $b \to 0$ and the configuration of the lowest energy state is uniquely determined by boundary conditions for the upper and left edges as shown in Fig. 16a. If we attach 0 and 1 to a vertical thick link and a thin link respectively, we obtain an array of 0, 1 sequence (Fig. 16b). Then, regarding the horizontal direction as the spatial direction and the vertical direction as the time direction, a

---

[9] It is not obvious (and not true in general) that $V_1 \otimes V_2$ is isomorphic to $V_2 \otimes V_1$ as U-modules, because the coproduct is not always symmetric under the switching of the tensor components.

[10] Usually we consider the evaluation module $V_\zeta = V \otimes \mathbb{C}[\zeta, \zeta^{-1}]$ where V is a finite dimensional module of the classical counter part of U, and $V_j = V_{\zeta_j}$ with an indeterminate ζ_i.

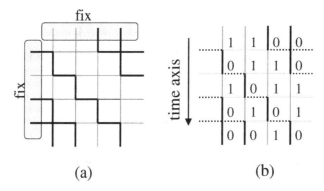

Fig. 16. a A boundary condition determines the zero temperature configuration of 6 vertex model. **b** Corresponding CA configuration

state of the 6-vertex model is considered to be a time evolution pattern of the CA which evolves according to the rule determined by the Bolzmann weight at zero temperature, or equivalently, by the R matrix at $q \to 0$. As in this example, 'crystallization' is the method to construct a CA by identifying a zero temperature configuration of a lattice model with the time evolution pattern of a CA. An initial condition of the CA corresponds to a boundary condition of the lattice model, and the time evolution rule of the CA corresponds to Boltzmann weights of the lattice model. Hikami-Inoue-Komori showed that the BBS is obtained from the zero temperature limit of the Bogoyavlensky lattice, which is a two-dimensional integrable lattice model, by crystallization[11]. The lattice model at zero temperature for the simple BBS is described as follows.

1. A vertical link takes one of the two states indexed by $\{0, 1\}$.
2. A horizontal link can take an infinite number of states indexed by non negative integers $\{0, 1, 2, 3, \dots\}$.
3. The allowed vertex configurations are shown in Fig. 17.
4. The nth horizontal link is in the state 0 for $|n| \gg 1$.

An example of the lattice pattern for the BBS is shown in Fig. 18. The pattern given by the states of vertical links coincides with the two soliton scattering '111'+'1' $\to$ '1'+ '111' in the BBS.

As explained in the previous subsection, we can construct an integrable lattice model if we know an R-matrix (intertwiner of tensor representations) of a quantum algebra. Combining this fact with the idea of crystallization, we notice that an 'integrable CA' can be constructed by using the R-matrix at zero temperature, which is the isomorphism of tensor representations of some quantum algebra at deformation parameter $q \to 0$ and is called the

---

[11] However the limit adopted by them is not continuous and it does not yield the BBS directly.

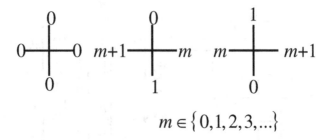

$$m \in \{0, 1, 2, 3, ...\}$$

Fig. 17. Allowed vertex configuration of the lattice model corresponding to the BBS

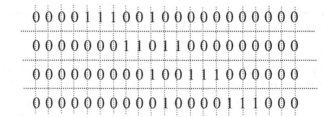

Fig. 18. A lattice configuration corresponding to a time evolution pattern of the BBS

'combinatorial R-matrix'. Hence we can associate an integrable CA with each representation of a quantum algebra at $q \to 0$. In particular one finds [17,18] that the box and ball system (M, θ_n, κ_t) discussed in the previous section, corresponds to the symmetric tensor representation of $U_q'(A_M^{(1)})$[12]. We shall now briefly discuss how to construct a BBS using crystals of $U_q'(A_M^{(1)})$.

Let B_k be the crystal [37] (a *good* representation at $q \to 0$) of $U_q'(A_M^{(1)})$ corresponding to the k-fold symmetric tensor representation. The crystal basis of B_k is characterized by the single row semistandard tableau of length k for letters $\{1, 2, ..., M + 1\}$. Hence, we denote a crystal base $b \in B_k$ by $b = \boxed{m_1 \, m_2 \cdots m_k}$ for $m_i \in \{1, \ldots, M + 1\}$. In particular, we express the vacuum (a highest weight vector) as $u_k := \boxed{1 \, 1 \cdots 1}$. We also denote by $R_{k,\ell}$ the combinatorial R-matrix from the representation $B_k \otimes B_\ell$ to $B_\ell \otimes B_k$. With these notations, we identify the array of vertical links of the square lattice model with an element

$$\cdots b_{n-2}^t \otimes b_{n-1}^t \otimes b_n^t \otimes b_{n+1}^t \otimes b_{n+2}^t \otimes \cdots$$

---

[12] $U_q'(A_M^{(1)})$ is a subalgebra of $U_q(A_M^{(1)})$, which does not include the elements q^d in $U_q(A_M^{(1)})$.

of the space
$$\cdots B_1 \otimes B_1 \otimes B_1 \otimes B_1 \otimes B_1 \cdots ,$$
where B_1 is the crystal of the vector representaion. We set $u_1 = \boxed{1}$ and we assume the boundary condition $b_n = u_1$ ($|n| \gg 1$). On the other hand, we identify the horizontal links with an element
$$\cdots v_{n-2}^t \otimes v_{n-1}^t \otimes v_n^t \otimes v_{n+1}^t \otimes v_{n+2}^t \otimes \cdots$$
of the space
$$\cdots B_\kappa \otimes B_\kappa \otimes B_\kappa \otimes B_\kappa \otimes B_\kappa \cdots .$$
Here we suppose $\kappa \gg 1$ and $v_n^t = u_\kappa$ ($|n| \gg 1$). Then, the Boltzmann weight at each vertex can be given by the combinatorial R matirix:
$$R_{\kappa,1} : v_n^t \otimes b_n^t \mapsto b_n^{t+1} \otimes v_{n+1}^t.$$
(for reasons of simplicity we do not consider the energy function, which is actually the analogue of a spectral parameter for such quantum integrable systems). Using the combinatorial R-matrices, we define a map (transfer matrix) T_κ
$$\cdots b_{n-1}^t \otimes b_n^t \otimes b_{n+1}^t \otimes b_{n+2}^t \otimes \cdots$$
$$\mapsto \cdots b_{n-1}^{t+1} \otimes b_n^{t+1} \otimes b_{n+1}^{t+1} \otimes b_{n+2}^{t+1} \otimes \cdots .$$

This map is actually an isomorphism determined by the combinatorial R-matrix:
$$B_\kappa \otimes (\cdots \otimes B_1 \otimes B_1 \otimes \cdots) \simeq (\cdots \otimes B_1 \otimes B_1 \otimes \cdots) \otimes B_\kappa$$
$$u_\kappa \otimes (\cdots \otimes b_n^t \otimes b_{n+1}^t \otimes \cdots) \simeq (\cdots \otimes b_n^{t+1} \otimes b_{n+1}^{t+1} \otimes \cdots) \otimes u_\kappa$$

Then, in the time evolution pattern:
$$\cdots\ b_{-2}^0\ b_{-1}^0\ b_0^0\ b_1^0\ b_2^0\ \cdots$$
$$\cdots\ b_{-2}^1\ b_{-1}^1\ b_0^1\ b_1^1\ b_2^1\ \cdots$$
$$\cdots\ b_{-2}^2\ b_{-1}^2\ b_0^2\ b_1^2\ b_2^2\ \cdots ,$$

if we regard $b_n^t = \boxed{m}$ ($m \geq 2$) as the nth box with a ball indexed $M + 2 - m$ at time t and $\boxed{1}$ as a vacant box, we find that the pattern coincides with that of the BBS with box capacity 1 and carrier capacity κ.

In Fig. 19, we show an example of a crystal lattice where $M = 3$, $\forall n$, $\theta_n = 1$, and $\forall t$, $\kappa_t = 2$. In order to compare a time evolution pattern of the BBS, we use 'j' and '$j_1 j_2$' instead of $\boxed{M + 2 - j}$ and $\boxed{M + 2 - j_1 M + 2 - j_2}$. Hence the vacuum states for the vertical and horizontal links are denoted by '4' and '44' respectively. When we observe only the vertical links and neglect the highest weight vector '4', this example presents the two soliton scattering, '13' + '2' $\longrightarrow$ '3'+ '12'.

Using this construction and by virtue of the commutativity of the transfer matrices $T_\kappa T_{\kappa'} = T_{\kappa'} T_\kappa$ we can prove the following properties for the general BBS (M, θ_n, κ_t).

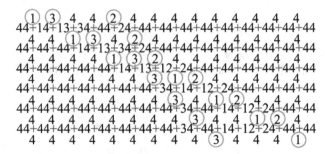

Fig. 19. A crystal lattice with parameters $M = 3, \theta_n = 1, \kappa_t = 2$

(1) By identifying solitons in the BBS with the crystal basis of $U_q(A_{M-1})$, the scattering of solitons is described by the action of the combinatorial R-matrix of $U_q(A_{M-1})$.
(2) The phase shifts of the solitons are determined by the energy functions associated with the R-matrix.
(3) The conserved quantities of the BBS are expressed by semistandard Young tableaux [17].

CA's with other symmetries (*i.e.*, corresponding to algebras of type B, C, etc.) have been discussed recently. However, at the time of writing, the relation of these CA's to the classical integrable systems is still an open problem.

6 Periodic BBS (PBBS)

In this section, we extend the original BBS to that with a periodic boundary condition [38,39,11]. Let us consider a one-dimensional array of N boxes. To be able to impose a periodic boundary condition, we assume that the Nth box is adjacent to the first one. (We may imagine that the boxes are arranged in a circle.) The box capacity is one for all the boxes, and each box is either empty or filled with a ball at any time step. We denote the number of balls by M, such that $M < \dfrac{N}{2}$. The balls are moved according to a deterministic time evolution rule. There are several equivalent ways to describe this rule. For example, as illustrated in Fig. 20,

1. In each filled box, create a copy of the ball.
2. Move all the copies once according to the following rules.
3. Choose one of the copies and move it to the nearest empty box on the right of it.
4. Choose one of the remaining copies and move it to the nearest empty box on the right of it.
5. Repeat the above procedure until all the copies have been moved.
6. Delete all the original balls.

Fig. 20. A rule for time evolution of a PBBS

It is not difficult to prove that the obtained result does not depend on the choice of the copies at each stage and that it coincides with the evolution rule of the original BBS when N goes to infinity. An example of the time evolution of the PBBS according to this rule is shown in Fig. 21.

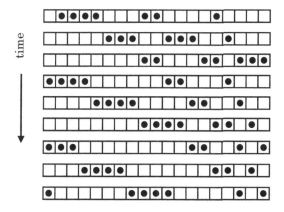

Fig. 21. An example of time evolution of a PBBS

In the following subsections, we will describe several important features of PBBS without giving proofs. The interested reader is referred to the original articles.

6.1 Boolean Formulae for PBBS

We show that the rules for PBBS introduced above can be formulated in terms of Boolean algebra [38]. Let N be the number of boxes. The space of the states of the PBBS is naturally regarded as $\mathbb{F}_2^N$ by denoting a filled box by $1 \in \mathbb{F}_2$ and a vacant box by $0 \in \mathbb{F}_2$ respectively. We denote a state $X(t)$ of the PBBS at time t by $X(t) = (x_1(t), x_2(t), \ldots, x_N(t)) \in \mathbb{F}_2^N$, where $x_i(t) = 0$ if the ith box is empty and $x_i(t) = 1$ if it is filled. Let $\wedge, \vee, \oplus$, be AND, OR and XOR respectively. These Boolean operators are realized in $\mathbb{F}_2^N$ as maps, $\mathbb{F}_2^N \times \mathbb{F}_2^N \to \mathbb{F}_2^N$. For $X = (x_1, x_2, \ldots, x_N)$, $Y = (y_1, y_2, \ldots, y_N)$, they are defined as

$$(X \wedge Y)_i := x_i \wedge y_i \equiv x_i y_i$$
$$(X \vee Y)_i := x_i \vee y_i \equiv x_i y_i + x_i + y_i$$
$$(X \oplus Y)_i := x_i \oplus y_i \equiv x_i + y_i.$$

We also define the rotate shift to the right, S,

$$SX = (x_N, x_1, x_2, \ldots, x_{N-1}).$$

The next theorem gives an expression of T,

$$T : X(t) \mapsto X(t+1),$$

in terms of these Boolean operators.

Theorem 6.1. *Suppose that $X(t) \in \mathbb{F}_2^N$ is the state of the PBBS at time step t. We consider the following recurrence equations,*

$$A^{(0)} = X(t), \quad B^{(0)} = SX(t), \tag{41}$$

$$\begin{cases} A^{(n+1)} := A^{(n)} \vee B^{(n)} \\ B^{(n+1)} := S(A^{(n)} \wedge B^{(n)}) \end{cases} \quad (n = 0, 1, 2, \ldots). \tag{42}$$

Then,

$$X(t+1) = A^{(N)} \oplus X(t), \quad \text{and} \quad B^{(N)} = \mathbf{0}, \tag{43}$$

where $\mathbf{0} := (0, 0, \ldots, 0)$.

This recurrence equation (43) is expressed with only three operations, AND, OR and SHIFT, and has a simple form. The SHIFT operator introduces the right-and-left symmetry breaking that comes from the definition of the direction of the movement of balls. From the Theorem, we also obtain the following corollary:

Corollary 6.1. *Suppose that $X(t) \in \mathbb{F}_2^N$ is given as the state at time t. Then the state at the next time step $X(t+1) = TX(t)$ is calculated by recurrence as follows.*

$$A^{(0)} := X(t), \quad B^{(0)} := SX(t) \tag{44}$$

$$\begin{cases} A^{(n+1)} := A^{(n)} \vee B^{(n)} \\ B^{(n+1)} := S^2(A^{(n)} \wedge B^{(n)}) \end{cases}, \tag{45}$$

and

$$X(t+1) = A^{(\lfloor N/2 \rfloor)} \oplus X(t), \quad B^{(\lfloor N/2 \rfloor)} = \mathbf{0}. \tag{46}$$

6.2 PBBS and Numerical Algorithm

The formulae for time evolution of the PBBS (42) have a simple and symmetric form, and we expect that they bear some relation to a good algorithm. In this subsection, we show that they have indeed the same structure as that of the algorithm to compute the pth root of a given number. Henceforth, let the truth values "0(false)" and "1(true)" be equivalent to the integers $0 \in \mathbb{Z}$ and $1 \in \mathbb{Z}$. Then we can replace $\wedge$ and $\vee$ with min and max as

$$\begin{cases} x \wedge y \iff \min{[x,y]} \\ x \vee y \iff \max{[x,y]} \end{cases}.$$

Following the notation in the previous subsection, we define that max and min act on $\mathbb{Z}^N$ bitwise. Then, (42) can be rewritten as an equation on integers as

$$\begin{cases} A^{(n+1)} = \max{\left[A^{(n)}, B^{(n)}\right]} \\ B^{(n+1)} = S \min{\left[A^{(n)}, B^{(n)}\right]} \end{cases}. \tag{47}$$

We construct the difference equations corresponding to (47) by means of inverse ultradiscretization. Noticing the identity:

$$\max{[x,y]} = \lim_{\epsilon \to +0} \epsilon \log{\left(e^{x/\epsilon} + e^{y/\epsilon}\right)} \quad (x, y \in \mathbb{R}),$$

and $\min{[x,y]} = -\max{[-x,-y]}$, we introduce the difference equations,

$$\begin{cases} a_i^{(n+1)} = \left\{ a_i^{(n)} + b_i^{(n)} \right\}/2 \\ b_i^{(n+1)} = 2 \left\{ \left(a_{i-1}^{(n)}\right)^{-1} + \left(b_{i-1}^{(n)}\right)^{-1} \right\}^{-1} \end{cases} \quad (1 \le i \le N). \tag{48}$$

The relation between (47) and (48) is obvious. When we replace $a_i^{(n)}$ and $b_i^{(n)}$ by $e^{(A^{(n)})_i/\epsilon}$ and $e^{(B^{(n)})_i/\epsilon}$ respectively, and take limit $\epsilon \to +0$, we obtain (47) from (48). The factor 2 in (48) is so chosen that the recurrence formulae do not diverge at $n \to \infty$.

When we disregard the space coordinates i in (48), or consider the case $N = 1$, we obtain the recurrence formulae,

$$\begin{cases} a^{(n+1)} = \dfrac{a^{(n)} + b^{(n)}}{2} \\ b^{(n+1)} = \dfrac{2a^{(n)}b^{(n)}}{a^{(n)} + b^{(n)}} \end{cases}, \tag{49}$$

which are the well-known arithmetic-harmonic mean algorithm and we have

$$\lim_{n\to\infty} a^{(n)} = \lim_{n\to\infty} b^{(n)} = \sqrt{a^{(0)}b^{(0)}}.$$

The recurrence formulae (48) for general N are also considered as a numerical algorithm to calculate the $2N$th root of a given number. To see this, first we note that (48) has a conserved quantity C with respect to the step n,

$$C^{(n)} := \prod_{i=1}^{N} a_i^{(n)} b_i^{(n)} = C^{(n-1)} = \cdots = C^{(0)} = \prod_{i=1}^{N} a_i^{(0)} b_i^{(0)} \equiv C, \tag{50}$$

where $\left\{a_i^{(0)}, b_i^{(0)}\right\}$ are the initial values. Then we can show the following proposition.

Proposition 6.1. *If all the initial values $\{a_i^{(0)}, b_i^{(0)}\}$ are positive, then sequences $a_k^{(n)}$ and $b_k^{(n)}$ converge to the same value*

$$\lim_{n\to\infty} a_k^{(n)} = \lim_{n\to\infty} b_k^{(n)} = \sqrt[2N]{\prod_{i=1}^{N} a_i^{(0)} b_i^{(0)}} = \sqrt[2N]{C} \quad (for\ all\ k).$$

Hence, the recurrence formula of the PBBS is regarded as a numerical algorithm for the $2N$th root.

Since the ultradiscrete model has to maintain the mathematical structures of a discrete model in the process of ultradiscretization, when we take the ultradiscrete limit of C, it is also a conserved quantity of the PBBS. In fact,

$$C = \prod_{i=1}^{N} a_i^{(0)} b_i^{(0)} \overset{\text{UD}}{\Longrightarrow} \sum_{i=1}^{N} \left\{A_i^{(0)} + B_i^{(0)}\right\} \tag{51}$$

gives twice the number of balls in the PBBS. The number of balls is, to be sure, a conserved quantity of the PBBS.

We can construct other conserved quantities of the recurrence formulae (42) by means of another inverse ultradiscretization.

From (42), we obtain

$$\begin{cases} A^{(n+1)} \wedge S^{-1} B^{(n+1)} = A^{(n)} \wedge B^{(n)} \\ A^{(n+1)} \vee S^{-1} B^{(n+1)} = A^{(n)} \vee B^{(n)} \end{cases}. \tag{52}$$

When we consider the inverse ultradiscretization of the above equation, we find

$$\begin{cases} a_i^{(n+1)} b_{i+1}^{(n+1)} = a_i^{(n)} b_i^{(n)} \\ a_i^{(n+1)} + b_{i+1}^{(n+1)} = a_i^{(n)} + b_i^{(n)} \end{cases} . \tag{53}$$

Thus, for arbitrary λ, $(\lambda + a_i^{(n)})(\lambda + b_i^{(n)}) = (\lambda + a_i^{(n+1)})(\lambda + b_{i+1}^{(n+1)})$ and we find that

$$C_n(\lambda) := \prod_{i=1}^{N} (\lambda + a_i^{(n)})(\lambda + b_i^{(n)}) \tag{54}$$

does not depend on n, which means that any symmetric polynomial with respect to $\{a_i^{(n)}\}$ and $\{b_i^{(n)}\}$ does not depend on n. Therefore, the ultra-discrete limit of such symmetric polynomials gives $2N$ conserved quantities $S_1, S_2, \dots, S_{2N}$ of (42). If we denote $B_i^{(n)} \equiv A_{N+i}^{(n)}$, these conserved quantities are explicitly given as

$$S_1 := \max_i \left[A_i^{(n)} \right]$$

$$S_2 := \max_{i<j} \left[A_i^{(n)} + A_j^{(n)} \right]$$

$$\cdots$$

$$S_{2N} := \sum_{i=1}^{2N} A_i^{(n)}.$$

6.3 PBBS as Periodic $A_M^{(1)}$ Crystal Lattice

In the previous sections, we showed that the BBS (with an infinite number of boxes) can be reformulated from the theory of crystal and the combinatorial R-matrix. The PBBS is also reformulated as a combinatorial R matrix lattice model with periodic boundary condition. For the original BBS, the time evolution is given by the isomorphism:

$$\mathcal{T}: B_\infty \otimes B_1^{\otimes N} \to B_1^{\otimes N} \otimes B_\infty$$
$$\mathcal{T}: |\{0\}\rangle \otimes |c(t)\rangle \mapsto |c(t+1)\rangle \otimes |\{0\}\rangle$$

where $|c(t)\rangle \in B_1^{\otimes N}$ is the state corresponding to the BBS at time t. For the PBBS, we have to take the trace of the vertical state, $i.e.$, by regarding $\mathcal{T} \in \mathrm{End}_{\mathrm{End}_{B_1^{\otimes N}}} B_\infty$, we define the matrix $T := \mathrm{Tr}_{B_\infty} \mathcal{T} \in \mathrm{End}_{\mathbb{C}} B_1^{\otimes N}$, which gives a time evolution as

$$T: B_1^{\otimes N} \to B_1^{\otimes N}$$
$$T: |c(t)\rangle \mapsto |c(t+1)\rangle.$$

At a glance, one may think that $|c(t+1)\rangle$ becomes a linear combination of many of the tensor products of B_1 crystals. However a tensor product of B_1

crystals is mapped to a unique tensor product of B_1 crystals and the resultant state exactly corresponds to the state of the PBBS at time step $t + 1$. Even for the PBBS with M ($M \geq 2$) kinds of balls and various box capacities, the above lattice model is also well defined as far as the dimension of the vertical crystal is large enough, that is, $\kappa_t \gg 1$ for the vertical crystal B_{κ_t} of $U'_q(A^{(1)}_{M-1})$. We will present a proof of this fact for the case with one kind of ball. Since the evolution rule for M kinds of balls is decomposed into M steps as far as $\kappa_t \gg 1$, and only one kind of ball is moved at each step according to the same evolution rule, the proof is also valid for the case with many kinds of balls. When κ is small, however, the above construction will not give a unique tensor product and will not define an evolution rule of the PBBS.

Since we treat only one kind of balls, the states are represented by a $U'_q(A^{(1)}_1)$ crystal. First we consider the isomorphism $B_\kappa \otimes B_\theta \simeq B_\theta \otimes B_\kappa$ given by the combinatorial R-matrix. A state b in B_κ is usually denoted by a single row semistandard Young tableau of length κ on letters 1 and 2. Instead, we denote $b = (y, \kappa - y)$ where y is the number of 1 in the Young tableau. For $(y, \kappa - y) \otimes (x, \theta - x) \simeq (x', \theta - x') \otimes (y', \kappa - y')$, we have the relation [35]

$$x' = y - \min[\kappa, x + y] + \min[\theta, x + y] \tag{55}$$

$$y' = x + \min[\kappa, x + y] - \min[\theta, x + y]. \tag{56}$$

For $\kappa > \theta$, the relation is explicitly written as

$$y' = \begin{cases} x & (x + y \leq \theta) \\ 2x + y - \theta & (\theta < x + y \leq \kappa) \\ x + \kappa - \theta & (\kappa < x + y) \end{cases} \tag{57}$$

$$x' = x + y - y' \tag{58}$$

Now let θ_n ($n = 1, 2, \ldots, N$) be the capacity of the nth box, and κ_t be the capacity of the carrying cart at time step t. The state at time step t is given by $|c(t)\rangle \in B_{\theta_1} \otimes B_{\theta_2} \otimes \cdots \otimes B_{\theta_N}$. Since B_{θ_n} is a $U'_q(A^{(1)}_1)$ crystal, a vector $b_n \in B_{\theta_n}$ is represented as $b_n = (x_n, \theta_n - x_n)$, where x_n corresponds to the number of the balls in the nth box. We denote a state $b_1 \otimes b_2 \otimes \cdots \otimes b_N$ by $[x_1, x_2, \ldots, x_N]$ for $b_i = (x_i, \theta_i - x_i)$ ($i = 1, 2, \ldots, N$). The combinatorial R-matrix of $U'_q(A^{(1)}_1)$ gives the isomorphism $\mathcal{T}$:

$$\mathcal{T} : B_{\kappa_t} \otimes (B_{\theta_1} \otimes B_{\theta_2} \otimes \cdots \otimes B_{\theta_N}) \simeq (B_{\theta_1} \otimes B_{\theta_2} \otimes \cdots \otimes B_{\theta_N}) \otimes B_{\kappa_t}$$
$$([y_0] \otimes [x_1, x_2, \ldots, x_N] \simeq [x'_1, x'_2, \ldots, x'_N] \otimes [y'_0])$$

From (58), we obtain the following recurrence equations:

$$y_n = F(y_{n-1}; x_n, \theta_n)$$

$$:= \begin{cases} x_n & (x_n + y_{n-1} \le \theta_n) \\ 2x_n + y_{n-1} - \theta_n & (\theta_n < x_n + y_{n-1} \le \kappa_t) \\ x_n + \kappa_t - \theta_n & (\kappa_t < x_n + y_{n-1}) \end{cases}$$

$$x'_n = \begin{cases} y_{n-1} & (x_n + y_{n-1} \le \theta_n) \\ \theta_n - x_n & (\theta_n < x_n + y_{n-1} \le \kappa_t) \\ y_n - \kappa_n + \theta_n & (\kappa_t < x_n + y_{n-1}) \end{cases} \qquad (59)$$

$$(n = 1, 2, \ldots, N)$$

$$y'_0 = y_N.$$

We see that the function $F(y; x, \theta)$ is a piecewise linear and monotonically increasing function of y which satisfies $F(y + 1; x, \theta) - F(y; x, \theta) = 0$ or 1 and $0 \le F(y; x, \theta) \le \kappa_t$. Since y'_0 is a function of y_0, we denote it by $y'_0 = F_N(y_0; \{x_i\}, \{\theta_i\}) := \underbrace{(F \circ F \circ \cdots \circ F)}_{N \text{ times}}(y_0)$. The function F_N is also a mono-tonically increasing piecewise linear function, and $F_N(y_0 + 1; \{x_i\}, \{\theta_i\}) - F_N(y_0; \{x_i\}, \{\theta_i\}) = 0$ or 1 and $0 \le F_N \le \kappa_t$. Thus, for $0 \le y_0 \le \kappa_t$, there is one and only one integer y^* or one and only one finite interval $[y_*, y^*]$ $(y_*, y^* \in \mathbb{Z})$ where the identity $y_0 = F_N(y_0; \{x_i\}, \{\theta_i\})$ holds. Furthermore, from (59), we find that the $\{x'_n\}$ do not vary for $y_* \le y_0 \le y^*$. Therefore we conclude that, for given $\{x_i\}$ and $\{\theta_i\}$, there is at least one y_0 $(0 \le y_0 \le \kappa_t)$ at which y'_0 is equal to y_0 and that $\{x'_i\}$ are uniquely determined and have the same values for y_0 which satisfies $y'_0 = y_0$. The above conclusion means that $T := \mathrm{Tr}_{B_{\kappa_t}} \mathcal{T}$, $\mathcal{T} \in \mathrm{End} B_{\theta_1} \otimes \cdots \otimes B_{\theta_N}$ maps a state $b_1 \otimes \cdots \otimes b_N$ to the state which is also described by a tensor product of crystal bases. We summarize the above statement as a Theorem.

Theorem 6.2. *If κ is greater than any θ_i $(i = 1, 2, \ldots, N)$, the map $T := \mathrm{Tr}_{B_\kappa} \mathcal{T} \in \mathrm{End} B_{\theta_1} \otimes B_{\theta_2} \otimes \cdots \otimes B_{\theta_N}$ sends a tensor product of crystal bases to a unique tensor product of crystal bases of $U'_q(A_1^{(1)})$. Furthermore, for sufficiently large κ, the statement holds for $U'_q(A_M^{(1)})$ with arbitrary positive integer M.*

From the construction of the map, it is clear that the PBBS discussed in the previous section corresponds to the case $\theta_i = 1$ $(i = 1, 2, \ldots, N)$, and we use this map to construct the PBBS with arbitrary box capacities and ball species.

6.4 PBBS as $A_{N-1}^{(1)}$ Crystal Chains

When there are M balls of the same kind and N boxes, we can also reformulate the PBBS in terms of the combinatorial R-matrix of $U'_q(A_{N-1}^{(1)})$ and the symmetric tensor product B_M and $B_{M'}$ where $M' := \sum_{i=1}^{N} \theta_i - M$ (Fig. 22).

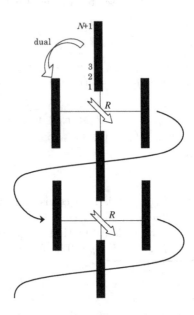

Fig. 22. Twisted chains of crystal $A_{N-1}^{(1)}$ and PBBS

A crystal $b \in B_M$ can be denoted by $b = (x_1, x_2, \ldots, x_N)$ with $0 \leq x_i \leq \theta_i$, $\sum_{i=1}^{N} x_i = M$. We associate a state of the PBBS with the crystal b in which x_i is the number of balls in the ith box of the state. For the crystal b, we define the dual crystal $\bar{b} = (\bar{x}_1, \bar{x}_2, \ldots, \bar{x}_N) \in B_{M'}$, where $\bar{x}_i = \theta_i - x_i$ $(i = 1, 2, \cdots, N)$. Then the crystal $b' \in B_M$ associated with the state at time $t + 1$ is given by the combinatorial R-matrix which gives the isomorphism $B_{M'} \otimes B_M \simeq B_M \otimes B_{M'}$ as

$$R : \quad \bar{b} \otimes b \longrightarrow b' \otimes \bar{b}'. \tag{60}$$

From [35], we see that this gives the same time evolution of the PBBS discussed above. As is shown in Fig. 22, the time evolution is described in two twisted chains of B_M and $B_{M'}$. Note that by changing the crystal b and/or $\bar{b}$ for another crystal (say B type crystal), we obtain other types of PBBS with a time evolution rule given by the isomorphism R. We may find interesting features in these CAs. However, the study of these PBBSs is a project for future research.

The isomorphism (60) has been shown to be expressed as an ultradiscrete KP equation (eqs. (22) and (23) in [18]), which is another reason why we claim the PBBS is an integrable CA. Here we do not repeat the results in [18], but we state a formula similar to (42). The space of the states, however, is no longer a finite field but $\mathbb{Z}^N$. For $X = (x_1, x_2, \ldots, x_N)$, $Y = (y_1, y_2, \ldots, y_N) \in \mathbb{Z}^N$, we define max and min: $\mathbb{Z}^N \times \mathbb{Z}^N \to \mathbb{Z}^N$ as

$$(\min[X, Y])_i = \min[x_i, y_i]$$
$$(\max[X, Y])_i = \max[x_i, y_i].$$

We also define the rotate shift to the right $S\colon \mathbb{Z}^N \to \mathbb{Z}^N$ as

$$SX = (x_N, x_1, x_2, \dots, x_{N-1}).$$

Let $\theta_i (\in \mathbb{Z}_{>0})$ be the capacity of the ith box and $x_i(t)$ $(0 \le x_i(t) \le \theta_i)$ be the number of balls in the ith box at time step t. We denote the state of the PBBS at t by $X(t) := (x_1(t), x_2(t), \dots, x_N(t))$. The state at $t+1$, $X(t+1)$, is obtained from the following Theorem.

Theorem 6.3. *Let* $A^{(0)} = X(t)$ *and* $B^{(0)} = SX(t)$. *We define* $A^{(n)}$ *and* $B^{(n)}$ $(n = 1, 2, \dots)$ *by the recurrence equations,*

$$\begin{cases} A^{(n+1)} := \min\left[A^{(n)} + B^{(n)}, \boldsymbol{\theta}\right] \\ B^{(n+1)} := S\max\left[A^{(n)} + B^{(n)} - \boldsymbol{\theta}, \mathbf{0}\right] \end{cases}, \tag{61}$$

where $\boldsymbol{\theta} := (\theta_1, \theta_2, \dots, \theta_N)$. *Then we obtain*

$$X(t+1) = A^{(N-1)} - X(t), \quad B^{(N-1)} = \mathbf{0}. \tag{62}$$

6.5 Fundamental Cycle of PBBS

Since the PBBS is composed of a finite number of cells, and it can only take on a finite number of patterns, the time evolution of the PBBS is necessarily periodic. In the present article, we investigate the fundamental cycle, *i.e.*, the shortest period of the discrete periodic motion of the PBBS.

As the original BBS, the PBBS has conserved quantities which are characterized by a Young diagram with M boxes. The Young diagram is constructed as follows. We denote an empty box by '0' and a filled box by '1'. Then the PBBS is represented as a $0, 1$ sequence in which the last entry is regarded as adjacent to the first entry. Let p_1 be the number of the 10 pairs in the sequence. If we eliminate these 10 pairs, we obtain a new $0, 1$ sequence. We denote by p_2 the number of 10 pairs in the new sequence. We repeat the above procedure until all the '1's are eliminated and obtain $p_2, p_3, \dots, p_l$.

Clearly $p_1 \ge p_2 \ge \cdots \ge p_l$ and $\displaystyle\sum_{i=1}^{l} p_i = M$. These $\{p_i\}_{i=1}^{l}$ are conserved in time evolution. Since $\{p_1, p_2, \dots, p_l\}$ is a weakly decreasing sequence of positive integers, we can associate it with a Young diagram with p_j boxes in the j-th column $(j = 1, 2, \dots, l)$. Then the lengths of the rows are also weakly decreasing positive integers, and we denote them

$$\underbrace{\{L_1, L_1, \dots, L_1,}_{n_1} \underbrace{L_2, L_2, \dots, L_2,}_{n_2} \cdots, \underbrace{L_s, L_s, \dots, L_s\}}_{n_s}$$

where $L_1 > L_2 > \cdots > L_s$. The set $\{L_j, n_j\}_{j=1}^s$ is an alternative expression of the conserved quantities of the system. In the limit $N \to \infty$, L_j means the length of the jth largest soliton and n_j is the number of solitons with length L_j.

The following two propositions are essential to our argument. Let $\ell_0 := N - 2M = N - \sum_{j=1}^l 2p_j = N - \sum_{j=1}^s 2n_j L_j$, $N_0 := \ell_0$, $L_{s+1} := 0$, and

$$\ell_j := L_j - L_{j+1} \quad (j = 1, 2, \ldots, s) \tag{63}$$

$$N_j := \ell_0 + 2n_1(L_1 - L_{j+1}) + 2n_2(L_2 - L_{j+1}) + \cdots + 2n_j(L_j - L_{j+1})$$

$$= \ell_0 + \sum_{k=1}^j 2n_k(L_k - L_{j+1}). \tag{64}$$

Then, for a fixed number of boxes N and conserved quantities $\{L_j, n_j\}$, the number of possible states of the PBBS $\Omega(N; \{L_j, n_j\})$ is given by the following formula.

Theorem 6.4.

$$\Omega(N; \{L_j, n_j\}) = \frac{N}{\ell_0} \binom{\ell_0 + n_1 - 1}{n_1} \binom{N_1 + n_2 - 1}{n_2} \binom{N_2 + n_3 - 1}{n_3}$$

$$\times \cdots \times \binom{N_{s-1} + n_s - 1}{n_s} \tag{65}$$

The fundamental cycle T is given as follows.

Theorem 6.5. *Let $\tilde{T}$ be defined as*

$$\tilde{T} := L.C.M. \left(\frac{N_s N_{s-1}}{\ell_s \ell_0}, \frac{N_{s-1} N_{s-2}}{\ell_{s-1} \ell_0}, \cdots, \frac{N_1 N_0}{\ell_1 \ell_0}, 1 \right), \tag{66}$$

where $L.C.M.(x, y) := 2^{\max[x_2, y_2]} 3^{\max[x_3, y_3]} 5^{\max[x_5, y_5]} \cdots$ for $x = 2^{x_2} 3^{x_3} 5^{x_5} \cdots$ and $y = 2^{y_2} 3^{y_3} 5^{y_5} \cdots$. Then T is a divisor of $\tilde{T}$. In particular, when there is no internal symmetry in the state $T = \tilde{T}$.

The definition of internal symmetry in the above proposition is rather complicated and we refer the reader to the original article [11]. However, for a given number of conserved quantities, we can always construct initial states which do not have any internal symmetry, in particular, if $^\forall i$, $n_i = 1$, the PBBS never has internal symmetry and $T = \tilde{T}$.

As for the asymptotic behaviour of the fundamental cycle T, we have the following theorems [40].

Theorem 6.6. *For $N \gg 1$ and $M = \rho N$ $(0 < \rho < 1/2)$, the maximum value of the fundamental cycle $T_{\max} \equiv T_{\max}(N; \rho)$ satisfies*

$$\exp\left[2\left(1 - \max[\sqrt{2 - 4\rho} - 1, 0]\right)\sqrt{N}\left(1 - \frac{c}{\log N}\right)\right]$$
$$< T_{\max} < \exp\left[2\sqrt{2\rho}\sqrt{N}\log N\right]. \tag{67}$$

Here c is a positive integer and $c \sim 0.1$ for $N \geq 10^{16}$.

Theorem 6.7. *Let $\bar{V}(N; \rho)$ be the number of initial states whose fundamental cycle is less than $\exp\left[\frac{2(\log N)^2}{-\log t_0}\right]$ with $t_0 := \frac{\rho}{1 - \rho}$ $(0 < t_0 < 1)$. Then, for fixed ρ,*

$$\lim_{N \to \infty} \frac{\bar{V}(N; \rho)}{V(N; \rho)} = 1. \tag{68}$$

From these theorems, a PBBS is shown to have no ergodicity in the sense that a trajectory leaves most of the states in the phase space unvisited. Although the maximum fundamental cycle $T_{\max} \gtrsim e^{\sqrt{N}}$ (Theorem 6.6), a generic state has fundamental cycle $T \lesssim e^{(\log N)^2}$ (Theorem 6.7).

7 Concluding Remarks

As was explained in detail, ultradiscretization is a limiting procedure which approximates an equation by a piecewise linear equation closed under a finite number of discrete values. This procedure can in principle be applied to any equation, for example discrete Painlevé equations [41]. However, in general, the naïve application of ultradiscretization to a non-integrable equation will not necessarily reproduce a specific feature of its solutions. In order to ultradiscretize non-integrable equations while retaining their features, we need some guiding principles. To find such principles is one of the important problems in the theory of ultradiscrete systems.

References

1. S. Wolfram: *Cellular Automata and Complexity* (Addison-Wesley, Reading, MA 1994)
2. S. Wolfram: Phys. Scr. **T9**, 170 (1985)
3. T. Tokihiro, D. Takahashi, J. Matsukidaira and J. Satsuma: Phys. Rev. Lett. **76**, 3247 (1996).
4. J. Matsukidaira, J. Satsuma, D. Takahashi, T. Tokihiro and M. Torii: Phys. Lett. A **255**, 287 (1997)
5. K. Park, K. Steiglitz, and W. P. Thurston: Physica D **19**, 423 (1986)
6. A. S. Fokas, E. P. Papadopoulou and Y. G. Saridakis: Physica D **41**, 297 (1990)
7. A. S. Fokas, E. P. Papadopoulou, Y. G. Saridakis and M. J. Ablowitz: Studies in Applied Mathematics **81**, 153 (1989)

8. M. J. Ablowitz, J. M. Keiser, L. A. Takhtajan: Quaestiones Math. **15**, 325 (1992)
9. D. Takahashi: 'On some soliton systems defined by boxes and balls'. In: *Proceedings of the International Symposium on Nonlinear Theory and Its Applications, NOLTA '93*, p.555 (1991)
10. D. Takahashi and J. Satsuma: J. Phys. Soc. Jpn. **59**, 3514 (1990)
11. D.Yoshihara, F.Yura and T.Tokihiro: J. Phys. A.FMath. Gen. **36**, 99 (2003)
12. A. Nagai, D. Takahashi and T. Tokihiro: Physics Letters A **255**, 265 (1999)
13. D. Takahashi and J. Matsukidaira: J. Phys. A.FMath. Gen. **30**, 733 (1997)
14. T. Tokihiro, A. Nagai and J. Satsuma: Inverse Probl. **15**, 1639 (1999)
15. T. Tokihiro, D. Takahashi and J. Matsukidaira: J. Phys. A.FMath. Gen. **33**, 607 (2000)
16. K. Hikami, R. Inoue, and Y. Komori: J. Phys. Soc. Jpn. **68**, 2234 (2000)
17. K. Fukuda, M. Okado, and Y. Yamada: Int. J. Mod. Phys. A **15**, 1379 (2000)
18. G. Hatayama, K. Hikami, R. Inoue, A. Kuniba, T. Takagi, and T. Tokihiro: J. Math. Phys. **42**, 274 (2001)
19. M. Bruschi, P. M. Santini and O. Ragnisco: Physics Letters A **169** 151 (1992)
20. A. Bobenko, M. Bordemann, C. Gunn, U. Pinkall: Comm. Math. Phys. **158**, 127 (1993)
21. R. Hirota: J. Phys. Soc. Jpn.**50**, 3785 (1981)
22. T. Miwa: Proceedings of the Japan Academy **58 A**, 9(1982)
23. E. Date, M. Jimbo, T. Miwa: J. Phys. Soc. Jpn. **51**, 4125 (1982)
24. R. Willox and J. Satsuma: Sato Theory and Transformation Groups. A Unified Approach to Integrable Systems, Lect. Notes Phys. 644, 17 (2004)
25. M. Sato: RIMS Kokyuroku **439**, 30 (1981).
26. E. Date, M. Jimbo, M. Kashiwara, T. Miwa: 'Transformation groups for soliton equations'. In: *Proceedings of RIMS symposium on Non-linear Integrable Systems-Classical Theory and Quantum Theory, Kyoto, Japan May 13 – May 16, 1981*, ed. by M. Jimbo, T. Miwa (World Scientific Publ. Co., Singapore 1983) pp. 39–119
27. T. Miwa, M. Jimbo and E. Date: *Solitons – Differential equations, symmetries and infinite dimensional algebras* (Cambridge University Press, UK 2000)
28. M. Toda: J. Phys. Soc. Jpn. **22**, 431 (1967)
29. A. Nagai, T. Tokihiro and J. Satsuma: Glasgow Math. J. **43A**,91 (2001)
30. M. Torii, D. Takahashi and J. Satsuma: Physica D **92**, 209 (1996)
31. W. Fulton: *Young Tableaux* (Cambridge University Press, UK, 1997)
32. C. N. Yang: Physical Review Letters **19**, 1312 (1967)
33. R. J. Baxter: Annals of Physics **70**, 193 (1972)
34. P. P. Kulish and E. K. Sklyanin: Journal of Soviet Mathematics **19**, 1596 (1982).
35. A. Nakayashiki and Y. Yamada: Selecta Mathematica, New Series **30**, 547 (1997)
36. See for example, M. Jimbo: 'Topics from representations of $U_q(g)$-an introductory guide to physicists'. In: *Nankai Lectures on Mathematical Physics* (World Scientific, Singapore, 1992), pp. 1-61.
37. M. Kashiwara: Communications in Mathematical Physics **133**, 249 (1990)
38. F. Yura and T. Tokihrio: J. Phys. A.FMath. Gen. **35**, 3787 (2002)
39. T. Kimijima and T. Tokihiro: Inverse Problems **18**, 1705 (2002)
40. J. Mada and T. Tokihiro: J. Phys. A.FMath. Gen.**36**, 7251 (2003)
41. A. Ramani, D. Takahashi, B. Grammaticos and Y. Ohta: Physica D **114** 185 (1998)

Time in Science:
Reversibility vs. Irreversibility

Yves Pomeau

Laboratoire de Physique Statistique de l'Ecole normale supérieure, 24 Rue Lhomond, 75231 Paris Cedex 05, France, pomeau@physique.ens.fr

Abstract. To discuss properly the question of irreversibility one needs to make a careful distinction between reversibility of the equations of motion and the 'choice' of the initial conditions. This is also relevant for the rather confuse philosophy of the 'wave packet reduction' in quantum mechanics. The explanation of this reduction requires also to make precise assumptions on what initial data are accessible in our world. Finally I discuss how a given (and long) time record can be shown in an objective way to record an irreversible or reversible process. Or: can a direction of time be derived from its analysis? This leads quite naturally to examine if there is a possible spontaneous breaking of the time reversal symmetry in many body systems, a symmetry breaking that would be put in evidence objectively by looking at certain specific time correlations.

1 Introduction

Scientists, as well as philosophers, have been always fascinated by time. The greatest of them all, Isaac Newton, made profound statements about the way time should enter into our rational understanding of the world. His deeply thought remarks (at the beginning of the Principia) are still valid today (for the nonrelativistic limit, relevant for most phenomena at human scale). One question, that is well discussed in the Principia too, concerns the initial conditions, clearly seen by Newton as something different from the laws of the motion, a notion foreign to many writers on the subject.

I believe this distinction between laws of motion and initial conditions is absolutely essential, and often not appreciated to its full extent. It is central to the discussion of two related issues in modern science:

1) the apparent opposition between reversibility of the laws of motion and everyday irreversibility in the behaviour of macroscopic systems,

2) the so-called reduction of the wave packet by measurements in quantum mechanics.

I am going to show that both issues have to do with the initial conditions of *our world*, and require us to make assumptions about them.

I felt that a visit to India was an opportunity to present some thoughts on questions of a more philosophical nature than usual, because it is a country of such long philosophical tradition. Later on I shall come back to more 'concrete' questions. In the spirit of [1], I shall discuss the following problem:

Y. Pomeau, Time in Science: Reversibility vs. Irreversibility, Lect. Notes Phys. **644**, 425–436 (2004)
http://www.springerlink.com/

given a time record, how is it possible to show in an objective way the fact that it records an irreversible process? In other terms, can a direction of time be derived from its analysis ? This has to do with 'practical' things as well as more fundamental ones, such as: could it be that the time in the Universe runs in different directions, depending on, say, the Galaxy one is inhabiting?

2 On the Phenomenon of Irreversibility in Physical Systems

A inexhaustible theme of discussions in physics and physics-related science is the apparent opposition between the reversible laws of motion and the irreversibility observed in everyday life. This discussion is much confused by attempts, conscious or not, to 'prove' the irreversibility of the motion of large groups of point particles, without stating clearly the assumptions made at the beginning. In some sense, this eludes one of the deepest message of Newton : the only way to do science is to derive a set of consequences, starting from explicitly stated 'laws' or 'axioms' (Newton used both terms interchangeably). The most complete derivation of irreversible behaviour from the law of mechanics was done in the kinetic theory of gases by Boltzmann. who based his theory on the so-called 'Stosszahlansatz' (meaning approximately 'assumption on the counting of hits'), which clearly means that, besides the laws of mechanics, an additional assumption is needed to prove irreversibility. This Stosszahlansatz says that, before colliding, two particles have never met before, and so have independent statistical properties. The difficulty with the Boltzmann Stosszahlansatz is that it cannot be exact. When expanding the collision operator beyond the lowest order in a density expansion, one finds the so-called ring collisions that yield a certain amount of correlation of two particles entering into a binary collision. But this correlation turns out to be small enough to make the Stosszahlansatz valid at low densities. In other terms, by running back in time, two particles should have no correlation at all at infinite negative times. The little amount of correlation created by the ring collisions is a short time effect, at least at low density (in my PhD thesis [2] I showed that this is not so at finite densities because of the occurence of slowly decaying hydrodynamical fluctuations). This implies that a given non equilibrium system to which the Boltzmann equation applies should have been made in the past of at least two completely separated sets, without any knowledge of each other, namely without correlation. The correlations brought in each system in the past by its own relaxation are actually negligible, because it can be assumed that they are at equilibrium, where particles are uncorrelated at low density. Therefore, at the end, the validity of Boltzmann Stosszahlansatz relies upon the absence of correlation between physically separated systems. This property cannot however be taken for granted. It has to be assumed, which can be done without violating any basic principle. That it must be assumed follows from the following idea.

Suppose that, once the system has decayed to equilibrium, one splits it in two parts. These two parts bear some special correlations: running backward in time, the system will follow the same trajectory. An outside observer looking at it with its time running in the opposite direction of the one of the sytem under consideration will have the impression that the Stosszahlansatz does not apply, because the H-theorem will certainly not apply. Therefore, for this particular observer, the correlation in the final state will be the correlations in its reverse time frame and will certainly not satisfy the condition of application of the Boltzmann theory. This implies that this type of correlation can exist, and have to be excluded from the real world by an explicit assumption, unprovable by any method.

Similar things could be said concerning the reduction of the wave packet in quantum mechanics. Its status is often quite ambiguous, it is even sometimes claimed to be a fundamental principle of quantum mechanics, although it is far closer to the Stosszahlansatz than many believe. That the reduction of the wave packet requires some irreversibility should be made obvious (hopefully) by the following gedanken experiment. Think of a quantum system with two possible states, A and B. To measure its state, one connects it to a macroscopic device such that it is in either state α or β (which can be seen as the two possible steady positions of a needle for instance). When connected to the quantum system, the energy of the state of the full system is the lowest in the joint states (α, A) or (β, B), so that the system reaches irreversibly in either of these states, with probability $1/2$ for instance. This irreversible step in the evolution is possible because α and β are states of a macroscopic system, with off-diagonal elements of the probability matrix that are very small by interference between various eigenstates. This step requires a hidden assumption about the possible correlations between the various components of the state of the system in its Hilbert space. If one reverses the time, the correlation introduced by the evolution will make the system return to a very unlikely initial state, the one preexisting the measure. Therefore the reduction of the wavepacket, again a process that requires assumptions about the initial conditions, cannot be proven just by looking at the equations of motion. To be a bit more specific, meaning more mathematical, let us introduce a quantum system on which the measurement is made with two quantum states A and B. The macroscopic (classical) measuring device is described by a decay equation,

$$\frac{dx}{dt} + \frac{\partial \Phi}{\partial x} = f(t). \tag{1}$$

In this generalized Langevin equation, x can be seen as the position of a needle in the measuring device, with two equilibria, $x = \alpha$ and $x = \beta$. These equilibria are the two minima of the function $\Phi(x)$. The thermal forcing is represented by the Gaussian white noise force $f(t)$, that is of zero average and is delta correlated in time,

$$\overline{f(t)f(t')} = T\delta(t - t'),$$

δ being the Dirac function, and T the absolute temperature. The device is a measuring device because the potential $\Phi(t)$ depends on the quantum state A or B, and on some coupling parameter. When the coupling is turned off, $\Phi(x)$ has minima at $x = \alpha$ and $x = \beta$, in such a way that the probability of a thermal jump above the barrier between the two states is negligible over macroscopic times. When the interaction is turned on, the potential $\Phi(x)$ has a single minimum. But the location of this minimum is either at $x = \alpha$ if the quantum system is in the A state or at $x = \beta$ if the quantum system is in the B state. Therefore, after the time needed to fall to the bottom of the potential Φ, the interaction brings an absolute correlation between the state of the quantum system and the macroscopic state of the measuring device. This correlation does not pose any problem as far as irreversibility is concerned. The state before the interaction is turned on, and the state after, may be different: for instance A for the quantum system and β for the measuring device is perfectly legitimate as an initial state. But after the interaction is turned on, the system eliminates states like (A, β). Once the measurement is made, the interaction is turned off, following the reversed time path of the turning-on process. Then, a reversibility paradox appears. No final state (A, β) is possible, although the system is perfectly time reversible since the turning-on and off of the interaction follows the same time dependance. Therefore it seems that the post-measurement states should be possibly in the same list as the pre-measurement states. This is not really a paradox, because of the small probability that, once the measurement has been initiated, the system may jump back by thermal fluctuations to a state like (A, β), starting from (B, β). Unlikely realizations of the noise $f(t)$ can do it. These unlikely realizations are precisely the ones that would be observed by someone looking at the system in the backward time direction, and that are excluded in the forward time direction by an argument similar to the Stosszahlansatz. It would be interesting to test this kind of idea by more detailed investigations on the time-dependent generalized Langevin equation. The philosophy here is that the so-called quantum discontinuity is mostly a classical, macroscopic phenomenon. To make a connection with the general theme of integrability, it is worth pointing out that, after all, the generalized Langevin equation rests upon the idea of the instability of the trajectories of the classical 'macroscopic measuring device'. Therefore the idea of quantum chaos should be linked in some way or another to the property of this 'device' of not behaving at all according to the generalized Langevin equation, that is to say, it should be integrable in some sense. This also brings to light the fact that quantum chaos should be linked to the behaviour of a system with many, if not infinitely many, degrees of freedom.

Below, as promised, I will review the following question: given some time-dependent signal, how is it possible to decide if it is time-reversible or not, and what does this mean?

3 Reversibility of Random Signals

Everyday life tells us if the magnetic tape from a record,for instance, is running forward or backward in time. This is obvious for a movie, because running backward shows people walking backward, filling glasses instead of drinking, etc. That a piece of music, especially a modern one is running backward is far less obvious to my untrained ears . One can even recognize classical pieces like Beethoven music when run backward. This brings me to my point: is it possible to make a clearcut and 'objective' difference between a signal that is 'time reversible' and another one that is not?

I assume that this signal, $x(t)$, a real and smooth function of time, lasts long enough to make it possible to 'measure' any kind of correlation function, like

$$\Psi(\tau) = \overline{(x(t) - \overline{x})(x(t + \tau) - \overline{x})}. \tag{2}$$

This correlation function is obtained by averaging over a stationary random process, so that Ψ depends on the time difference between the two functions $(x(.) - \overline{x})$, averages being denoted by an overbar. For a stationary random process, by its very definition, function $\Psi(\tau)$ has the same value if τ is changed into $-\tau$. In other words no difference can be made between the two possible directions of time by looking at the pair correlation $\Psi(\tau)$. Therefore, part of the information in the signal $x(.)$, if it is not time reversible, is lost by looking at pair correlations as given in (2).

Consider now the product $x(t)x(t + 2\tau)x(t + 3\tau)$. It is a function of τ only, like the pair-correlation $\Psi(\tau)$ that has been just defined. The center of gravity of the three arguments is at $t + \frac{5}{3}\tau$, so that the three times in arguments of $x(.)$ are not symmetrical with respect to this barycenter. Looking backward in time, one would replace the triple product $x(t)x(t + 2\tau)x(t + 3\tau)$ by $x(t)x(t + \tau)x(t + 3\tau)$. Unless something special happens, the two averages have no reason to be the same. Therefore, it is relevant to introduce the correlation of the difference between the two cubic averages,

$$\Psi'(\tau) = \overline{x(t)x(t + 2\tau)x(t + 3\tau) - x(t)x(t + \tau)x(t + 3\tau)}. \tag{3}$$

Function $\Psi'(\tau)$ is exactly zero for a time reversible process, that is, for a process such that no time direction can be derived from its analysis. Indeed, an infinite number of functions vanishing for time-symmetric process and non-zero otherwise can be imagined. For instance, it can be that the plus and minus value of x have equal probability, so that $\Psi'(\tau)$ is zero, even for a non-time reversible process. In such a case an even function like

$$\Psi''(\tau) = \overline{x^3(t)x(t + \tau) - x(t)x^3(t + \tau)} \tag{4}$$

can be used to discriminate between a time-symmetric and a non-symmetric process.

Both Ψ' and Ψ'' are odd functions of τ. If signal $x(.)$ is smooth enough, the Taylor expansions of Ψ' and Ψ'' near $\tau = 0$ are

$$\Psi'(\tau) \approx -\frac{10}{3}\tau^3 \overline{\left(\frac{dx}{dt}\right)^3}$$

and

$$\Psi''(\tau) \approx -\tau^3 \overline{x(t)\left(\frac{dx}{dt}\right)^3}.$$

In both expression, the average is a single time-average.

Now I am going to examine various situations where this question of time symmetry is relevant.

The standard Ornstein-Uhlenbeck process is given formally by the solution of the Langevin equation,

$$\frac{dx}{dt} + x = f(t), \tag{5}$$

where $f(.)$ is a Gaussian white noise of temperature T such that $\overline{f} = 0$ and $\overline{f(t)f(t')} = T\delta(t - t')$. The stochastic process $x(.)$ is time reversible because of a balance between the damping term, the $+x$ on the left side of (5), and the special noise term on the right-hand side. Consider, for instance, the correlation function entering into Ψ'', namely the average $\overline{x^3(t)x(t + \tau)}$. Since $x(.)$ is a Gaussian variable of zero average, the usual rules for Gaussian variables yield at once:

$$\overline{x^3(t)x(t + \tau)} = 3(\overline{x})^3\overline{x(t)x(t + \tau)}$$

and

$$\overline{x(t)x^3(t + \tau)} = 3(\overline{x})^3\overline{x(t)x(t + \tau)}.$$

Therefore $\Psi''(\tau) = 0$ for this process.

This symmetry can be shown in a more general way by looking at the auto-correlation function of the Ornstein-Uhlenbeck process, namely the function $P(x(t), x(t + \tau))$ such that any correlation function $\overline{h(x(t))g(x(t + \tau))}$ is yielded by

$$\overline{h(x(t))g(x(t + \tau))} = \int_{-\infty}^{+\infty} dx \int_{-\infty}^{+\infty} dx_+ h(x)g(x_+)P(x, x_+). \tag{6}$$

In this expression, x is for $x(t)$ and x_+ for $x(t + \tau)$. Because $x(t)$ is a linear function of a Gaussian variable, $f(t)$, both x and x_+ are Gaussian variables. Therefore the autocorrelation function, $P(x, x_+)$, is a Gaussian joint probability of the general form

$$P(x, x_+) = \frac{1}{Z(\tau)}e^{-\left[a(\tau)(x^2 + x_+^2) + 2b(\tau)xx_+\right]}. \tag{7}$$

The factor a is the same for x^2 and x_+^2 because the averages over x_+ and over x, independent of the knowledge of the other, have to be the same. The two functions of τ, a and b, as well as the normalization factor Z, are derived from a knowledge of the various averages. The average value of 1 yields

$$\int_{-\infty}^{+\infty} dx \int_{-\infty}^{+\infty} dx_+ P(x, x_+) = 1.$$

That yields:

$$Z = \frac{\pi}{\sqrt{a^2 - b^2}}.$$

The standard properties of the Ornstein-Uhlenbeck process yield

$$\overline{x^2} = \frac{T}{2}$$

$$\overline{xx_+} = \frac{T}{2} e^{-|\tau|}.$$

Although the same quantities computed with $P(x, x_+)$ are

$$\overline{x^2} = \frac{a}{2(a^2 - b^2)},$$

$$\overline{xx_+} = -\frac{b}{2(a^2 - b^2)}.$$

By identification, this yields after some algebraic manipulations,

$$P(x, x_+) = \frac{1}{\pi T(1 - e^{-2|\tau|})} e^{-\frac{1}{T(1 - e^{-2|\tau|})} \left[x^2 + x_+^2 - 2xx_+ e^{-|\tau|} \right]}. \tag{8}$$

Note that because this expression depends on the absolute value of τ only, and because of its symmetry under the exchange of x and x_+, the cross correlation of the Ornstein-Uhlenbeck process is actually time symmetric. No measurement of a functions depending on the signal will be able to distinguish between the forward and backward directions of time.

The same symmetry can be shown to hold true for more general Langevin equations. Take the function of time $x(t)$ solution of the equation

$$\frac{dx}{dt} + \frac{d\Phi}{dx} = f(t), \tag{9}$$

where $f(t)$ is the same Gaussian noise as before, and Φ a generalization of the potential $\frac{x^2}{2}$ in the familiar linear Langevin equation. It can be shown that the same property of time-reversal symmetry holds for the random process given by (9). The proof is based upon a formal solution of the Chapman-Kolmogoroff equation for the pair-correlation $P(x(t), x(t+\tau))$, a proof given in [1]. This proof extends to situations where x is actually a vector, denoted by

x, and where $\frac{d\Phi}{dx}$ is replaced by the gradient with respect to **x** of a dissipation function, $\nabla\Phi$, which depend on the various components of **x**. However the proof does not work for non-gradient systems, that is for systems such as

$$\frac{d\mathbf{x}}{dt} + \mathbf{F}(\mathbf{x}) = \mathbf{f}$$

with **f** Gaussian white noise, and **F** non-potential field, with a non-zero rotational. This remark is interesting in the modelization of thermal fluctuations added to the equations of fluid mechanics. Those equations, when linearized (the so-called Stokes limit of the Navier-Stokes equations) can be written in the gradient form (a consequence of the Rayleigh-Prigogine principle of minimum production of entropy). With the non-linear terms added, it has been known for a long time that the full Navier-Stokes equations cannot be written in such a gradient form, so that an extenal noise source cannot be added to describe the equilibrium and weakly non-equilibrium thermal fluctuations. At equilibrium these fluctuations must satisfy the constraint of reversibility.

This brings me to my next point. The constraint of reversibility is relevant for equilibrium situations only. This is not completely obvious. One might have the impression that it is a direct consequence of the reversibility of the fundamental equations of classical dynamics. Actually it also requires that the system is at equilibrium. To show this, it is enough to find a counter-example, namely a correlation that fails the test of time reversal symmetry as soon as the system is out of equilibrium. The example I choose is the following one. Suppose we have a gas in a 2D system, with four possible values of molecular velocities, of modulus one, and directed along four directions at right angle of each other. In this system, the velocity space is discrete, with index i, between $i = 1$ and $i = 4$. The convention will be that the sum $i + j$, i and j integers less than 4, will be taken modulo 4. The direction i and $i + 2$ are opposite. The Boltzmann equation for this discrete gas is

$$\frac{dN_i}{dt} = \nu(N_{i+1}N_{i+3} - N_{i+2}N_i). \tag{10}$$

The velocity distribution N_i is a set of four time-dependent positive numbers (I assume that the gas is homogeneous in space), normalized by the condition

$$\Sigma_{i=1}^{i=4} N_i = 1.$$

Moreover the quantity ν that appears in (10) is a frequency of collision. Indeed, (10) is actually a list of four equations. Let us consider a Couette shear flow. In this flow, and near the center of the flow, there is no mean velocity and an Enskog-Hilbert expansion shows that the local stationary distribution takes the form

$$N_i = \frac{1}{4} + \epsilon(-1)^i. \tag{11}$$

In this expression, $\epsilon(-1)^i$ denotes the deviation of the velocity distribution from equilibrium. It would result from the shear flow. In this model, if i, for instance, is in the y direction, and the x coordinate is orthogonal to y, the shear flow must be a function of $(x + y)$, the velocity being oriented in the direction of the bissectrix of the axis such that $x = y$, which is because this model has spurious collisional invariants.

The time correlation functions are quantities of the form

$$\overline{F(i,t)G(j,t+\tau)} = \Sigma_{i,j=1,4}N_i^0 M(j,\tau;i)G(j)F(i). \tag{12}$$

By definition in this expression, $F(i,t)$ is the value of an arbitrary function of the velocities at time t in a given microscopic state of the gas reached at this time, although $F(i)$ is the function of i that is averaged over all the particles to obtain this microscopic value. Moreover, N_i^0 is the velocity distribution in the stationary state under study, and $M(j,\tau;i)$ is the time-correlation function. This correlation function is a solution of the linearized kinetic equation with a delta-like initial condition,

$$\frac{dM(j,\tau;i)}{d\tau} = \nu(N_{j+1}^0 M_{j+3} + N_{j+3}^0 M_{j+1} - N_j^0 M_{j+2} + N_{j+2}^0 M_j) \tag{13}$$

In this expression, M_{j+3} on the right-hand side stands for $M(j+3,\tau;i)$. The initial condition is $M(j,\tau = 0;i) = \delta_{i,j}$, where $\delta_{i,j}$ stands for the discrete Kronecker delta, equal to 1 if $i = j$ and to zero otherwise.

From the correlation of F and G at different times, as given in (12), one can define a new correlation-function that should be zero for a time-reversible process, by subtracting its time reversed expression. Since the speed changes sign under time-reversal, $\overline{F(i,t)G(j,t+\tau)}$ becomes $\overline{G(i+2,t)F(j+2,t+\tau)}$ under time-reversal, reversing speeds amounting to adding two to the indices. Therefore the time-correlation that should vanish for a time-reversal invariant process is $\overline{[F(i,t)G(j,t+\tau) - G(i+2,t)F(j+2,t+\tau)]}$. Take a function N_l^0 such that $N_l^0 = N_{l+2}^0$, and functions F and G such that $G(l) = G(l+2)$ and $F(l) = F(l+2)$. Elementary but rather long calculations yields the solution of the linear problem, (13), with the result

$$\overline{[F(i,t)G(j,t+\tau) - G(i+2,t)F(j+2,t+\tau)]} =$$

$$\frac{exp - 2\left[|\nu\tau|(N_1^0 + N_2^0)\right]}{(N_1^0 + N_2^0)}(G(2)F(1) - G(1)F(2))\left[(N_1^0)^2 - (N_2^0)^2\right]. \tag{14}$$

With $N_1^0 = \frac{1}{4} - \epsilon$ and $N_2^0 = \frac{1}{4} + \epsilon$, this yields:

$$\overline{[F(i,t)G(j,t+\tau) - G(i+2,t)F(j+2,t+\tau)]} =$$
$$2\epsilon\,(G(2)F(1) - G(1)F(2))\,e^{-|\nu\tau|}$$

This last expression shows that the test function for time-reversal symmetry vanishes at equilibrium (for $\epsilon = 0$). It does not vanish for $\tau = 0$ because the velocity fluctuations of the individual particles do not depend smoothly on time.

Another interesting property of the time fluctuations of non-time-symmetric systems has to do with the phase of the Fourier transform of the signal. Let $X(t)$ be a fluctuating signal or noise. The Wiener-Khinchin theorem relates the time correlation of this signal to the modulus of its Fourier transform. Let

$$\tilde{X}_\tau(\omega) = \int_{t_0}^{t_0+\tau} d\tau' e^{i\omega\tau'} X(\tau')$$

be the Fourier tranform in the time-window $[t_0, t_0 + \tau]$. By definition, the autocorrelation of the signal is

$$S_a(t) = \lim_{\tau\to\infty} \frac{1}{\tau} \int_{t_0}^{t_0+\tau} d\tau' (X(\tau') - \overline{X})(X(t+\tau') - \overline{X}). \tag{15}$$

The Wiener-Khinchin theorem states that

$$\lim_{\tau\to\infty} \frac{1}{\tau} |\tilde{X}_\tau(\omega)|^2 = \int_{-\infty}^{+\infty} dt e^{i\omega t} S_a(t)$$

This shows that the spectral function is independant on the phase of $\tilde{X}_\tau(\omega)$. But this spectral function is insensitive to time-reversal, as is the autocorrelation function $S_a(t)$. Therefore, any information related to the time reversal symmetry of the signal is stored in the phase of its Fourier transform. This leads to a rather curious property of the phase of various transform of a time symmetric signal. To show it, consider two functions of the signal $X(\tau)$, like $A(X(\tau))$ and $B(X(\tau))$. Consider now their Fourier transform,

$$\tilde{A}_\tau(\omega) = \int_{t_0}^{t_0+\tau} d\tau' e^{i\omega\tau'} A(X(\tau'))$$

$$\tilde{B}_\tau(\omega) = \int_{t_0}^{t_0+\tau} d\tau' e^{i\omega\tau'} B(X(\tau'))$$

Let us try now to compute the part of correlation function between A and B that vanishes for a time-symmetric signal. By definition, this is the following correlation function,

$$S_{AB}(t) = \lim_{\tau\to\infty} \frac{1}{\tau} \int_{t_0}^{t_0+\tau} d\tau' \left[A(X(\tau'))B(X(t+\tau')) - B(X(\tau'))A(X(t+\tau')) \right].$$
$$\tag{16}$$

If the noise is time symmetric, this function should be zero for any choice of A and B. An obvious extension of the derivation of the Wiener-Khinchin theorem shows that

$$\lim_{\tau\to\infty} \frac{2i}{\tau} \sin(\varphi_A(\omega) - \varphi_B(\omega)) |\tilde{A}_\tau(\omega)\tilde{B}_\tau(\omega)| = \int_{-\infty}^{+\infty} dt e^{i\omega t} S_{AB}(t).$$

In this expression, $\varphi_{A,B}$ are the phases of the Fourier transform, while $A_\tau(\omega)$ and $B_\tau(\omega)$ are seen as complex numbers. Therefore one obtains the non-obvious result that the phases of the Fourier transform of any function of $X(t)$ are the same if the noise is time-reversible. Indeed this phase has a meaning for a well defined and unique choice of the integration bounds for the Fourier transform.

A final remark on this question of the analysis of time-reversal symmetry is that 'time' there is only to mean a real variable, going from minus to plus infinity. Therefore it makes sense to try to extend the same idea to situations where the time is replaced by another continuous variable, typically a position in space. In [1] I suggested this possibility, by introducing the minimal field theoretic model showing this breaking of symmetry under reflection. The idea is actually to consider an interaction between various points that has the same lack of symmetry as, for instance, the triple correlations non vanishing in a non-reversible system. It is even thinkable that this symmetry under reflection is broken spontaneously, for instance as temperature is lowered in a system of interacting spins on a lattice.

4 Conclusion and Perspectives

This essay intended to show that the time-dependance of physical phenomena remains an active subject of investigations. Particularly, in the mathematical-physical approach, it is still full of yet poorly understood questions. The question of integrability is in some sense behind the whole subject. Any irreversible behaviour is due not to the lack of reversibility of the equations, but to their lack of integrability. I pointed out the fact that quantum chaos is actually far closer to the usual type of chaos than expected, if one looks at the phenomenon responsible for the reduction of the wave packet by measurements. From this point of view, again, a fundamental understanding of the issues at stake seems to be still lacking.

A final comment: part of the oral presentation concerned dynamical models with discrete time (various cellular automata models, some developed even before the word became known). I refer the interested reader to the original publications [3]. From the point of view of the topic of this school, I think that the most relevant issue I have raised is the possible existence of conserved quantities in reversible cellular automata (CA), something I leave as a a suggestion to the readers of the present notes. To formulate the problem in the quickest possible way, consider a Boolean CA on a square lattice. Each site is indexed by a pair of integers, positive or negative (i, j). The value of the Boolean variable there is at time n (discrete) $\sigma_{i,j}^n$. It takes either the value 0 or +1. The law of evolution is reversible (the same in backward and forward direction of time),

$$\sigma_{i,j}^{n+1} + \sigma_{i,j}^{n-1} = F(\sigma_{i',j'}^n). \tag{17}$$

Equation (17) is written in Boolean algebra, that is such that $0 + 0 = 0$, $0 + 1 = 1$ and $1 + 1 = 0$. Moreover its right-hand side is a function of the σ's in the square lattice at sites that are neighbours of $\sigma_{i,j}$, that are (for instance) the four nearest neighbours, with $i' = i \pm 1$ and $j' = j \pm 1$. Since the function F has two possible values, 0 or 1, there are 2^{2^K} such functions, K being the numbers of neighbours, 4 in the present case. This gives $2^{16} = 65536$ possible functions. The $Q2R$ model is one possible choice for a function F. It has a conserved quantity proportional to the size of the system (and found by trial and error, not by using any Noether-like argument). As far as I am aware, no study has been undertaken of other possible choices of function F giving an invariant quantity (or eventually more than one), which makes it an interesting topic of study.

Acknowledgements

My travel to the CIMPA school on "Discrete integrable systems" in Pondicherry (India) was made possible thank to the support of the Direction des relations internationales de l'Académie des sciences, which is warmly thanked.

References

1. Y. Pomeau, J. de Physique (Paris), **43**, (1982) 859.
2. Y. Pomeau, Phys. Letters **26A**, (1968) 34.
3. J. Hardy, de Pazzis O. and Pomeau Y., Phys. Rev. Letters **31**, (1973) 276; D. d'Humières, Lallemand P. and Pomeau Y., C. R. Ac. Sci. **301**, (1985) 1391; U. Frisch, Hasslacher B. and Pomeau Y., Phys. Rev. Lett. **56**, (1986) 1505; Y. Pomeau, J. of Phys. **A17**(1984) L415; Y. Pomeau, Vichniac G., Comment, J. of Physics **A21**, (1988) 3297.

Index

Lecture Notes in Physics

For information about Vols. 1–599
please contact your bookseller or Springer-Verlag
LNP Online archive: springerlink.com